Earthquake Engineering

Earthquake Engineering

Theory and Implementation with the 2015 International Building Code

Nazzal S. Armouti, Ph.D., P.E.
Professor of Civil Engineering
University of Jordan
International Director
International Code Council, ICC

Third Edition

McGraw Hill Education

New York Chicago San Francisco Athens
London Madrid Mexico City Milan
New Delhi Singapore Sydney Toronto

Cataloging-in-Publication Data is on file with the Library of Congress.

Earthquake Engineering: Theory and Implementation with the 2015 International Building Code, Third Edition

1 2 3 4 5 6 7 8 9 0 QVS/QVS 1 2 0 9 8 7 6 5 4

ISBN 978-1-25-958712-2
MHID 1-25-958712-6

The pages within this book were printed on acid-free paper.

Sponsoring Editor
Lauren Poplawski

Editorial Supervisor
Donna M. Martone

Project Manager
Namita Gahtori, Cenveo® Publisher Services

Production Supervisor
Lynn M. Messina

Copy Editor
Cenveo Publisher Services

Proofreader
Cenveo Publisher Services

Composition
Cenveo Publisher Services

Art Director, Cover
Jeff Weeks

Contents

Foreword

This is the third edition of a one-of-a-kind textbook. This book explains the fundamental concepts of structural dynamics and earthquake engineering with an exceptional clarity and an unprecedented quantity of numerical examples that help the reader fully understand the concepts being discussed.

Professor Armouti has done a phenomenal job of explaining the difficult concepts of linear and nonlinear dynamics and structural response to earthquake excitations. The presentation style, simplicity of language, and the vast number of examples help make the concepts presented easily understandable even to those who face them for the first time.

This is an ideal textbook for teaching a first undergraduate or graduate course in earthquake engineering. It not only explains the structural dynamics theories necessary for understanding linear and nonlinear response to earthquake excitations, but also covers the basic design of earthquake-resistant steel and reinforced concrete buildings, bridges, and isolated systems, in accordance with the latest codes of the United States. The provisions of ASCE 7 standard as well as those in the International Building Code (IBC), ACI-318, and AISC seismic provisions are clearly explained and illustrated through numerical examples.

Students of the subject will find this book easy to follow and will appreciate the wealth of numerical examples presented for every small and large issue discussed. The instructors will find this book useful because of the simplicity of the presentation, the extensive number of solved examples, and the problems contained at the end of the first five chapters. To aid instructors in using the book effectively for teaching the subject, an Instructor's Manual containing solutions to end of chapter problems and a set of Powerpoint presentation slides are made available to qualified instructors. Last, but not the least, engineering practitioners will find this book to be an invaluable source of information regarding response of various systems and components to earthquake excitations.

When I was first presented with the manuscript of the first edition of this book by the International Code Council, which was seeking my opinion regarding potential publication in the United States, the first thought that crossed my mind was: an earthquake engineering book from Jordan for the U.S. market? This initial reaction, however, rapidly faded when I went over the contents and the presentation of the book.

I did strongly recommend publication of the first and second editions of this textbook for the U.S. market and I am very pleased to have done the same for this third edition of the book.

Farzad Naeim, Ph.D., S.E., Esq.
President,
Farzad Naeim, Inc.
Irvine, California
Past President,
Earthquake Engineering Research Institute

Acknowledgment

I gratefully acknowledge the help, support, and encouragement received from my family, friends, and colleagues throughout Jordan and abroad which have converted my exhaustion into motivation.

THANK YOU ALL !!!

Earthquake Engineering

CHAPTER 1
Introduction

Earthquake engineering is the science that studies the behavior of structures under earthquake excitation and provides the rules on how to design structures to survive seismic shocks. Earthquakes are wild and violent events that can have dramatic effects on structures. In fact, many structures have collapsed during earthquakes because earthquake-induced forces or displacements exceeded the ultimate capacity of the structures. Therefore, the study of structural behavior at full capacity is a necessary element of earthquake engineering.

Earthquakes are extremely random and oscillatory in nature (as shown in Fig. 1-1). Because earthquakes cause structures to largely deform in opposite directions, earthquake engineering also requires an understanding of structural behavior under cyclic loading. Figure 1-2 shows an example of cyclic loading in the inelastic range. Furthermore, the extreme randomness and uncertain occurrence of earthquakes also require the use of a probability approach in the analysis and design of structures that may experience seismic excitation.

The civil engineering field is basically distinguished by two design philosophies: elastic design and plastic (inelastic) design. In the elastic design philosophy, structures are designed to remain elastic, and no internal force redistribution is permitted during the lifetime of the structure. Elastic design can be characterized by its reversibility: the uniqueness between stress-strain or load-displacement (F-Δ) relationships and whether those relationships are linear or nonlinear, as seen in Fig. 1-3(a) and (b). This behavior implies the recovery of the work done by the external loads after their removal. However, in the plastic design philosophy, the relationship between stress and strain or between load and displacement is not unique. In this philosophy, plastic deformations are defined as those deformations that remain permanent after the removal of forces. Such deformations result in a hysteresis loop as shown in Fig. 1-3(c). This behavior also implies that dissipation of energy has occurred. In addition, because redistribution of internal forces is permitted, the restoring force (resistance of the structure) depends on both material properties and loading history.

Earthquake-resistant structures can be designed to remain elastic under large earthquakes. However, such design requires high strength and, in turn, high cost, which might not be economically feasible. As shown in Fig. 1-4, and as will be described later, the inelastic response of structures under seismic excitation permits them to be designed with strength less than their elastic strength demand. Experience from previous earthquakes and inelastic dynamic analysis shows that elastic and inelastic displacements remain within the same range.

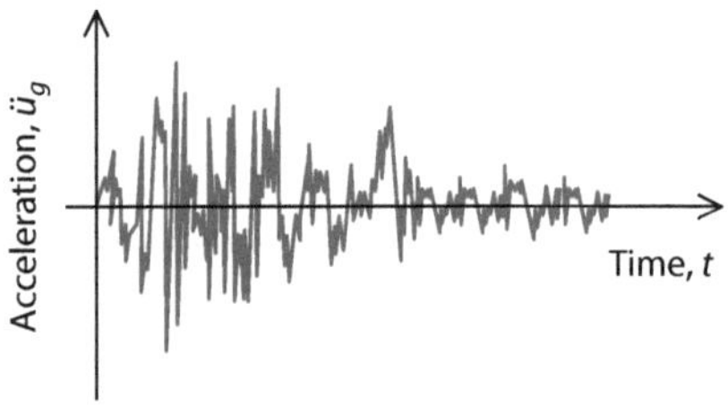

Figure 1-1 Earthquake record.

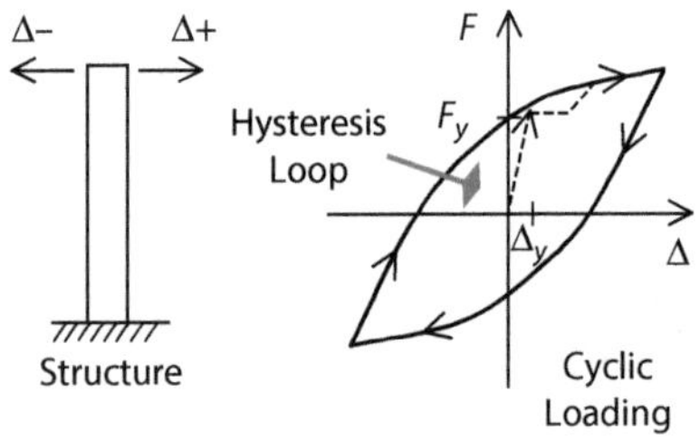

Figure 1-2 Cyclic behavior.

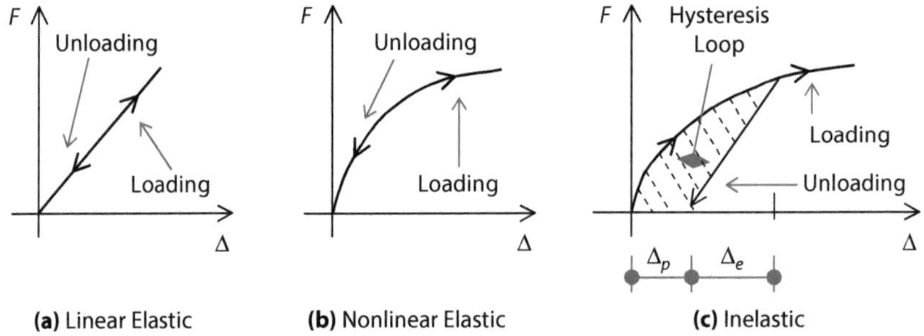

Figure 1-3 Elastic versus inelastic behavior.

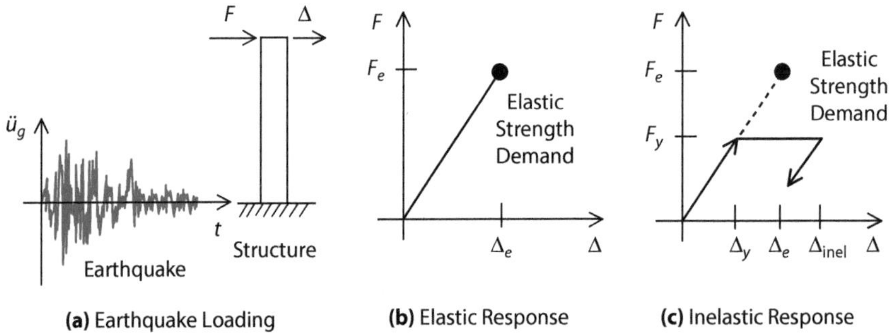

Figure 1-4 Elastic versus inelastic response.

Inelastic dynamic analysis shows that structures with strength less than their elastic strength demand can survive earthquake excitation if they possess enough capacity to deform in the inelastic region. Therefore, we rely on inelastic behavior (yielding and ductility) to design structures with strength less than their elastic strength demand. This implies that designing structures with strength less than their elastic strength demand imposes requirements on the structure other than the traditional requirements in structural design. These requirements include good ductility, good energy dissipation, and good self-centering capacity of the structure.

Therefore, the objective of this book is to provide an understanding of the behavior of structures under earthquake excitation, the characteristics of earthquakes, and the relationship between the force reduction factor and ductility demand. This understanding will provide the knowledge needed to realize the requirements for achieving ductility capacity to meet ductility demand and, eventually, for designing cost-effective structures that can survive earthquake excitation.

Modern seismic codes have set the following objectives as their ultimate goals of earthquake-resistant design:

1. Prevent nonstructural damage caused by minor and frequent earthquakes.
2. Prevent structural damage during moderate and less frequent earthquakes.
3. Prevent collapse of structures during major and rare earthquakes (ultimate goal is to protect human life).
4. Maintain functionality of essential facilities during and after any earthquake (i.e., hospitals, fire departments, and police stations). This would also include lines of transportation (e.g., bridges).

Achieving the first and second objectives in reference to Fig. 1-5 requires that a minimum stiffness, k, and minimum strength, F_y, be provided to keep structural response within acceptable performance limits, usually within the elastic range. Although minimum stiffness requirements limit the elastic deformations needed to achieve objective 1 (prevent nonstructural damage), minimum strength requirements will ensure the achievement of objective 2 (prevent structural damage). Objectives 3 and 4 are

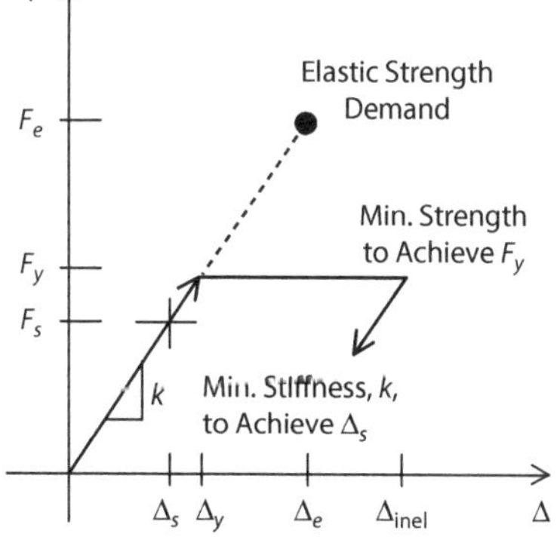

Figure 1-5 Seismic design objectives.

achieved by using the reserved capacity of structures that are due to the inelastic response of structures.

To achieve these objectives, one must understand structural analysis, structural dynamics, inelastic behavior of structures, and earthquake characteristics. This book covers these essential areas through detailed analysis of the characteristics of earthquakes, elastic and inelastic response of structures to dynamic loading (earthquake excitation), behavior of structures under earthquake excitation, and design of earthquake-resistant structures.

In addition to the introduction, the contents of this book may be grouped into four main topics: nature and properties of earthquakes, theory and analysis, practical application and treatment by seismic codes, and special topics. The 12 chapters of the book are organized in a logical sequence of topics, and therefore the reader is advised to start the book with Chap. 1 and proceed chapter by chapter. To receive the maximum benefit of this book, it is highly recommended that the basics (Chaps. 1 through 5) be studied before going to practical applications and special topics. Brief descriptions of the main topics and pertinent chapters is given in the following paragraphs.

A general introduction to the subject is given in Chap. 1. The nature and characteristics of earthquakes are treated in Chap. 2. This chapter does not intend to provide the reader with the geological base of earthquakes; rather, it intends to provide a basic understanding of earthquakes necessary for their direct incorporation in the analysis and design of structures.

The basic theory and analysis of earthquake engineering from a structural viewpoint is treated in Chaps. 3 through 5. Elastic dynamic analysis is treated in Chap. 3; inelastic dynamic analysis in Chap. 4. Dynamic analysis and vibration properties of structures subjected to earthquakes only are treated in these chapters. The behavior of structures under seismic excitation is treated in Chap. 5. The core of modern earthquake engineering is presented in this chapter. The basic seismic parameters and definitions, such as ductility, energy dissipation, and others are also given in this chapter. Ample figures and examples are included in these chapters to further illustrate concepts and ideas.

Practical applications and treatment of seismic codes are covered in Chaps. 6 through 9. Chapter 6 treats seismic provision and design requirements of buildings using the *International Building Code®* (IBC®), and includes the ASCE 7 Standard as the book's reference code for seismic provisions. Force calculations, drift limitations, selection of seismic systems, methods of analysis, and other pertinent issues are presented in this chapter. Comprehensive examples are used to illustrate concepts as deemed necessary.

Chapter 7 treats, in depth, provisions and design of reinforced concrete buildings according to ACI-318. This chapter is used as a model of practical application of seismic design, and hence, comprehensive treatment and design of various systems such as types of frames and types of shear walls are provided. Comprehensive and detailed examples are presented for illustration and practice. Chapter 8 introduces seismic provisions of structural steel buildings according to the American Institute of Steel Construction (AISC) seismic provisions. Basic provisions and identification of systems are illustrated in this chapter without giving examples. This chapter intends to give the reader a flavor of the variation of seismic requirements when the material is changed. Masonry and wood structures are not addressed in this book.

Practical applications and treatment of bridges are covered in Chap. 9, which treats seismic provision and design requirements according to the American Association of State Highway and Transportation Officials (AASHTO) Code. This chapter presents, in detail, the methods used in AASHTO to calculate seismic forces in bridges with comprehensive and illustrative examples. New simplified and efficient methods are also presented in this chapter, which are based on published developments of the author.

The last group of chapters, Chaps. 10 through 12, presents special topics pertinent to earthquake engineering. Concise presentation of geotechnical aspects, such as liquefaction and popular problems in foundations pertinent to earthquake effects are presented in Chap. 10. The basis and development of synthetic earthquake records, which are usually needed in explicit inelastic dynamic analysis, are presented in Chap. 11. The modern technique of seismic isolation is covered in Chap. 12. Seismic isolation is usually used to alleviate harmful effect of earthquakes, and to control and protect structures from damage under seismic excitation.

CHAPTER 2

Characteristics of Earthquakes

2.1 Causes of Earthquakes

As mentioned in Chap. 1, it is important to understand the nature of earthquakes in order to further understand their effect on structures and consequently to develop rational methods of their analysis. However, this book does not intend to study earthquakes from the geological point of view; rather, it intends to highlight their characteristics from an engineering point of view that would be closely related to engineering analysis. Therefore, only a brief introduction will be given on the origin, location, and subjective measurements, such as the Richter and Mercalli scales, of earthquakes. More attention will be given to their instrumental measurements such as acceleration, power spectral density, and response spectrum. Such measurements are directly used to evaluate and analyze the direct impact of earthquakes on structures.

Earthquakes can occur as the result of activities such as tectonic movements, volcanic activities, cave collapses, natural and man-made explosions, and filling of reservoirs. The reservoir-induced Koyna earthquake in India in 1967 resulted in the death of about 180 people. The collapse of the World Trade Center towers in New York on September 11, 2001, generated an earthquake of 5.2 magnitude on the Richter scale. However, earthquakes can most reliably be explained by tectonic movements, which generate 90 percent of all earthquake phenomena.

2.2 Plate Tectonic Theory

The crust of the earth is divided into several large tectonic plates that sometimes encompass more than one continent. These plates include the Eurasian Plate, African Plate, North American Plate, South American Plate, Australian Plate, and Pacific Plate, among others. Figure 2-1 shows the extent of some of these plates.

Major geological faults are formed along the boundaries of these plates, and these faults are the main source of earthquakes.

The tectonic plates are in continuous movement against each other. However, friction forces between these plates prevent differential displacements at the boundaries of the plates. This action generates energy buildup along plate boundaries in a form of strain energy that is stored in the plates.

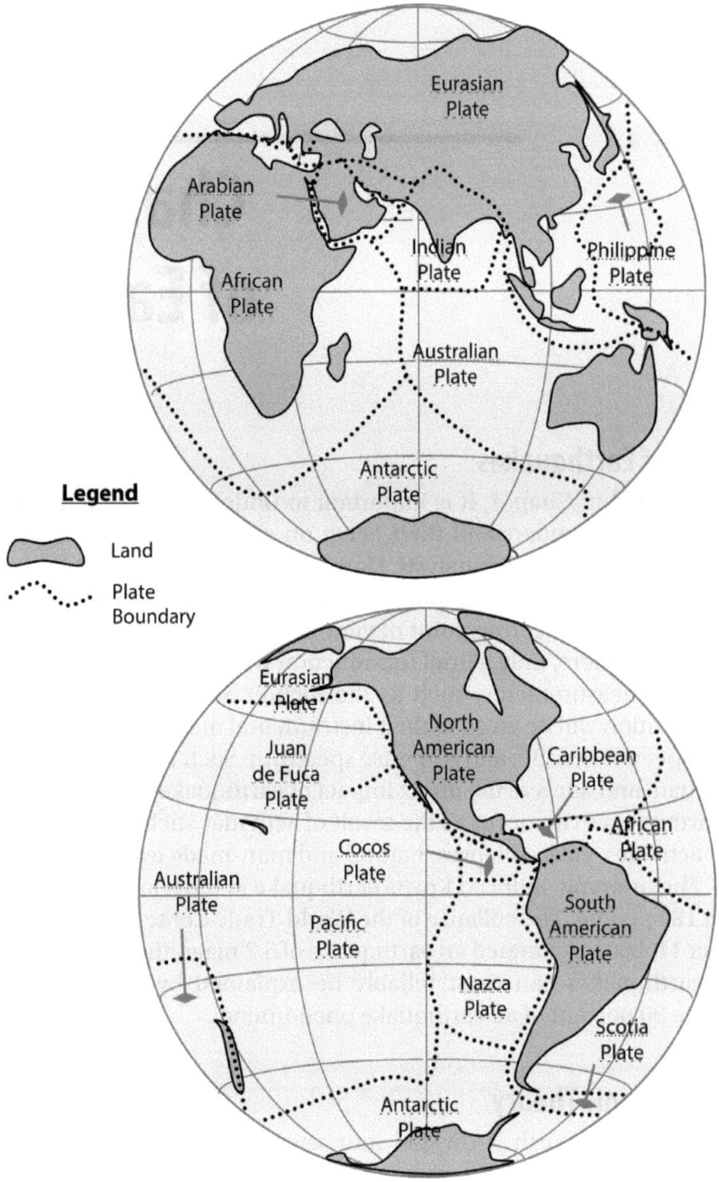

FIGURE 2-1 Major tectonic plates.

When the stored energy increases to levels that exceed the ability of the friction forces to hold the plate boundaries together, sliding along those boundaries occurs, creating a phenomenon known as elastic rebound. Elastic rebound releases the stored energy in the form of seismic strain waves in all directions. This marks the onset of an earthquake event. Figure 2-2 shows the different types of seismic strain waves that are generated by earthquakes.

Strain waves are classified into two main groups: body waves and surface waves. Body waves are classified as fast primary waves or P-waves, and slow shear waves or

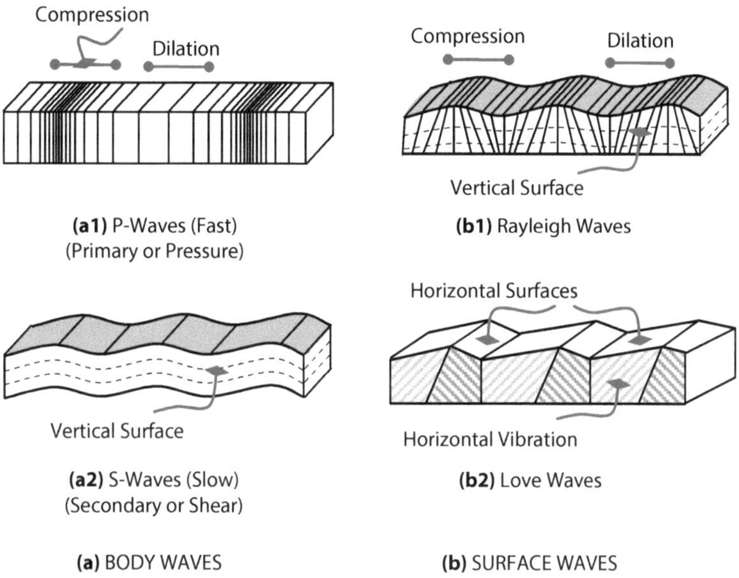

FIGURE 2-2 Types of seismic strain waves.

S-waves. Body waves can be used to estimate the distance of the site of measurement from the earthquake source. Because P-waves are faster than S-waves, a measurement of the time difference between their arrivals at the site can be converted into a distance.

Surface waves are classified as Rayleigh waves, or R-waves, and Love waves, or L-waves. The shape of their oscillation is shown in Fig. 2-2.

Note that these seismic strain waves travel through random media that modify the waves with an infinite number of effects, which include filtering, amplification, attenuation, reflection, and refraction. These random modifications give each site a different wave profile, making it very difficult to predict the characteristics of an earthquake at a specific site.

The extreme randomness and uncertainty of earthquake characteristics require the use of probabilistic means in the treatment and design of structures. Therefore, the design and survival of structures is also based on probabilistic consideration of earthquake excitations.

The following sections summarize additional measures that are used to describe earthquakes.

2.3 Measures of Earthquakes

Earthquake measures quantify the size and effect of earthquakes. The size of an earthquake is measured by the amount of energy released at the source, its magnitude, whereas the effect of an earthquake at different locations is measured by its intensity at a specific site. Figure 2-3 defines the relevant components of an earthquake with its measures at the source and at any site.

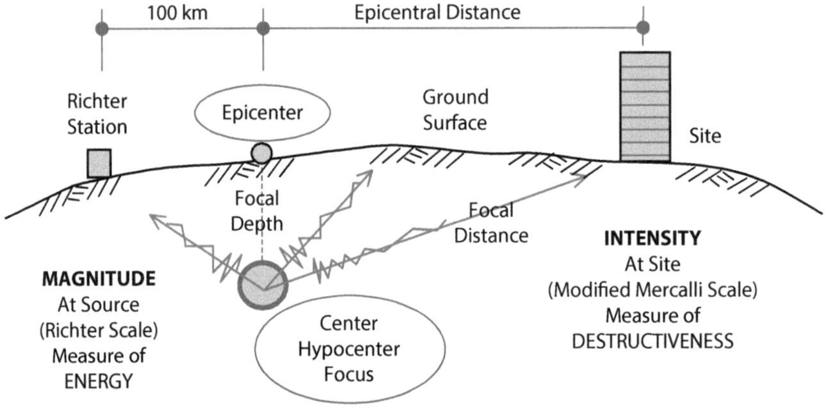

FIGURE 2-3 Measures of earthquakes.

2.3.1 Magnitude

The size of an earthquake at its source is known as the magnitude of the earthquake and is measured by the Richter scale. The magnitude, M, is given as

$$M = \log A$$

where A is the amplitude in μm measured on a Wood-Anderson–type seismometer.

The earthquake may be described in general terms according to the value of M as follows:

$M = 1$ to 2: earthquake is barely noticeable.
$M < 5$: earthquake is not expected to cause structural damage.
$M > 5$: earthquake is expected to cause structural damage.
$M = 8, 9$: earthquake causes the most structural damage recorded.

Note that ground motion intensity decreases with the distance from focus. Therefore, M does not measure local destructiveness of earthquakes. The M-value is only an indication of the energy released.

2.3.2 Intensity

Intensity is a subjective measure of the local destructiveness of an earthquake at a given site. Intensity scales are based on human feelings and observations of the effect of ground motion on natural and man-made objects.

The most popular scale of intensity is called the Modified Mercalli (MM) scale. This scale is divided into 12 grades (I to XII) as follows:

I. Not felt except under exceptionally favorable circumstances.
II. Felt by persons at rest.
III. Felt indoors; may not be recognized as an earthquake.
IV. Windows, dishes, and doors disturbed; standing motor cars rock noticeably.
V. Felt outdoors; sleepers wakened; doors swung.
VI. Felt by all; walking unsteady; windows and dishes broken.

VII. Difficult to stand; noticed by drivers; fall of plaster.

VIII. Steering of motor cars affected; damage to ordinary masonry.

IX. General panic; weak masonry destroyed, ordinary masonry heavily damaged.

X. Most masonry and frame structures destroyed with foundations; rails bent slightly.

XI. Rails bent greatly; underground pipes broken.

XII. Damage total; objects thrown into the air.

Note that MM depends on the magnitude of the earthquake and on the distance between the site and the source. This scale may be expressed as a function of magnitude and distance from the epicenter by the following expression:

$$\text{MM} = 8.16 + 1.45\,M - 2.46 \ln r$$

where MM = Modified Mercalli scale intensity grade
M = earthquake magnitude (Richter scale)
r = distance from epicenter (km)

A plot of this relation is shown in Fig. 2-4.

Example 1 A maximum amplitude of 1.2 m is recorded on a Wood-Anderson seismometer at a standard Richter station. Describe the expected damage in a city located at 160 km from the earthquake epicenter.

Solution

(1) Richter scale magnitude

$$M = \log A = \log 1.2 \times 10^6 = 6.08$$

(2) MM scale intensity

$$\text{MM} = 8.16 + 1.45\,M - 2.46 \ln r$$

$$\text{MM} = 8.16 + 1.45\,(6.08) - 2.46 \ln (160)$$

$$\text{MM} = 4.5 \approx \text{grade V}$$

Therefore, the damage in the city may be described as grade V according to the MM scale: the earthquake was felt outdoors, sleepers wakened, and doors swung.

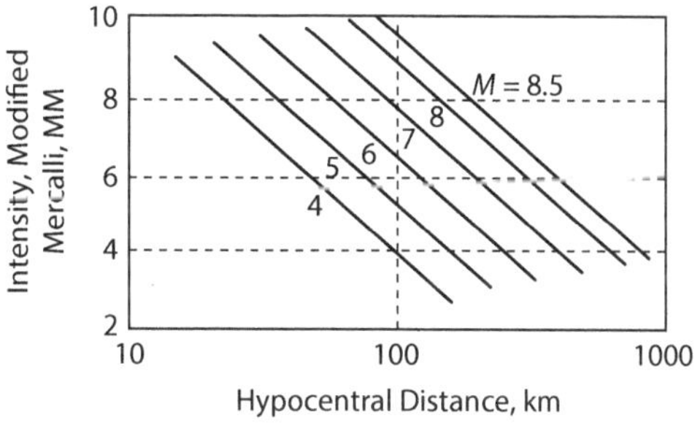

Figure 2-4 Relationship between magnitude and seismic intensity.

2.3.3 Instrumental Scale

In general, both magnitude and intensity scales of earthquakes are useful in estimating the size and severity of earthquakes. However, they are not useful for engineering purposes, especially in structural engineering. Structural engineers need a quantitative measure that can be used in analysis and design. This measure is provided in an accelerogram, which is a record of the ground acceleration versus time.

Figure 2-5 shows a sample accelerogram record of the famous El Centro earthquake that occurred in May 1940, killing nine people and damaging 80 percent of the buildings in Imperial, California. The accelerogram contains important parameters of the earthquake, such as peak ground acceleration (PGA), total duration, and length of continuous pulses. The accelerogram can be mathematically analyzed to obtain other important parameters of an earthquake, such as frequency content, peak ground velocity, peak ground displacement, and power spectral density. Accelerograms are also used to construct response spectra.

A correlation between the peak ground acceleration and the MM scale has been observed. Statistical analysis shows that the following approximate relations may be used for estimate purposes.

On the average, PGA (in *g*) is given as

$$PGA_{avg} = 0.1 \times 10^{-2.4 + 0.34\,MM}$$

A conservative value of PGA (in *g*) is given as

$$PGA_{design} = 0.1 \times 10^{-1.95 + 0.32\,MM}$$

Table 2-1 can be used as a guideline to estimate the relationship between the shown earthquake parameters and its effects.

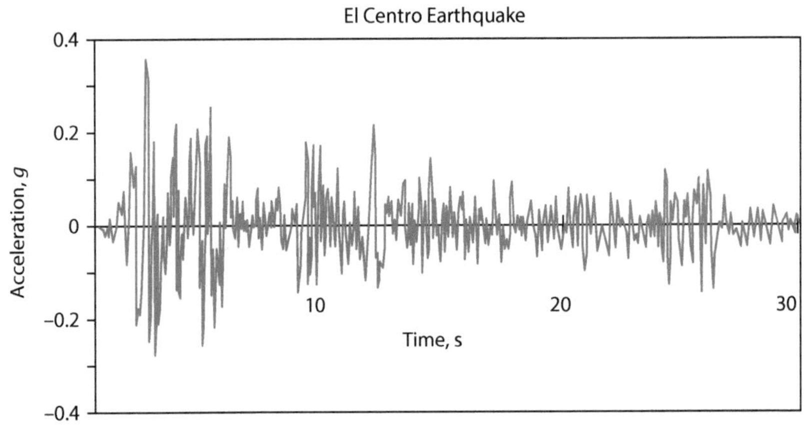

Figure 2-5 Sample earthquake record.

MM	PGA (*g*)	UBC zone (*Z*)	IBC: Mapped spectral acceleration at short period (S_s)	IBC: Mapped spectral acceleration at 1-s period (S_1)
IV	< 0.03	1	around 30	around 12
V	0.03–0.08			
VI	0.08–0.15			
VII	0.15–0.25	2A	around 60	around 25
		2B	around 75	around 30
VIII	0.25–0.45	3	around 115	around 45
IX	0.45–0.60			
X	0.60–0.80	4	around 150	around 60
XI	0.80–0.90			
XII	> 0.9			

Table 2-1 Approximate Relationships between PGA, MM, Uniform Building Code (UBC) Zoning, and International Building Code (IBC) Zoning Criterion

2.3.4 Fourier Amplitude Spectrum

Any time function, $f(t)$, can be transformed into its frequency domain as a function of frequency, $F(j\omega)$, and vice versa using Fourier transform pairs

$$F(j\omega)=\int_{-\infty}^{\infty} f(t)\,e^{-j\omega t}\,dt$$

$$f(t)=\frac{1}{2\pi}\int_{-\infty}^{\infty} F(j\omega)\,e^{j\omega t}\,d\omega$$

An earthquake accelerogram is composed of an infinite number of harmonics. The frequencies of these harmonics, known as frequency content, can be found. The power associated with these frequencies can also be obtained from Fourier analysis because the power spectral density is directly proportional to the square of Fourier amplitude,

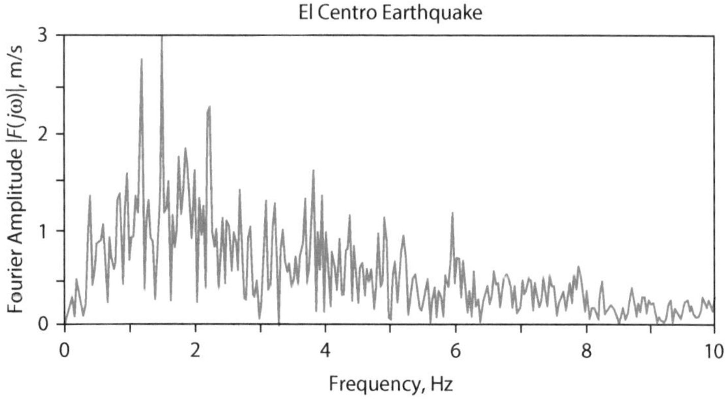

Figure 2-6 Fourier amplitude spectrum.

$|F(j\omega)|$. Figure 2-6 shows a Fourier amplitude spectrum for the El Centro earthquake. More information on this subject is given in Chap. 12.

2.3.5 Power Spectral Density

The power spectral density, $S(\omega)$, is another measure of the energy associated with each frequency contained in the earthquake. Power spectral density is directly proportional to the square of Fourier amplitude in the following form:

$$S(\omega) = \lim_{T\to\infty} \frac{|F(j\omega)|^2}{2\pi T}$$

Power spectral density is widely used in constructing synthetic earthquakes. Synthetic earthquake records can be used as standard records in design, similar to the use of actual earthquake records, as shown in Chap. 12.

2.3.6 Response Spectrum

Response spectrum is the most useful measure of earthquakes for engineers. Response spectrum is a chart that plots the response of a single degree of freedom oscillator to a specific earthquake. By varying the frequency (or the period) and the damping ratio of the system, the maximum structural response quantities can be evaluated in terms of maximum displacement, maximum velocity, and maximum acceleration of the system. A detailed analysis of response spectrum is provided in Chap. 3. As an example, Fig. 2-7 shows a sample response spectrum of the El Centro earthquake.

The different measurements and characteristics introduced in the previous sections serve different purposes in the chapters that follow. For example, some earthquake measurements such as acceleration profile and response spectrum are necessary and will be used in the later chapters to study and quantify the engineering effect of earthquakes on structures, whereas other measurements such as Fourier spectrum and power spectral density function are necessary and will be used to generate synthetic earthquakes as another necessary component for analysis of structures under seismic excitations.

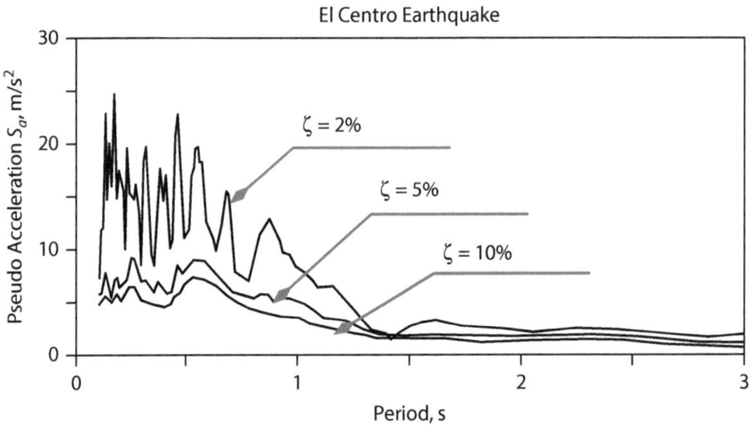

FIGURE 2-7 Response spectrum.

CHAPTER 3

Linear Elastic Dynamic Analysis

3.1 Introduction

Structural dynamics is a well-established science with a rich and complete foundation of literature. The objective of this book is to focus on the aspects of support excitation that are due to earthquakes. For extensive study on the many other aspects of structural dynamics, the reader may refer to any textbook on structural dynamics.

3.2 Single Degree of Freedom System

3.2.1 System Formulation

An idealized single degree of freedom (SDOF) dynamic system is shown in Fig. 3-1. The system consists of a concentrated mass, m, subjected to a force varying with time, $p(t)$, where its motion is resisted by its inertial force, f_I, a viscous dashpot, c, and an elastic spring, k. The mass can move in its only degree of freedom, in direction u.

Dynamic equilibrium may be established with the help of the D'Alembert principle, which states that dynamic equilibrium can be established similar to static equilibrium if the inertial force, f_I, is considered a reaction to its motion in a direction opposite to the direction of its acceleration. As shown in Fig. 3-2, this principle is expressed in mathematical form as follows.

Since,

$$p(t) = m\ddot{u} \qquad \text{Newton second law}$$

or,

$$p(t) - m\ddot{u} = 0$$

or,

$$p(t) - f_I = 0 \qquad \text{dynamic equilibrium state}$$

then,

$$f_I = m\ddot{u}$$

where f_I is inertial force equals to $m\ddot{u}$ and opposite to $\ddot{u}$-direction.

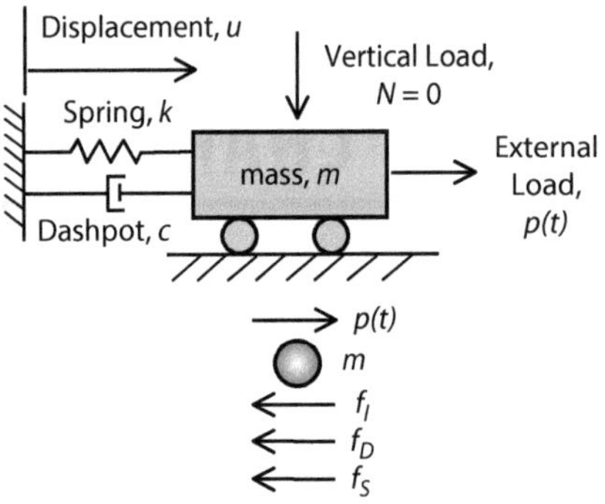

FIGURE 3-1 Single degree of freedom system.

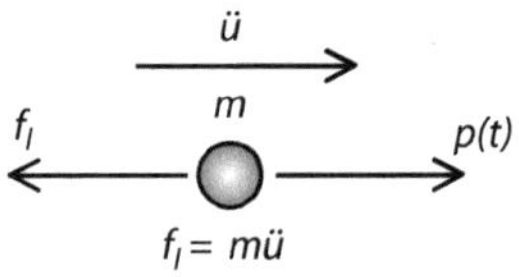

FIGURE 3-2 D'Alembert principle.

Using the D'Alembert principle, and in reference to Fig. 3-1, dynamic equilibrium for a SDOF system subjected to an external force, $p(t)$, can be established as follows:

$$\sum F_X = 0 \rightarrow f_I + f_D + f_S = p(t)$$

$$m\ddot{u} + c\dot{u} + ku = p(t)$$

where f_I = inertial force
f_D = damping force, which is modeled as a viscous force proportional to the velocity
f_S = spring force
$p(t)$ = any external force acting upon the mass
m = mass
c = coefficient of viscous damping
k = spring stiffness
u = displacement in the shown direction
$\dot{u}$ = mass velocity
$\ddot{u}$ = mass acceleration

For an SDOF structure with properties m, c, and k as shown in Fig. 3-3, and subjected to ground motion $u_g(t)$ where $p(t) = 0$ and $N = 0$, dynamic equilibrium requires that

$$f_I + f_D + f_S = 0$$

$$m\ddot{u}_{\text{tot}} + c\dot{u} + ku = 0$$

$$m(\ddot{u}_g + \ddot{u}) + c\dot{u} + ku = 0$$

$$m\ddot{u} + c\dot{u} + ku = -m\ddot{u}_g$$

Note that in the case of support excitation, or earthquake loading, the equation of motion will be identical to the regular SDOF system shown earlier, with ground acceleration acting as an external force, $p(t) = -m\ddot{u}_g$. Remember that only relative displacements cause internal straining actions.

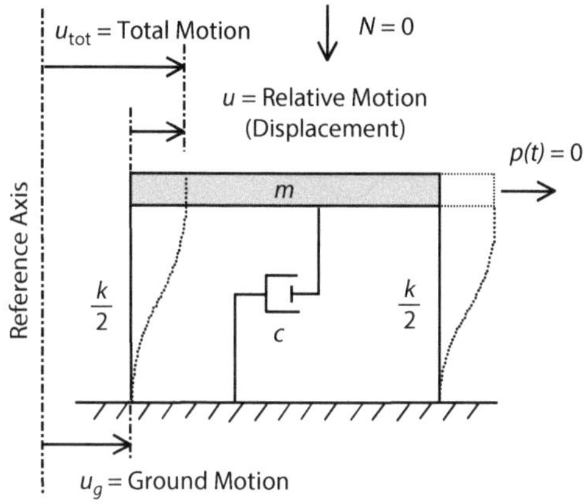

Figure 3-3 SDOF structure.

In structural dynamics, the equation of motion may be divided by the mass, m, to yield the following:

$$\ddot{u}+\frac{c}{m}\dot{u}+\frac{k}{m}u=-\ddot{u}_g$$

or

$$\ddot{u}+2\zeta\omega\dot{u}+\omega^2u=-\ddot{u}_g$$

where ω = circular frequency (rad/s)
ζ = damping ratio, dimensionless quantity, $\zeta=\frac{c}{c_{cr}}$
c_{cr} = critical coefficient of damping, $c_{cr}=2\sqrt{km}$

For $c < c_{cr}$, or undercritically damped structures, the free vibration solution ($\ddot{u}_g = 0$) is given as

$$u(t) = e^{-\zeta\omega t}\{A \cos \omega_d t + B \sin \omega_d t\}$$

or

$$u(t) = e^{-\zeta\omega t}\{A \cos (\omega_d t + \theta)\}$$

where A and B are constants determined from initial conditions. θ is the phase angel. ω_d is the damped frequency, which is equal to

$$\omega_d=\omega\sqrt{1-\zeta^2}$$

However, for most structural engineering applications

$$\zeta \lll 1.0(2\%, 5\%)$$

and, therefore,

$$\omega_d \approx \omega$$

A plot of the above solution is shown in Fig. 3-4.

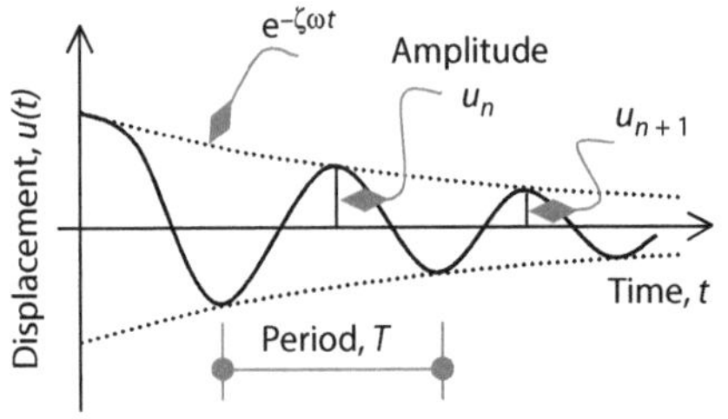

FIGURE 3-4 Displacement history, SDOF.

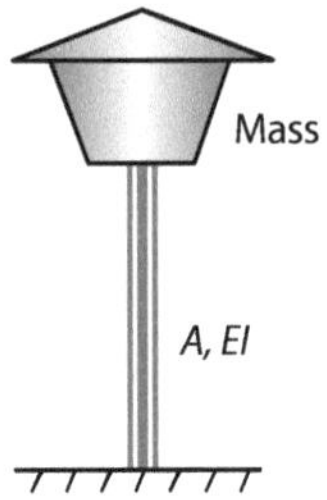

FIGURE 3-5 Elevated tank.

Note that for free vibration, damping decreases by the so-called logarithmic decrement, δ, which is given by the following expression:

$$\delta = \ln \frac{u_n}{u_{n+1}} \approx 2\pi\zeta \qquad (\text{for } \zeta << 1.0)$$

Few examples that have SDOF exist in practice. For example, the elevated water tower tank shown in Fig. 3-5 is considered an SDOF system because the mass is concentrated in one location inside the tank.

3.2.2 Response Spectrum of Elastic Systems

The solution of the equation of motion under the excitation of a specific excitation is obtained by numerical analysis and will be shown later. During the excitation time of the earthquake, the solution yields the response displacement history of the system as shown in Fig. 3-6.

For a specific ground motion, $u_g(t)$, we are only interested in the maximum response of the structure, or $|u_{max}|$, which will be denoted by S_d. S_d is known as the spectral displacement that is defined as the maximum relative displacement of an SDOF system subjected to a specific ground motion (earthquake). Note that at the time of maximum displacement, the force in the spring reaches its maximum value, $f_{s,max}$, because the velocity equals zero and the damping force, f_D, will be equal to zero. Therefore,

$$f_{s,max} = kS_d$$

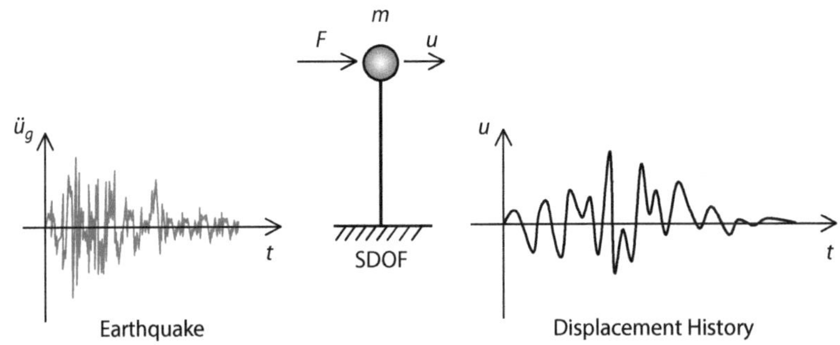

FIGURE 3-6 Earthquake excitation of SDOF system.

If both sides of the above equation are divided by m, then

$$\frac{f_{s,\max}}{m} = \left(\frac{k}{m}\right) S_d$$

Because $\omega^2 = k/m$, and if the quantity $f_{s,\max}/m$ is defined as S_a, then

$$S_a = \omega^2 S_d$$

S_a is known as the spectral pseudoacceleration as it is not the actual acceleration of the mass. Instead, it is an acceleration look-alike parameter that produces the maximum spring force when multiplied by the mass. This maximum spring force is known as the base shear of the structure.

Knowing S_a, the force in the spring may now be given in the following form:

$$f_{s,\max} = mS_a$$

A third quantity of interest in the response spectrum analysis is known as the spectral pseudovelocity, S_v, which is related to the maximum kinetic energy of the system and is defined as the maximum total velocity of the system. When the maximum strain energy, U, of the system is equated with the maximum kinetic energy, T, of the system, then

$$U_{\max} = T_{\max}$$

$$\left(\frac{1}{2}\right) k u_{\max}^2 = \left(\frac{1}{2}\right) m \dot{u}_{\max}^2$$

(Note that at $\dot{u}_{\max}$, $\ddot{u} = 0$.)

$$kS_d^2 = mS_v^2$$

Similar to the case of spectral acceleration, if both sides of the equation above are divided by m and we notice that k/m is the square of the frequency, then the equation above becomes

$$S_v = \omega S_d$$

S_v is also referred to as the maximum earthquake response integral.

For a specific system and a specific earthquake, the quantities S_d, S_v, and S_a are related through the system frequency or the system period. The response spectrum is constructed for a specific earthquake by varying the system period (T_1, T_2, T_3, ...) and the damping ratio (ζ_1, ζ_2, ζ_3, ...). For each combination of T and ζ, the system response is evaluated in terms of S_d, S_v, or S_a. The results are then plotted in a form similar to Fig. 3-7, which shows a spectral velocity response spectrum plot for the El Centro earthquake.

Because S_d, S_v, and S_a are related through the system frequency (period), they can be plotted on a single chart using the log-log scale. The resulting chart is called a tripartite chart where S_d and S_a appear with constant lines at 45°. Figure 3-8 shows such a chart for the El Centro earthquake.

Plotted on tripartite paper, response spectra exhibit general shapes as shown in Fig. 3-9. If the maximum ground parameters, maximum ground displacement, d, maximum ground velocity, v, and maximum ground acceleration, a, are also plotted on the same chart, they basically follow the same trend.

Figure 3-9 shows that for very short-period systems where period tends to zero ($T \to 0$ and $k \to \infty$), the spectral acceleration approaches peak ground acceleration, or $S_a \to a$. For very long period systems where period tends to infinity ($T \to \infty$ and $k \to 0$), the

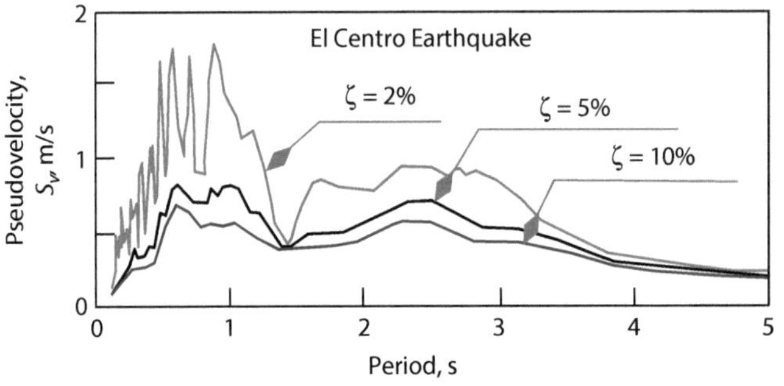

Figure 3-7 Sample response spectrum.

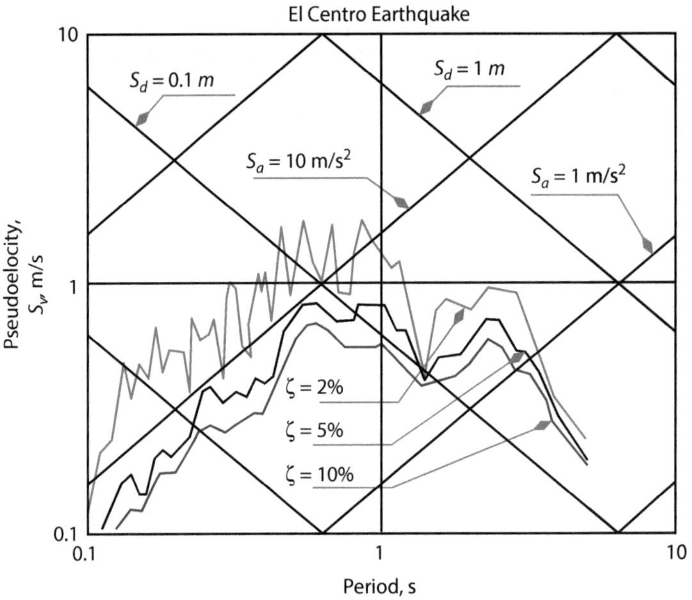

Figure 3-8 Sample response spectrum, tripartite chart.

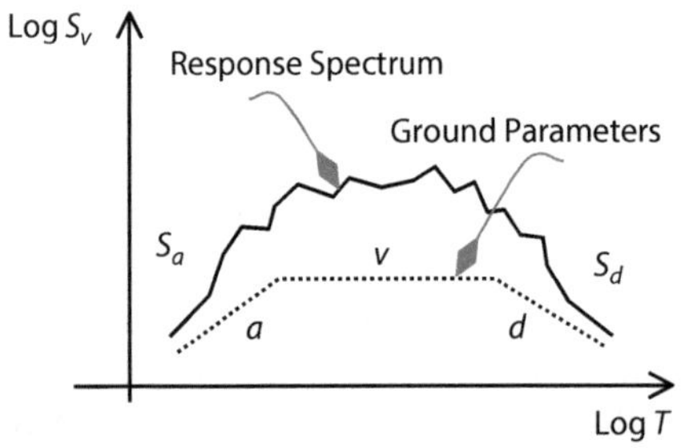

Figure 3-9 General response spectrum.

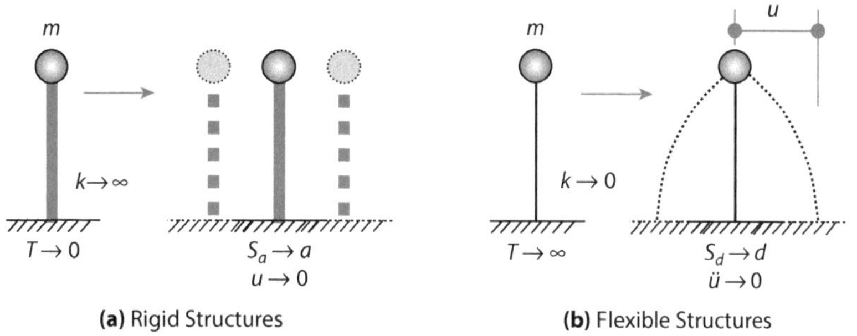

Figure 3-10 Limiting structural response.

spectral displacement approaches peak ground displacement, or $S_d \to d$. This behavior is depicted in Fig. 3-10.

The relationships shown in Fig. 3-10 constitute the basic concept behind base isolation, or seismic isolation, where the structures can be isolated from the ground at the base or isolated internally to reduce the amount of acceleration imparted to the structure. Remember that seismic isolation induces large displacements in the structure, which must be accommodated by the system. Therefore, seismic isolation is generally a simple tradeoff between large forces and large displacements under earthquake excitations.

Many seismic isolation systems have already been developed and installed in buildings and bridges. One popular seismic isolation system is the lead-rubber bearing system shown in Fig. 3-11. In this system, the lead core restricts the structure displacement under normal loads (nonseismic). Under earthquake excitation, the lead yields and gives control to the rubber to accommodate large displacement. This yielding of the lead core also serves as an energy dissipater. Chapter 13 provides a comprehensive discussion of this subject.

In addition to the measures of earthquakes given in Chap. 2, spectral velocity is also used as a measure of earthquakes by defining a quantity known as the response spectrum integral, SI. This integral is an energy measure because spectral velocity is related to kinetic energy. SI is expressed as the area under the response spectrum

$$SI = \int S_v \, dT$$

where T is the system period.

3.2.3 Design Response Spectrum

As illustrated in Chap. 2, the shape of a specific response spectrum is extremely jagged and erratic with peaks and valleys as the period varies.

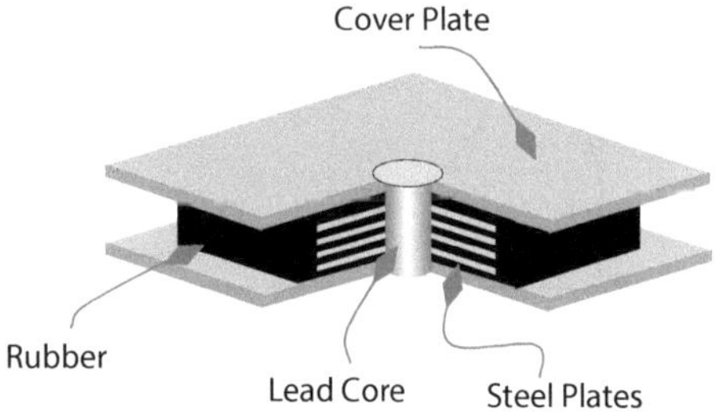

Figure 3-11 Lead-rubber bearing.

In real structures, exact calculations of the period are very difficult as the period is a function of both mass and stiffness ($T = 2\pi\sqrt{m/k}$). Whereas the stiffness is affected by nonstructural elements, which are usually not considered in the analysis, the mass is simply a random quantity that depends on the occupancy of the structure at the time of the excitations.

For design purposes, many response spectra are usually added together and smoothened to yield a smooth design response spectrum as an upper limit as shown in Fig. 3-12.

Response spectrum analysis is usually accepted for regular structures and for simple irregular structures. For complex structures, especially in high-risk regions of special importance, response spectrum procedures are not acceptable. Instead, the explicit inelastic dynamic analysis that is necessary requires many numbers from actual or synthetic acceleration records. This very involved task requires a special understanding of structure stiffness (hysteresis loops) and special skills in modeling of structures.

As discussed before, response spectrum can be used as a measure of earthquakes because it reflects the amplifications of ground motions parameters *d*, *v*, and *a*. A study of actual earthquake records conducted by Seed et al. in the United States used the response spectrum in different site conditions.

Figure 3-13 shows the results of the Seed study, which represent lines of average results of many records. Each site condition (type of soil characteristics) exhibits a

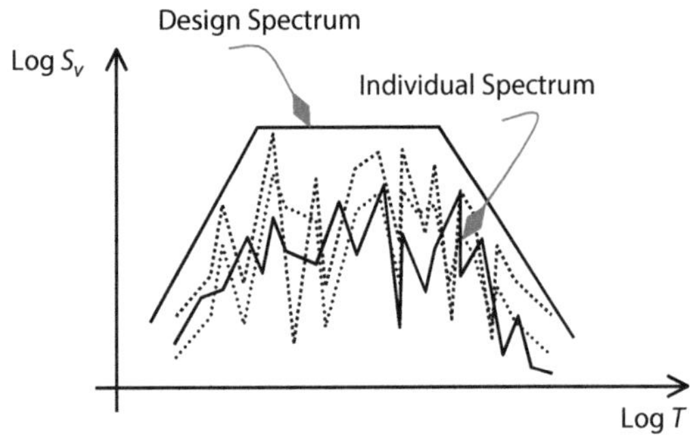

FIGURE 3-12 Design response spectrum.

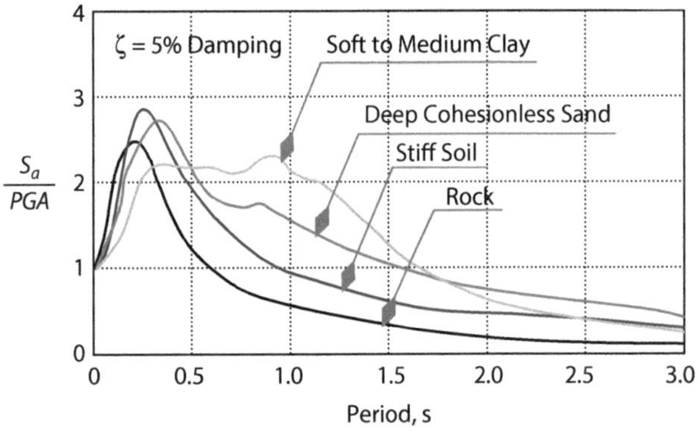

FIGURE 3-13 Elastic response spectrum, Seed et al. (Averaged field measurements.)

different response shape. In general, harder sites are characterized by high energy at higher frequencies.

In seismic codes in the United States, these shapes constitute the basic response spectrum shapes for design. A more in-depth discussion is provided in the design Chaps. 6 and 9.

Example 3-1

1. The SDOF structure shown in Example 3-1, Fig. 1, has the following properties:

$$m = 600 \text{ kN}\cdot\text{s}^2/\text{m}$$
$$EI_c = 20 \times 10^3 \text{ kN}\cdot\text{m}^2$$
$$c = 220 \text{ kN}\cdot\text{s/m}$$

Determine the maximum displacement and base shear force if the structure is excited by the El Centro earthquake.

2. Repeat requirements in (1) if the stiffness of the structure is increased to

$$EI = 60 \times 10^3 \text{ kN}\cdot\text{m}^2$$

Solution

Part (1)

Stiffness: If column stiffness is designated k_c, then

$$k_c = \frac{12EI}{L^3} = \frac{12(20\times 10^3)}{3^3} = 8{,}889 \text{ kN/m}$$

$$k_{tot} = 2(8{,}889) = 17{,}778 \text{ kN/m}$$

Frequency:
$$\omega = \sqrt{\frac{k}{m}} = \sqrt{\frac{17{,}778}{600}} = 5.44 \text{ rad/s}$$

Damping:
$$c_{cr} = 2\sqrt{km} = 2\sqrt{17{,}778(600)} = 6{,}532 \text{ kN}\cdot\text{s/m}$$

$$\zeta = \frac{c}{c_{cr}} = \frac{220}{6{,}532} = 0.034 \approx 3\%$$

$$T = \frac{2\pi}{\omega} = \frac{2\pi}{5.44} = 1.15 \text{ s}$$

knowing T and ζ, the response spectrum given in Fig. 3-8 may be used to directly read the response spectrum quantities S_d, S_v, and S_a, which read

$$S_d = 0.15 \text{ m},\qquad S_v = 0.8 \text{ m/s},\qquad S_a = 4 \text{ m/s}^2$$

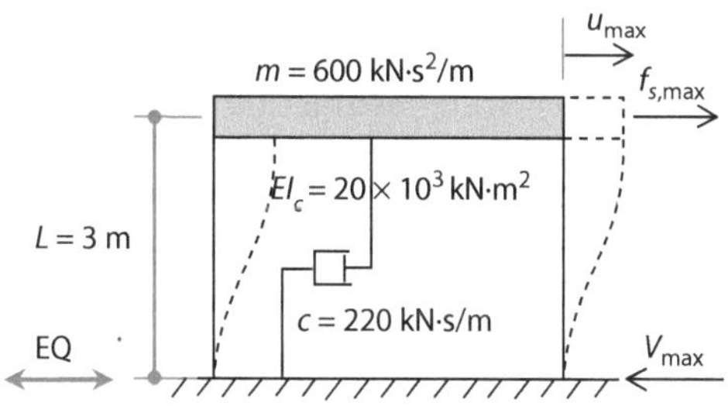

Example 3-1, Figure 1

Therefore, displacement and base shear are given as

$$u_{max} = S_d = 0.15 \text{ m}$$
$$V_{max} = f_{s,max} = mS_a = 600(4) = 2{,}400 \text{ kN}$$

Part (2)
Stiffness: If column stiffness is k_c, then

$$k_c = \frac{12EI}{L^3} = \frac{12(60\times10^3)}{3^3} = 26{,}667 \text{ kN/m}$$
$$k_{tot} = 2(26{,}667) = 53{,}334 \text{ kN/m}$$

Frequency: $$\omega = \sqrt{\frac{k}{m}} = \sqrt{\frac{53{,}334}{600}} = 9.43 \text{ rad/s}$$

Damping: $$c_{cr} = 2\sqrt{km} = 2\sqrt{53{,}334(600)} = 11{,}313 \text{ kN·s/m}$$

$$\zeta = \frac{c}{c_{cr}} = \frac{220}{11{,}313} = 0.019 \approx 2\%$$
$$T = \frac{2\pi}{\omega} = \frac{2\pi}{9.43} = 0.66 \text{ s}$$

knowing T and ζ, the response spectrum given in Fig. 3-8 may be used to directly read the response spectrum quantities S_d, S_v, and S_a, which read

$$S_d = 0.09 \text{ m}, \qquad S_v = 0.9 \text{ m/s}, \qquad S_a = 7 \text{ m/s}^2$$

Therefore, displacement and base shear are given as

$$u_{max} = S_d = 0.09 \text{ m}$$
$$V_{max} = f_{s,max} = mS_a = 600(7) = 4{,}200 \text{ kN}$$

Note from Part (1) and (2) that increasing stiffness attracts more earthquake force. Therefore, increasing strength, which in most cases inherently increases the stiffness of the structure, may not be conducive to earthquake design.

3.3 Generalized Single Degree of Freedom

The dynamic solution to a structural system can be simplified and treated as an SDOF if its vibration can be expressed in a single quantity. Doing so will restrict all the displacements of the structure to a specified deflected shape, the shape function. The results of such analysis are only as good as the conformance of the assumed shape with the exact one.

For an assumed shape function, $\psi(x)$, the displacement of the structure, $v(x,t)$, which is a function of time and space, may be expressed as a product of two functions, one function of space, $\psi(x)$, and one function of time, $z(t)$. This technique is known in mathematics as a separation of variables. Accordingly, the displacement is written as

$$v(x,t) = \psi(x)z(t)$$

where $z(t)$ is defined as a generalized coordinate. The principle of virtual work may be used to formulate such problems. For example, this principle can be applied to the vertical cantilever (shear wall) shown in Fig. 3-14. In this example, the cantilever has a uniformly distributed mass, m, uniform damping, c, and uniform flexure stiffness, EI.

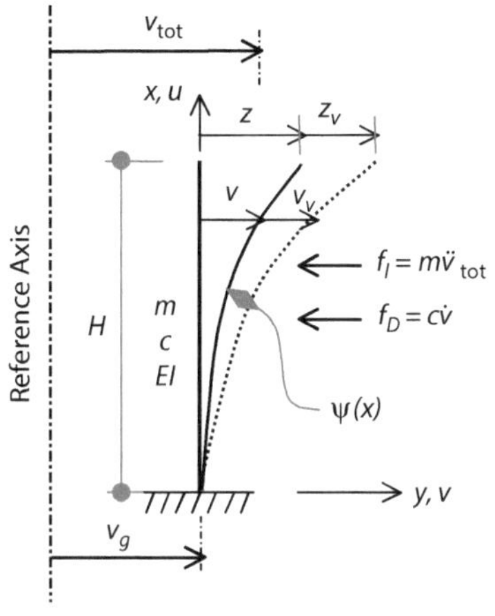

FIGURE 3-14 Generalized SDOF.

Using the virtual work principle, we let the external forces go through a virtual displacement, v_v. Hence, the virtual work done by external forces, f_I and f_D, is equated with the internal work stored by the system (strain energy).

A review of flexure properties reveals that the following relations hold.

Because,

$$v(x,t) = \psi(x)z(t)$$

then

$$v = \psi\, z; \qquad \rightarrow \qquad v_v = \psi\, z_v$$

and

$$\dot{v} = \psi\, \dot{z}; \qquad \ddot{v} = \psi\, \ddot{z}; \qquad v'' = \psi''\, z;$$

and also

$$\ddot{v}_{\text{tot}} = \ddot{v} + \ddot{v}_g = \psi\, \ddot{z} + \ddot{v}_g$$

For beam displacement-curvature, or v-κ relations, the following relations also hold

$$\kappa = -v'' \;\rightarrow\; \kappa_v = -v_v''$$
$$M = EI\kappa = -EI\,v''$$

Knowing the above relations, and with reference to Fig. 3-14, and if $r(x)$ represents the component of ground motion applied to each mass element (DOF) in the direction of motion, the virtual work equation is expressed as follows:

$$W_{\text{ext}} = W_{\text{int}}$$

where $W_{\text{ext}} = -\int (f_I + f_D) v_v dx$

$$= -\int (m\ddot{v}_{\text{tot}} + c\dot{v}) v_v dx = -\int (m\ddot{v} + mr(x)\ddot{v}_g + c\dot{v}) v_v dx$$

$$= -\int (m\psi\ddot{z} + mr\ddot{v}_g + c\psi\dot{z})\psi z_v dx$$

$$= -z_v \int (m\psi^2\ddot{z} + m\psi r\ddot{v}_g + c\psi^2\dot{z}) dx$$

where $r(x)$ is a function known as earthquake directional vector which represents the component of ground motion applied to each infinitesimal mass in the direction of all DOFs. In this example, $r(x)$ is taken 1 because the ground motion is acting parallel to axis v, the direction of mass freedom.

$$\begin{aligned} W_{\text{int}} &= \int M\, d\alpha \\ &= \int M\kappa_v dx = \int (EI\,\kappa)\kappa_v dx = \int (-EI\,v'')(-v_v'')dx \\ &= \int (EI\,\psi'' z)(\psi'' z_v)dx = z_v \int EI(\psi'')^2 z\, dx \end{aligned}$$

By setting $W_{\text{ext}} = W_{\text{int}}$ and noting that z_v cancels out, the equation of motion of the system may be written as

$$\ddot{z}\int m\psi^2 dx + \dot{z}\int c\psi^2 dx + z\int EI(\psi'')^2 dx = -\ddot{v}_g \int m\psi r\, dx$$

or

$$m^*\ddot{z} + c^*\dot{z} + k^* z = -£\ddot{v}_g$$

where m^* is defined as generalized mass, c^* as generalized damping, and k^* as generalized stiffness. The quantity £ is defined as the earthquake excitation factor.

Similar to the SDOF case considered earlier, if the above equation is divided by m^*, the equation of motion reduces to the following form:

$$\ddot{z} + 2\zeta\omega\dot{z} + \omega^2 z = -\Gamma\ddot{v}_g$$

where Γ is defined as the participation factor

$$\Gamma = -\frac{£}{m^*}$$

Note that the above equation is identical to an SDOF system, except that the acceleration is modified by a participation factor Γ. Therefore, for a given ζ and $\omega = \sqrt{k^*/m^*}$, or $T = 2\pi/\omega$, the response spectrum of an earthquake can be used to find the maximum response values of the system S_d, S_v, and S_a. Note also that because $\ddot{v}_g$ is multiplied by Γ in the equation of motion, the response spectrum must also be multiplied by Γ as shown in Fig. 3-15. Thus, the generalized displacement, $z_{\max}$ is given as

$$z_{\max} = \Gamma S_d,$$

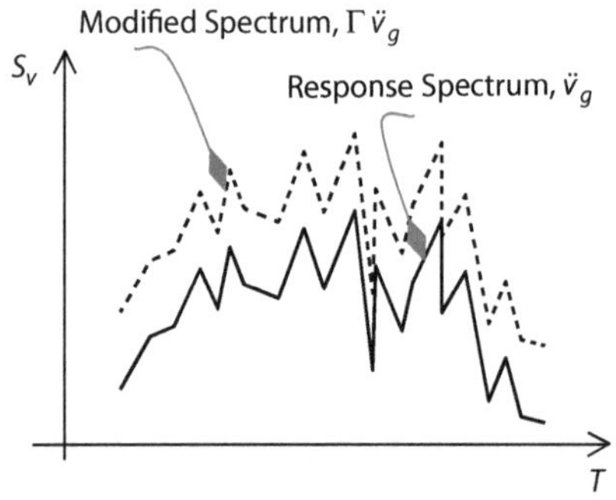

FIGURE 3-15 Modified response spectrum.

and since $v_{max} = \psi z_{max}$, the maximum displacement v_{max} is given as

$$v_{max} = \psi \Gamma S_d$$

The accuracy of this solution largely depends on the choice of the deflected shape ψ. Because ψ is only an approximation of the true shape, its derivatives will be less accurate. Therefore, finding forces from derivatives is considered a poor approximation. For example, if ψ is chosen as $\psi = 1 - \cos \pi x/2H$, for the cantilever example given in Fig. 3-14, the shear force at the base can be given in terms of derivatives as $V = -EIv'''$. Note that the shear at the base is maximum, whereas the derivative gives the shear at the base as zero ($\propto v'''$), a completely erroneous result!

A more accurate approximation is to find the force, f_s, first and then evaluate shears and moments accordingly. The force, f_s, may be found by noting that the true vibration mode shape must satisfy the differential equation ($m\ddot{u} + EIu^{iv} = 0$) in undamped harmonic-free vibration. From structural dynamics, the solution to this equation is given as ($u = u_o \cos\omega t$). Differentiating the solution twice shows that ($\ddot{u} = -\omega^2 u$). Therefore, the differential equation may be written as

$$m\ddot{u} + EIu^{iv} = 0$$
$$m(-\omega^2 u) + f_s = 0$$

or

$$f_s = m\omega^2(\psi z)$$

hence,

$$f_{s,max} = m\omega^2 \psi z_{max}$$
$$= m\psi\omega^2(\Gamma S_d)$$
$$f_{s,max} = m\psi\Gamma S_a$$

This relationship gives the earthquake-induced maximum forces in the structure. Note that the force distribution is proportional to the assumed shape function as shown in Fig. 3-16. The reaction of this maximum force at the base, or the total earthquake-induced force that is known as the base shear, V_B, may be expressed as follows:

$$V_B = \int f_{s,max}\, dx$$

The required shape function for this analysis may be approximated with various functions. A good guess for these shapes is to use the deflected shape from static loadings, for example, top concentrated force, uniformly distributed load, or triangle loading as shown in Fig. 3-17.

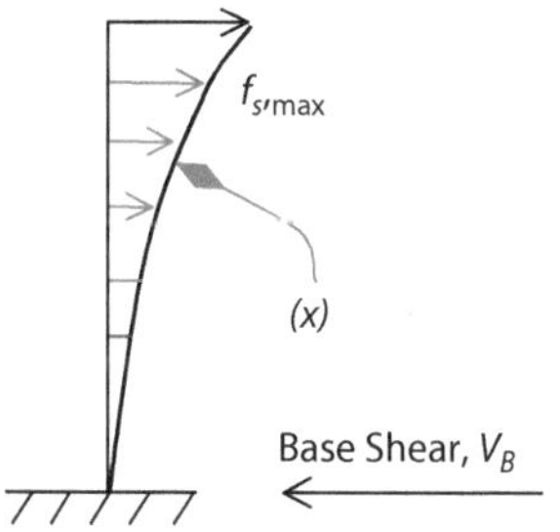

FIGURE 3-16 Earthquake equivalent force.

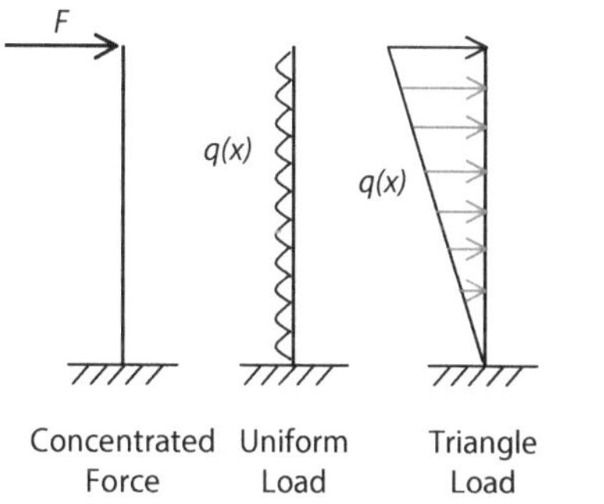

FIGURE 3-17 Static loading.

Example 3-2 The cantilever wall shown in Example 3-2, Fig. 1 has the following properties:

Uniform mass: $m = 0.77\ \text{kN·s}^2/\text{m}^2$

Uniform inertia: $EI = 0.4 \times 10^6\ \text{kN·m}^2$

This figure also shows that the wall is carrying two concentrated masses at the top and at midheight. The deflected shape of this structure is given as

$$\psi(x) = 1 - \cos\left(\frac{\pi x}{2H}\right)$$

The structure is excited by the El Centro earthquake with $\zeta = 5\%$. It is required to analyze this wall under two conditions:

(A) If the wall does not carry any concentrated masses.
(B) If the wall carries both concentrated masses, M_1 and M_2, as shown in the same figure.

For both cases (A) and (B), determine:

1. The maximum top displacement, base moment, and base shear.
2. The maximum displacement, moment, and shear at midheight.

Solution
Case (A)
Wall without concentrated masses

Part (A-1)
Shape function

$$\psi(x) = 1 - \cos\left(\frac{\pi x}{2H}\right)$$

$$\psi'(x) = \left(\frac{\pi}{2H}\right)\sin\left(\frac{\pi x}{2H}\right)$$

$$\psi''(x) = \left(\frac{\pi}{2H}\right)^2 \cos\left(\frac{\pi x}{2H}\right)$$

Generalized parameters

$$m^* = \int_0^H \psi m \psi \, dx$$

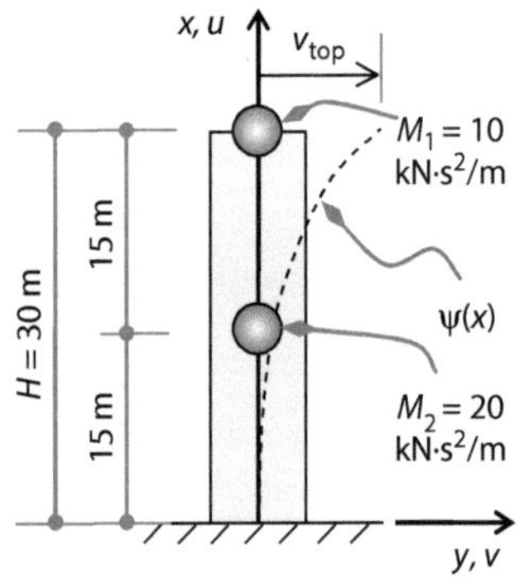

Example 3-2, Figure 1

$$= m\int_0^H \left[1-\cos\left(\frac{\pi x}{2H}\right)\right]^2 dx$$

$$= 0.227 \text{ mH}$$

$$= 0.227(0.77)(30) = 5.244 \text{ kN·s}^2\text{/m}$$

$$k^* = \int_0^H EI(\psi'')^2 dx$$

$$= \int_0^H EI\left[\left(\frac{\pi}{2H}\right)^2 \cos\left(\frac{\pi x}{2H}\right)\right]^2 dx$$

$$= EI\left(\frac{\pi}{2H}\right)^4 \left(\frac{H}{2}\right)$$

$$= 400\times 10^3 \left(\frac{\pi}{2\,(30)}\right)^4 \left(\frac{30}{2}\right) = 45.096 \text{ kN/m}$$

$$\pounds = \int_0^H \psi m r dx$$

$$= m\int_0^H \left[1-\cos\left(\frac{\pi x}{2H}\right)\right](1) dx$$

$$= 0.363\, mH = 0.363\,(0.77)(30) = 8.385 \text{ kN·s}^2\text{/m}$$

$$\Gamma = \frac{\pounds}{m^*} = \frac{8.386}{5.244} = 1.599$$

$$\omega = \sqrt{\frac{k^*}{m^*}} = \sqrt{\frac{45.096}{5.244}} = 2.93 \text{ rad/s}$$

$$T = \frac{2\pi}{\omega} = \frac{2\pi}{2.932} = 2.14 \text{ s}$$

With $T = 2.14$ s and $\zeta = 5\%$, the El Centro earthquake response spectrum given in Fig. 3-8 yields

$$S_d = 0.23 \text{ m}, \qquad S_a = 2.0 \text{ m/s}^2$$

Therefore, the resulting displacements and forces are given as

$$v_{\text{top}} = \psi \Gamma S_d$$

$$= 1(1.599)(0.23) = 0.37 \text{ m}$$

$$f_{s,\text{top}} = m\psi\Gamma S_a$$

$$= 0.77(1)(1.599)(2.0)$$

$$= 2.462 \text{ kN/m}$$

Base shear, V_B, in Example 3-2, Fig. 2

$$V_B = \int_0^H f_{s,\max} dx$$

$$= \int_0^H m\psi\Gamma S_a dx$$

$$= m\Gamma S_a \int_0^H \left[1-\cos\left(\frac{\pi x}{2H}\right)\right] dx$$

$$= m\Gamma S_a (0.363H)$$

$$= 0.77\,(1.599)(2.0)(0.363)(30) = 26.82 \text{ kN}$$

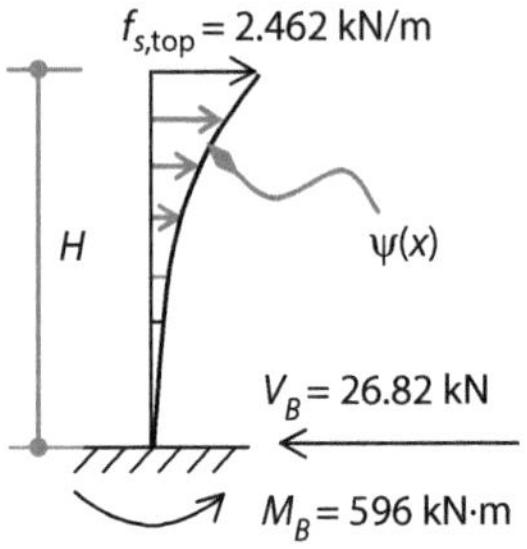

EXAMPLE 3-2, FIGURE 2 Base reactions.

Base Moment, M_B, in Example 3-2, Fig. 2

$$
\begin{aligned}
M_B &= \int_0^H f_{s,\max}\, x dx \\
&= \int_0^H m\psi\Gamma S_a x dx \\
&= m\Gamma S_a \int_0^H \left[1-\cos\left(\frac{\pi x}{2H}\right)\right] x dx \\
&= m\Gamma S_a (0.269\, H^2) \\
&= 0.77(1.599)(2.0)(0.269)(30^2) = 596 \text{ kN·m}
\end{aligned}
$$

Part (A-2)

Similarly, displacements and forces at the midheight of 15 m can be found by integration as follows:

$$
\begin{aligned}
v_{\text{mid}} &= \psi\Gamma S_d = \left[1-\cos\left(\frac{\pi x}{2H}\right)\right]\Gamma S_d \\
&= \left[1-\cos\left(\frac{\pi(15)}{2(30)}\right)\right](1.599)(0.23) = 0.108 \text{ m}
\end{aligned}
$$

$$
\begin{aligned}
V_{\text{mid}} &= \int_{15}^{30} f_{s,\max}\, dx \\
&= \int_{15}^{30} m\psi\Gamma S_a dx \\
&= m\Gamma S_a (0.313\, H) \\
&= 0.77\,(1.599)(2.0)(0.313)(30) = 23.12 \text{ kN}
\end{aligned}
$$

$$
\begin{aligned}
M_{\text{mid}} &= \int_{15}^{30} f_{s,\max}(x-15)dx \\
&= \int_{15}^{30} m\psi\Gamma S_a (x-15)dx \\
&= m\Gamma S_a (0.093\, H^2) \\
&= 0.77(1.599)(2.0)(0.093)(30^2) = 206 \text{ kN·m}
\end{aligned}
$$

Case (B):

Wall with concentrated masses

Results from case (A) can be used in this case.

Part (B-1)

Shape function values at concentrated mass locations

$$\psi(x) = 1 - \cos\left(\frac{\pi x}{2H}\right)$$

$$\psi(15) = 1 - \cos\left[\frac{\pi(15)}{2(30)}\right] = 0.293$$

$$\psi(30) = 1 - \cos\left[\frac{\pi(30)}{2(30)}\right] = 1.0$$

Generalized parameters

$$m^* = \int_0^H \psi m \psi \, dx + \sum \psi_i M_i \psi_i$$

$$= \text{same as case A} + \psi_1 M_1 \psi_1 + \psi_2 M_2 \psi_2$$

$$= 5.244 + 10(1)^2 + 20(0.293)^2$$

$$= 5.244 + 10 + 1.717 = 16.961 \text{ kN·s}^2\text{/m}$$

$$k^* = \text{same as case A} = 45.096 \text{ kN/m}$$

$$£ = \int_0^H \psi m r \, dx + \sum \psi_i M_i r$$

$$= \text{same as case A} + \psi_1 M_1 r + \psi_2 M_2 r$$

$$= 8.385 + 10(1)(1) + 20(0.293)(1)$$

$$= 8.385 + 10 + 5.86 = 24.245 \text{ kN·s}^2\text{/m}$$

$$\Gamma = \frac{£}{m^*} = \frac{24.245}{16.961} = 1.429$$

$$\omega = \sqrt{\frac{k^*}{m^*}} = \sqrt{\frac{45.096}{16.961}} = 1.63 \text{ rad/s}$$

$$T = \frac{2\pi}{\omega} = \frac{2\pi}{1.63} = 3.85 \text{ s}$$

With $T = 3.85$ s and $\zeta = 5\%$, the El Centro earthquake response spectrum given in Fig. 3-8 yields

$$S_d = 0.20 \text{ m} \qquad S_a = 0.53 \text{ m/s}^2$$

Therefore, displacements and forces are given as

$$v_{\text{top}} = \psi \Gamma S_d$$

$$= 1\,(1.429)(0.20) = 0.286 \text{ m}$$

$$f_{s,\text{top}} = m\,\psi \Gamma S_a$$

$$= 0.77\,(1)(1.429)(0.53)$$

$$= 0.583 \text{ kN/m}$$

$$F_{s1} = M_1 \psi_1 \Gamma S_a$$

$$= 10\,(1)(1.429)(0.53)$$

$$= 7.57 \text{ kN}$$

$$F_{s2} = M_2 \psi_2 \Gamma S_a$$

$$= 20\,(0.293)(1.429)(0.53)$$

$$= 4.44 \text{ kN}$$

Base shear, V_B, in Example 3-2, Fig. 3

$$V_B = \int_0^H f_{s,\max}\,dx + F_{s1} + F_{s2}$$
$$= m\Gamma S_a(0.363\,H) + 7.57 + 4.44$$
$$= 0.77(1.429)(0.53)(0.363)(30) + 7.57 + 4.44$$
$$= 6.35 + 7.57 + 4.44 = 18.36 \text{ kN}$$

Base moment, M_B, in Example 3-2, Fig. 3

$$M_B = \int_0^H f_{s,\max}\,x\,dx + F_{s1}(H) + F_{s2}(H/2)$$
$$= m\Gamma S_a(0.269\,H^2) + F_{s1}(H) + F_{s2}(H/2)$$
$$= 0.77(1.429)(0.53)(0.269)(30^2) + 7.57(30) + 4.44(15)$$
$$= 141 + 227 + 67 = 435 \text{ kN·m}$$

Part (B-2)
Similarly, displacements and forces at the midheight of 15 m can be found by integration as follows:

$$v_{\text{mid}} = \psi\Gamma S_d = \left[1 - \cos\left(\frac{\pi x}{2H}\right)\right]\Gamma S_d$$
$$= \left[1 - \cos\left(\frac{\pi(15)}{2(30)}\right)\right](1.429)(0.20) = 0.084 \text{ m}$$

The shear force at midheight will have a jump at the location of F_{s2}.
Thus, shear force above F_{s2} equals to

$$V_{\text{mid}} = \int_{15}^{30} f_{s,\max}\,dx + F_{s1}$$
$$= m\Gamma S_a(0.313\,H) + F_{s1}$$
$$= 0.77(1.429)(0.53)(0.313)(30) + 7.57$$
$$= 5.48 + 7.57 = 13.05 \text{ kN}$$

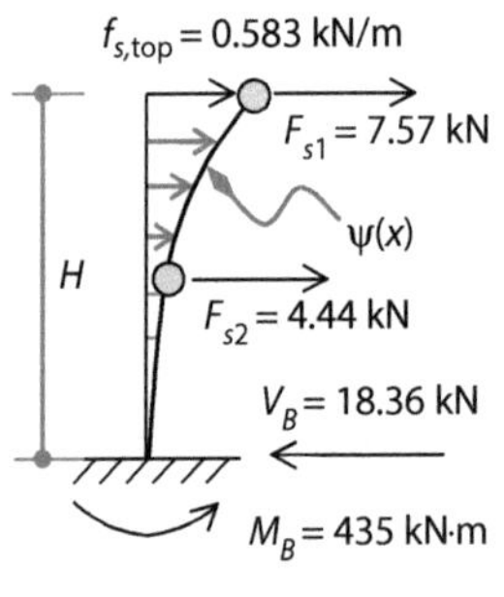

Example 3-2, Figure 3 Base reactions.

Shear force below F_{s2} equals to

$$\begin{aligned}V_{\text{mid}} &= \int_{15}^{30} f_{s,\max}\,dx + F_{s1} + F_{s2}\\ &= m\Gamma S_a(0.313\,H) + F_{s1} + F_{s2}\\ &= 0.77(1.429)(0.53)(0.313)(30) + 7.57 + 4.44\\ &= 5.48 + 7.57 + 4.44 = 17.49\ \text{kN}\end{aligned}$$

$$\begin{aligned}M_{\text{mid}} &= \int_{15}^{30} f_{s,\max}(x-15)\,dx + F_{s1}\left(\frac{H}{2}\right)\\ &= m\Gamma S_a(0.093\,H^2) + F_{s1}\left(\frac{H}{2}\right)\\ &= 0.77(1.429)(0.53)(0.093)(30^2) + 7.57(15)\\ &= 49 + 114 = 163\ \text{kN·m}\end{aligned}$$

3.4 Multiple Degrees of Freedom System

Multiple degrees of freedom (MDOF) systems will be presented in the following sections for the case of two-dimensional (2D) analysis, 2D systems, and for the case of three-dimensional (3D) analysis, 3D systems.

3.4.1 Multiple Degrees of Freedom System in 2D Analysis

An MDOF system consists of multiple lumped (concentrated) masses (m_1, m_2, etc.), where these masses are subjected to varying forces with time as shown in Fig. 3-18. The movement of each mass is resisted by its inertial force, f_I, damping force, f_D, and elastic stiffness, f_S. By establishing the dynamic equilibrium of each mass in the direction of its movement, Fig. 3-19 shows the equilibrium of mass, m_2, which requires that

$$\sum F_{x2} = 0$$

$$m_2\,\ddot{u}_{\text{tot}} + c_2(\dot{u}_2 - \dot{u}_1) + k_2(u_2 - u_1) - c_3(\dot{u}_3 - \dot{u}_2) - k_3(u_3 - u_2) = 0$$

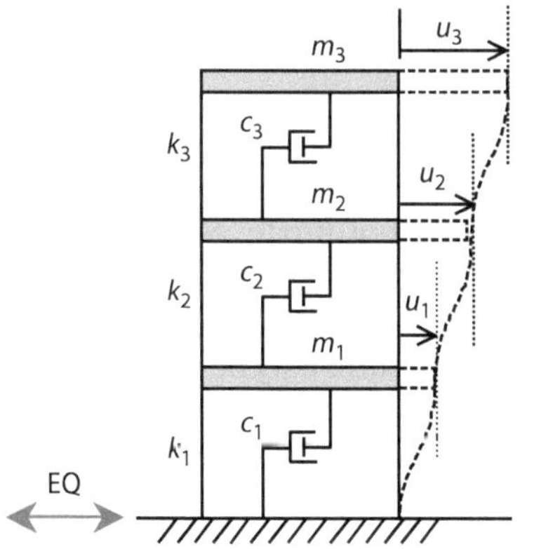

FIGURE 3-18 MDOF structure.

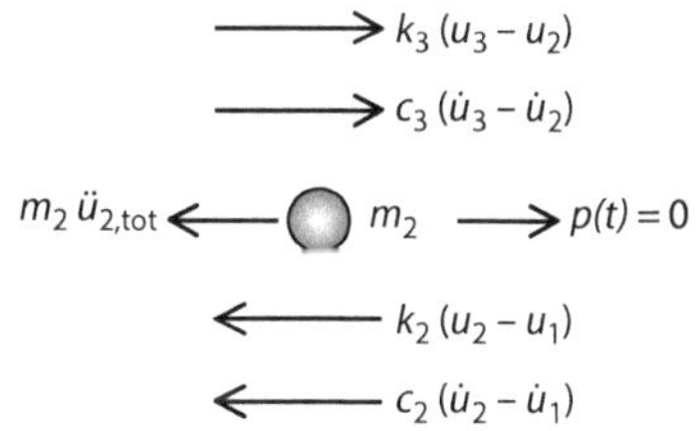

FIGURE 3-19 Dynamic equilibrium of mass m_2.

Similarly, by taking $\Sigma F_{xi} = 0$ at each mass, substituting $\ddot{u}_{tot} = \ddot{u} + \ddot{u}_g$ and rearranging the equations, they can be put in matrix form as follows:

$$\begin{bmatrix} m_1 & 0 & 0 \\ 0 & m_2 & 0 \\ 0 & 0 & m_3 \end{bmatrix} \begin{Bmatrix} \ddot{u}_1 \\ \ddot{u}_2 \\ \ddot{u}_3 \end{Bmatrix} + \begin{bmatrix} c_{11} & c_{12} & c_{13} \\ c_{21} & c_{22} & c_{23} \\ c_{31} & c_{32} & c_{33} \end{bmatrix} \begin{Bmatrix} \dot{u}_1 \\ \dot{u}_2 \\ \dot{u}_3 \end{Bmatrix} + \begin{bmatrix} k_{11} & k_{12} & k_{13} \\ k_{21} & k_{22} & k_{23} \\ k_{31} & k_{32} & k_{33} \end{bmatrix} \begin{Bmatrix} u_1 \\ u_2 \\ u_3 \end{Bmatrix}$$

$$= -\begin{bmatrix} m_1 & 0 & 0 \\ 0 & m_2 & 0 \\ 0 & 0 & m_3 \end{bmatrix} \begin{Bmatrix} 1 \\ 1 \\ 1 \end{Bmatrix} \ddot{u}_g$$

or,

$$[M]\{\ddot{U}\}+[C]\{\dot{U}\}+[K]\{U\}=-[M]\{R\}\ddot{u}_g$$

where $[M]$ = mass matrix (square matrix)
$[C]$ = damping matrix (square matrix)
$[K]$ = stiffness matrix (square matrix)
$\{U\}, \{\dot{U}\}, \{\ddot{U}\}$ = displacement, velocity, and acceleration matrices (column matrices)
$\{R\}$ = earthquake directional vector, or influence vector, which defines the component of ground motion in the direction of the DOFs
= vector of ground motion components in the direction of each DOF resulting from unit support displacement in direction of ground motion
$\{\ddot{u}_g\}$ = ground acceleration (m/s²)

In general, for N-degrees of freedom, there are square matrices, $N \times N$, and column matrices, $N \times 1$, of order. The complete formulation from the equilibrium equation yields N-coupled linear differential equations, which can be solved by numerical analysis. However, for a linear elastic system, the solution may be simplified using the technique of modal analysis, also known as modal superposition.

Modal Analysis

Consider the case of an undamped MDOF system in free vibration

$$[M]\{\ddot{U}\} + [K]\{U\} = \{0\}$$

By separation of variables x and t, the displacement, $U(x,t)$, will be shown to be given as $\{U(x,t)\} = [\Phi(x)]\{Z(t)\}$, where $\phi_n(x)$ is the nth mode of vibration associated with an nth frequency, ω_n. Z_n is defined as the nth modal coordinate. Note that ϕ_n has a fixed shape that vibrates with frequency ω_n and has an amplitude Z_n.

The mode shape, $\phi_n(x)$, is found by solving the eigenvalue problem, which is also known as the characteristic-value problem. The solution may be obtained by considering the solution of the free vibration of the time function, $Z(t)$.

Since the solution of free vibration time function, $Z(t)$, is given as

$$Z = A_o \sin\omega t$$

then,

$$\dot{Z} = \omega A_o \cos\omega t$$

$$\ddot{Z} = -\omega^2 A_o \sin\omega t = -\omega^2 Z$$

Substitution in free vibration equation yields

$$[M]\{\ddot{U}\} + [K]\{U\} = \{0\}$$

$$[M]\{\phi\}\ddot{Z} + [K]\{\phi\}Z = \{0\}$$

From above

$$-\omega^2 [M]\{\phi\} Z + [K]\{\phi\}Z = \{0\}$$

$$[-\omega^2 [M]\{\phi\} + [K]\{\phi\}]Z = \{0\}$$

For nontrivial solution, that is $Z \neq 0$, then

$$[K]\{\phi\} - \omega^2[M]\{\phi\} = \{0\} \quad \rightarrow \quad \text{Note that } [K]\{\phi\} = \omega^2 [M]\{\phi\}$$

$$[[K] - \omega^2 [M]]\{\phi\} = \{0\}$$

The last equation above represents a set of N-linear homogeneous equations. The solution can only be obtained for relative values of $\{\phi\}$. Therefore, for $\{\phi\} \neq \{0\}$, Kramer rule requires that the following determinant equals to zero.

$$|\{[K] - \omega^2 [M]\}| = 0$$

The equation above is known as the frequency equation, where ω^2 is its eigenvalue or characteristic value. Solving an eigenvalue problem, the above frequency equation yields N frequencies ($\omega_1, \omega_2, \ldots \omega_N$) and N-mode shapes ($\{\phi_1\}, \{\phi_2\}, \ldots \{\phi_N\}$), where the lowest frequency is called the fundamental frequency of the system and is associated with the fundamental mode, or the first mode of vibration.

Orthogonality of Mode Shapes

The mode shapes, $\phi_n(x)$, obtained from the eigenvalue solution exhibit the property of orthogonality

$$\{\phi_m\}^T [M]\{\phi_n\} = \{0\} \qquad \text{for } n \neq m$$

$$\{\phi_m\}^T [K]\{\phi_n\} = \{0\} \qquad \text{for } n \neq m$$

Orthogonality may be proved as follows:
Because

$$\{[K] - \omega_n^2 [M]\}\{\phi_n\} = \{0\}$$

$$[K]\{\phi_n\} - \omega_n^2 [M]\{\phi_n\} = \{0\}$$

Premultiply by $\{\phi_m\}^T$ leads to

$$\{\phi_m\}^T [K]\{\phi_n\} - \omega_n^2 \{\phi_m\}^T [M]\{\phi_n\} = \{0\} \tag{3-1}$$

Switching subscripts leads to

$$\{\phi_n\}^T [K]\{\phi_m\} - \omega_m^2 \{\phi_n\}^T [M]\{\phi_m\} = \{0\} \tag{3-2}$$

Transpose of (2)

$$\{\phi_m\}^T [K]\{\phi_n\} - \omega_m^2 \{\phi_m\}^T [M]\{\phi_n\} = \{0\} \tag{3-3}$$

Note that,

$$K^T = K \ldots \text{symmetry}$$

and

$$M^T = M \ldots \text{symmetry}$$

$$(3\text{-}1) - (3\text{-}3) \quad \rightarrow \quad \{\phi_m\}^T[M]\{\phi_n\}\{\omega_m^2 - \omega_n^2\} = \{0\}$$

Note that if $m \neq n$, then $\omega_m \neq \omega_n$ and hence we get

$$\left.\begin{array}{ll} \{\phi_m\}^T[M]\{\phi_n\} = \{0\} & \text{if } m \neq n \\ \{\phi_m\}^T[K]\{\phi_n\} = \{0\} & \text{if } m \neq n \end{array}\right\} \text{Orthogonality condition}$$

Likewise (second line above)

Therefore, if $\{U(x,t)\}$ is given as $\{U(x,t)\} = [\Phi(x)]\{Z(t)\}$, the equation of motion may be written as

$$[M]\{\ddot{U}\}+[C]\{\dot{U}\}+[K]\{U\}=-[M]\{R\}\ddot{u}_g$$

$$[M][\Phi]\{\ddot{Z}\}+[C][\Phi]\{\dot{Z}\}+[K][\Phi]\{Z\}=-[M]\{R\}\ddot{u}_g$$

Premultiply by $[\Phi(x)]^T$ results in

$$[\Phi]^T[M][\Phi]\{\ddot{Z}\}+[\Phi]^T[C][\Phi]\{\dot{Z}\}+[\Phi]^T[K][\Phi]\{Z\}=-[\Phi]^T[M]\{R\}\ddot{u}_g$$

Because of orthogonality, the equation above results in diagonal matrices in the form

$$\begin{bmatrix} m_1^* & 0 & 0 \\ 0 & m_2^* & 0 \\ 0 & 0 & m_3^* \end{bmatrix} \begin{Bmatrix} \ddot{z}_1 \\ \ddot{z}_2 \\ \ddot{z}_3 \end{Bmatrix} + \begin{bmatrix} c_1^* & 0 & 0 \\ 0 & c_2^* & 0 \\ 0 & 0 & c_3^* \end{bmatrix} \begin{Bmatrix} \dot{z}_1 \\ \dot{z}_2 \\ \dot{z}_3 \end{Bmatrix} + \begin{bmatrix} k_1^* & 0 & 0 \\ 0 & k_2^* & 0 \\ 0 & 0 & k_3^* \end{bmatrix} \begin{Bmatrix} z_1 \\ z_2 \\ z_3 \end{Bmatrix} = -\begin{Bmatrix} £_1 \\ £_2 \\ £_3 \end{Bmatrix} \ddot{u}_g$$

These are N-uncoupled equations in the form

$$m_n^*\ddot{Z}_n + c_n^*\dot{Z}_n + k_n^*Z_n = -£_n\ddot{u}_g$$

The equation above is known as the modal equation of motion,

where $m_n^* = \{\phi_n\}^T[M]\{\phi_n\}$, modal mass of mode n
$c_n^* = \{\phi_n\}^T[C]\{\phi_n\}$, modal damping of mode n
$k_n^* = \{\phi_n\}^T[K]\{\phi_n\}$, modal stiffness of mode n
$£_n = \{\phi_n\}^T[M]\{R\}$, modal earthquake excitation factor of mode n
$\{R\}$ = earthquake directional vector as defined earlier

Dividing the modal equation by m_n^* results in

$$\ddot{Z}_n + 2\zeta\omega_n\dot{Z}_n + \omega_n^2 Z_n = -\Gamma_n\ddot{u}_g$$

where Γ_n is the modal participation factor for mode n which is equal to

$$\Gamma_n = \frac{£_n}{m_n^*} = \frac{\{\phi_n\}^T[M]\{R\}}{\{\phi_n\}^T[M]\{\phi_n\}}$$

Note the similarity of the modal equation of motion and the SDOF system. Using the same procedure for an SDOF solution, the maximum modal response values of S_{dn}, S_{vn}, and S_{an} may be found from the response spectrum using $T_n = 2\pi/\omega_n$. Therefore, $Z_{n,\max}$ and $U_{n,\max}$ are given as $Z_{n,\max} = \Gamma_n S_{dn}$, and since $\{U_{n,\max}\} = \{\phi_n\}Z_{n,\max}$, the modal displacements are given as

$$\{U_{n,\max}\} = \{\phi_n\}Z_{n,\max}$$

$$\{U_{n,\max}\} = \{\phi_n\}\Gamma_n S_{dn}$$

Because the frequency equation in previous sections yields the following relation

$$[K]\{\phi\} = \omega^2 [M]\{\phi\}$$

the modal forces are given as

$$\begin{aligned}\{f_{sn,\max}\} &= [K]\{U_{n,\max}\}\\ &= [K]\{\phi_n\}Z_{n,\max}\\ &= \omega^2 [M]\{\phi_n\}Z_{n,\max}\\ &= \omega^2 [M]\{\phi_n\}\Gamma_n S_{dn}\\ \{f_{sn,\max}\} &= [M]\{\phi_n\}\Gamma_n S_{an}\end{aligned}$$

In summary, the displacements and mode shapes are given in forms similar to those of the generalized SDOF system by replacing the continuous functions by their counterpart matrices

$$\{U_{n,\max}\} = \{\phi_n\}\Gamma_n S_{dn}$$

$$\{f_{sn,\max}\} = [M]\{\phi_n\}\Gamma_n S_{an}$$

The procedures above are performed for N-modes yielding N-displacements and N-modal forces. Of course, the modal forces result in modal shears and modal moments. The total response of the structure will be equal to the summation of all modal quantities (modal superposition). However, on account of the difference of modal frequencies (periods), the vibration modes do not usually vibrate in phase. Therefore, the summation of their absolute values will be conservative. The summation of absolute values is abbreviated as SABS and is given in the following form:

$$\{U\}_{\max} = \sum \left|\{U_{n,\max}\}\right|$$

$$\{F_s\}_{\max} = \sum \left|\{F_{sn,\max}\}\right|$$

A more realistic summation procedure is the square root of the sum of squares (SRSS) which is given by the following form:

$$\{U_{\max}\} = \sqrt{\sum_{1}^{N} \{U_{n,\max}^2\}}$$

$$\{F_{s,\max}\} = \sqrt{\sum_{1}^{N} \{F_{sn,\max}^2\}}$$

The SRSS method gives satisfactory answers if the natural periods of the structure are well separated (if they are unlikely to vibrate in phase). However, because periods might come close to each other in 3D structures, there is a likelihood that the modes will vibrate in phase. If this occurs, the SABS method should be used. Remember that the SABS is always an upper bound solution.

Better formulas exist and are incorporated in commercial programs. For example, a more general complete quadratic combinations (CQC) method could be used. This method is based on probabilistic correlation of the periods of the mode shapes and requires statistical analysis as described by Clough (1993).

Caution

One must observe the right sequence of summation before adding the modal quantities. For example, the SRSS values of the interstory drift, Δ_i, should be obtained by adding the squares of the interstory drift from each mode and then by taking the square root of the summation. It would be wrong to obtain the interstory drift by taking the difference of the SRSS of the displacement of the adjacent stories. Figure 3-20 illustrates the summation sequence if the displacements of stories 1 and 2 are given in the first and second mode as follows:

Mode (1): $$\{U_{1,\max}\} = \begin{Bmatrix} 2 \\ -1 \end{Bmatrix}$$

Mode (2): $$\{U_{2,\max}\} = \begin{Bmatrix} 5 \\ 3 \end{Bmatrix}$$

SRSS of story 1: $$u_1 = \sqrt{2^2 + 5^2} = 5.39$$

SRSS of story 2: $$u_2 = \sqrt{1^2 + 3^2} = 3.16$$

SRSS of interstory: $$\Delta_{1-2} = \sqrt{3^2 + 2^2} = 3.6$$

The correct interstory drift is the SRSS of the interstory drift, or 3.6. It would be wrong to consider the interstory drift as the difference between the SRSS of the displacement of adjacent stories, for example (5.39 – 3.16 = 2.23), or even (5.39 + 3.16 = 8.55).

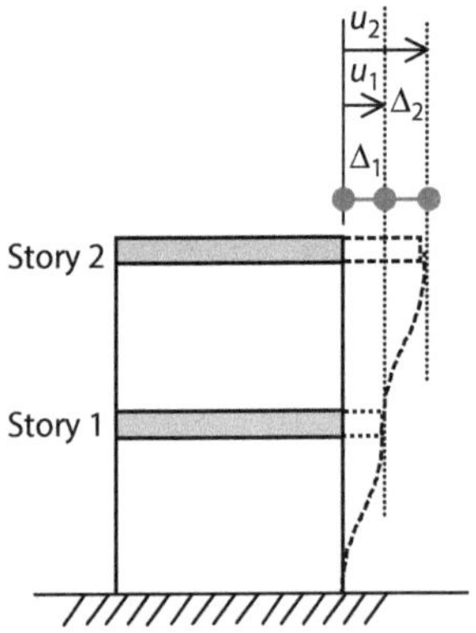

FIGURE 3-20 Interstory drift.

Example 3-3 The structure shown in Example 3-3, Fig. 1, is idealized as 2DOFs, v_1 and v_2. The structure properties are given as shown in the figure as

$$m = 15 \text{ kN·s}^2/\text{m}$$
$$EI = 1{,}500 \text{ kN·m}^2$$
$$L = 1 \text{ m}$$

Calculate the maximum SRSS base shear and base moment if the structure is excited by the El Centro earthquake along its axis, *Z-Z*. Take $\zeta = 2\%$. Note that the members are considered axially rigid.

Solution
Define relevant matrices

1. Displacement vector

$$\{U\} = \begin{Bmatrix} v_1 \\ v_2 \end{Bmatrix}$$

2. Mass matrix
The mass matrix is constructed by applying unit acceleration in each DOF and calculating the corresponding resulting inertial forces. Remember that the acceleration of all other DOFs must be kept zero, hence
Applying unit acceleration along v_1 results in matrix element, $m_{11} = 2\,m$.
Applying unit acceleration along v_2 results in matrix element, $m_{22} = 1\,m$.
Therefore,

$$[M] = m\begin{bmatrix} 2 & 0 \\ 0 & 1 \end{bmatrix}$$

3. Stiffness matrix
The stiffness matrix can be constructed in the usual procedures either by the stiffness method or the flexibility method. If the stiffness method is used, the matrix must be condensed by kinematic condensation procedures to eliminate the rotational DOF. In this case, it is easier to use the flexibility method where the flexibility matrix can be directly constructed without the rotational DOF. The inverse of the flexibility matrix will result in the required stiffness matrix.

Flexibility Method
Apply unit force in the direction of each DOF. Then find displacements accordingly. For example

$$\begin{Bmatrix} v_1 \\ v_2 \end{Bmatrix} = \begin{bmatrix} f_{11} & f_{12} \\ f_{21} & f_{22} \end{bmatrix} \begin{Bmatrix} F_1 \\ F_2 \end{Bmatrix}$$

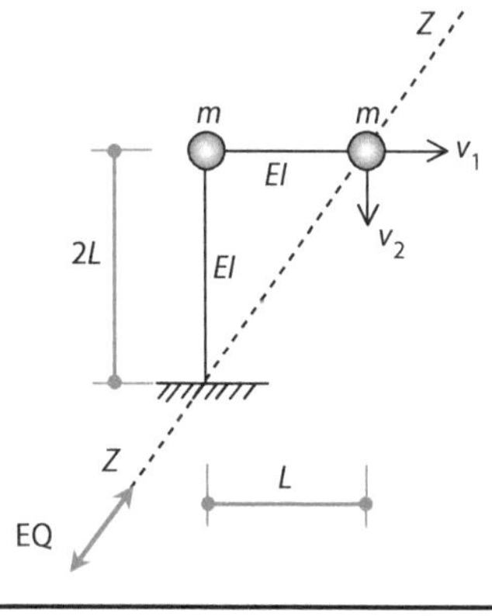

EXAMPLE 3-3, FIGURE 1

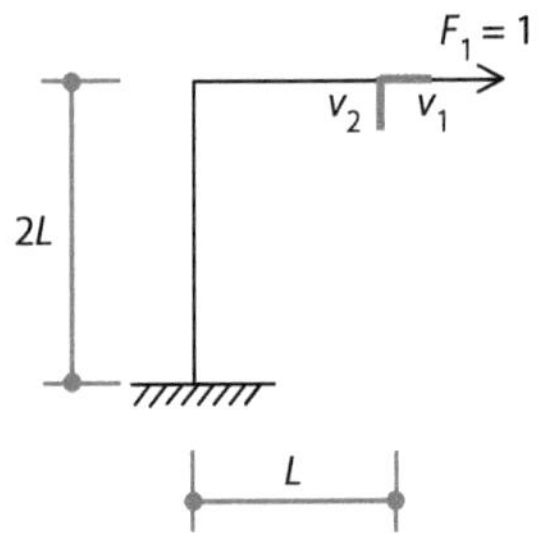

EXAMPLE 3-3, FIGURE 2 Unit force in Direction 1.

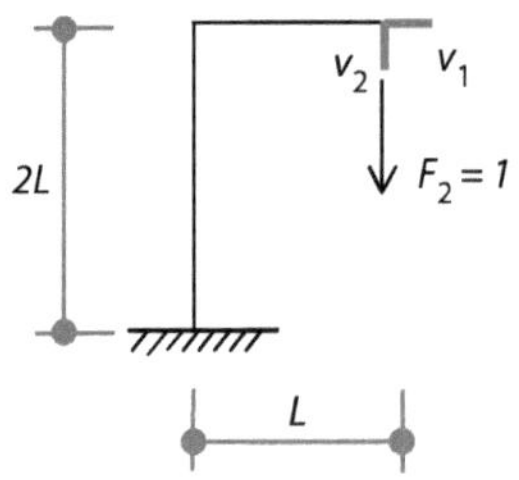

EXAMPLE 3-3, FIGURE 3 Unit force In Direction 2.

By applying unit force in direction 1 as shown in Example 3-3, Fig. 2, the corresponding displacements may be found using any method in structural analysis (e.g., virtual work). The resulting displacements are calculated as follows:

$$v_1 = f_{11} = \frac{8L^3}{3EI}$$
$$v_2 = f_{21} = \frac{2L^3}{EI}$$

Similarly, the corresponding displacements may also be found by applying unit force in direction 2 as shown in Example 3-3, Fig. 3. The resulting displacements are calculated as follows:

$$v_1 = f_{12} = \frac{2L^3}{EI}$$
$$v_2 = f_{22} = \frac{7L^3}{3EI}$$

The results above can be arranged to yield the flexibility matrix as follows:

$$[F] = \frac{L^3}{3EI}\begin{bmatrix} 8 & 6 \\ 6 & 7 \end{bmatrix}$$

The stiffness matrix can now be obtained by taking the inverse of [F] to yield

$$[K] = \frac{3EI}{20L^3}\begin{bmatrix} 7 & -6 \\ -6 & 8 \end{bmatrix}$$

Stiffness Method with Kinematic Condensation

When we use the stiffness formulation in reference to Example 3-3, Fig. 4, the given structure has 4DOFs, two displacements (v_1, v_2), and two rotations (θ_1, θ_2), provided the structure has

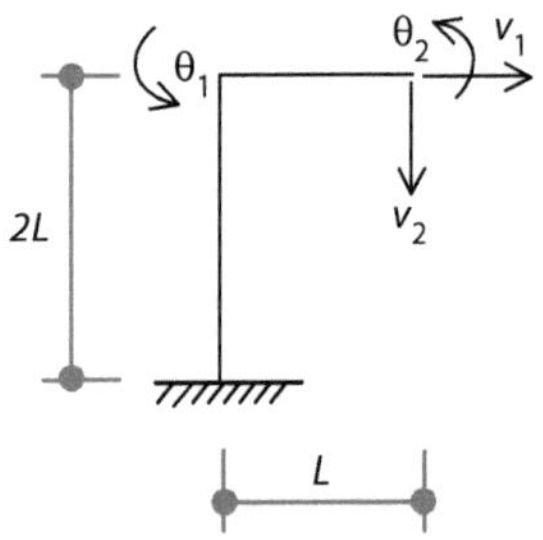

EXAMPLE 3-3, FIGURE 4 Stiffness method.

axial rigidity. The global stiffness matrix may be assembled with the usual procedures and arranged by separation of the displacement DOFs from the rotational DOFs. If the matrices are partitioned, they can be given in the following form:

$$\begin{Bmatrix} \{F\} \\ \{M\} \end{Bmatrix} = \begin{bmatrix} [K_{11}] & [K_{12}] \\ [K_{21}] & [K_{22}] \end{bmatrix} \begin{Bmatrix} \{v\} \\ \{\theta\} \end{Bmatrix}$$

where $\{F\}$ and $\{v\}$ are the force and displacement submatrices, and $\{M\}$ and $\{\theta\}$ are the moment and rotational submatrices. Insofar as the external moments at the rotational DOFs will always be zero, the rotational DOFs may be eliminated as follows:

Multiplying the bottom row in the matrix above and noting that $\{M\} = 0$ yields

since $$\{M\} = [K_{21}]\{v\} + [K_{22}]\{\theta\} = \{0\}$$

then $$\{\theta\} = -[K_{22}]^{-1}[K_{21}]\{v\}$$

Multiplying the first row of the original matrix yields

$$\{F\} = [K_{11}]\{v\} + [K_{12}]\{\theta\}$$

substitution of $\{\theta\}$, as previously obtained from the equation above, yields

$$\{F\} = [K_{11}]\{v\} + [K_{12}]\{-[K_{22}]^{-1}[K_{21}]\{v\}\}$$
$$\{F\} = \{[K_{11}] - [K_{12}][K_{22}]^{-1}[K_{21}]\}\{v\}$$

Using the sign convention given in Example 3-3, Fig. 4, the relevant matrices are constructed as follows:

Element stiffness matrix is given for the beam element shown in Example 3-3, Fig. 5, as follows:

$$\{V\} = [k_e]\{v\}$$

or

$$\begin{Bmatrix} V_i \\ M_i \\ V_j \\ M_j \end{Bmatrix} = \frac{EI}{L^3} \begin{bmatrix} 12 & 6L & -12 & 6L \\ 6L & 4L^2 & -6L & 2L^2 \\ -12 & -6L & 12 & -6L \\ 6L & 2L^2 & -6L & 4L^2 \end{bmatrix} \begin{Bmatrix} v_i \\ \theta_i \\ v_j \\ \theta_j \end{Bmatrix}$$

Global matrix $$\rightarrow \begin{Bmatrix} V_1 \\ V_2 \\ M_1 \\ M_2 \end{Bmatrix} = \frac{EI}{L^3} \left[\begin{array}{cc|cc} 1.5 & 0 & 1.5L & 0 \\ 0 & 12 & 6L & 6L \\ \hline 1.5L & 6L & 6L^2 & 2L^2 \\ 0 & 6L & 2L^2 & 4L^2 \end{array}\right] \begin{Bmatrix} v_1 \\ v_2 \\ \theta_1 \\ \theta_2 \end{Bmatrix}$$

Partitioned global matrix,

$$[K_{11}] = \frac{EI}{L^3}\begin{bmatrix} 1.5 & 0 \\ 0 & 12 \end{bmatrix} \qquad [K_{12}] = \frac{EI}{L^3}\begin{bmatrix} 1.5L & 0 \\ 6L & 6L \end{bmatrix}$$

$$[K_{21}] = \frac{EI}{L^3}\begin{bmatrix} 1.5L & 6L \\ 0 & 6L \end{bmatrix} \qquad [K_{22}] = \frac{EI}{L^3}\begin{bmatrix} 6L^2 & 2L^2 \\ 2L^2 & 4L^2 \end{bmatrix}$$

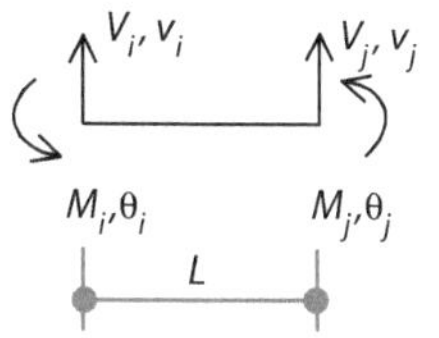

Example 3-3, Figure 5 Element stiffness matrix.

The inverse is,

$$[K_{22}]^{-1} = \frac{L^3}{20EI}\begin{bmatrix} \frac{4}{L^2} & \frac{-2}{L^2} \\ \frac{-2}{L^2} & \frac{6}{L^2} \end{bmatrix}$$

After performing the matrix multiplication above, the final matrix will be

$$\{F\} = [K]\{v\}$$

where

$$[K] = \frac{3EI}{20L^3}\begin{bmatrix} 7 & -6 \\ -6 & 8 \end{bmatrix}$$

which is the same result obtained by the force method.

4. Eigenvalue solution

$$|[K] - \omega^2[M]| = 0$$

$$\left| \frac{3EI}{20L^3}\begin{bmatrix} 7 & -6 \\ -6 & 8 \end{bmatrix} - \omega^2 m \begin{bmatrix} 2 & 0 \\ 0 & 1 \end{bmatrix} \right| = 0$$

If λ is set to ($\lambda = \omega^2 m\,(20L^3)/3EI$), the equation above becomes

$$\begin{vmatrix} (7-2\lambda) & -6 \\ -6 & (8-\lambda) \end{vmatrix} = 0$$

expansion of the determinant yields

$$(7 - 2\lambda)(8 - \lambda) - 36 = 0$$

or

$$\lambda^2 - 11.5\lambda + 10 = 0$$

Solution of the quadratic equation above yields two roots of λ, ($\lambda_1 = 0.948$ and $\lambda_2 = 10.55$).

For each value of λ, there will be a frequency and an associated mode shape, which are obtained as follows:

For $\lambda_1 = 0.948$,

Frequency:

$$\omega_1^2 = \lambda_1 \frac{3EI}{20mL^3} = 0.948\left(\frac{3EI}{20mL^3}\right)$$

$$\omega_1 = 0.377\sqrt{\frac{EI}{mL^3}}$$

Mode shape: let v_1 arbitrarily be 1, then

$$\begin{bmatrix} 7-2\,(0.948) & -6 \\ -6 & 8-1(0.948) \end{bmatrix}\begin{Bmatrix} 1 \\ v_2 \end{Bmatrix} = \begin{Bmatrix} 0 \\ 0 \end{Bmatrix}$$

Multiplication of the first row yields

$$7 - 2\,(0.948)(1) - 6\,v_2 = 0 \quad \rightarrow \quad v_2 = 0.85$$

Therefore, mode (1) becomes

$$\{\phi_1\}\begin{Bmatrix} 1.0 \\ 0.85 \end{Bmatrix}$$

Similarly for mode (2), we have

For $\lambda_2 = 1.258$,

Frequency:

$$\omega_2^2 = \lambda_2\left(\frac{3EI}{20mL^3}\right) = 10.55\frac{3EI}{20mL^3}\left(\frac{3EI}{20mL^3}\right)$$

hence,

$$\omega_2 = 1.258\sqrt{\frac{EI}{mL^3}}$$

Mode shape: let v_1 arbitrarily be 1, then

$$\begin{bmatrix} 7-2(10.55) & -6 \\ -6 & 8-1(10.55) \end{bmatrix} \begin{Bmatrix} 1 \\ v_2 \end{Bmatrix} = \begin{Bmatrix} 0 \\ 0 \end{Bmatrix}$$

Multiplication of the first row yields

$$7 - 2(10.55)(1) - 6\,v_2 = 0$$

hence,

$$v_2 = -2.35$$

Therefore, mode (2) becomes

$$\{\phi_2\} = \begin{Bmatrix} 1.0 \\ -2.35 \end{Bmatrix}$$

Summary of results

$$\{\omega\} = \begin{Bmatrix} 0.377 \\ 1.258 \end{Bmatrix}\sqrt{\frac{EI}{mL^3}},\ [\Phi] = [\ \phi_1 \ \vdots\ \phi_2\ ] = \begin{bmatrix} 1 & 1 \\ 0.85 & -2.35 \end{bmatrix}$$

The mode shapes are shown in Example 3-3, Fig. 6.

5. Modal analysis

The earthquake directional vector, $\{R\}$, is constructed by giving the structure unit rigid body motion in the direction of the earthquake as shown in Example 3-3, Fig. 7. Then calculate the displacements of the DOFs accordingly. For example,

$$\{R\} = \begin{Bmatrix} 0.447 \\ -0.894 \end{Bmatrix}$$

Because the construction of vector $\{R\}$ completes the construction of all required matrices for modal analysis, the analysis can be started mode by mode to find earthquake-induced forces in the form

$$\{f_{s,\max}\} = [M]\{\phi\}\Gamma S_a$$

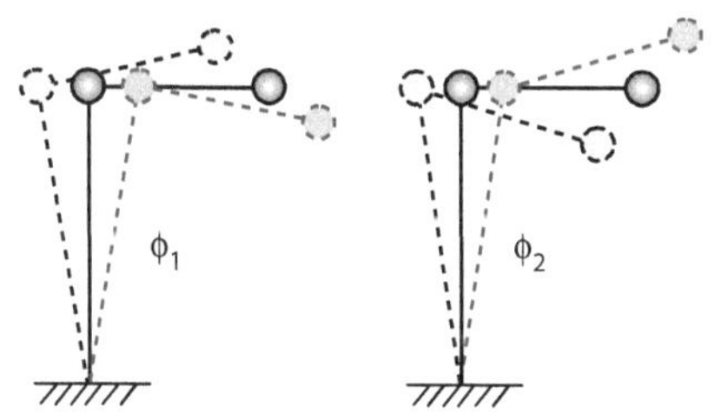

Example 3-3, Figure 6 Mode shapes.

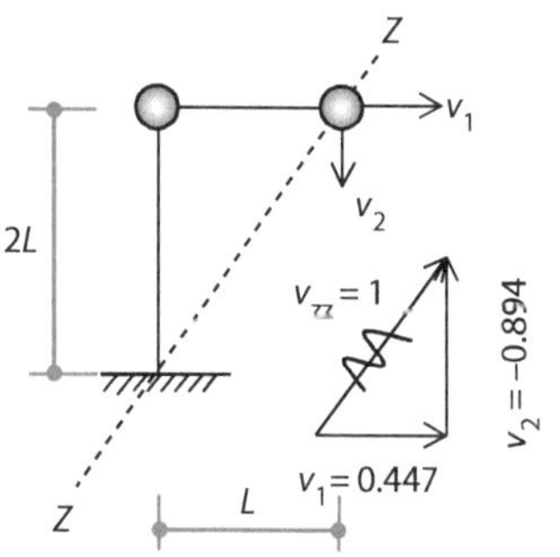

Example 3-3, Figure 7 Rigid body motion.

Mode (1)

$$\{f_{s1}\} = [M]\{\phi_1\}\Gamma_1 S_{a1}$$

$$£_1 = \{\phi_1\}^T[M]\{R\}$$

$$= \{1.0 \;\; 0.85\}(15)\begin{bmatrix} 2 & 0 \\ 0 & 1 \end{bmatrix}\begin{Bmatrix} 0.447 \\ -0.894 \end{Bmatrix} = 2.01\ \text{kN·s}^2/\text{m}$$

$$m_1^* = \{\phi_1\}^T[M]\{\phi_1\}$$

$$= \{1.0 \;\; 0.85\}(15)\begin{bmatrix} 2 & 0 \\ 0 & 1 \end{bmatrix}\begin{Bmatrix} 1.0 \\ 0.85 \end{Bmatrix} = 40.84\ \text{kN·s}^2/\text{m}$$

$$\Gamma_1 = \frac{£_1}{m_1^*} = \frac{2.01}{40.84} = 0.049$$

$$\omega_1 = 0.377\sqrt{\frac{EI}{mL^3}} = 0.377\sqrt{\frac{1500}{15(1)}} = 3.77\ \text{rad/s}$$

$$T_1 = \frac{2\pi}{\omega_1} = \frac{2\pi}{3.77} = 1.67\ \text{s}$$

With $T_1 = 1.67$ s and $\zeta_1 = 2\%$, S_{a1} can be directly read from the El Centro response spectrum curves given in this chapter to yield $S_{a1} = 2.2\ \text{m/s}^2$. Thus, the modal forces are given as

$$\{f_{s1}\} = [M]\{\phi_1\}\Gamma_1 S_{a1}$$

$$= (15)\begin{bmatrix} 2 & 0 \\ 0 & 1 \end{bmatrix}\begin{Bmatrix} 1.0 \\ 0.85 \end{Bmatrix}(0.049)(2.2) = \begin{Bmatrix} 3.23 \\ 1.37 \end{Bmatrix}\text{kN}$$

With these forces applied as shown in Example 3-3, Fig. 8, the reactions at the base may be calculated by statics to yield the base shear V_1, and base moment M_1.

$$V_1 = 3.23\ \text{kN}$$

$$M_1 = 3.23(2) + 1.37(1) = 7.83\ \text{kN·m}$$

Similarly, mode (2) can be evaluated as follows:

Mode (2)

$$\{f_{s2}\} = [M]\{\phi_2\}\Gamma_2 S_{a2}$$

$$£_2 = \{\phi_2\}^T[M]\{R\}$$

$$= \{1.0 \;\; -2.35\}(15)\begin{bmatrix} 2 & 0 \\ 0 & 1 \end{bmatrix}\begin{Bmatrix} 0.447 \\ -0.894 \end{Bmatrix} = 44.85\ \text{kN·s}^2/\text{m}$$

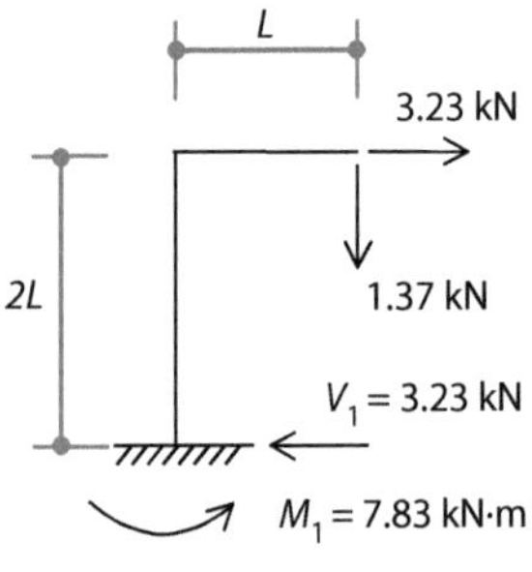

Example 3-3, Figure 8 Modal forces, mode (1).

$$m_2^* = \{\phi_2\}^T[M]\{\phi_2\}$$

$$= \{1.0 \;\; -2.35\}(15)\begin{bmatrix} 2 & 0 \\ 0 & 1 \end{bmatrix}\begin{Bmatrix} 1.0 \\ -2.35 \end{Bmatrix} = 112.98 \text{ kN·s}^2\text{/m}$$

$$\Gamma_2 = \frac{£_2}{m_2^*} = \frac{44.85}{112.98} = 0.398$$

$$\omega_2 = 1.258\sqrt{\frac{EI}{mL^3}} = 1.258\sqrt{\frac{1500}{15(1)}} = 12.58 \text{ rad/s}$$

$$T_2 = \frac{2\pi}{\omega_2} = \frac{2\pi}{12.58} = 0.5 \text{ s}$$

With $T_2 = 0.5$ s and $\zeta_2 = 2\%$, S_{a2} can be directly read from the El Centro response spectrum curves given in Chap. 2 to yield $S_{a2} = 10$ m/s^2. Thus, the modal forces are given as

$$\{f_{s2}\} = [M]\{\phi_2\}\Gamma_2 S_{a2}$$

$$= (15)\begin{bmatrix} 2 & 0 \\ 0 & 1 \end{bmatrix}\begin{Bmatrix} 1.0 \\ -2.35 \end{Bmatrix}(0.398)(10.0) = \begin{Bmatrix} 120 \\ -140 \end{Bmatrix}\text{kN}$$

With these forces applied as shown in Example 3-3, Fig. 9, the reactions at the base may be calculated by statics to yield the base shear, V_2, and base moment, M_2

$$V_2 = 120 \text{ kN}$$

$$M_2 = 120(2) - 140(1) = 100 \text{ kN·m}$$

The final results are obtained by adding values from mode (1) and mode (2), including the base shear and base moments. If the SRSS method of summation is used, the base shear and base moments are calculated as follows:

$$V_{\max} = \sqrt{V_1^2 + V_2^2}$$

$$= \sqrt{3.23^2 + 120^2} = 120.04 \text{ kN}$$

$$M_{\max} = \sqrt{M_1^2 + M_2^2}$$

$$= \sqrt{7.83^2 + 100^2} = 100.31 \text{ kN·m}$$

In this example, mode (1) has little effect on the final force. This is not necessarily the usual case in analysis.

The earthquake-induced displacements may be found for each mode by statics that are due to modal forces or may be found using the modal displacement equations. The required S_{dn} may be read directly from the response spectrum or may be calculated from S_{an}. Using modal displacement procedure results in

$$\{U_{n,\max}\} = \{\phi_n\}\Gamma_n S_{\text{dn}}$$

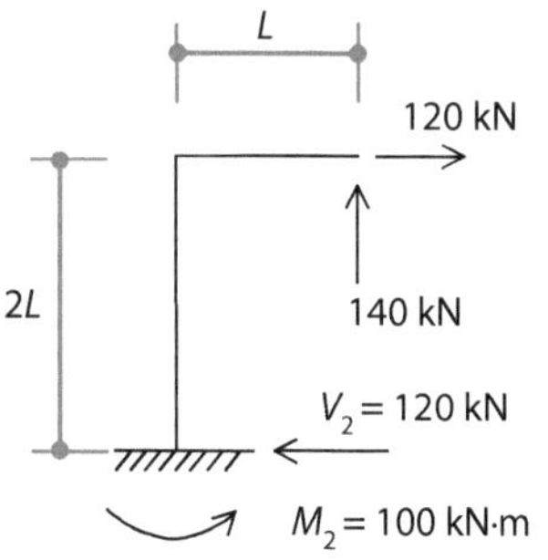

Example 3-3, Figure 9 Modal forces, mode (2).

Mode (1)

$$\{U_1\} = \{\phi_1\}\Gamma_1 S_{d1}$$

where

$$S_{d1} = \frac{S_{a1}}{\omega_1^2} = \frac{2.2}{2.77^2} = 0.15\,\text{m}$$

Therefore,

$$\begin{Bmatrix} v_1 \\ v_2 \end{Bmatrix} = \begin{Bmatrix} 1.0 \\ 0.85 \end{Bmatrix}(0.049)(0.15) = \begin{Bmatrix} 0.00735 \\ 0.00625 \end{Bmatrix}\text{m}$$

Mode (2)

$$\{U_2\} = \{\phi_2\}\Gamma_2 S_{d2}$$

where

$$S_{d2} = \frac{S_{a2}}{\omega_2^2} = \frac{10}{12.58^2} = 0.063\,\text{m}$$

Therefore,

$$\begin{Bmatrix} v_1 \\ v_2 \end{Bmatrix} = \begin{Bmatrix} 1.0 \\ -2.35 \end{Bmatrix}(0.398)(0.063) = \begin{Bmatrix} 0.0251 \\ -0.0589 \end{Bmatrix}\text{m}$$

The SRSS displacements become,

$$v_{1,\max} = \sqrt{0.00735^2 + 0.0251^2} = 0.026\,\text{m}$$
$$v_{2,\max} = \sqrt{0.00625^2 + 0.0589^2} = 0.059\,\text{m}$$

Example 3-4

Because the structure shown in Example 3-4, Fig. 1, has very stiff beams, it will be idealized as two degrees of freedom. The structure properties are given as follows:

Mass: $m_1 = 0.2$ kN·s²/mm
$m_2 = 0.4$ kN·s²/mm

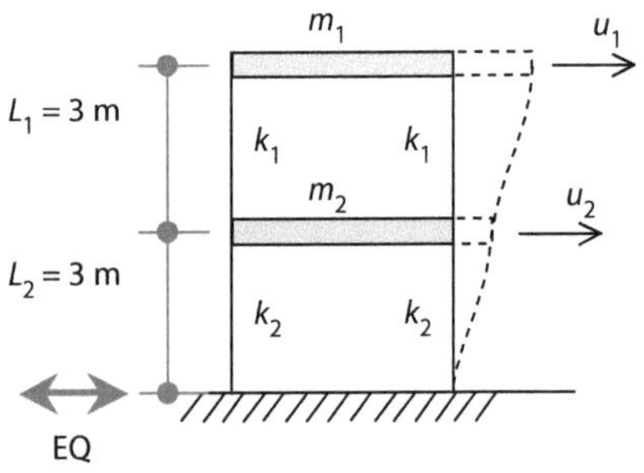

Example 3-4, Figure 1

Stiffness: $$k_1 = \frac{12EI_1}{L_1^3} = 50\,\text{kN/mm}$$

$$k_2 = \frac{12EI_2}{L_2^3} = 100\,\text{kN/mm}$$

Damping: $$\zeta = 5\%$$

If the structure is excited by the El Centro earthquake in the direction shown in the same figure, calculate the maximum displacements, maximum story forces, and maximum reactions (the base shear and base moment using the SRSS combining method).

Solution

1. Displacement vector

$$\{U\} = \begin{Bmatrix} u_1 \\ u_2 \end{Bmatrix}$$

2. Mass matrix
   The mass matrix may be found as in the previous examples by applying unit acceleration along each degree of freedom as shown in Example 3-4, Fig. 2, and then by calculating the resulting forces. Accordingly, the mass matrix is given as

$$[M] = 0.2\begin{bmatrix} 1 & 0 \\ 0 & 2 \end{bmatrix} \text{kN·s}^2\text{/mm}$$

3. Stiffness matrix
   Because the beams are stiff enough to prevent joint rotations, the stiffness matrix can be constructed by simply applying unit displacements in the direction of each degree of freedom (as shown in Example 3-4, Fig. 3) and by calculating the resulting forces. Accordingly, the stiffness matrix is given as

$$[K] = 100\begin{bmatrix} 1 & -1 \\ -1 & 3 \end{bmatrix} \text{kN/mm}$$

4. Eigenvalue solution

$$|[K] - \omega^2 [M]| = 0$$

$$\left| 100\begin{bmatrix} 1 & -1 \\ -1 & 3 \end{bmatrix} - \omega^2 (0.2)\begin{bmatrix} 1 & 0 \\ 0 & 2 \end{bmatrix} \right| = 0$$

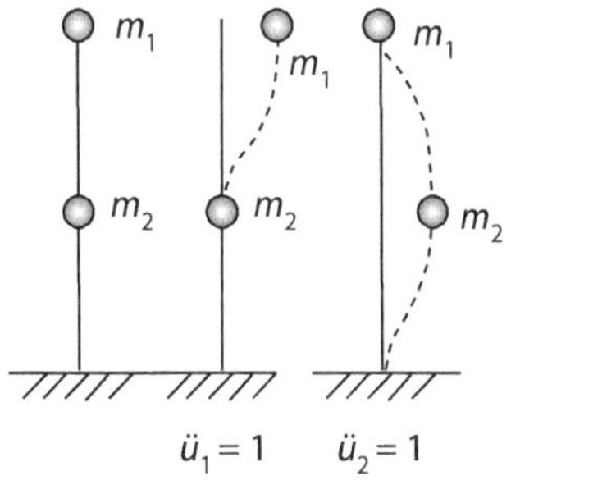

Example 3-4, Figure 2

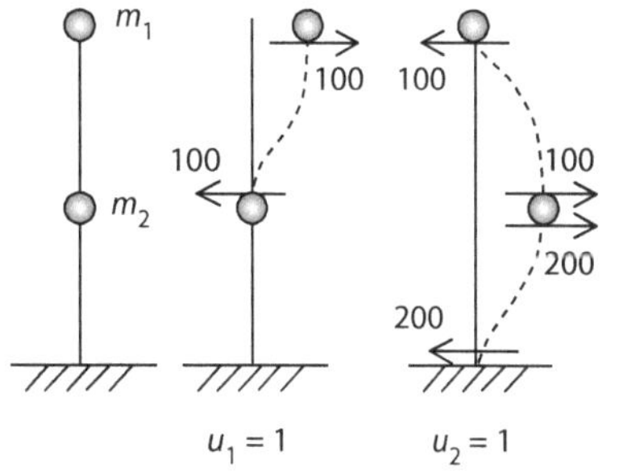

Example 3-4, Figure 3

If λ is set to ($\lambda = \omega^2 (0.2)/(100) = \omega^2/500$), the equations above become

$$\begin{vmatrix} (1-\lambda) & -1 \\ -1 & (3-2\lambda) \end{vmatrix} = 0$$

Expansion of the determinant yields

$$(1-\lambda)(3-2\lambda) - 1 = 0$$

or

$$\lambda^2 - 2.5\lambda + 1 = 0$$

Solution of the quadratic equation above yields two roots of λ

$$\lambda_1 = 0.5$$

and

$$\lambda_2 = 2.0$$

For each value of λ there will be a frequency and an associated mode shape which are obtained as follows:

For $\lambda_1 = 0.5$

Frequency: $\omega_1^2 = \lambda_1(500) = 0.5(500) = 250$

$$\omega_1 = \sqrt{250} = 15.81 \text{ rad/s}$$

Mode shape: let u_1 arbitrarily be 1, then

$$\begin{bmatrix} 1-(0.5) & -1 \\ -1 & 3-2(0.5) \end{bmatrix} \begin{Bmatrix} 1 \\ u_2 \end{Bmatrix} = \begin{Bmatrix} 0 \\ 0 \end{Bmatrix}$$

Multiplication of the first row yields

$$0.5 - u_2 = 0 \quad \rightarrow \quad u_2 = 0.5$$

Therefore, mode (1) becomes

$$\{\phi_1\} = \begin{Bmatrix} 1.0 \\ 0.5 \end{Bmatrix}$$

Similarly for mode (2), we have

For $\lambda_2 = 2$

Frequency: $\omega_2^2 = \lambda_2(500) = 2(500) = 1{,}000$

$$\omega_2 = \sqrt{1{,}000} = 31.62 \text{ rad/s}$$

Mode shape: let v_1 arbitrarily be 1, then

$$\begin{bmatrix} 1-(2) & -1 \\ -1 & 3-2(2) \end{bmatrix} \begin{Bmatrix} 1 \\ u_2 \end{Bmatrix} = \begin{Bmatrix} 0 \\ 0 \end{Bmatrix}$$

Multiplication of the first row yields

$$-1 - u_2 = 0 \quad \rightarrow \quad u_2 = -1$$

Therefore, mode (2) becomes

$$\{\phi_2\} = \begin{Bmatrix} 1.0 \\ -1.0 \end{Bmatrix}$$

The results summary is given below and shown graphically in Example 3-4, Fig. 4.

$$\{\omega\} = \begin{Bmatrix} 15.81 \\ 31.62 \end{Bmatrix}$$

$$[\Phi] = \left[\ \phi_1 \ \vdots \ \phi_2\ \right] = \left[\begin{array}{c:c} 1 & 1 \\ 0.5 & -1 \end{array}\right]$$

5. Modal analysis
The earthquake directional vector, $\{R\}$, is constructed by giving the structure unit rigid body motion in the direction of the earthquake and then calculating the displacements of the DOFs accordingly

$$\{R\} = \begin{Bmatrix} 1 \\ 1 \end{Bmatrix}$$

Mode (1)

$$£_1 = \{\phi_1\}^T[M]\{R\} = \{1.0 \quad 0.5\}(0.2)\begin{bmatrix} 1 & 0 \\ 0 & 2 \end{bmatrix}\begin{Bmatrix} 1 \\ 1 \end{Bmatrix} = 0.4 \text{ kN}\cdot\text{s}^2/\text{mm}$$

$$m_1^* = \{\phi_1\}^T[M]\{\phi_1\} = \{1.0 \quad 0.5\}(0.2)\begin{bmatrix} 1 & 0 \\ 0 & 2 \end{bmatrix}\begin{Bmatrix} 1.0 \\ 0.5 \end{Bmatrix}$$
$$= 0.3 \text{ kN}\cdot\text{s}^2/\text{mm}$$

$$\Gamma_1 = \frac{£_1}{m_1^*} = \frac{0.4}{0.3} = 1.333$$
$$T_1 = \frac{2\pi}{\omega_1} = \frac{2\pi}{15.81} = 0.4 \text{ s}$$

With $T_1 = 0.4$ s and $\zeta_1 = 5\%$, S_{d1} and S_{a1} can be directly read from the El Centro response spectrum curves (Fig. 3-8) to yield, $S_{d1} = 0.022$ m and $S_{a1} = 6$ m/s^2. Thus, the modal response is given as

$$\{U_1\} = \{\phi_1\}\Gamma_1 S_{d1} = \begin{Bmatrix} u_1 \\ u_2 \end{Bmatrix} = \begin{Bmatrix} 1.0 \\ 0.5 \end{Bmatrix}(1.333)(22) = \begin{Bmatrix} 29 \\ 15 \end{Bmatrix} \text{mm}$$

$$\{f_{s1}\} = [M]\{\phi_1\}\Gamma_1 S_{a1} = (0.2)\begin{bmatrix} 1 & 0 \\ 0 & 2 \end{bmatrix}\begin{Bmatrix} 1.0 \\ 0.5 \end{Bmatrix}(1.333)(6{,}000) = \begin{Bmatrix} 1{,}600 \\ 1{,}600 \end{Bmatrix} \text{kN}$$

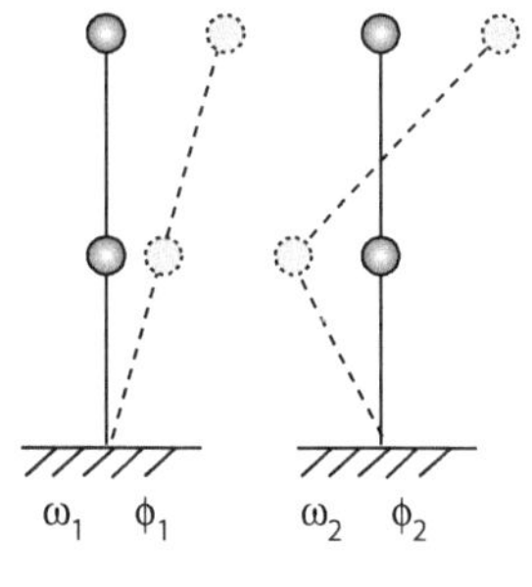

Example 3-4, Figure 4

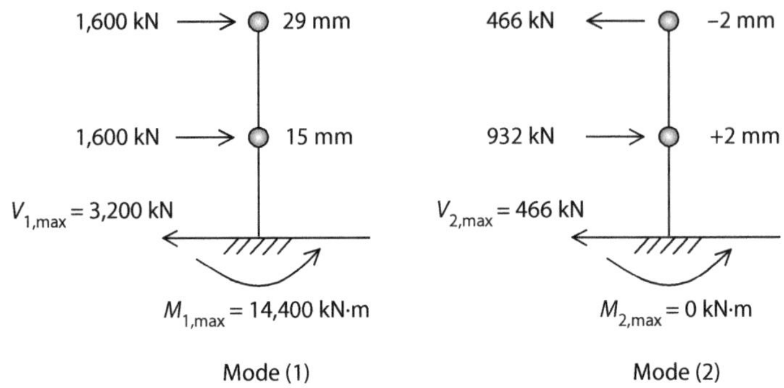

Example 3-4, Figure 5

Mode (2)

$$£_2 = \{\phi_2\}^T[M]\{R\} = \{1 \quad -1\}(0.2)\begin{bmatrix} 1 & 0 \\ 0 & 2 \end{bmatrix}\begin{Bmatrix} 1 \\ 1 \end{Bmatrix} = -0.2\ \text{kN·s}^2/\text{mm}$$

$$m_2^* = \{\phi_2\}^T[M]\{\phi_2\} = \{1 \quad -1\}(0.2)\begin{bmatrix} 1 & 0 \\ 0 & 2 \end{bmatrix}\begin{Bmatrix} 1 \\ -1 \end{Bmatrix} 0.6\ \text{kN·s}^2/\text{mm}$$

$$\Gamma_2 = \frac{£_2}{m_2^*} = \frac{-0.2}{0.6} = -0.333$$

$$T_2 = \frac{2\pi}{\omega_2} = \frac{2\pi}{31.62} = 0.2\ \text{s}$$

With $T_2 = 0.2$ s and $\zeta_2 = 5\%$, S_{d2} and S_{a2} can be directly read from the El Centro response spectrum curves (Fig. 3-8) to yield $S_{d2} = 0.007$ m and $S_{a2} = 7$ m/s². Thus, the modal response is given as

$$\{U_2\} = \{\phi_2\}\Gamma_2 S_{d2}\begin{Bmatrix} u_1 \\ u_2 \end{Bmatrix} = \begin{Bmatrix} 1 \\ -1 \end{Bmatrix}(-0.333)(7) = \begin{Bmatrix} -2 \\ 2 \end{Bmatrix}\text{mm}$$

$$\{f_{s2}\} = [M]\{\phi_2\}\Gamma_2 S_{a2} = (0.2)\begin{bmatrix} 1 & 0 \\ 0 & 2 \end{bmatrix}\begin{Bmatrix} 1 \\ -1 \end{Bmatrix}(-0.333)(7{,}000) = \begin{Bmatrix} -466 \\ 932 \end{Bmatrix}\text{kN}$$

Finally, the total response is found using the SRSS method. Summary of modal response of modes (1) and (2) is shown in Example 3-4, Fig. 5, hence

$$u_{1,\max} = \sqrt{29^2 + 2^2} = 29.07\ \text{mm}$$

$$u_{2,\max} = \sqrt{15^2 + 2^2} = 15.13\ \text{mm}$$

$$V_{\max} = \sqrt{3{,}200^2 + 466^2} = 3{,}233\ \text{kN}$$

$$M_{\max} = \sqrt{14{,}400^2 + 0^2} = 14{,}400\ \text{kN·m}$$

Example 3-5 A two-story building is idealized as a prismatic stick model with 2DOFs as shown in Example 3-5, Fig. 1(a). The mass and properties of the model are given as follows:

Mass: $m_1 = 100$ kN·s²/m

$m_2 = 200$ kN·s²/m

Stiffness: $EI = 3 \times 10^6$ kN·m²

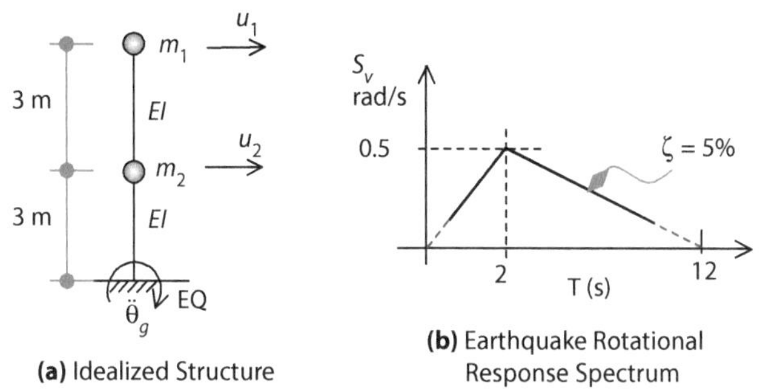

(a) Idealized Structure

(b) Earthquake Rotational Response Spectrum

EXAMPLE 3-5, FIGURE 1

The building is excited by the rotational component of an earthquake as shown in the same figure. The earthquake has a rotational response spectrum as shown in Example 3-5, Fig. 1(b). Calculate the maximum displacements, maximum story forces, and maximum reactions using the SRSS combining method.

Solution

1. Displacement vector

$$\{U\} = \begin{Bmatrix} u_1 \\ u_2 \end{Bmatrix}$$

2. Mass matrix
As in the previous examples, the mass matrix may be found by applying unit acceleration along each degree of freedom (as shown in Example 3-5, Fig. 2) and then calculating the resulting forces. Accordingly, the mass matrix is given as

$$[M] = \begin{bmatrix} 100 & 0 \\ 0 & 200 \end{bmatrix} \text{kN·s}^2/\text{m}$$

3. Stiffness matrix
Because the beam is a cantilever with rotational DOF, it will be easier to construct the stiffness matrix from the flexibility matrix than constructing the general stiffness matrix and kinematically condensing the rotational DOF as illustrated in Example 3-3.
The flexibility matrix can be constructed by simply applying unit displacements in the direction of each degree of freedom. In the following analysis, the conjugate beam method is used to calculate the required displacements.

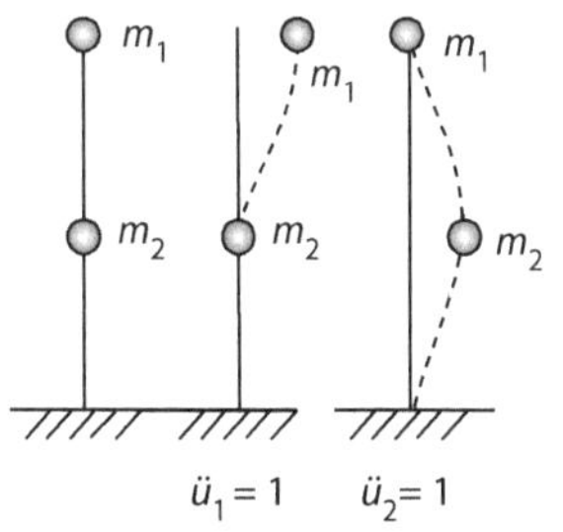

EXAMPLE 3-5, FIGURE 2

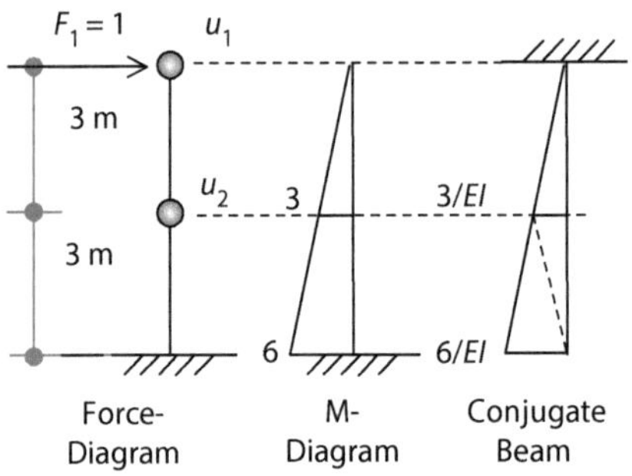

Example 3-5, Figure 3

u_1-direction
Apply unit force at joint 1 as shown in Example 3-5, Fig. 3. The resulting bending moment and conjugate beam will be as shown. Accordingly, the resulting displacements at joints 1 and 2 are calculated as follows:

$$f_{11} = \frac{qL^2}{3} = \frac{6}{3\times10^6}\frac{6^2}{3} = 24\times10^{-6}\,\text{m}$$

$$f_{21} = \text{two triangles} = \frac{q_1L^2}{3} + \frac{q_2L^2}{6}$$

$$f_{21} = \frac{6}{3\times10^6}\frac{3^2}{3} + \frac{3}{3\times10^6}\frac{3^2}{6} = 7.5\times10^{-6}\,\text{m}$$

u_2-direction
Apply unit force at joint 2 as shown in Example 3-5, Fig. 4. The resulting bending moment and conjugate beam will be as shown. Accordingly, the resulting displacements at joints 1 and 2 are calculated as follows:

$$\begin{aligned} f_{12} &= \frac{1}{2}q(L)\left(L + \frac{2}{3}L\right) \\ &= \frac{1}{2}\frac{3}{3\times10^6}(3)(3+2) \\ &= 7.5\times10^{-6}\,\text{m} \end{aligned}$$

$$f_{22} = \frac{3}{3\times10^6}\frac{3^2}{3} = 3\times10^{-6}\,\text{m}$$

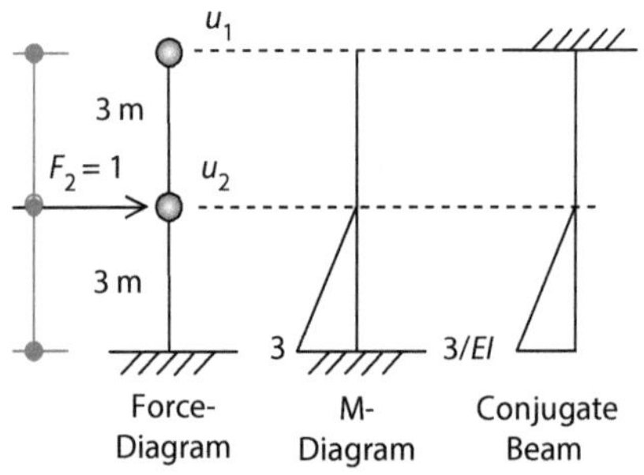

Example 3-5, Figure 4

Accordingly, the flexibility matrix, $[F]$, is given as

$$[F]=\begin{bmatrix} f_{11} & f_{12} \\ f_{21} & f_{22} \end{bmatrix}=10^{-6}\begin{bmatrix} 24 & 7.5 \\ 7.5 & 3 \end{bmatrix}\text{m/kN}$$

The stiffness matrix is obtained by taking the inverse of matrix $[F]$. Because the determinant of $\|[F]\| = [24\,(3) - (7.5)^2] \times 10^{-12} = 15.75 \times 10^{-12}$, $[K]$ will be

$$[K]=\frac{10^{-6}}{15.75\times10^{-12}}\begin{bmatrix} 3 & -7.5 \\ -7.5 & 24 \end{bmatrix}=63,492\begin{bmatrix} 3 & -7.5 \\ -7.5 & 24 \end{bmatrix}\text{kN/m}$$

4. Eigenvalue solution

$$\|[K] - \omega^2[M]\| = 0$$

$$\left|63,492\begin{bmatrix} 3 & -7.5 \\ -7.5 & 24 \end{bmatrix}-\omega^2(100)\begin{bmatrix} 1 & 0 \\ 0 & 2 \end{bmatrix}\right|=0$$

If λ is set to ($\lambda = \omega^2\,(100)/(63{,}492) = \omega^2/634.92$), the equation above becomes

$$\begin{vmatrix} (3-\lambda) & -7.5 \\ -7.5 & (24-2\lambda) \end{vmatrix}=0$$

expansion of the determinant yields

$$(3 - \lambda)(24 - 2\lambda) - (7.5)^2 = 0$$

or,

$$2\lambda^2 - 30\lambda + 15.75 = 0$$

Solution of the above quadratic equation yields two roots of λ

$$\lambda_1 = 0.545$$

and

$$\lambda_2 = 14.455$$

For each value of λ, there will be a frequency and an associated mode shape that are obtained as follows:

For $\lambda_1 = 0.545$

Frequency: $\omega_1^2 = \lambda_1\,(634.92) = 0.545(634.92) = 346$

$$\omega_1 = \sqrt{346} = 18.60\text{ rad/s}$$

Mode shape: let u_1 arbitrarily be 1, then

$$\begin{bmatrix} 3-(0.545) & -7.5 \\ -7.5 & 24-2(0545) \end{bmatrix}\begin{Bmatrix} 1 \\ u_2 \end{Bmatrix}=\begin{Bmatrix} 0 \\ 0 \end{Bmatrix}$$

Multiplication of the first row yields

$$2.455 - 7.5\,u_2 = 0 \quad \rightarrow \quad u_2 = 0.327$$

Therefore, mode (1) becomes

$$\{\phi_1\}=\begin{Bmatrix} 1.0 \\ 0.327 \end{Bmatrix}$$

Similarly for mode (2), we have

For $\lambda_2 = 14.455$

Frequency: $\omega_2^2 = \lambda_2(634.92) = 14.455\,(634.92) = 9{,}178$

$$\omega_2 = \sqrt{9{,}178} = 95.80\ \text{rad/s}$$

Mode shape: let u_1 arbitrarily be 1, then

$$\begin{bmatrix} 3-(14.455) & -7.5 \\ -7.5 & 24-2(14.455) \end{bmatrix} \begin{Bmatrix} 1 \\ u_2 \end{Bmatrix} = \begin{Bmatrix} 0 \\ 0 \end{Bmatrix}$$

Multiplication of the first row yields

$$-11.455 - 7.5\,u_2 = 0 \quad \rightarrow \quad u_2 = -1.527$$

Therefore, mode (2) becomes

$$\{\phi_2\} = \begin{Bmatrix} 1.0 \\ -1.527 \end{Bmatrix}$$

The summary of results is given below and shown graphically in Example 3-5, Fig. 5.

$$\{\omega\} = \begin{Bmatrix} 18.60 \\ 95.80 \end{Bmatrix}$$

$$[\Phi] = \begin{bmatrix} \phi_1 & \vdots & \phi_2 \end{bmatrix} = \begin{bmatrix} 1 & \vdots & 1 \\ 0.327 & \vdots & -1.527 \end{bmatrix}$$

5. Modal analysis
The earthquake directional vector, $\{R\}$, is constructed by giving the structure a unit rigid body motion in the direction of the earthquake and then calculating the displacements of the DOFs accordingly. In this example, the rigid body motion will be a unit rotation, ($\theta = 1$), of the member around its base as shown in Example 3-5, Fig. 6. Thus, the displacement of joint 1 and 2 will be equal to its distance from the base times the base rotation.
It should be pointed out that the rotational response spectrum is built up by normalizing the linear acceleration to the rotational acceleration. For example, if the mass is elevated one unit from the base, that is, $h = 1$, the equation of motion becomes

$$m\ddot{u} + c\dot{u} + ku = -m\ddot{u}_g$$

$$m\ddot{u} + c\dot{u} + ku = -m(1)\ddot{\theta}_g$$

$$m\ddot{u} + c\dot{u} + ku = -m\ddot{\theta}_g$$

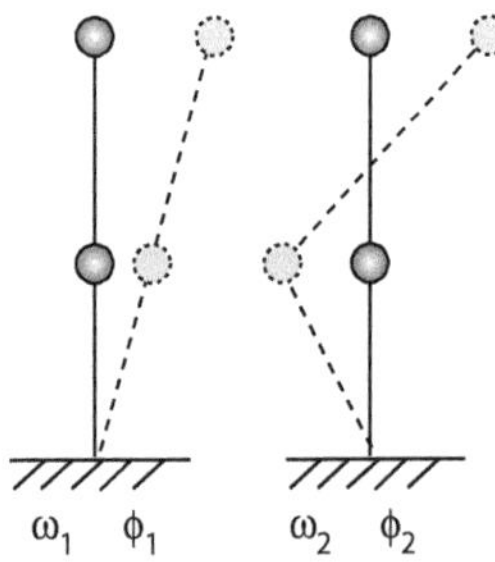

Example 3-5, Figure 5

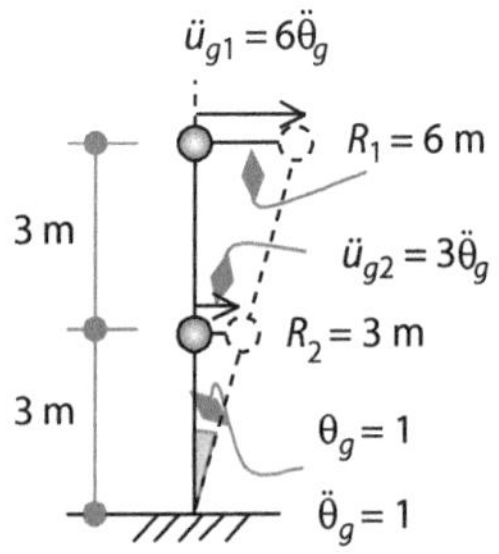

Example 3-5, Figure 6 Rigid body rotation.

Consequently, the vector $\{R\}$ is given as

$$R_1 = 1\,(6) = 6\text{ m}$$
$$R_2 = 1\,(3) = 3\text{ m}$$
$$\{R\} = \begin{Bmatrix} 6 \\ 3 \end{Bmatrix}\text{m}$$

Mode (1)

$$£_1 = \{\phi_1\}^T[M]\{R\}$$
$$= \{1.0 \quad 0.327\}(100)\begin{bmatrix} 1 & 0 \\ 0 & 2 \end{bmatrix}\begin{Bmatrix} 6 \\ 3 \end{Bmatrix} = 796.2\text{ kN·s}^2$$

$$m_1^* = \{\phi_1\}^T[M]\{\phi_1\} = \{1.0 \quad 0.327\}(100)\begin{bmatrix} 1 & 0 \\ 0 & 2 \end{bmatrix}\begin{Bmatrix} 1.0 \\ 0.327 \end{Bmatrix}$$
$$= 121.39\text{ kN·s}^2\text{/m}$$

$$\Gamma_1 = \frac{£_1}{m_1^*} = \frac{796.2}{121.39} = 6.559\text{ m}$$

$$T_1 = \frac{2\pi}{\omega_1} = \frac{2\pi}{18.6} = 0.34\text{ s}$$

Note that, unlike the displacement spectrum, the participation factor of the rotational response spectrum has a unit of meter. Refer to the rotational response spectrum in Example 3-5, Fig. 1, at $T_1 = 0.34$ s. S_{v1} is read as 0.085 rad/s. Thus

$$S_{d1} = \frac{S_{v1}}{\omega_1} = \frac{0.085}{18.6} = 0.005\text{ rad}$$
$$S_{a1} = S_{v1}\omega_1 = 0.085(18.6) = 1.581\text{ rad/s}^2$$

The modal response is given as

$$\{U_1\} = \{\phi_1\}\Gamma_1 S_{d1} = \begin{Bmatrix} u_1 \\ u_2 \end{Bmatrix} = \begin{Bmatrix} 1.0 \\ 0.327 \end{Bmatrix}(6.559)(0.005) = \begin{Bmatrix} 0.033 \\ 0.011 \end{Bmatrix}\text{m}$$

$$\{f_{s1}\} = [M]\{\phi_1\}\Gamma_1 S_{a1}$$
$$= (100)\begin{bmatrix} 1 & 0 \\ 0 & 2 \end{bmatrix}\begin{Bmatrix} 1.0 \\ 0.327 \end{Bmatrix}(6.559)(1.581) = \begin{Bmatrix} 1{,}037 \\ 678 \end{Bmatrix}\text{kN}$$

Mode (2)

$$£_2 = \{\phi_2\}^T[M]\{R\} = \{1 \quad -1.527\}(100)\begin{bmatrix} 1 & 0 \\ 0 & 2 \end{bmatrix}\begin{Bmatrix} 6 \\ 3 \end{Bmatrix} = -316.2\text{ kN·s}^2$$

$$m_2^* = \{\phi_2\}^T[M]\{\phi_2\} = \{1 \quad -1.527\}(100)\begin{bmatrix} 1 & 0 \\ 0 & 2 \end{bmatrix}\begin{Bmatrix} 1 \\ -1.527 \end{Bmatrix}$$
$$= 566.35\text{ kN·s}^2\text{/m}$$

$$\Gamma_2 = \frac{£_2}{m_2^*} = \frac{-316.2}{566.35} = -0.558\text{ m}$$

$$T_2 = \frac{2\pi}{\omega_2} = \frac{2\pi}{95.8} = 0.066\text{ s}$$

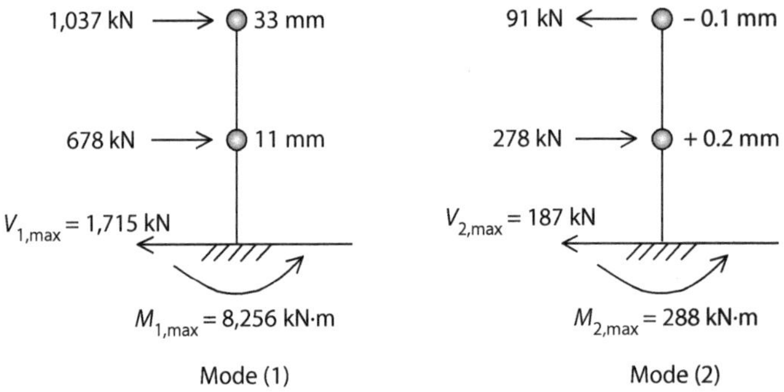

EXAMPLE 3-5, FIGURE 7

Similar to first mode, refer to the response spectrum in Example 3-5, Fig. 1, at $T_2 = 0.066$ s, S_{v2} is read as 0.017 rad/s, hence

$$S_{d2} = \frac{S_{v2}}{\omega_2} = \frac{0.017}{95.8} = 0.0002 \text{ rad}$$

$$S_{a2} = S_{v2}\omega_2 = 0.017(95.8) = 1.629 \text{ rad/s}^2$$

$$\{U_2\} = \{\phi_2\}\Gamma_2 S_{d2} = \begin{Bmatrix} u_1 \\ u_2 \end{Bmatrix} = \begin{Bmatrix} 1 \\ -1.527 \end{Bmatrix}(-0.558)(0.0002) = \begin{Bmatrix} -0.0001 \\ 0.0002 \end{Bmatrix} \text{m}$$

$$\{f_{s2}\} = [M]\{\phi_2\}\Gamma_2 S_{a2} = (100)\begin{bmatrix} 1 & 0 \\ 0 & 2 \end{bmatrix}\begin{Bmatrix} 1 \\ -1.527 \end{Bmatrix}(-0.558)(1.629) = \begin{Bmatrix} -91 \\ 278 \end{Bmatrix} \text{kN}$$

Finally, the total response is found using the SRSS method. The summary of modal response of modes (1) and (2) is shown in Example 3-5, Fig. 7, which is evaluated as follows:

$$u_{1,\max} = \sqrt{33^2 + 0.1^2} = 33 \text{ mm}$$

$$u_{2,\max} = \sqrt{11^2 + 0.2^2} = 11.002 \text{ mm}$$

$$V_{\max} = \sqrt{1{,}715^2 + 187^2} = 1{,}725 \text{ kN}$$

$$M_{\max} = \sqrt{8{,}256^2 + 288^2} = 8{,}261 \text{ kN·m}$$

Importance of Modes

The advantage of modal analysis with response spectrum technique is that only a few modes significantly contribute to the response of the structure. Experience in dynamic analysis shows that the participation factor, Γ, diminishes rapidly with higher modes.

For low-rise buildings (short period), the higher modes pick up less forces than lower modes as shown in Fig. 3-21. For tall buildings (long period), higher modes pick up more forces than lower modes. Therefore, higher modes should be investigated with care in high-rise buildings.

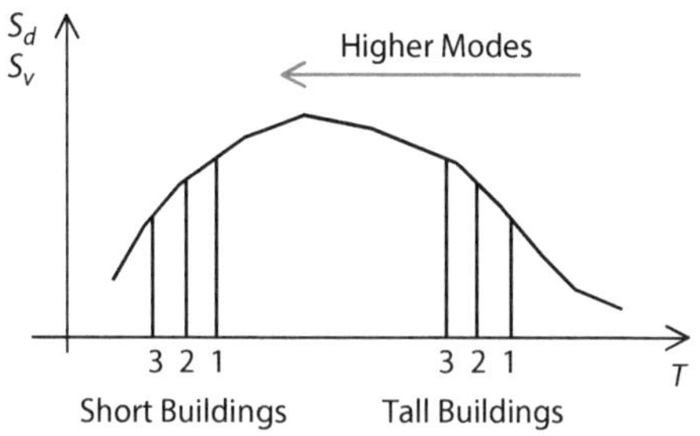

FIGURE 3-21 Importance of higher modes.

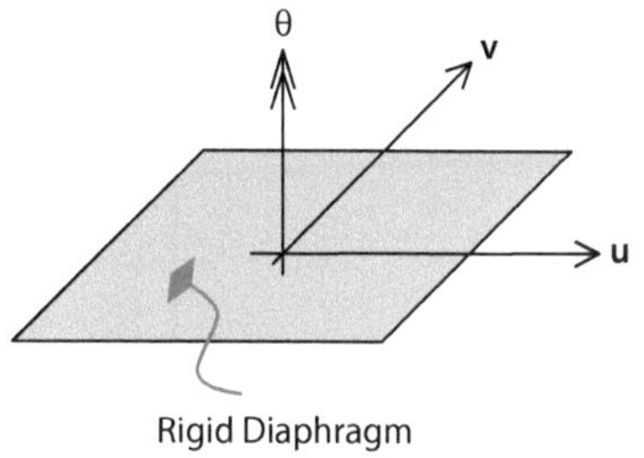

FIGURE 3-22 Rigid diaphragm degrees of freedom.

Seismic codes include this behavior in their response spectra by lifting up the curves of the response in the long period range to accommodate the effect of higher modes in the long period range.

3.4.2 Multiple Degrees of Freedom System in 3D Analysis

Buildings with rigid floors (diaphragms) where vertical acceleration of the ground is neglected can be modeled with at least three dynamic DOFs, the three mass movements u, v, and, θ as shown in Fig. 3-22.

If the center of mass (CM) coincides with the center of resistance (CR), also known as center of stiffness or center of rigidity, the resulting mass and stiffness matrices will be diagonals. For example, if a rectangular diaphragm with constant thickness (uniform mass) is supported by four circular columns with stiffness k as shown in Fig. 3-23, the resulting matrices will be as follows.

For a displacement vector, $\{U\} = \{u \ \ v \ \ \theta\}^T$, the mass and stiffness matrices take the following form:

$$[M] = \begin{bmatrix} m_u & 0 & 0 \\ 0 & m_v & 0 \\ 0 & 0 & m_\theta \end{bmatrix}, [K] = \begin{bmatrix} k_u & 0 & 0 \\ 0 & k_v & 0 \\ 0 & 0 & k_\theta \end{bmatrix}$$

where m_u and m_v are the translational mass coefficients and $m_\theta = I_p$ is the mass polar moment of inertia, k_u and k_v are the translational stiffness coefficients, and k_θ is the rotational stiffness coefficient.

Eigenvalue formulation yields

$$[K - \omega^2 M]\{\phi\} = \{0\}$$

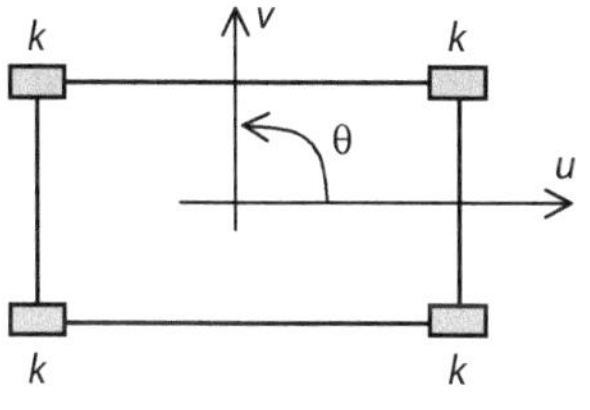

FIGURE 3-23 Symmetric diaphragm.

Frequency equation is given as

$$\begin{vmatrix} (k_u - \omega^2 m_u) & 0 & 0 \\ 0 & (k_v - \omega^2 m_v) & 0 \\ 0 & 0 & (k_\theta - \omega^2 m_\theta) \end{vmatrix} = 0$$

The results are three independent equations of the form $(k_i - \omega_i^2 m_i = 0)$, which yields three frequencies

$$\omega_i^2 = \sqrt{\frac{k_i}{m_i}}$$

Eigenvectors can be obtained as follows.
Mode (1) is given in the u-direction with frequency and mode shape ω_u, ϕ_u, and by setting u arbitrarily to 1, the mode shape is obtained as follows:

$$\begin{bmatrix} (k_u - \omega_u^2 m_u) & 0 & 0 \\ 0 & (k_v - \omega_u^2 m_v) & 0 \\ 0 & 0 & (k_\theta - \omega_u^2 m_\theta) \end{bmatrix} \begin{Bmatrix} 1 \\ v \\ \theta \end{Bmatrix} = \begin{Bmatrix} 0 \\ 0 \\ 0 \end{Bmatrix}$$

The second row yields: $(k_v - \omega_u{}^2 m_v)\, v = 0$
The third row yields: $(k_\theta - \omega_u{}^2 m_\theta)\, \theta = 0$

Because $(k_u - \omega_u^2 m_u = 0)$ and consequently $(k_v - \omega_u^2 m_v \neq 0)$, $(k_\theta - \omega_u^2 m_\theta \neq 0)$, the two equations above require that $v = 0$ and $\theta = 0$.

Similarly, in the v-direction for $v = 1$, the other two displacements are $u = 0$ and $\theta = 0$. Similarly, in the θ-direction for $\theta = 1$, the other two displacements are $u = 0$ and $v = 0$. Therefore, the three mode shapes are given as

$$[\Phi] = \left[\ \phi_u \ \vdots\ \phi_v \ \vdots\ \phi_\theta\ \right] = \begin{bmatrix} 1 & 0 & 0 \\ 0 & 1 & 0 \\ 0 & 0 & 1 \end{bmatrix}$$

Using the earthquake directional vector, $\{R\}$, it can be shown that the participation factor Γ only exists for the mode shape in the direction of excitation, whereas they vanish for the other modes. The implication of these results is that structures in which the center of mass and center of resistance coincide have uncoupled degrees of freedom. It is also implied that their modes of vibration are only excited by a ground motion in the direction of their freedom. In other words, $\{\phi_u\}$ is excited only by $\ddot{u}_g$, $\{\phi_v\}$ is excited only by $\ddot{v}_g$, and $\{\phi_\theta\}$ is excited only by $\ddot{\theta}_g$.

This important result states that the torsional modes, which are dangerous modes, will not be excited by translational ground motion. However, many symmetric buildings fail in torsional response during earthquakes because torsional response can occur for many reasons, including accidental asymmetry (or the coupling of the system because of local failures during an excitation). Furthermore, even though we design for translational earthquake components, earthquakes also contain some rotational components. For these reasons, even if structural systems are selected symmetric, seismic

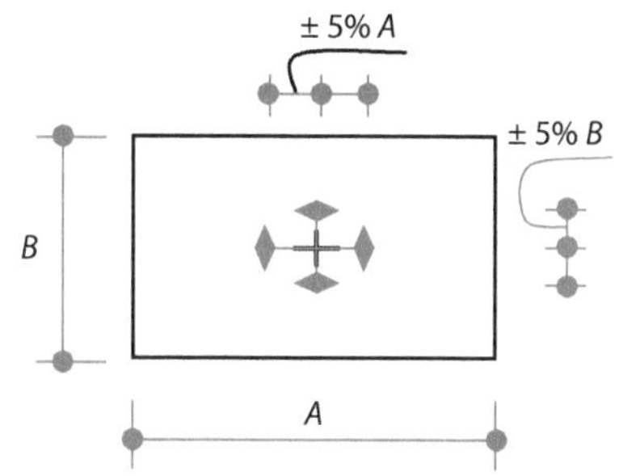

Figure 3-24 Accidental eccentricity.

codes prescribe a minimum additional amount of eccentricity known as *accidental eccentricity* for design. For example, the UBC and IBC codes require a minimum accidental eccentricity by displacing the mass 5 percent as shown in Fig. 3-24. The mass should be displaced ±5 percent alongside *B* if the earthquake excitation is in the longitudinal direction alongside *A*. Remember that this mass displacement is added to any existing eccentricity.

In general, buildings may be modeled as space frames with rigid diaphragms using standard structural analysis software. The mass is allocated at the nodes as concentrated masses as shown in Fig. 3-25.

Experience and skills in modeling are important aspects of dynamic analysis because of the effort involved. For example, the dynamic DOFs can be taken different from static DOFs. Although the rotational DOFs are very important in static analysis, they may not contribute much to dynamic analysis. These rotational DOFs might be eliminated in dynamic analysis. The kinematic condensation illustrated in Example 3-1 may be used for this purpose.

Mass distribution is also an important step in the modeling process. Mass allocation must be distributed in a manner that captures the location of the center of mass with respect to the center of resistance because this location represents the rotational degree

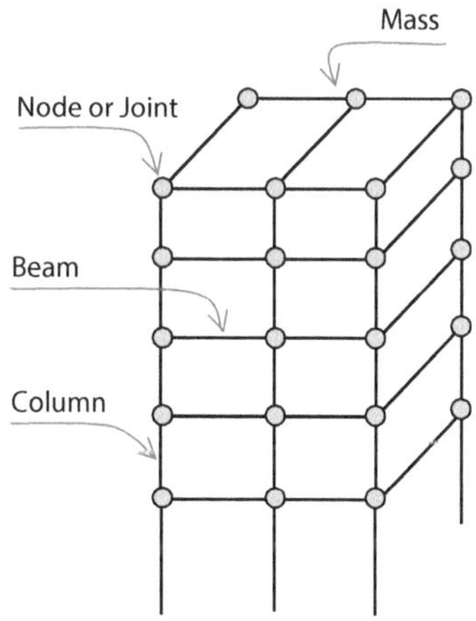

Figure 3-25 3D frame model.

of freedom of the entire floor. This allocation is also important because the torsional effect of the earthquake is related to this distribution. Example 3-7 illustrates this point.

It is important to point out that the mass allocation at the nodes is performed for structures with rigid diaphragms. In a structure with flexible diaphragms, the floors must at least be explicitly modeled as 2D elements. As an example, a 3D shell element may be used to model the floor where the mass is directly applied to the surface of the elements.

Very complex structures, such as nuclear power plants, may require finite element analysis. Many commercial software programs are available for this purpose.

Example 3-6 A 150-mm concrete slab is supported by four circular columns, which are located as shown in Example 3-6, Fig. 1. The slab is considered rigid, and the columns are weightless and totally fixed to the slab. The clear height of the columns is 3 m. The concrete mass is given as $\gamma_c = 25$ kN/m³.

1. Express the mass matrix and mode shapes in terms of the slab centroid coordinates u, v, and θ, as shown in Example 3-6, Fig. 1.
2. If the El Centro earthquake acts in the direction of coordinate u, determine the maximum dynamic displacement at the top of each column in the first mode of vibration. Damping ratio is taken as $\zeta = 2\%$.

Solution

Part (1)

1. Displacement vector.

$$\{U\} = \begin{Bmatrix} u \\ v \\ \theta \end{Bmatrix}$$

2. Mass matrix is given only for the slab since columns are weightless

Weight per m²: $w = \gamma h = 25(0.15) = 3.75 \text{ kN/m}^2$

Mass per m²: $m = \dfrac{w}{g} = \dfrac{3.75}{9.81} = 0.3823 \text{ kN}\cdot\text{s}^2/\text{m/m}^2$

Area: $A = 9(6) = 54 \text{ m}^2$

Polar moment of inertia:

$$I_p = I_x + I_y = \frac{6(9)^3}{12} + \frac{9(6)^3}{12} = 526.5 \text{ m}^4$$

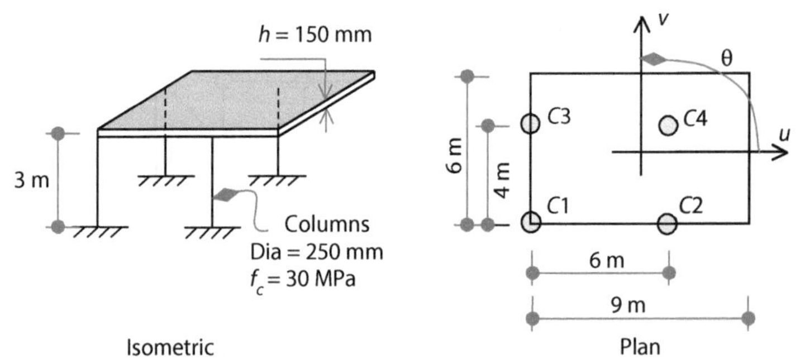

Example 3-6, Figure 1

Mass matrix: $$[M]=m\begin{bmatrix}54 & 0 & 0\\ 0 & 54 & 0\\ 0 & 0 & 526.5\end{bmatrix}, \text{(kN, m, s units)}$$

3. Stiffness matrix
   Stiffness of concrete columns

Modulus of elasticity: $$E_c = 4{,}700\sqrt{f_c'} = 4{,}700\sqrt{30} = 25{,}700\ \text{MPa}$$

Moment of inertia: $$I_c = \frac{\pi D^4}{64} = \frac{\pi(250)^4}{64} = 192\times10^6\ \text{mm}^4$$

$$EI = 25{,}700(192\times10^6) = 4{,}928\times10^9\ \text{N·m}^2$$

Column stiffness: $$k_c = \frac{12EI}{L^3} = \frac{12(4{,}928)}{(3)^3} = 2{,}190\ \text{kN/m}$$

The stiffness matrix is constructed by giving the slab unit displacements in the direction of each degree of freedom, u, v, and θ, and then calculating the resulting forces and moments accordingly.

Unit displacement in u-direction as shown in Example 3-6, Fig. 2, that is, $u = 1$, $v = 0$, and $\theta = 0$ result in the following:

$$F_u = 4k_c(1) = 4k_c\ \text{kN/m}$$
$$F_v = 0$$
$$F_\theta = (2\times3 - 2\times1)k_c(1) = 4k_c\ \text{kN·m/m}$$

Unit displacement in v-direction as shown in Example 3-6, Fig. 3, that is, $v = 1$, $u = 0$, and $\theta = 0$ result in the following:

$$F_u = 0$$
$$F_v = 4k_c(1) = 4\,k_c\ \text{kN/m}$$
$$F_\theta = (2\times1.5 - 2\times4.5)k_c(1)$$
$$= -6k_c\ \text{kN·m/m}$$

Unit displacement (rotation) in θ-direction as shown in Example 3-6, Fig. 4, that is, $\theta = 1$, $u = 0$, and $v = 0$ result in the third column of the stiffness matrix. For unit rotation, calculations become easier if the concept of the instantaneous center is utilized. The displacement of any point in any direction equals rotation times the perpendicular distance to the instantaneous center. In this case, the centroid of the slab is the instantaneous center.

Accordingly, and in reference to Example 3-6, Fig. 4, the forces in column 3 may be calculated as follows.

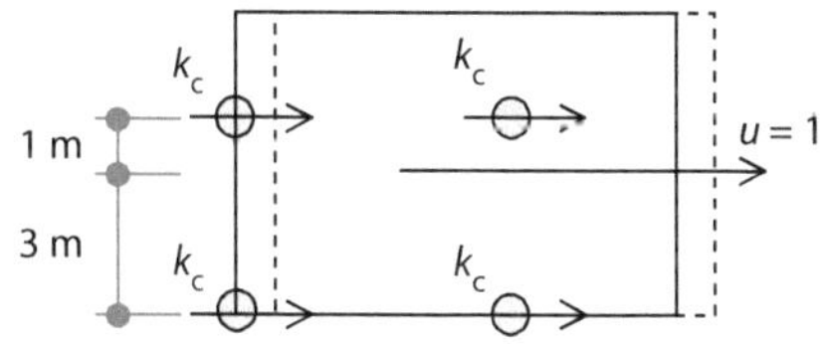

EXAMPLE 3-6, FIGURE 2

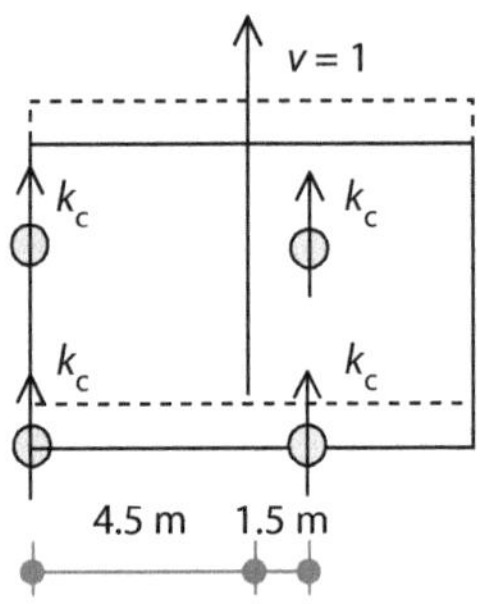

Example 3-6, Figure 3

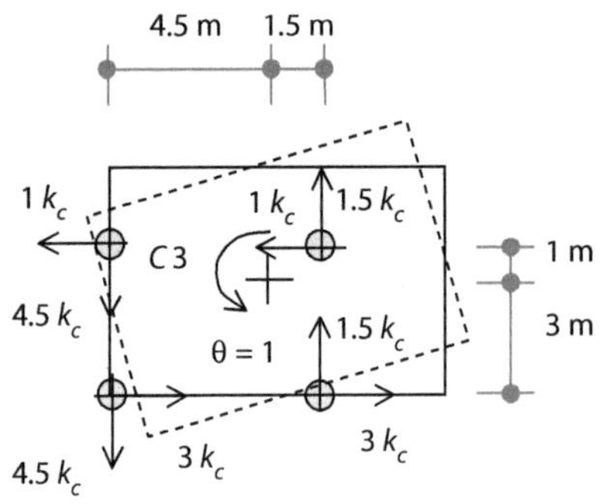

Example 3-6, Figure 4

For the horizontal component in column C3,

$$
\begin{aligned}
F_u &= k_c(1\text{ m})\,\theta \\
&= k_c\,(1)(1) \\
&= k_c\text{ kN/rad}
\end{aligned}
$$

For the vertical component in column C3,

$$
\begin{aligned}
F_v &= k_c(4.5\text{ m})\,\theta \\
&= k_c(4.5)(1) \\
&= 4.5\,k_c\text{ kN/rad}
\end{aligned}
$$

Similarly, the rest of the force components in all columns are calculated with resulting values as shown in Example 3-6, Fig. 4.

With these values known, the total forces needed for the stiffness matrix can be evaluated by equilibrium as follows:

$$
\begin{aligned}
F_u &= 4\,k_c\text{ kN/rad} \\
F_v &= -\,6\,k_c\text{ kN/rad} \\
F_\theta &= [(4.5\times 4.5\times 2)+(1.5\times 1.5\times 2)+(3\times 3\times 2)+(1\times 1\times 2)]\,k_c \\
&= 65\,k_c\text{ kN·m/rad}
\end{aligned}
$$

Stiffness matrix: $[K]=k_c\begin{bmatrix} 4 & 0 & 4 \\ 0 & 4 & -6 \\ 4 & -6 & 65 \end{bmatrix}$, (kN, m, and s units)

Part (2)

1. Eigenvalue solution

$$|[K] - \omega^2 [M]| = 0$$

$$\left| k_c \begin{bmatrix} 4 & 0 & 4 \\ 0 & 4 & -6 \\ 4 & -6 & 65 \end{bmatrix} - \omega^2 m \begin{bmatrix} 54 & 0 & 0 \\ 0 & 54 & 0 \\ 0 & 0 & 526.5 \end{bmatrix} \right| = 0$$

If λ is set to ($\lambda = \omega^2 \text{ m}/\text{k}_c$), the equation above becomes

$$\begin{vmatrix} 4-54\lambda & 0 & 4 \\ 0 & 4-54\lambda & -6 \\ 4 & -6 & 65-526.5\lambda \end{vmatrix} = 0$$

expansion of the determinant yields

$$(4 - 54\lambda)\cdot\{(4 - 54\lambda)(65 - 526.5\lambda) - 36 - 16\} = 0$$

or

$$(4 - 54\lambda)\cdot\{2{,}8431\lambda^2 - 5{,}616\lambda + 208\} = 0$$

Solution of the equation above yields three roots of λ:

$$\lambda_1 = 0.049{,}4, \qquad \lambda_2 = 0.074{,}1, \qquad \lambda_3 = 0.148{,}1$$

For each value of λ, there will be a frequency and an associated mode shape which are obtained as follows.

For $\lambda_1 = 0.0494$

Frequency: $$\omega_1^2 = \lambda_1 \frac{k_c}{m} = 0.0494\left(\frac{2{,}190}{0.3823}\right)$$

$$\omega_1 = 16.8 \text{ rad/s}$$

Mode shape: let u arbitrarily be 1, then

$$\begin{bmatrix} 4-54(0.0494) & 0 & 4 \\ 0 & 4-54(0.0494) & -6 \\ 4 & -6 & 65-526.5(0.0494) \end{bmatrix} \begin{Bmatrix} 1 \\ v \\ \theta \end{Bmatrix} = \begin{Bmatrix} 0 \\ 0 \\ 0 \end{Bmatrix}$$

or

$$\begin{bmatrix} 1.333 & 0 & 4 \\ 0 & 1.333 & -6 \\ 4 & -6 & 39 \end{bmatrix} \begin{Bmatrix} 1 \\ v \\ \theta \end{Bmatrix} = \begin{Bmatrix} 0 \\ 0 \\ 0 \end{Bmatrix}$$

The first row yields: $1.333 + 4\theta = 0, \quad \rightarrow \quad \theta = -0.333$

The second row yields: $1.333\,v - 6\theta = 0, \quad \rightarrow \quad v = -1.5$

Therefore, mode (1) becomes

$$\{\phi_1\} = \begin{Bmatrix} 1.0 \\ -1.5 \\ -0.333 \end{Bmatrix}$$

Although second and third modes can be found with identical procedures; however, they are not required for this example. Note that the mode shape includes a rotational component because the center of mass and center of resistance do not coincide as explained earlier.

2. Modal analysis
   Earthquake directional vector, $\{R\}$
   This vector is constructed by giving the structure unit rigid body motion in the direction of the earthquake. For an earthquake in direction, u, the displacements will be as shown in Example 3-6, Fig. 2

$$\{R\} = \begin{Bmatrix} 1 \\ 0 \\ 0 \end{Bmatrix}$$

Mode (1)

$$£_1 = \{\phi_1\}^T[M]\{R\}$$

$$= \{1.0 \ \ -1.5 \ \ -0.333\}(0.3823)\begin{bmatrix} 54 & 0 & 0 \\ 0 & 54 & 0 \\ 0 & 0 & 526.5 \end{bmatrix}\begin{Bmatrix} 1 \\ 0 \\ 0 \end{Bmatrix} = 20.6 \text{ kN·s}^2/\text{m}$$

$$m_1^* = \{\phi_1\}^T[M]\{\phi_1\}$$

$$= \{1.0 \ \ -1.5 \ \ -0.333\}(0.3823)\begin{bmatrix} 54 & 0 & 0 \\ 0 & 54 & 0 \\ 0 & 0 & 526.5 \end{bmatrix}\begin{Bmatrix} 1 \\ -1.5 \\ -0.333 \end{Bmatrix}$$

$$= 89.27 \text{ kN·s}^2/\text{m}$$

Therefore

$$\Gamma_1 = \frac{£_1}{m_1^*} = \frac{20.6}{89.27} = 0.231$$

$$T_1 = \frac{2\pi}{\omega_1} = \frac{2\pi}{16.8} = 0.374 \text{ s}$$

With $T_1 = 0.374$ s and $\zeta_1 = 2\%$, S_{d1} can be directly read from the El Centro response spectrum curves given in Fig. 3-8 to yield $S_{d1} = 0.04$ m. Thus, the first mode displacements are given as

$$\{U\} = \begin{Bmatrix} u \\ v \\ \theta \end{Bmatrix} = \{\phi\}\Gamma_1 S_{d1}$$

$$= \begin{Bmatrix} 1 \\ -1.5 \\ -0.333 \end{Bmatrix}(0.231)(0.04)$$

$$= \begin{Bmatrix} 0.009 \\ -0.014 \\ -0.003 \end{Bmatrix} \begin{matrix} \text{m} \\ \text{m} \\ \text{rad} \end{matrix}$$

Knowing these displacements, the two components of the displacement u_i and v at the top of each column can be found by instantaneous center. With reference to Example 3-6, Fig. 5

C1: $u_1 = u + 3\theta, \qquad v_1 = v - 4.5\theta$

$$u_1 = 0.009 + 3(-0.003) = 0$$

$$v_1 = -0.014 - 4.5(-0.003) = -0.0005 \text{ m}$$

$$\Delta_{max} = \sqrt{0^2 + 0.0005^2} = 0.0005 \text{ m}$$

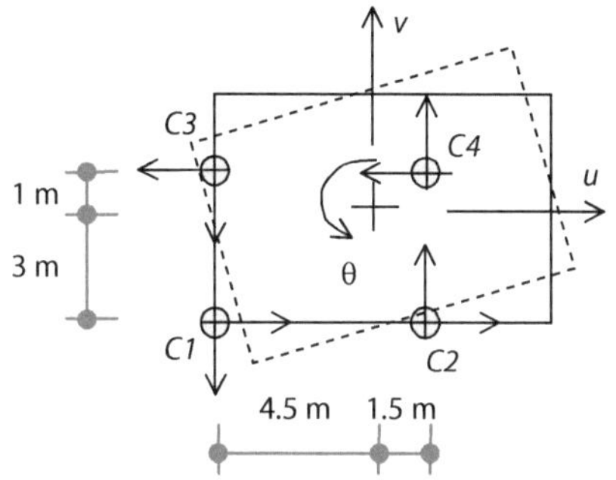

Example 3-6, Figure 5

C2: $u_2 = u + 3\theta, \quad v_2 = v + 1.5\theta$

$u_2 = 0.009 + 3(-0.003) = 0$

$v_2 = -0.014 + 1.5(-0.003) = -0.0185 \text{ m}$

$$\Delta_{\max} = \sqrt{0^2 + 0.0185^2} = 0.0185 \text{ m}$$

C3: $u_3 = u - 1\theta, \quad v_3 = v - 4.5\theta$

$u_3 = 0.009 - 1(-0.003) = 0.012 \text{ m}$

$v_3 = -0.014 - 4.5(-0.003) = -0.0005 \text{ m}$

$$\Delta_{\max} = \sqrt{0.012^2 + 0.0005^2} = 0.0120 \text{ m}$$

C4: $u_4 = u - 1\theta, \quad v_4 = v + 1.5\theta$

$u_4 = 0.009 - 1(-0.003) = 0.012 \text{ m}$

$v_4 = -0.014 + 1.5(-0.003) = -0.0185 \text{ m}$

$$\Delta_{\max} = \sqrt{0.012^2 + 0.0185^2} = 0.022 \text{ m}$$

The force can be found by multiplying the top displacement of each column by its stiffness, k_c. Alternatively, the force is obtained by using similar procedures using the proper equations for modal force calculations, namely, two forces and a moment in the form $\{f_{s1}\} = [M]\{\phi_1\}\Gamma_1 S_{a1}$.

Example 3-7 Use standard structural analysis software to find the frequencies and mode shapes of the structure given in Example 3-6, which is also shown in Example 3-7, Fig. 1. For this purpose, consider two schemes of modeling:

1. Allocation of mass to frame nodes (joints).
2. Allocation of mass to a single location at the center of mass.

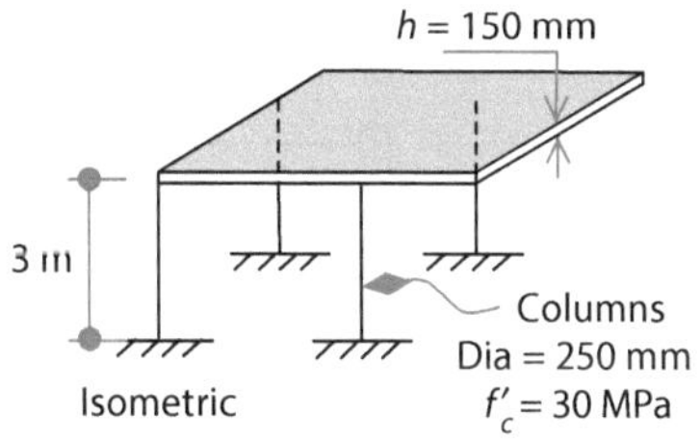

Example 3-7, Figure 1

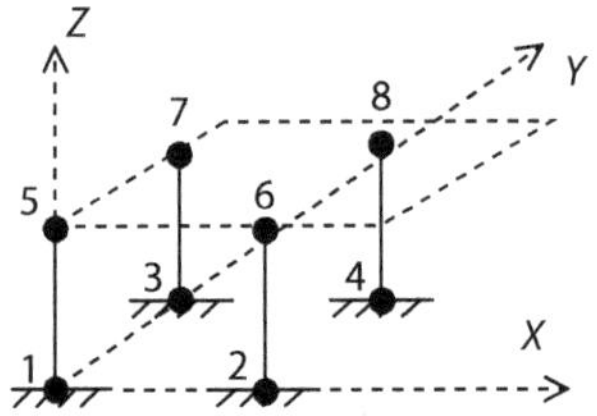

Example 3-7, Figure 2 Space frame model.

Solution

Part (1)

The structure may be modeled as a space frame (3D) with eight nodes and four members as shown in Example 3-7, Fig. 2. The nodes are numbered 1 through 8. The supports are input as totally fixed nodes. The floor nodes, 5 to 7, are also restrained against rotation. The rigidity of the diaphragm may be introduced by the concept of the master node where all nodes are restricted to the displacement of one node in the plane of the diaphragm. For example, node 5 may be declared as a master node for nodes 6, 7, and 8.

Some software programs can explicitly declare the rigid diaphragm feature. In this case, the four nodes 5, 6, 7, and 8, are declared as constrained to a rigid diaphragm at the level of the slab.

For this example, and if the input is given to the computer in kN-m units, the material should be given as, $E_c = 25.7 \times 10^6$ kN·m². Section properties are given to the computer for the four members as the circular section of diameter = 0.25.

Because the columns are assumed to be weightless, their density is assigned to zero. The mass of the slab is allocated to four nodes 5, 6, 7, and 8.

The mass must be allocated to the nodes while preserving the center of mass of the group at the original, CM, as shown in Example 3-7, Fig. 3.

The modeling of the mass must be considered in the two horizontal directions, X and Y. This is analogous to directions u and v in Example 3-6. The total mass of the slab is calculated as given in Example 3-6 as

$$
\begin{aligned}
m_{\text{tot}} &= mA \\
&= 0.3823(54) \\
&= 20.644 \text{ kN·s}^2/\text{m}
\end{aligned}
$$

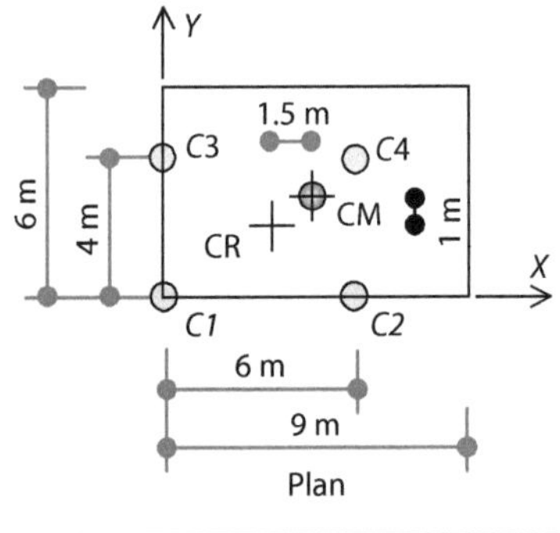

Example 3-7, Figure 3

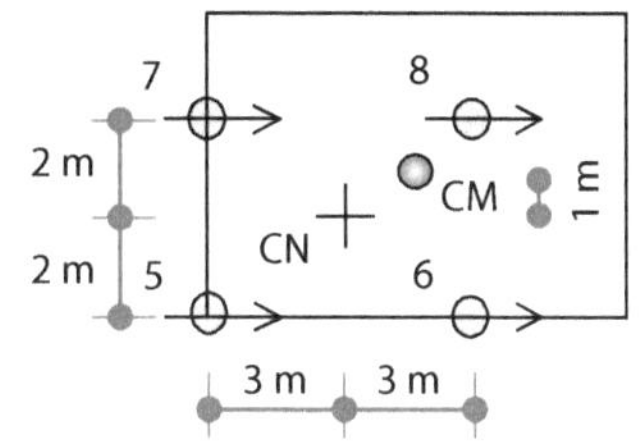

Example 3-7, Figure 4 Mass in direction X.

Direction X

The eccentric moment of masses, $M_{m,\text{cn}}$, about the center of nodes, CN, is given as the total mass multiplied by the distance between CM and CN, or e_{mn}, as shown in Example 3-7, Fig. 4

$$M_{m,\text{cn}} = m_{\text{tot}}\, e_{\text{mn}}$$
$$= 20.644(1) = 20.644 \text{ kN·s}^2$$

The share of each node from the total mass may be obtained by any appropriate method as long as the mass distribution preserves both magnitude of the total mass and moment of mass about CN. One convenient method is considered in this context by using procedures similar to those used in finding forces in piles. This may be given by the following expression:

$$m_i = \frac{m_{\text{tot}}}{N} + \frac{M_{m,\text{cn}}}{\sum r_j^2} r_i$$

where $\sum r_j^2 = 2(2)^2 + 2(2)^2 = 16 \text{ m}^2$

$$m_{7,8} = \frac{20.644}{4} + \frac{20.644}{16}(2) = 5.161 + 2.581 = 7.742 \text{ kN·s}^2\text{/m}$$

$$m_{5,6} = \frac{20.644}{4} - \frac{20.644}{16}(2) = 5.161 - 2.581 = 2.580 \text{ kN·s}^2\text{/m}$$

Note that in this case, the center of nodes, CN, coincides with the center of resistance, CR. This coincidence is not necessarily the case in general.

Alternative procedures that can be used to allocate the mass to nodes are to consider uniform mass distribution on both sides of the center of nodes for any selected number of nodes. Since the moment of mass about CN for uniform masses $\bar{m}_j$ acting in opposite directions is given as follows:

$$M_{m,\text{cn}} = \sum m_j |r_j| = \bar{m}_j \sum |r_j|$$

then,

$$\bar{m}_j = \pm \frac{M_{m,\text{cn}}}{\sum |r_j|}$$

Note that the summation of $\sum \bar{m}$ in one direction does not always vanish, and hence, only the difference between m_{tot} and $\sum \bar{m}$ is equally distributed to all mass locations. Consequently, the final distribution of mass m_i becomes

$$\Delta m = m_{\text{tot}} \pm \sum \bar{m}_j$$

$$m_i = \frac{\Delta m}{N} \pm \bar{m}_i$$

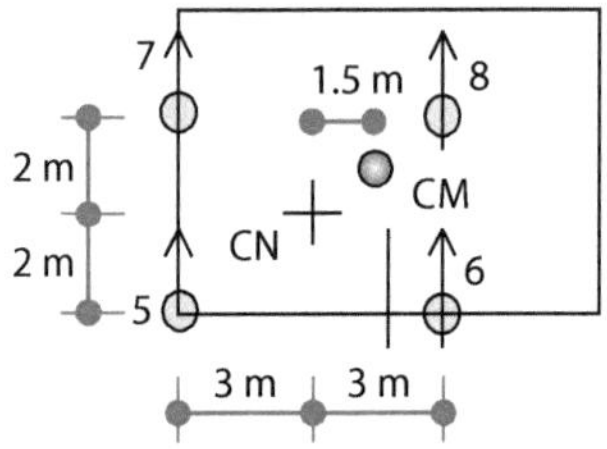

Example 3-7, Figure 5 Mass in direction *Y*.

The application of this procedure is illustrated in the calculations of mass allocation in the *Y*-direction as follows.

Direction Y

The mass will be allocated as shown in Example 3-7, Fig. 5.

Moment of mass about CN

$$M_{m,\mathrm{cn}} = m_{\mathrm{tot}}\, e_{\mathrm{mn}}$$

$$= 20.644(1.5) = 30.966\ \mathrm{kN{\cdot}s^2}$$

$$\Sigma\left|r_j\right| = 2(3)+2(3) = 12\ \mathrm{m}$$

$$\bar{m}_j = \pm\frac{M_{m,\mathrm{cn}}}{\Sigma\left|r_j\right|} = \pm\frac{30.966}{12} = \pm 2.581$$

$$\Delta m = m_{\mathrm{tot}} \pm \Sigma\,\bar{m}_j$$

$$= 20.644 + 2(2.581) - 2(2.581)$$

$$= 20.644\ \mathrm{kN{\cdot}s^2/m}$$

$$m_i = \frac{\Delta m}{N} \pm \bar{m}_i$$

$$m_{6,8} = \frac{20.644}{4} + 2.581 = 7.742\ \mathrm{kN{\cdot}s^2/m}$$

$$m_{5,7} = \frac{20.644}{4} - 2.581 = 2.580\ \mathrm{kN{\cdot}s^2/m}$$

The final distribution of the mass to the four nodes is shown in Example 3-7, Fig. 6. These masses can be input directly in any convenient software programs such as SAP2000. They are given as joint mass in directions *X* and *Y*.

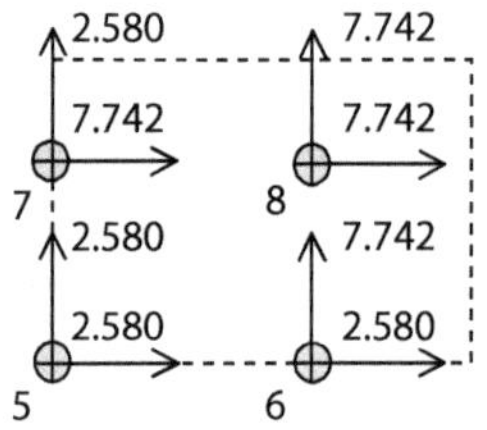

Example 3-7, Figure 6 Final mass distribution ($\mathrm{kN{\cdot}s^2/m}$).

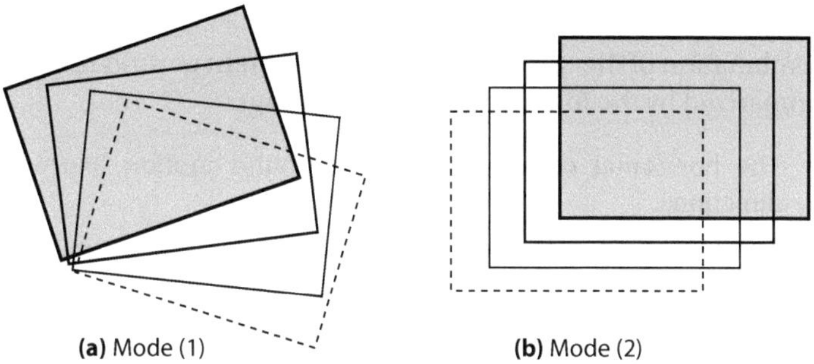

EXAMPLE 3-7, FIGURE 7 Frequencies and mode shapes, SAP2000 output.

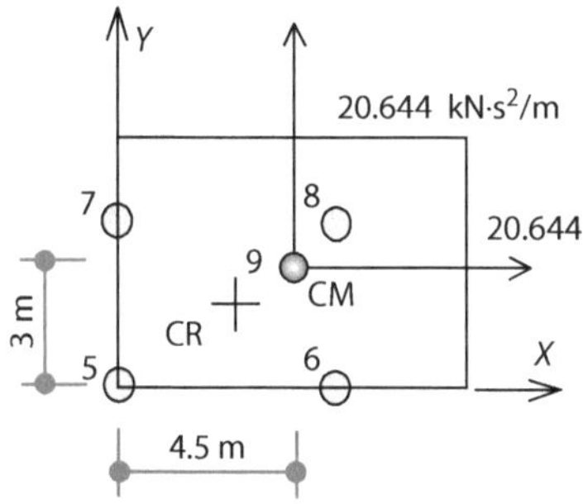

EXAMPLE 3-7, FIGURE 8 Single node mass.

Solution Scheme	Period
Rigid floor with two displacements and rotation, manual (Example 3-6)	0.374 s
Mass allocated to frame nodes, SAP2000 (Example 3-7)	0.365 s
Mass allocated to single node, SAP2000 (Example 3-7)	0.342 s

EXAMPLE 3-7, TABLE 1 Comparison Between Fundamental Periods

The input data above were fed to SAP2000 to yield the results shown in Example 3-7, Fig. 7, as animated by SAP2000.

Part (2)
In this scheme, the total mass will be input to the program at a single node. Similar to Part (1), the mass must be input in both directions X and Y. To do this, a new node must be created at the center of mass and must also be attached to the rigid diaphragm. This task is accomplished by introducing node number 9 as shown in Example 3-7, Fig. 8.
This model was run through SAP2000 to yield results as in Part (1). Example 3-7, Table 1, shows the comparison between the periods that resulted from different solution schemes of Example 3-6 and Example 3-7.

Combination Effect of Different Ground Motions

The combination of the directional effect of different ground motion components may be summarized by the following research findings:

- The horizontal component of the ground motion has equal intensity in all directions.
- The vertical component of the ground motion is about two-thirds its intensity in the horizontal direction.
- The horizontal components in any two orthogonal directions and the vertical component are approximately uncorrelated; they do not take place simultaneously.

Consequently, the same design response spectrum can be used for any two perpendicular directions. The vertical response spectrum may be obtained by multiplying the horizontal response spectrum by two-thirds, or any other factor deemed appropriate.

Combining the values from the three motions in the three directions may be accomplished using the SRSS combining method; for any quantity *A*, the maximum value is given as

$$(A_{\max})^2 = (A_x)^2 + (A_y)^2 + (A_z)^2$$

where A_x, A_y, and A_z are the resulting values of any event *A* from the earthquake citations in directions *x*, *y*, and *z*, respectively.

3.4.3 Mass Participation in Buildings

Mass participation is defined as the contribution of each mode to the total base shear due to unit acceleration in each of the global directions *x*, *y*, and *z*. Seismic codes usually require a number of modes to be used in the analysis to assure at least 90 percent of the participating mass is included in the dynamic response for each principal horizontal direction.

Before the treatment of MDOF system, the response of SDOF to unit acceleration is considered. The dynamic equation due to unit acceleration for an SDOF is given as

$$m\ddot{u} + c\dot{u} + ku = p(t)$$

$$m\ddot{u} + c\dot{u} + ku = -m\ddot{u}_g$$

$$m\ddot{u} + c\dot{u} + ku = -m(1)$$

Dividing by mass, then $\rightarrow$ $\ddot{u} + 2\zeta\omega\dot{u} + \omega^2 u = -1$

For undamped system $\rightarrow$ $\ddot{u} + \omega^2 u = -1$

Solution is given as, displacement $\rightarrow$ $u = \dfrac{-1}{\omega^2}(1 - \cos\omega t)$

By differentiating twice, acceleration $\rightarrow$ $\ddot{u} = -\cos\omega t$

Hence, the system maximum acceleration $\rightarrow$ $|\ddot{u}_{\max}| = 1$

Since the spectral acceleration $S_a = |\ddot{u}_{\max}|$, the spectral acceleration of the system becomes $S_a = 1$.

Accordingly, the maximum force in the system, $f_{s,\max}$, is calculated as

$$f_{s,\max} = mS_a = m(1) = m$$

It can be seen from the equation above that the maximum force in the SDOF system due to unit acceleration equals to the total mass of the system m.

The same concept may now be applied to the general case of MDOF system as follows. General equation of motion

$$[M]\{\ddot{U}\}+[C]\{\dot{U}\}+[K]\{U\}=-[M]\{R\}\ddot{u}_g$$

Equation of motion due to unit acceleration in one direction, say x-direction, becomes

$$[M]\{\ddot{U}\}+[C]\{\dot{U}\}+[K]\{U\}=-[M]\{R_x\}(1)$$

where $\{R_x\}$ = earthquake directional vector, or, influence vector in direction x
= { I }, a unit vector = $\{1\ 1\ 1\ .....\}^T$, in this case

Modal equations

$$m_n^*\ddot{Z}_n + c_n^*\dot{Z}_n + k_n^* Z_n = -£_n(1)$$

Dividing the modal equation by m_n^*, and considering undamped system result in

$$\ddot{Z}_n + \omega_n^2 Z_n = -\Gamma_{n,x}(1)$$

where Γ_n is the modal participation factor for mode n which is equal to

$$\Gamma_{n,x} = \frac{£_n}{m_n^*} = \frac{\{\phi_n\}^T[M]\{R_x\}}{\{\phi_n\}^T[M]\{\phi_n\}}$$

Comparing results with SDOF solution, then $\ddot{Z}_{n,\max}$ is given as

$$\ddot{Z}_{n,\max} = \Gamma_{n,x}$$

The modal forces are given as

$$\{f_{sn,\max}\} = [M]\{\ddot{U}_{n,\max}\}$$

$$\{f_{sn,\max}\} = [M]\{\phi_n\}\ddot{Z}_{n,\max}$$

$$\{f_{sn,\max}\} = [M]\{\phi_n\}\Gamma_{n,x}$$

The base shear due to first mode is the summation of $\{f_{sn,\max}\}$ components, that is

$$V_{n,x} = \{I\}^T\{f_{sn,\max}\}$$

$$V_{n,x} = \{I\}^T[M]\{\phi_n\}\Gamma_{n,x}$$

Taking the transpose of the equation above and noting that $\{R_x\} = \{\mathrm{I}\}$, in this case, then

$$V_{n,x}^T = [\{I\}^T[M]\{\phi_n\}\Gamma_{n,x}]^T$$

$$V_{n,x} = \{\phi_n\}^T[M]^T\{I\}\Gamma_{n,x}$$

or,

$$V_{n,x} = \{\phi_n\}^T [M]\{R_x\}\Gamma_{n,x}$$
$$V_{n,x} = \left(\Gamma_{n,x} m_n^*\right)\Gamma_{n,x}$$
$$V_{n,x} = \Gamma_{n,x}^2 m_n^*$$

Since the ratio of the mass participation in the base shear is proportional to the base shear force due to mode n, which is given as, $V_{n,x}/V_{\text{total}}$, and since the acceleration is unity, the contribution of the mode expressed in a mass participation ratio MP becomes

$$MP_{n,x} = \frac{V_{nx}}{V_{\text{tot}}} = \frac{\Gamma_{n,x}^2 m_n^*}{m_{\text{tot},x}} = \Gamma_{n,x}^2 \frac{m_n^*}{m_{\text{tot},x}}$$

or,

$$MP_{n,x} = \Gamma_{n,x}^2 \frac{m_n^*}{m_{\text{tot},x}}$$

The same procedures can be demonstrated for the other principal axes y and z, hence we have

$$MP_{n,y} = \Gamma_{n,x}^2 \frac{m_n^*}{m_{\text{tot},y}}$$

$$MP_{n,z} = \Gamma_{n,x}^2 \frac{m_n^*}{m_{\text{tot},z}}$$

Example 3-8 The structure shown in Example 3-8, Fig. 1, has been solved in Example 3-4 earlier. It is required to use the data obtained from Example 3-4 to find the mass participation ratio in direction x. Refer to Example 3-4 for the data and the results obtained, which are as follows.

1. Mass and stiffness (given)

$$m_1 = 0.2 \text{ kN·s}^2\text{/mm}$$
$$m_2 = 0.4 \text{ kN·s}^2\text{/mm}$$
$$k_1 = 50 \text{ kN/mm}$$
$$k_2 = 100 \text{ kN/mm}$$

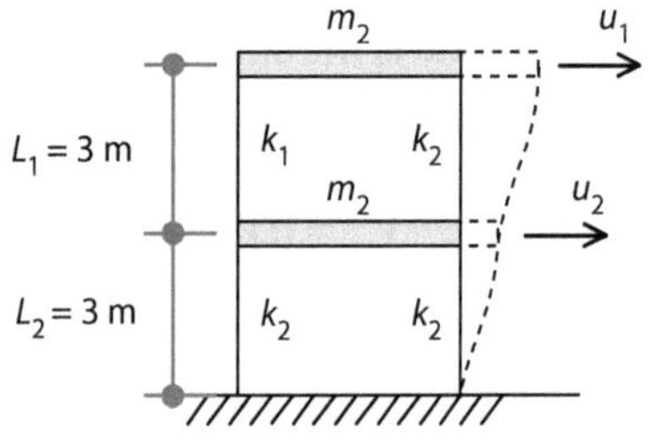

EXAMPLE 3-8, FIGURE 1

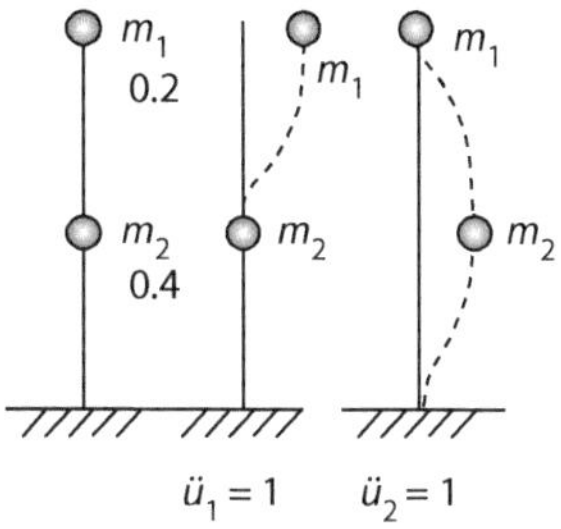

Example 3-8, Figure 2

2. Displacement vector
   Refer to Example 3-8, Fig. 2

$$\{U\} = \begin{Bmatrix} u_1 \\ u_2 \end{Bmatrix}$$

3. Mass matrix results
   Refer to Example 3-8, Fig. 2

$$[M] = 0.2\begin{bmatrix} 1 & 0 \\ 0 & 2 \end{bmatrix} \text{kN·s}^2\text{/mm}$$

4. Stiffness matrix results
   Refer to Example 3-8, Fig. 3

$$[K] = 100\begin{bmatrix} 1 & -1 \\ -1 & 3 \end{bmatrix} \text{kN/mm}$$

5. Eigenvalue solution results

$$|[K] - \omega^2[M]| = 0$$

The results summary is given below and shown graphically in Example 3-8, Fig. 4.

$$\{\omega\} = \begin{Bmatrix} \omega_1 \\ \omega_2 \end{Bmatrix} = \begin{Bmatrix} 15.81 \\ 31.62 \end{Bmatrix}$$

$$[\Phi] = \begin{bmatrix} \phi_1 & \phi_2 \end{bmatrix} = \begin{bmatrix} 1 & 1 \\ 0.5 & -1 \end{bmatrix}$$

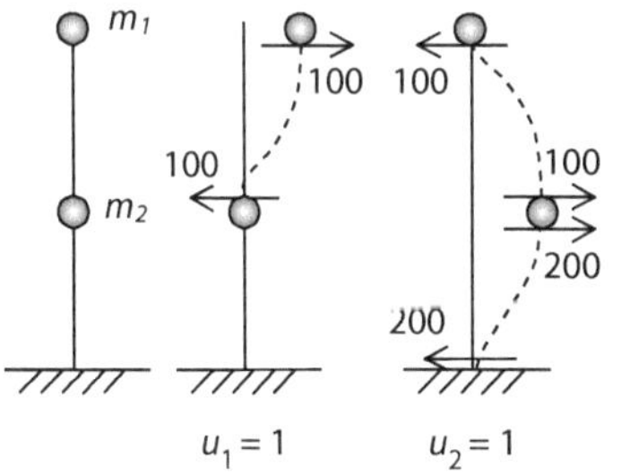

Example 3-8, Figure 3

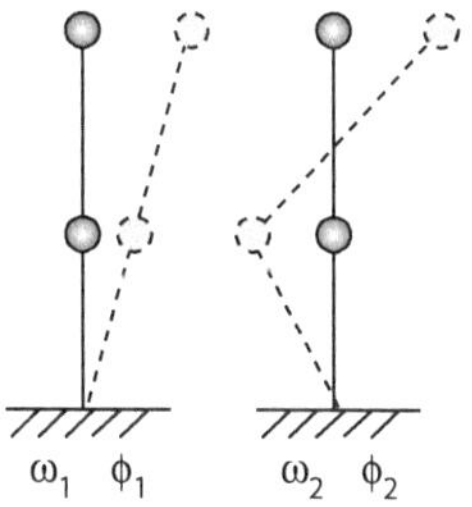

Example 3-8, Figure 4

6. Modal analysis results
   The earthquake directional vector,

$$\{R_x\} = \begin{Bmatrix} 1 \\ 1 \end{Bmatrix}$$

Mode (1): $£_1 = 0.4 \text{ kN·s}^2/\text{mm},\ m_1^* = 0.3 \text{ kN·s}^2/\text{mm}$

$$\Gamma_1 = \frac{£_1}{m_1^*} = \frac{0.4}{0.3} = 1.333$$

Mode (2): $£_2 = -0.2 \text{ kN·s}^2/\text{mm},\ m_2^* = 0.6 \text{ kN·s}^2/\text{mm}$

$$\Gamma_2 = \frac{£_2}{m_2^*} = \frac{-0.2}{0.6} = -0.333$$

7. Mass participation ratios

Mode (1): $$MP_{1,x} = \Gamma_{n,x}^2 \frac{m_1^*}{m_{\text{tot},x}} = (1.333)^2 \frac{(0.3)}{0.6} = 0.888(89\%)$$

Mode (2): $$MP_{2,x} = \Gamma_{n,x}^2 \frac{m_2^*}{m_{\text{tot},x}} = (0.333)^2 \frac{(0.6)}{0.6} = 0.111(11\%)$$

8. Total participating mass

$$\begin{aligned} MP_{\text{total},x} &= MP_{1,x} + MP_{2,x} \\ &= 89\% + 11\% \\ &= 100\% \text{ of the total mass} \end{aligned}$$

This conclusion indicates that the first mode contributes to 89 percent of the participating mass in the dynamic response, which is the ratio that is referred to in seismic codes.

3.5 Shear Beam

The deflection of a shear beam is due to sliding strips (planes) that remain parallel to each other as shown in Fig. 3-26. In this deformation pattern, no rotation or curvature is allowed. Recall that curvature is defined as rotation per unit length, or ($\kappa = M/EI$).

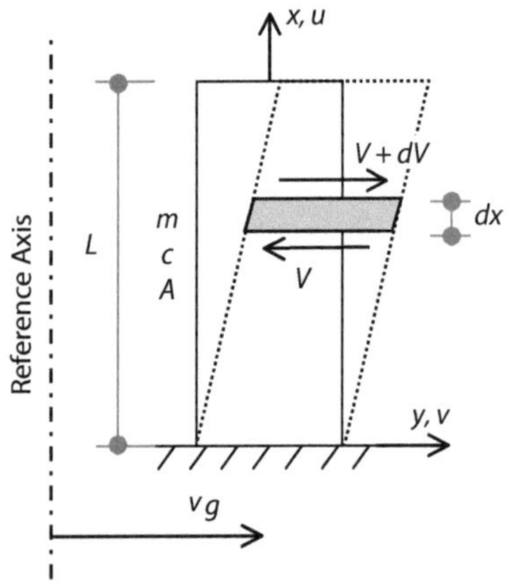

FIGURE 3-26 Shear beam model.

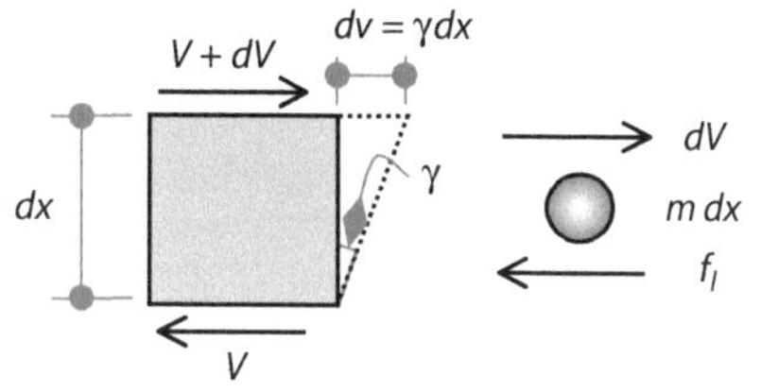

FIGURE 3-27 Shear differential element.

For a shear beam, the curvature, κ, tends to zero as the flexural rigidity, *EI*, of the member approaches ∝.

Figure 3-27 shows the deformation of a shear element *dx*, which is given as follows. Shear strain

$$\gamma = \frac{dv}{dx} = v'$$

Shear force

$$\begin{aligned} V &= \tau A \\ &= G\gamma A = GA\,v' \end{aligned}$$

Equilibrium of the differential element with constant mass per unit length, *m*, is given as shown in Fig. 3-27

$$dV - f_I = 0$$

by substitution from above

$$\frac{\partial V}{\partial x}dx - (m\,dx)\ddot{v} = 0$$

$$\frac{\partial}{\partial x}(GAv') - m\ddot{v} = 0$$

$$m\ddot{v} - GAv'' = 0$$

The equation above is the differential equation of a prismatic beam with constant area, *A*, and with constant mass per unit length, *m*, that vibrates in pure shear mode. For forced vibration with earthquake excitation, the differential equation becomes

$$m\ddot{v} - G\,Av'' = -mr\ddot{v}_g$$

The frequencies and mode shapes can be extracted similar to the MDOF procedures. Because the shear beam is a continuous system, it is considered an MDOF with infinite number of degrees of freedom. Consequently, the beam has an infinite number of frequencies and an infinite number of mode shapes that can be found as follows.

Consider the free vibration equation

$$m\ddot{v} - GAv'' = 0$$

$$\ddot{v} - \frac{GA}{m}v'' = 0, \quad \text{by setting } c^2 = GA/m$$

then

$$\ddot{v} - c^2\,v'' = 0$$

Using separation of variables technique, the displacement function $v(x,t)$ may written as

$$v(x,t) = \phi(x)\, Z(t)$$

Substitution in the differential equation yields

$$\phi\ddot{Z} - c^2\phi'' Z = 0$$

$$\frac{\ddot{Z}}{Z} = c^2 \frac{\phi''}{\phi}$$

Because $\phi(x)$ and $Z(t)$ are two independent quantities, the above equation can only be satisfied if

$$\frac{\ddot{Z}}{Z} = \text{constant}$$

and

$$c^2 \frac{\phi''}{\phi} = \text{constant}$$

For convenience, let this constant be equal to $-\omega^2$. Thus,

$$\frac{\ddot{Z}}{Z} = -\omega^2 \quad \rightarrow \quad \ddot{Z} + \omega^2 Z = 0$$

$$c^2 \frac{\phi''}{\phi} = -\omega^2 \quad \rightarrow \quad \phi'' + \frac{\omega^2}{c^2}\phi = 0$$

with

$$k^2 = \omega^2/c^2$$

the second equation becomes

$$\phi'' + k^2\phi = 0$$

and the solution is

$$\phi = A\cos kx + B\sin kx$$

where A and B are constants determined by boundary conditions (BC). For the vertical cantilever given earlier, the boundary conditions are given as zero displacement at the base and zero shearing force at the top as shown in Fig. 3-28.

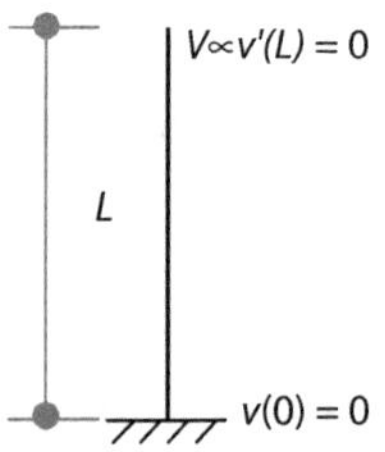

Figure 3-28 Boundary conditions.

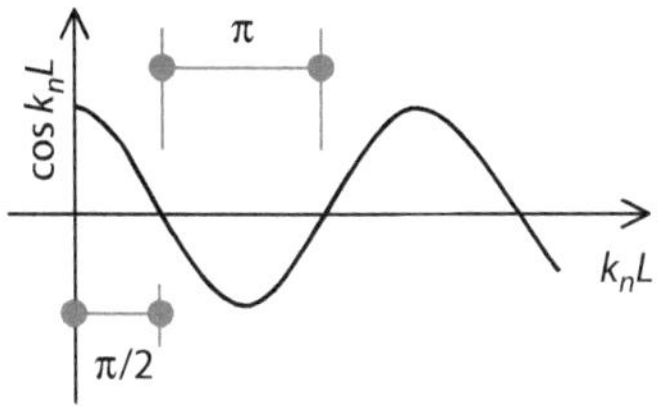

FIGURE 3-29 Cos k_nL solution.

BC #1: $v(0) = 0, \quad \rightarrow \quad \phi(0) = 0 = A\cos 0 = 0,$

therefore,

$$A = 0$$

BC #2: $v'(L) = 0$

hence,

$$\phi'(L) = 0$$
$$kB\cos kL = 0$$

Therefore, and in reference to Fig. 3-29

$$\cos kL = 0$$
$$k_n L = \left(\frac{\pi}{2}\right)(2n-1);$$
$$n = 1, 2, 3, \ldots$$

From the results above, frequencies and mode shapes are obtained as follows.
Frequencies:

$$k_n L = \left(\frac{\pi}{2}\right)(2n-1); \qquad \text{let } \lambda_n = \frac{\pi}{2}(2n-1)$$
$$\left(\frac{\omega_n}{c}\right)L = \lambda_n$$
$$\omega_n = \lambda_n \frac{c}{L}$$

where $\lambda_n = \frac{\pi}{2}(2n-1)$

since $$c = \sqrt{\frac{GA}{m}}$$

the frequency ω_n is given as

$$\omega_n = \lambda_n \sqrt{\frac{GA}{mL^2}}, \quad \lambda_n = \frac{\pi}{2}(2n-1)$$

Mode shapes: with constant B taken arbitrarily equal to 1, the mode shapes are given as

$$\phi_n = B \sin k_n x = \sin\left(\frac{\pi}{2}\right)(2n-1)\left(\frac{x}{L}\right)$$

$$\phi_n = \sin\lambda_n \frac{x}{L}, \qquad \lambda_n = \frac{\pi}{2}(2n-1)$$

With these given frequencies and mode shapes, modal analysis can be used in the usual manner for finding modal mass, m^*, modal earthquake excitation factor $£_n$, and modal participation factor Γ_n. The earthquake-induced forces and displacements for each mode are found from the response spectrum as given before

$$v_{n,\max} = \phi_n \Gamma_n S_{dn}$$

$$f_{sn,\max} = m\phi_n \Gamma_n S_{an}$$

The final forces and displacements are found by combining modal quantities by any applicable method such SABS, SRSS, or CQC.

The shear beam can be utilized to analyze regular buildings that have constant uniform distribution of mass and stiffness and rigid diaphragms. For this analysis to be valid, the girder needs to be stiff enough to prevent nodal rotations of the columns, hence simulating shear deformations. This condition is usually satisfied if $EI/L_{\text{girder}} > 2\ EI/L_{\text{column}}$. The relevant parameter of the shear beam may be obtained from such a building by analogy. The following analogy can be made in reference to Fig. 3-30.

Because it is required to find the parameter $c\left(c = \sqrt{GA/m}\right)$ in order to find the frequencies and mode shapes, it will be required to find the equivalent shear beam axial rigidity, GA, and equivalent uniform mass distribution per unit length, m. These quantities can be found with reference to Fig. 3-30 as

Equivalent mass,

$$m = \frac{M}{h}$$

Equivalent shear strain,

$$\gamma = \frac{\Delta}{h}$$

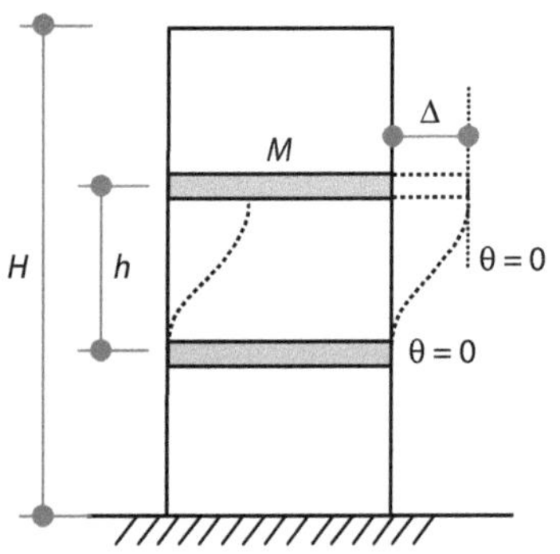

FIGURE 3-30 Shear beam building.

where M is the story mass and Δ is the relative displacement between adjacent stories. Because the shear force for a shear beam is given as

$$\begin{aligned} V &= \tau A \\ &= G\gamma A = GA\gamma \end{aligned}$$

or

$$V = GA\left(\frac{\Delta}{h}\right)$$

and because the shear force (story shear) for a building is given as

$$\begin{aligned} V &= \frac{\Sigma 12\, EI}{h^3}\Delta \\ &= \frac{\Sigma 12\, EI}{h^2}\left(\frac{\Delta}{h}\right) \end{aligned}$$

A comparison of the two equations above shows that the equivalent GA of the buildings is given as

$$GA = \frac{\Sigma 12EI}{h^2}$$

Hence,

$$c = \sqrt{\frac{\sum 12\, EI}{mh^2}}$$

$$\omega_n = \lambda_n \frac{c}{H}, \lambda_n = \frac{\pi}{2}(2n-1), \qquad n = 1, 2, 3, \ldots$$

$$\phi_n = \sin\left(\lambda_n \frac{x}{H}\right)$$

The first three mode shapes are shown in Fig. 3-31.

The modal participation factor, modal base shear, and modal base moments are given as follows.

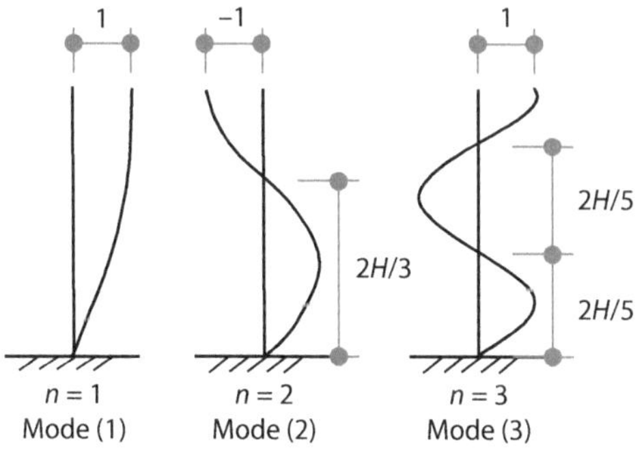

FIGURE 3-31 Shear beam mode shapes.

Participation factor, Γ_n

$$\pounds = \int_H^H \phi m r dx$$

$$= m\int_0^H \phi(1) dx$$

$$= m\int_0^H \left[\sin\left(\lambda_n \frac{x}{H}\right)\right] dx = -m\frac{H}{\lambda_n}\left(\cos\lambda_n \frac{x}{H}\right)_0^H = m\frac{H}{\lambda_n}$$

$$m^* = \int_0^H \phi m \phi dx = m\int_0^H \phi^2 dx$$

$$= m\int_0^H \left[\sin^2\left(\lambda_n \frac{x}{H}\right)\right] dx = m\left(\frac{H}{2}\right)_0^H = m\frac{H}{2}$$

$$\Gamma_n = \frac{\pounds}{m^*} = \frac{2}{\lambda_n}$$

Base shear, V_B

$$V_B = \int_0^H f_s\, dx = \int m\phi\Gamma_n S_a\, dx = m\Gamma_n S_a \int \phi\, dx = m\left(\frac{2}{\lambda_n}\right) S_a \left(\frac{H}{\lambda_n}\right)$$

$$= mHS_a\left(\frac{2}{\lambda_n^2}\right)$$

Base moment, M_B

$$M_B = \int_0^H f_s x\, dx = \int m\phi\Gamma_n S_a x\, dx = m\Gamma_n S_a \int \phi x\, dx$$

since

$$\int \phi x\, dx = \int x\left[\sin\left(\lambda_n \frac{x}{H}\right)\right] dx = \text{integration by parts} = \frac{H^2}{\lambda_n^2}$$

then

$$M_B = m\left(\frac{2}{\lambda_n}\right) S_a \left(\frac{H^2}{\lambda_n^2}\right)$$

$$= mH^2 S_a \left(\frac{2}{\lambda_n^3}\right)$$

Example 3-9 For preliminary design purposes, the tall building shown in Example 3-9, Fig. 1 will be assumed to behave as a uniform shear beam.

1. Determine the frequency, ω, the mode shape, ϕ, and the participation factor, Γ, for each of the first five modes. Note that, $\phi(x)$ is a continuous function.

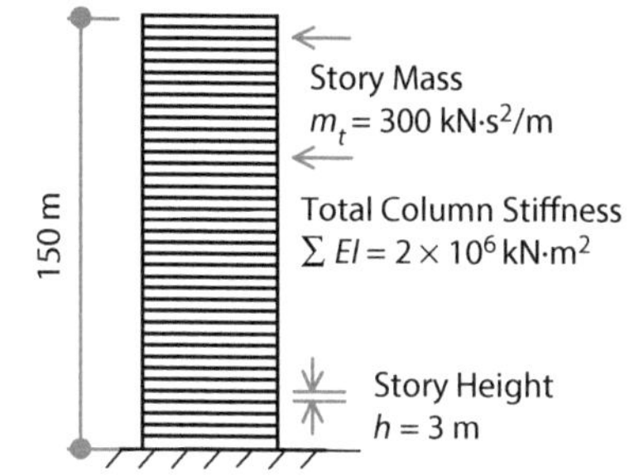

Example 3-9, Figure 1

2. Compute the approximate maximum top displacement base shear and base overturning moment by the SRSS method, including the first five modes. Assume that the velocity response spectrum value for each mode is 0.5 m/s.

Solution

Part (1)
Frequencies:

$$\omega_n = \lambda_n \frac{c}{H}$$

$$\lambda_n = \frac{\pi}{2}(2n-1), \quad n = 1,2,3,\ldots$$

where,

$$c = \sqrt{\frac{\Sigma 12EI}{mh^2}} = \sqrt{\frac{12(2\times10^6)}{100(3)^2}} = 163.3 \text{ m/s}$$

$$H = 150 \text{ m}$$

hence,

$$\omega_n = \frac{163.3}{150}\lambda_n = 1.089\left(\frac{\pi}{2},\frac{3\pi}{2},\frac{5\pi}{2},\frac{7\pi}{2},\frac{9\pi}{2}\right)$$

$$\omega_n = 1.71,\ 5.13,\ 8.55,\ 11.97,\ 15.39 \text{ rad/s}$$

mode shapes,

$$\phi_n = \sin\lambda_n \frac{x}{H}$$

$$\phi_n = \sin\lambda_n \frac{x}{150} = \sin\frac{x}{150}\left(\frac{\pi}{2},\frac{3\pi}{2},\frac{5\pi}{2},\frac{7\pi}{2},\frac{9\pi}{2}\right)$$

See Fig. 3-31 for plots of the first three modes.
participation factor,

$$\Gamma_n = \frac{2}{\lambda_n} = \frac{2}{\left(\frac{\pi}{2}\right)(1,3,5,7,9)} = \frac{4}{\pi},\frac{4}{3\pi},\frac{4}{5\pi},\frac{4}{7\pi},\frac{4}{9\pi}$$

$$= 1.273,\ 0.424,\ 0.255,\ 0.182,\ 0.141$$

Part (2)

Top displacement: $u_{top} = \Gamma_n S_{dn}$

where,

$$\Gamma_n = \frac{2}{\lambda_n} = 1.273, 0.424, 0.255, 0.182, 0.141 \qquad \text{see Part (1)}$$

$$S_{dn} = \frac{S_{vn}}{\omega_n} = \frac{0.5}{(1.71, 5.13, 8.55, 11.97, 15.39)} \qquad \text{see Part (1)}$$

$$= 0.29,\ 0.097,\ 0.058,\ 0.042,\ 0.032\ \text{m}$$

Therefore, the top displacements that are due to first five modes are

$$u_{top} = 0.369,\ 0.041,\ 0.015,\ 0.008,\ 0.005\ \text{m}$$

Percent of the first mode: $u_{top} = 100\%, 11\%, 4\%, 2\%, 1\%$

SRSS: $u_{top} = 0.372$ m
$= (100.8\%$ of the first mode$)$

base shear,

$$V_B = mHS_{an}\left(\frac{2}{\lambda_n^2}\right)$$

where

$$mH = 100(150) = 15{,}000\ \text{kN·s}^2/\text{m}$$

$$S_{an} = S_{vn}\omega_n = 0.5(1.71,\ 5.13,\ 8.55,\ 11.97,\ 15.39) \qquad \text{see Part (1)}$$

$$= 0.855,\ 2.565,\ 4.275,\ 5.985,\ 7.695\ \text{m/s}^2$$

$$\frac{2}{\lambda_n^2} = \frac{2}{\left(\frac{\pi}{2}\right)^2 (1^2, 3^2, 5^2, 7^2, 9^2)}$$

hence,

$$V_B = 10{,}394,\ 3{,}462,\ 2{,}082,\ 1{,}486,\ 1{,}151\ \text{kN}$$

Percent of the first mode: $V_B = 100\%, 33\%, 20\%, 14\%, 11\%$

SRSS: $V_B = 11{,}309$ kN
$= (109\%$ of the first mode$)$

base moment,

$$M_B = mH^2 S_{an}\left(\frac{2}{\lambda^3}\right)$$

where

$$mH^2 = 100(150)^2 = 2.25 \times 10^6\ \text{kN·s}^2$$

$$S_{an} = S_{vn}\omega_n = 0.5(1.71,\ 5.13,\ 8.55,\ 11.97,\ 15.39) \qquad \text{see Part (1)}$$

$$= 0.855,\ 2.565,\ 4.275,\ 5.985,\ 7.695\ \text{m/s}^2$$

$$\frac{2}{\lambda_n^3} = \frac{2}{\left(\frac{\pi}{2}\right)^3 (1^3, 3^3, 5^3, 7^3, 9^3)}$$

hence,

$$M_B = 992{,}515,\ 110{,}193,\ 39{,}762,\ 20{,}271,\ 12{,}215\ \text{kN·m}$$

Percent of the first mode: $M_B = 100\%,\ 11\%,\ 4\%,\ 2\%,\ 1\%$

SRSS: $M_B = 999{,}685\ \text{kN·m}$

$= (100.7\%\ \text{of the first mode})$

3.6 Cantilever Flexure Beam

Cantilever flexure beams can be found in most buildings. If the structural system of the building consists mainly of shear walls, or if the building behavior is dominated by flexure modes, the behavior of the cantilever flexure beam is an appropriate model for analyzing building behavior.

The deflection of a beam in flexure is due to rotation of its plane sections. Curvature, κ, is defined as the rotation per unit length ($\kappa = d\alpha/dx$).

Figure 3-32 shows a vertical cantilever with uniformly distributed mass, m, damping c, and flexure stiffness, EI.

Figure 3-33 shows the well-known deformation of a flexure element, dx, which is summarized as follows.

Flexure strain; curvature: $\kappa = -v''$

Moment: $M = EI\kappa = -EIv''$

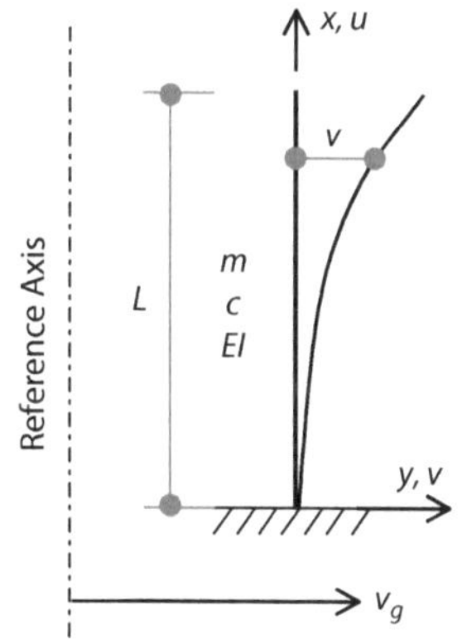

FIGURE 3-32 Cantilever flexure beam.

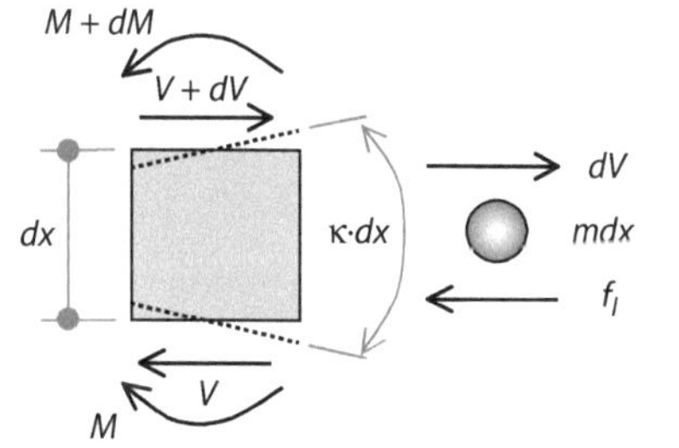

FIGURE 3-33 Flexure differential element.

Static equilibrium of the infinitesimal element dx requires that

$$dM = Vdx$$

or

$$V = M'$$

Dynamic equilibrium of the infinitesimal mass ($m \cdot dx$) requires that

$$dV - f_I = 0$$

by substitution from above

$$\frac{\partial V}{\partial x} dx - (mdx)\ddot{v} = 0$$

$$\frac{\partial}{\partial x}(M') - m\ddot{v} = 0$$

$$M'' - m\ddot{v} = 0$$

$$m\ddot{v} + EIv^{iv} = 0$$

Because the earthquake loading can be applied as external load, as shown in the previous chapter, the equation of motion in this case becomes

$$m\ddot{v} + EIv^{iv} = -mr\ddot{v}_g$$

Because the frequencies and mode shapes can be extracted in a way similar to the shear beam system, the cantilever flexure beam is considered an MDOF with an infinite number of degrees of freedom. Therefore, the beam has an infinite number of frequencies and an infinite number of mode shapes that can be found as follows.

To find the frequencies and mode shapes, consider the free vibration equation

$$m\ddot{v} + EIv^{iv} = 0$$

$$\ddot{v} + \frac{EI}{m} v^{iv} = 0, \qquad \text{by setting } c^2 = \frac{EI}{m}$$

then

$$\ddot{v} + c^2 v^{iv} = 0$$

Using the separation of variables technique, the displacement function $v(x,t)$ may be written as

$$v(x,t) = \phi(x)\, Z(t)$$

Substitution in the differential equation yields

$$\phi \ddot{Z} + c^2 \phi^{iv} Z = 0$$

$$\frac{\ddot{Z}}{Z} = -c^2 \frac{\phi^{iv}}{\phi}$$

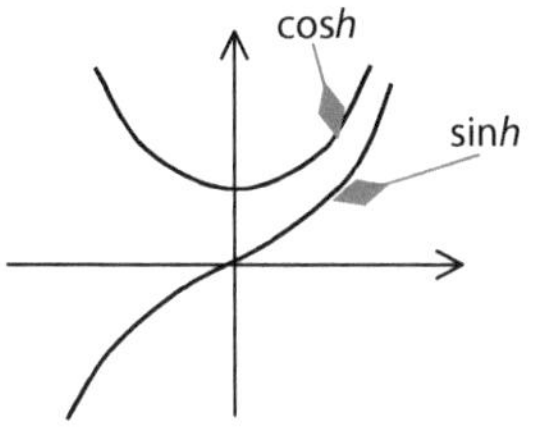

Figure 3-34 Hyperpolic functions.

Because $\phi(x)$ and $Z(t)$ are two independent quantities, the equation above can only be satisfied if

$$\frac{\ddot{Z}}{Z} = \text{constant}$$

and

$$-c^2 \frac{\phi^{iv}}{\phi} = \text{constant}$$

For convenience, let this constant $= -\omega^2$.
Therefore,

$$\frac{\ddot{Z}}{Z} = -\omega^2 \quad \rightarrow \quad \ddot{Z} + \omega^2 Z = 0$$

$$-c^2 \frac{\phi^{iv}}{\phi} = -\omega^2 \quad \rightarrow \quad \phi^{iv} - \frac{\omega^2}{c^2}\phi = 0$$

With $k^4 = \omega^2/c^2$, the second equation becomes

$$\phi^{iv} - k^4\phi = 0$$

Solution,

$$\phi = A \cos kx + B \sin kx + C \cosh kx + D \sinh kx$$

where A, B, C, and D are constants determined by boundary conditions. Recall that the hyperbolic functions take the shape shown in Fig. 3-34. For the vertical cantilever given earlier, the boundary conditions are given as zero displacement and zero slope at the base, and zero shear and zero moment at the top as shown in Fig. 3-35. Applying these four boundary conditions leads to the following:

$$\cosh kL \cos kL + 1 = 0$$

A graphical solution to this equation is shown in Fig. 3-36. Values of kL that satisfy this equation are

$$k_1 L = 1.875$$

$$k_n L \approx \left(\frac{\pi}{2}\right)(2n-1),\ n = 2,\ 3,\ \ldots$$

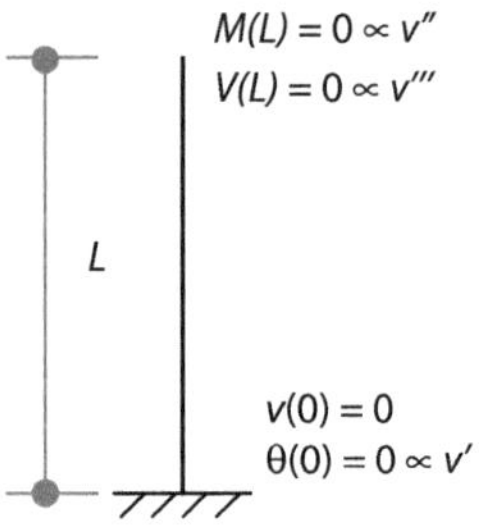

FIGURE 3-35 Boundary conditions.

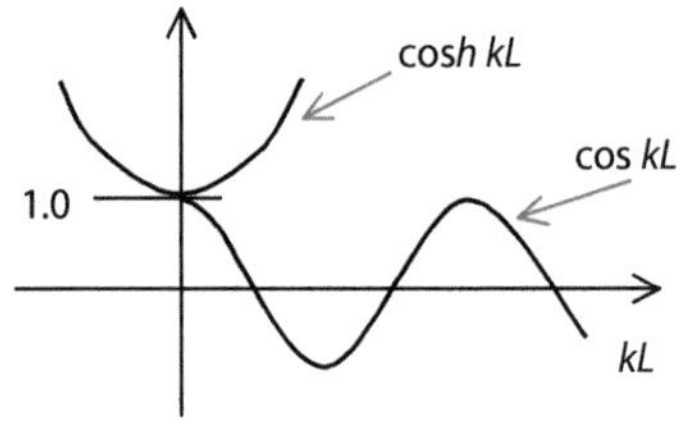

FIGURE 3-36 Solution condition.

From the results above, frequencies and mode shapes are obtained as follows.
Frequencies:

Because

$$k_n^4 = \left(\frac{\omega_n}{c}\right)^2$$

and letting

$$k = \frac{\lambda}{L}$$

then

$$\frac{\lambda_n{}^4}{L^4} = \left(\frac{\omega_n}{c}\right)^2$$

or,

$$\omega_n^2 = \lambda_n^4 \frac{c^2}{L^4} = \lambda_n^4 \frac{EI}{mL^4}$$

or,

$$\omega_n = \lambda_n^2 \sqrt{\frac{EI}{mL^4}}, \begin{bmatrix} \lambda_1 = 1.875 \\ \lambda_n = \frac{\pi}{2}(2n-1); \quad n = 2, 3, 4, \ldots \end{bmatrix}$$

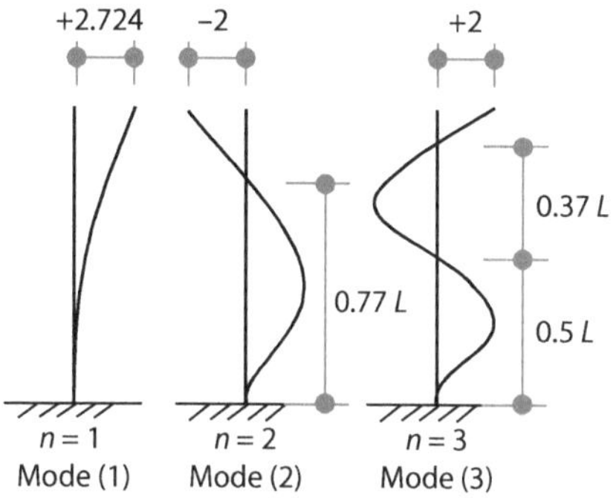

Figure 3-37 Mode shapes of cantilever flexure beam.

Recall that the mode shapes are given with relative values. Solving for the mode shapes, they are given in the following expression.

$$\phi_n = \left(\sin\lambda_n \frac{x}{L} - \sinh\lambda_n \frac{x}{L}\right) + K_n\left(\cosh\lambda_n \frac{x}{L} - \cos\lambda_n \frac{x}{L}\right)$$

where $$K_n = \frac{\sin\lambda_n + \sinh\lambda_n}{\cos\lambda_n + \cosh\lambda_n}, \quad \begin{bmatrix} \lambda_1 = 1.875 \\ \lambda_n = \dfrac{\pi}{2}(2n-1); \quad n = 2, 3, 4, \ldots \end{bmatrix}$$

The first three mode shapes are shown in Fig. 3-37.

The modal participation factor, modal base shear, and modal base moments are obtained by integration as in the case of the shear beam.

Because of the complicated procedures of integration, only the resulting expressions are given here.

Participation factor, Γ_n

$$\Gamma_n = \frac{£}{m^*} = \frac{4}{\lambda_n K_n}$$

Base shear, V_B

$$V_B = mHS_a\left(\frac{4}{(\lambda_n K_n)^2}\right)$$

Base moment, M_B

$$M_B = mH^2 S_a\left(\frac{4}{\lambda_n^3 K_n}\right)$$

where λ_n and K_n are as defined before.

Once frequencies and mode shapes are defined, modal analysis can be used as in all previous cases, that is, modal quantities m_n^*, $£_n$, and Γ_n can be calculated for each mode. With a specified response spectrum, the response values for each mode can be evaluated as

$$v_{n,\max} = \phi_n \Gamma_n S_{dn}$$
$$f_{sn,\max} = m\phi_n \Gamma_n S_{an}$$

	Mode	Participation Factor, Γ_n	Base Shear	Base Moment
Shear Beam		$\frac{2}{\lambda_n}$	$\frac{2}{\lambda_n^2}$	$\frac{2}{\lambda_n^3}$
	1	1.273 (100%)	0.811 (100%)	0.516 (100%)
	2	0.424 (33%)	0.0901 (11%)	0.0191 (4%)
	3	0.255 (20%)	0.0324 (4%)	0.0041 (1%)
Cantilever Flexure Beam		$\frac{4}{\lambda_n K_n}$	$\frac{4}{(\lambda_n K_n)^2}$	$\frac{4}{\lambda_n^3 K_n}$
	1	1.566 (100%)	0.613 (100%)	0.445 (100%)
	2	0.868 (55%)	0.188 (31%)	0.0394 (9%)
	3	0.509 (33%)	0.0648 (11%)	0.00825 (2%)

TABLE 3-1 Comparison between Shear Beam and Cantilever Flexure Beam

The total response is then combined from all modes using any combination method, SABS, SRSS, or CQC as applicable.

The flexure beam can be used to model regular buildings with shear walls as their lateral force resisting elements because shear walls behave primarily as vertical cantilever with flexure domination. Note that the shear beam can be used to idealize frame buildings as discussed earlier. In general, because buildings usually contain both systems (shear walls and frames), their response will be a mixture of both beams. Note that the curvature of the first mode in each beam is reversed; therefore, the combined effect of the beams tends to follow a straight line.

In most seismic codes, an equivalent lateral static force procedure is offered for regular buildings. This force distribution is usually taken close to a straight line distribution, which is compatible with the combined effect of both shear beam and flexure beam.

Comparison between Shear Beam and Cantilever Flexure Beam

Table 3-1 shows a comparison that may be drawn between the shear beam and the cantilever flexure beam if the mode shapes shown in Fig. 3-37 are normalized to a value of 1.0 at the top of the beam. The comparison considers the participation factor at the top, which is proportional to the lateral displacement $v_{\text{top}} = \Gamma_n S_d$, base shear, which is proportional to, $2/\lambda_n^2$ and $4/(\lambda_n K_n)^2$, and base moment, which is proportional to $2/\lambda_n^3$ and $4/(\lambda_n^3 K_n)$ for shear and flexure beam as given in previous equations.

Example 3-10 A tall steel building is designed to resist lateral forces by a series of core concentric frames, a core of vertical trusses, as shown in Example 3-10, Fig. 1.

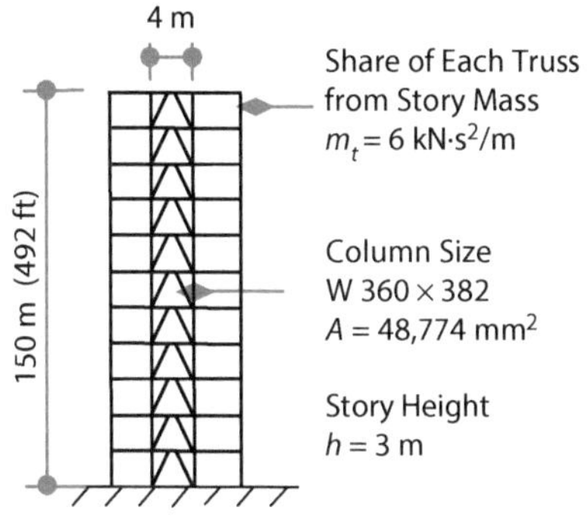

Example 3-10, Figure 1

If the building is assumed to behave as a flexure beam and is subjected to an earthquake having the response spectrum of $S_a = \frac{1}{T} \le 1.0\,\text{m/s}^2$.

(*) Compute the top maximum displacement, base shear, and base moment for the first and second mode.

Solution

1. Structural properties

Modulus of elasticity: $E_s = 200 \times 10^3$ MPa

Mass: $m = \frac{m_t}{3} = \frac{6}{3} = 2\,\text{kN·S}^2/\text{m/m}$

Moment of inertia:

Because the core truss resists lateral loads, and because it consists of concentric bracing, the moment of inertia is idealized as a cross section that consists of two beams spaced 4 m apart, accordingly

$$I \approx \sum Ad^2$$
$$I \approx 48{,}774(2{,}000)^2(2) = 390{,}192 \times 10^6\,\text{mm}^4$$

Therefore,

$$EI = 200 \times 10^3 (390{,}192 \times 10^6)$$
$$= 78 \times 10^{15}\,\text{N·mm}^2 (78 \times 10^6\,\text{kN·m}^2)$$

Modal analysis:

Mode (1)

Frequency: $$\omega_1 = \lambda_1^2 \sqrt{\frac{EI}{mL^4}} = (1.875)^2 \sqrt{\frac{78 \times 10^6}{2(150)^4}} = 0.98\,\text{rad/s}$$

Period: $$T_1 = \frac{2\pi}{\omega_1} = \frac{2\pi}{0.98} = 6.41\,\text{s}$$

Spectrum acceleration: $$S_{a1} = \frac{1}{T_1} = \frac{1}{6.41} = 0.156\,\text{m/s}^2$$

Spectrum displacement: $$S_{d1} = \frac{S_{a1}}{\omega_1^2} = \frac{0.156}{0.98^2} = 0.159\,\text{m}$$

Top displacement: $v_{top,1} = \Gamma_1 S_{d1}$ (see Table 3-1 for Γ_1)

$= 1.566\,(0.159) = 0.25$ m

Base shear: $V_{B1} = mHS_{a1}\dfrac{4}{(\lambda_1 K_1)^2}$ $\left(\text{see Table 3-1 for } \dfrac{4}{(\lambda_1 K_1)^2}\right)$

$= (2)(150)(0.156)(0.613) = 29$ kN

Base moment: $M_{B1} = mH^2\, S_{a1}\dfrac{4}{\lambda_1^3 K_1}$ $\left(\text{see Table 3-1 for} \dfrac{4}{(\lambda_1^3 K_1)^2}\right)$

$= (2)(150)^2(0.156)(0.445) = 3{,}124$ kN·m

Mode (2)

Frequency: $\omega_2 = \lambda_2^2\sqrt{\dfrac{EI}{mL^4}} = (4.712)^2\sqrt{\dfrac{78\times10^6}{2\,(150)^4}} = 6.16$ rad/s

Period: $T_2 = \dfrac{2\pi}{\omega_2} = \dfrac{2\pi}{6.16} = 1.019$ s

Spectrum acceleration: $S_{a2} = \dfrac{1}{T_2} = \dfrac{1}{1.019} = 0.98$ m/s^2

Spectrum displacement: $S_{d2} = \dfrac{S_{a1}}{\omega_1^2} = \dfrac{0.98}{6.16^2} = 0.026$ m

Top displacement: $v_{top,2} = \Gamma_2 S_{d2}$ (see Table 3-1 for Γ_2)

$= 0.868(0.026) = 0.023$ m

Base shear: $V_{B2} = mH\; S_{a2}\dfrac{4}{(\lambda_2 K_2)^2}$ $\left(\text{see Table 3-1 for } \dfrac{4}{(\lambda_2 K_2)^2}\right)$

$= (2)(150)(0.98)(0.188) = 55$ kN

Base moment: $M_{B2} = mH^2 S_{a2}\dfrac{4}{\lambda_2^3 K_2}$ $\left(\text{see Table 3-1 for} \dfrac{4}{\lambda_2^3 K_2}\right)$

$= (2)(150)^2(0.98)(0.039) = 1{,}720$ kN·m

SRSS results

$$v_{top} = \sqrt{0.249^2 + 0.023^2} = 0.250\text{ m}$$

$$V_B = \sqrt{29^2 + 55^2} = 62\text{ kN}$$

$$M_B = \sqrt{3{,}124^2 + 1{,}720^2} = 3{,}566\text{ kN·m}$$

3.7 Simple Flexure Beam

Simple flexure beams are usually found in bridges. The deck of a single-span bridge behaves as a simple flexure beam simply supported at the ends as shown in Fig. 3-38. For more details on bridges, see Chap. 9.

The governing differential equation of the simple flexure beam is the same differential equation for the cantilever flexure beam

$$m\ddot{v} + EIv^{iv} = 0$$

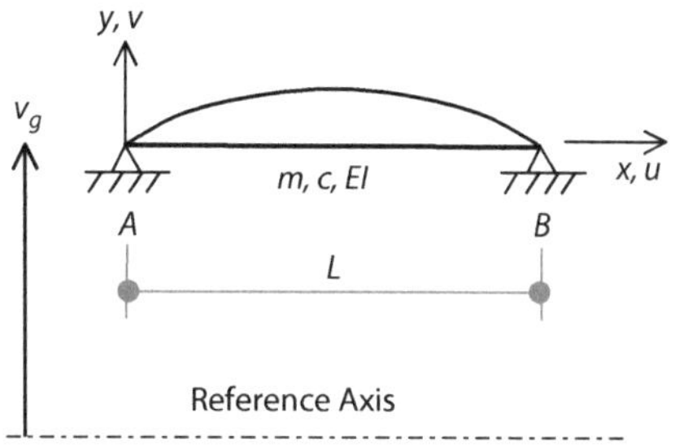

FIGURE 3-38 Simple flexure beam.

With earthquake loading applied as external load, the equation of motion in this case becomes

$$m\ddot{v} + EIv^{iv} = -mr\ddot{v}_g$$

The frequencies and mode shapes can be extracted in a way that is similar to the cantilever flexure beam. Therefore, for an infinite number of DOF, the beam will have an infinite number of frequencies and an associated infinite number of mode shapes.

The solution to the beam differential equation is obtained for the vertical flexure beam as follows:

$$\phi = A \cos kx + B \sin kx + C \cosh kx + D \sinh kx$$

where $k^4 = \omega^2/c^2$, and where A, B, C, and D are constants determined by boundary conditions. The boundary conditions are given as zero displacement and zero moment at both ends of the beam as shown in Fig. 3-39. Applying these four boundary conditions leads to zero values for constants A, C, and D. Thus, the mode shape is given by the following expression:

$$\phi = B \sin kx$$

For the nontrivial solution ($B \neq 0$), the above equation can be satisfied if $\sin kx = 0$. Since $\phi(L) = 0$, then

$$\sin(kL) = 0$$

or

$$(kL) = n\pi, \qquad n = 0, 1, 2, 3, \ldots$$

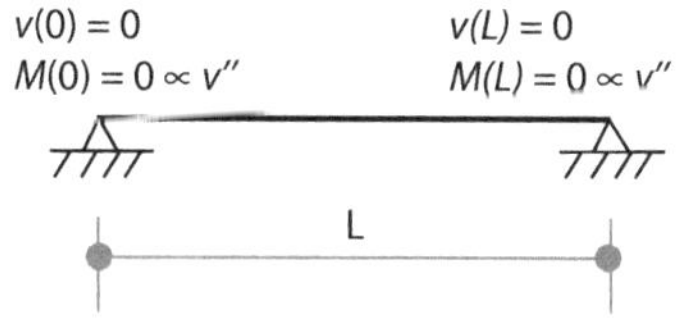

FIGURE 3-39 Boundary conditions.

From the above results, frequencies and mode shapes are obtained as follows.
Frequencies:

Since

$$k_n^4 = \left(\frac{\omega_n}{c}\right)^2$$

then

$$k_n^4 L^4 = \left(\frac{\omega_n}{c}\right)^2 L^4 = (n\pi)$$

or

$$\omega_n^2 = (n\pi)^4 \left(\frac{c^2}{L^4}\right) = (n\pi)^4 \frac{EI}{mL^4}$$

$$\omega_n = (n\pi)^2 \sqrt{\frac{EI}{mL^4}}, \quad n = 1, 2, 3, \ldots$$

Remember that mode shapes are given with relative values. Thus, by selecting constant, $B = 1$, and substituting, $k_n L = n\pi$, the mode shapes are given in the following expression:

$$\phi_n = \sin\left(n\pi \frac{x}{L}\right)$$

$$n = 1, 2, 3, \ldots$$

The first three mode shapes are shown in Fig. 3-40.

The modal participation factor, modal base shear, and modal base moments are obtained by integration as in the case of the shear beam.

Participation factor Γ_n

$$\Gamma_n = \frac{£_n}{m_n^*} = \frac{\int \phi m r \, dx}{\int \phi m \phi \, dx}$$

$$= \frac{\int \sin\left(n\pi \frac{x}{L}\right) dx}{\int \sin^2\left(n\pi \frac{x}{L}\right) dx}, \quad n = 1, 2, 3, \ldots$$

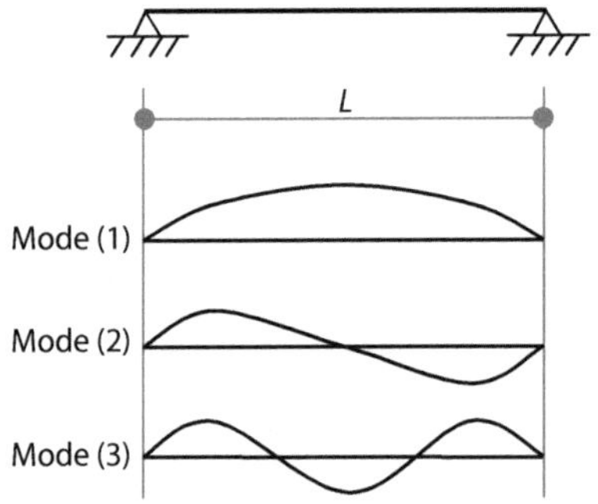

Figure 3-40 Mode shapes of simple flexure beam.

$$\Gamma_n = \frac{4}{n\pi}, \qquad n = 1, 3, 5, \ldots$$

$$\Gamma_n = 0, \qquad n = 2, 4, 6, \ldots$$

Once frequencies and mode shapes are defined, modal analysis can be used as in all previous cases, that is, modal quantities—m_n^*, L_n, and, Γ_n, can be calculated for each mode. With a specified response spectrum, response values for each mode can be evaluated as

$$v_{n,\max} = \phi_n \Gamma_n S_{\text{dn}}$$
$$f_{\text{sn,max}} = m\phi_n \Gamma_n S_{\text{an}}$$

The total response is then combined from all modes using any combination method, SABS, SRSS, or CQC as applicable.

3.8 Axial Beam

An axial beam is a beam in axial vibration, or vibration along its centroidal axis. A vertical column and a shear wall are considered axial beams under earthquake vertical components.

The deflection of an axial beam is due to elongation of its axial elements as shown in Fig. 3-41. In this deformation pattern, no rotation or curvature is allowed.

Figure 3-42 shows deformation of an axial element, *dx,* which is given as follows:

Axial strain: $$\varepsilon = \frac{du}{dx} = u'$$

Axial force: $$N = \sigma A = E\varepsilon A = EAu'$$

Equilibrium of a differential element with constant mass is given in Fig. 3-42. For constant mass intensity per unit length, *m*, equilibrium requires

$$dN - f_I = 0$$

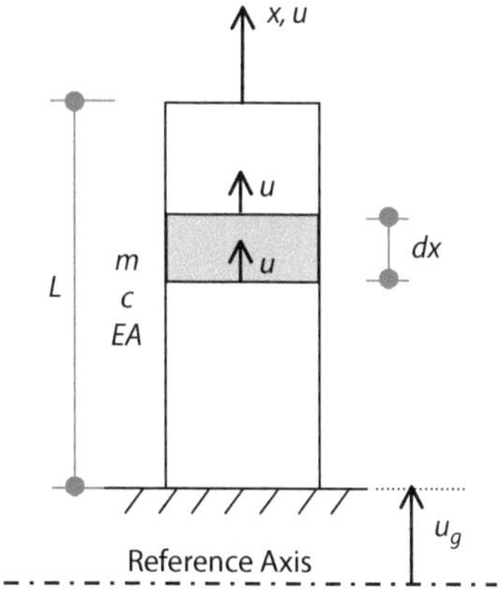

Figure 3-41 Axial beam.

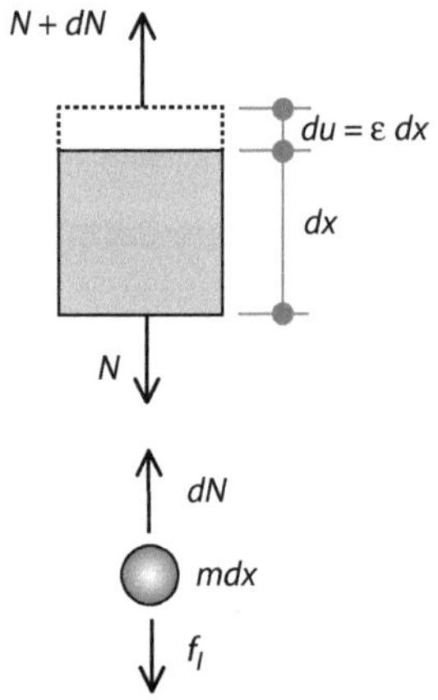

Figure 3-42 Axial Differential Element.

by substitution from above

$$\frac{\partial N}{\partial x}dx-(mdx)\ddot{u}=0$$

$$\frac{\partial}{\partial x}(EAu')-m\ddot{u}=0$$

$$m\ddot{u}-EAu''=0$$

The equation above is the differential equation of a prismatic beam that has a constant cross-sectional area, A, and constant mass per unit length, m, and that vibrates in pure axial mode. For forced vibration with earthquake excitation, the differential equation becomes

$$m\ddot{u}-EAu''=-mr\ddot{u}_g$$

Because the equation above is analogous to the differential equation of the shear beam derived in the previous section, the same expression for frequencies and mode shapes apply for both cases

Frequencies:

$$\omega_n=\lambda_n\sqrt{\frac{EA}{mL^2}}$$

$$\lambda_n=\frac{\pi}{2}(2n-1)$$

Mode shapes:

$$\phi_n=\sin\lambda_n\frac{x}{L}$$

$$\lambda_n=\frac{\pi}{2}(2n-1)$$

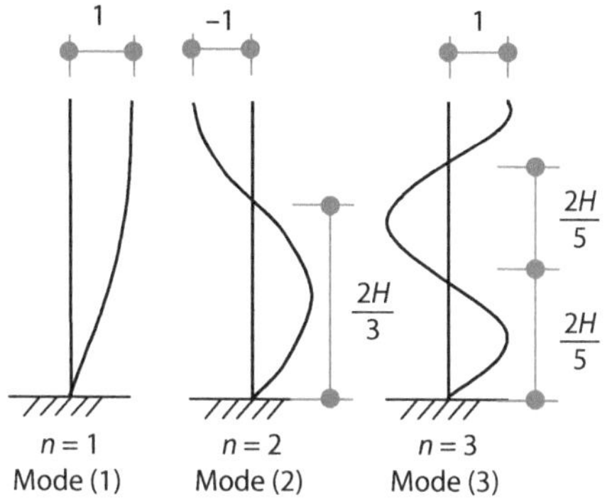

FIGURE 3-43 Axial beam mode shapes.

The first three mode shapes shown in Fig. 3-43 are exactly the same as those for shear beam modes.

With these given frequencies and mode shapes, modal analysis can be used in the usual manner, that is, finding modal mass, m^*, modal earthquake excitation factor, $£_n$, and modal participation factor, Γ_n. The earthquake-induced forces and displacements for each mode are found from the response spectrum as given before

$$v_{n,\max} = \phi_n \Gamma_n S_{\mathrm{dn}}$$
$$f_{\mathrm{sn},\max} = m\phi_n \Gamma_n S_{\mathrm{an}}$$

The final forces and displacements are then combined using any applicable method such as SABS, SRSS, or CQC.

3.9 Finite Element Method

Finite element methods are the most general and powerful methods for structural analysis. They offer an approximate (next to exact) solution for practically any kind of problem in structural mechanics, dynamics, and many other fields of engineering.

The application of finite element analysis in structural dynamics is well recognized and appreciated because it offers a solution for any continuous system, including 1D, 2D, and 3D problems.

The major advantage of finite element analysis is that the differential equation of the system is not required as in the case of the finite differences method. Furthermore, boundary conditions can also be described without any complications as in the finite differences method.

Finite element formulation can be developed by many methods such as virtual work, least squares, or the Galerkin method. The two most powerful approaches that are frequently applied in finite element formulation are the virtual work method and Galerkin method. The Galerkin formulation requires the differential equation of the system, whereas the virtual work method can be used to formulate the problem can be formulated without the differential equation.

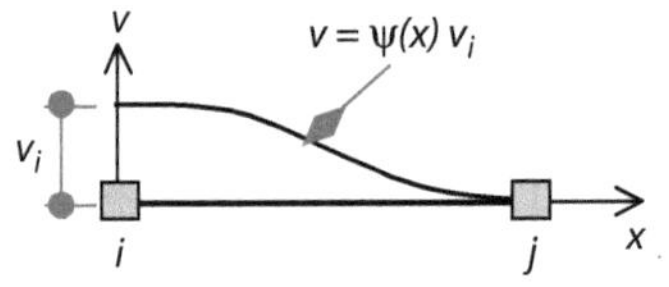

FIGURE 3-44 Beam deformation.

3.9.1 Finite Element Concept in Structural Engineering

The virtual work concept can be used to illustrate finite element analysis. Virtual work is a well-known and well-established concept and tool in structural engineering. When virtual work is used to develop mechanics problems, this approach usually requires evaluation of the external work and the internal work represented by strain energy of a deformed system. In order to evaluate the work quantities, the deformation characteristics of the structure are required.

Because the exact deformation characteristics of structures are not always known, finite element analysis can then be used to assume a deformed shape known as the shape function. The finite element concept states that if the displacements at the ends of any element are given, the displacements between the end points can be interpolated using some shape functions. For example, if the displacement of the beam end, i, as shown in Fig. 3-44 is given as v_i, the displacement $v(x)$ of any point inside the beam between ends i and j may be interpolated if the deflected shape is known or approximated by the function $\psi(x)$ such that

$$v = \psi(x)v_i$$

For this reason, the shape functions are also known as interpolating functions. In the finite element field, the end points of the elements are known as nodes and the displacement in between these nodes is known as field displacement.

Restricting the field displacement to the assumed shape function results in converting a continuous system of an infinite number of DOFs to a discrete system with a finite number of DOFs at the defined nodes of the system. Because the shape function plays an important role in finite element analysis, the accuracy of the solution will be as good as the accuracy of the shape function in representing the exact deformed shape.

The literature is rich in describing the various shape functions for different systems. The shape functions used in structural analysis are usually of the Hermitian polynomial type. For example, linear shape functions are used for truss elements, whereas cubic polynomials are used for beam elements.

Although a lumped mass may be used in dynamic analysis, a more accurate solution is obtained if the mass is considered to be distributed. Consideration of distributed mass in dynamic analysis may be developed using finite element analysis. The following sections will use the virtual work approach and the Galerkin approach to demonstrate such development for beam elements, which are used for most practical structural engineering problems.

3.9.2 Stiffness Matrix (Virtual Work Approach)

The beam element formulation may be simplified if it is remembered that beam behavior is dominated by flexure. The well-known element stiffness matrix [4 × 4] excluding axial deformation is usually given as a function of nodal displacements and

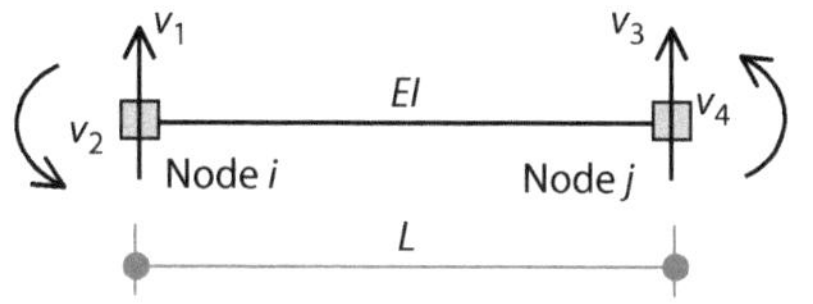

FIGURE 3-45 Classical beam element.

nodal rotations. This matrix is constructed using basic structural analysis theories and may be found in any textbook on structural engineering. Figure 3-45 shows a reference numbering of the classical beam element that designates a nodal displacement and a nodal rotation v_1 and v_2 at node i with a nodal displacement and a nodal rotation v_3 and v_4 at node j. For each node displacement, there will be a consistent end-member force associated with these displacements. For example, there will be a nodal force and a nodal moment p_1 and p_2 at node i with a nodal force and a nodal moment p_3 and p_4 at node j. This matrix may be expressed as follows:

$$\{P\} = [k_m]\{V\}$$

where

$$[k_m]=\begin{bmatrix} \frac{12EI}{L^3} & \frac{6EI}{L^2} & \frac{-12EI}{L^3} & \frac{6EI}{L^2} \\ \frac{6EI}{L^2} & \frac{4EI}{L} & \frac{-6EI}{L^2} & \frac{2EI}{L} \\ \frac{-12EI}{L^3} & \frac{-6EI}{L^2} & \frac{12EI}{L^3} & \frac{-6EI}{L^2} \\ \frac{6EI}{L^2} & \frac{2EI}{L} & \frac{-6EI}{L^2} & \frac{4EI}{L} \end{bmatrix}$$

The constants E and I are the modulus of elasticity and the moment of inertia of the cross section, respectively.

In this section, the beam element stiffness matrix will be developed using the finite element approach for demonstration purposes.

For the beam nodal displacements as defined in Fig. 3-45, a shape function is associated with each nodal displacement as shown in Fig. 3-46. These shape functions are the well-known Hermitian polynomials, which may be expressed mathematically as follows:

$$\psi_1 = +1-3\left(\frac{x}{L}\right)^2+2\left(\frac{x}{L}\right)^3$$

$$\psi_2 = +x\left(1-\frac{x}{L}\right)^2$$

$$\psi_3 = +3\left(\frac{x}{L}\right)^2-2\left(\frac{x}{L}\right)^3$$

$$\psi_4 = -\left(\frac{x^2}{L}\right)\left(1-\frac{x}{L}\right)$$

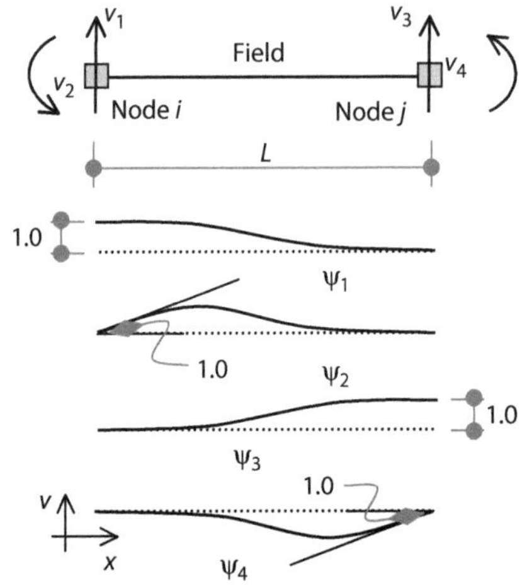

FIGURE 3-46 Shape functions of beam element.

In general, any stiffness element in the stiffness matrix can be found by giving the force under consideration a virtual displacement in the direction associated with that element.

As an example, the element k_{12} in the matrix above, which represents the nodal force, p_1, due to unit rotation, v_2, may be evaluated by giving the system of forces due to v_1 a virtual nodal displacement of v_{2v}.

Recall that the curvature, κ, is the negative of the second derivative of displacement, v'', and the moment is proportional to curvature through its flexure stiffness. Recall also that the rotation, α, is the integration of curvature over length

$$\kappa = -v''$$

Hence,

$$\kappa_v = -v_v''$$

and

$$M = EI\kappa = -EIv''$$

and

$$d\alpha = \kappa dx$$

Consequently, if the forces p_{11} and p_{21} due to displacement v_1, given in Fig. 3-47(a), are given a virtual displacement v_{2v}, as shown in Fig. 3-47(b), the principle of virtual work requires that the virtual work done by external loads must equal the virtual internal work (strain energy) stored by the system. Thus

$$W_{\text{ext}} = W_{\text{int}}$$

$$p_{21}v_{2v} = \int M_{\text{int}} d\alpha_v = \int (EI\kappa)(\kappa_v\, dx)$$

$$= \int (-EIv'')(-v_v'')dx$$

$$= \int (EI\psi_1'' v_1)(\psi_2'' v_{2v})dx$$

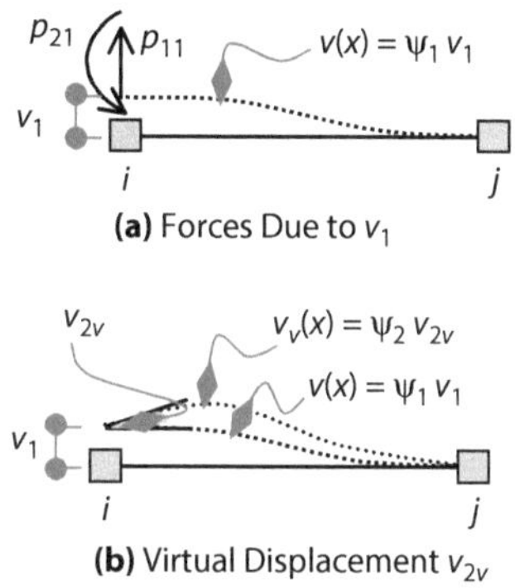

FIGURE 3-47 Virtual displacement.

Canceling v_{2v} on both sides yields

$$p_{21} = \int \{EI\, \psi_2'' \psi_1'' dx\} v_1$$

since $p_{21} = k_{21} v_1$, the stiffness term is given as

$$k_{21} = \int EI\, \psi_2'' \psi_1'' dx$$

Note that the subscripts of the force, p, match the subscripts of the shape functions in the integrand. Therefore, using the same procedures, it can be shown that the general element, k_{ij}, of the stiffness matrix of the beam is given by the following general expression:

$$k_{ij} = \int EI\, \psi_i'' \psi_j'' dx$$

As an example, the element k_{11} may be evaluated as follows:

$$k_{11} = \int EI\, \psi_1'' \psi_1'' dx$$

where

$$y_1 = 1 - 3\left(\frac{x}{L}\right)^2 + 2\left(\frac{x}{L}\right)^3$$

$$\psi_1' = -6\left(\frac{x}{L}\right)\left(\frac{1}{L}\right) + 6\left(\frac{x}{L}\right)^2\left(\frac{1}{L}\right)$$

$$\psi_1'' = -\frac{6}{L^2} + 12\frac{x}{L^3}$$

hence,

$$k_{11} = EI\int(\psi_1'')^2 dx$$

$$= EI\int\left(-\frac{6}{L^2}+12\frac{x}{L^3}\right)^2 dx$$

$$k_{11} = \frac{12EI}{L^3}$$

The answer ($12EI/L^3$) is an exact answer of the beam stiffness because the assumed shape function is the same exact deflected shape of the beam under these deformations. This is not usually the case in general finite element formulation.

Another example of interest in this context is the truss element. The truss shape functions are given as straight lines, as shown in Fig. 3-48, and may be mathematically expressed as

$$\psi_1 = 1 - \frac{x}{L}$$

$$\psi_2 = \frac{x}{L}$$

Recall that the strain, ε, is the first derivative of displacement, u', and the axial force is proportional to axial strain through its axial stiffness. In addition, the displacement, u, is the integration of strain over length

$$\varepsilon = u'$$

Hence,

$$\varepsilon_v = u_v'$$

and,

$$P = EA\varepsilon = EAu'$$

and,

$$du = \varepsilon dx$$

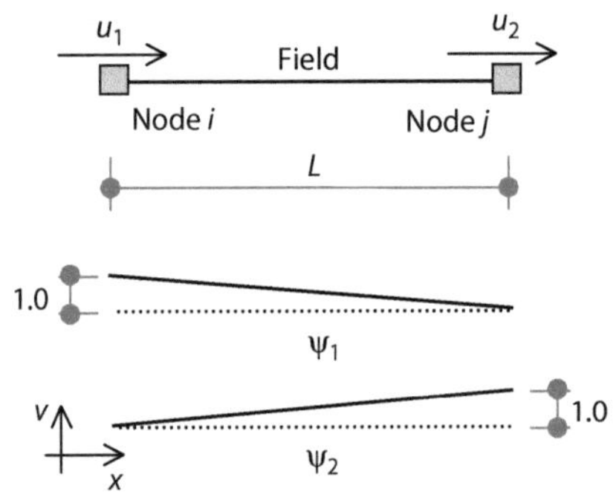

FIGURE 3-48 Shape functions of truss element.

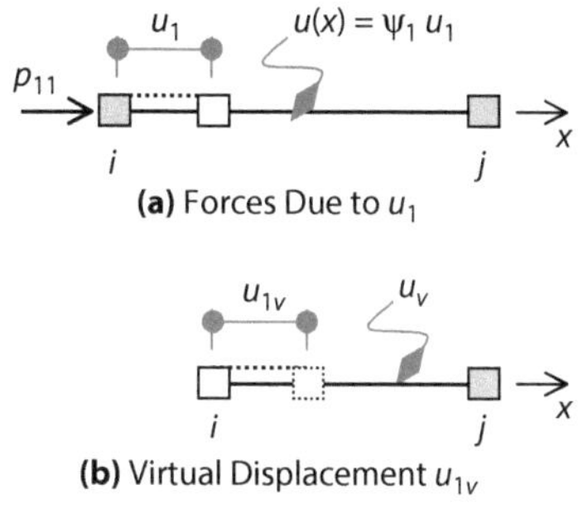

FIGURE 3-49 Virtual displacement.

Consequently, if the force p_{11} due to the displacement u_1, given in Fig. 3-49(a), is given a virtual displacement v_{1v}, as shown in Fig. 3-49(b). The principle of virtual work requires that the virtual work done by external loads must equal the virtual internal work (strain energy) stored by the system

$$W_{\text{ext}} = W_{\text{int}}$$

$$\begin{aligned} p_{11}u_{1v} &= \int P_{\text{int}}\, du_v \\ &= \int (EA\varepsilon)(\varepsilon_v dx) \\ &= \int (EAu')(u_v')dx \\ &= \int (EA\psi_1'u_1)(\psi_1'u_{1v})dx \end{aligned}$$

Canceling u_{1v} on both sides yields

$$p_{11} = \int \{EA\psi_1'\psi_1'\,dx\}u_1$$

Because $(p_{11} = k_{11}u_1)$, the stiffness term is given as

$$k_{11} = \int EA\psi_1'\psi_1'dx$$

Similar to the beam element, the general element, k_{ij}, of the stiffness matrix of the truss is given in the following expression:

$$k_{ij} = \int EA\,\psi_i'\,\psi_j'dx$$

The same procedure used for the beam element can be used to easily show that the element (k_{11}) in the stiffness matrix of the truss is given as

$$k_{11} = \int EA\psi_1'\psi_1'dx = \frac{EA}{L}$$

It should be noted again that the answer (*EI/A*) is an exact answer because the assumed shape function is the exact deflected shape of the truss. For the nodal numbering given by Fig. 3-48, the truss element stiffness matrix is given as

$$[k_m]=\frac{EA}{L}\begin{bmatrix} 1 & -1 \\ -1 & 1 \end{bmatrix}$$

3.9.3 Mass Matrix (Virtual Work Approach)

The mass matrix can be developed by similar procedures given in the previous section for stiffness matrix. Using virtual work principles, the inertial nodal forces can be evaluated by adding the inertial nodal forces to the stiffness nodal forces given in Fig. 3-47. If the inertial nodal forces are designated, p_i, with the field inertial differential force designated, f_i, as shown in Fig. 3-50(a); and if the beam is given a virtual nodal displacement, v_{2v}, as shown in Fig. 3-50(b), principle of virtual work requires

$$W_{\text{ext}} = W_{\text{int}}$$

$$p_{21}v_{2v} + p_{i,21}v_{2v} - f_{i,}v_v = \int M_{\text{int}}d\alpha_v$$

Since, $p_{21}v_{2v} = \int M_{\text{int}}d\alpha$, from the previous section, they both cancel out on both sides of the equation, hence,

$$p_{i,21}v_{2v} - f_{i,}v_v = 0$$

$$p_{i,21}v_{2v} - \int (m\,dx)\ddot{v}v_v = 0$$

or,

$$\begin{aligned} p_{i,21}v_{2v} &= \int (m\,dx)\ddot{v}v_v \\ &= \int (m\,dx)\psi_1\ddot{v}_1\psi_2 v_{2v} \end{aligned}$$

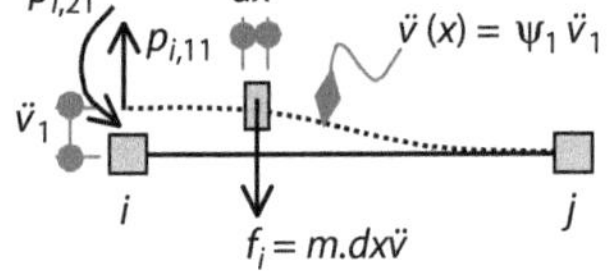

(a) Forces Due to $\ddot{v}_1$

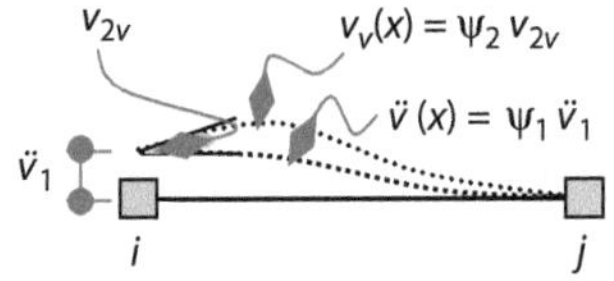

(b) Virtual Displacement v_{2v}

FIGURE 3-50 Virtual displacement.

Canceling v_{2v} on both sides yields

$$p_{i,21} = \int \{m\psi_2 \psi_1 dx\} \ddot{v}_1$$

since $p_{i,21} = m_{21}\, \ddot{v}_1$, the mass term is given as

$$m_{21} = \int m\psi_2 \psi_1 dx$$

Note that the subscripts of the force, p, match the subscripts of the shape functions in the integrand. Therefore, using the same procedures, it can be shown that the general element, m_{ij}, of the mass matrix of the beam is given by the following general expression:

$$m_{ij} = \int m\psi_i \psi_j dx$$

This integral is known as the consistent mass matrix because it is based on shape functions that are consistent with the stiffness shape function. If this integration is carried out for the four nodal quantities of the beam, it yields the known consistent mass matrix of the beam element which is given as

$$[m_m] = \frac{\bar{m}L}{420} \begin{bmatrix} 156 & 22L & 54 & -13L \\ 22L & 4L^2 & 13L & -3L^2 \\ 54 & 13L & 156 & -22L \\ -13L & -3L^2 & -22L & 4L^2 \end{bmatrix}$$

where $\bar{m}$ is the mass density per unit length of the member.

The same procedures may be applied to the truss element to yield the 2×2 mass consistent matrix of the truss as

$$m_{ij} = \int m\psi_i \psi_j dx$$

In matrix form,

$$[m_m] = \frac{\bar{m}L}{420} \begin{bmatrix} 140 & 70 \\ 70 & 140 \end{bmatrix}$$

3.9.4 Stiffness and Mass Matrices (Galerkin Approach)

The Galerkin approach is another useful method to formulate the finite element problems if the differential equation of the system is known. The Galerkin method falls under the general concept of error minimization, especially the concept of weighted residuals. Because the shape functions are an approximate representation of the actual displacement fields, the Galerkin method uses the concept of weighted residuals to minimize the error by assuming the weights of residuals as the shape functions themselves. The Galerkin method is considered the most accurate one among all methods in this category.

Application of the method will be demonstrated by using a simple case of a truss element. Recall that the differential equation of truss element under dynamic loading as developed under the axial beam section is given as

$$m\ddot{u} - EAu'' = -mr\ddot{u}_g$$

Assume an approximate displacement field, $\bar{u}(x)$, such that

$$\bar{u}(x,t) = \sum_j \psi_j(x) u_j(t)$$

where $\psi(x)$ is a shape function and $u_i(t)$ is the nodal displacement. The Galerkin method minimizes the error over the entire domain so that the integration of the error over the entire domain is zero. This may be expressed as follows.

If the function $u(x,t)$ is defined as $u(x,t) = m\ddot{u} - EAu''$, the differential equation given above can be written as

$$u(x,t) = -mr\ddot{u}_g$$

The approximate function is given as

$$\bar{u}(x,t) = m\ddot{\bar{u}} - EA\bar{u}''$$

The error between the two functions is given as the difference between the original and the approximate function such that

$$\text{error} = \bar{u}(x,t) - u(x,t) = \left(m\ddot{\bar{u}} - EA\bar{u}''\right) - (-mr\ddot{u}_g)$$

The Galerkin method states that the summation of the error components multiplied by their weights is set to zero. The weights are considered to be the shape functions themselves

$$\int \text{error}\cdot\text{weight}\, dx = 0$$

$$\int (m\ddot{\bar{u}} - EA\bar{u}'' + mr\ddot{u}_g)\psi_i dx = 0, i = 1,2$$

$$\int (m\ddot{\bar{u}} - EA\bar{u}'')\psi_i dx = -\int (mr\ddot{u}_g)\psi_i\, dx, i = 1,2$$

substituting $\bar{u}(x,t) = \Sigma_j \psi_j(x) u_j(t)$,

$$\sum_j \int (m\psi_j \ddot{u}_j - EA\psi_j'' u_j)\psi_i dx = -\int (mr\ddot{u}_g)\psi_i dx$$

$$\sum_j \int (m\psi_i \psi_j \ddot{u}_j - EA\psi_i \psi_j'' u_j) dx = -\int (mr\ddot{u}_g)\psi_i dx$$

Using integration by parts, the second integration can be converted to symmetrical integration with reduction of the second derivative to the first derivative

$$\int -EA\psi_i\psi_j''u_j dx = \int -EA\psi_i u_j d\psi_j' = 0 + \int EA\psi_i'\psi_j' u_j dx$$

Therefore, the above integration becomes

$$\sum_j \int (m\psi_i\psi_j\ddot{u}_j + EA\psi_i'\psi_j' u_j)dx = -\int (mr\ddot{u}_g)\psi_i dx$$

The integration above yields three quantities in terms of mass, m, axial stiffness, EA, and ground acceleration, $\ddot{u}_g$. These quantities are defined as the consistent mass matrix $[m_m]$, the axial stiffness matrix, $[k_m]$, and the excitation matrix [£], such that

$$m_{\text{ij}} = \int m\psi_i\psi_j dx$$

$$k_{\text{ij}} = \int EA\psi_i'\psi_j' dx$$

$$£_{\text{ij}} = \int mr\psi_j dx$$

Note that the first matrix and second matrix are the mass matrix and the stiffness matrix, which are the same matrices obtained by the virtual work approach given in the previous section.

The same procedures may be applied to the beam element to yield the 4 × 4 mass consistent matrix of the beam as

$$m_{\text{ij}} = \int m\psi_i\psi_j dx$$

3.9.5 Other Matrices

Other important element matrices may be developed using the same concept and procedures of the Galerkin approach. Examples of such matrices include the damping matrix, $[C]$, and the well-known geometric stiffness matrix, $[K_G]$, that is utilized for stability analysis of frames. These matrices are given as follows:

Damping matrix: $$c_{\text{ij}} = \int c\psi_i\psi_j dx$$

Geometric stiffness matrix: $$k_{\text{Gij}} = \int P\psi_i'\psi_j' dx$$

In matrix form

$$[k_{\text{Gm}}] = \frac{P}{30L}\begin{bmatrix} 36 & 3L & -36 & 3L \\ 3L & 4L^2 & -3L & -L^2 \\ -36 & -3L & 36 & -3L \\ 3L & -L^2 & -3L & 4L^2 \end{bmatrix}$$

The geometric stiffness matrix also has special importance in second-order analysis. Since the beam nodal forces that are due to any change in geometry are related to the nodal displacement through the geometric stiffness matrix, $[k_{\mathrm{Gm}}]$, and because the beam nodal forces that are due to transverse loads are related to the nodal displacement through the ordinary stiffness matrix, $[k_m]$, the total nodal forces of the beam may be expressed as

$$\{Q\} = [[k_m] - [k_{\mathrm{Gm}}]]\{U\}$$

The effective stiffness matrix of the beam element, including the effect of axial load, may be expressed as follows:

$$[k_{\mathrm{eff}}] = [k_m] - [k_{\mathrm{Gm}}]$$

where $[k_{\mathrm{eff}}]$ = element effective stiffness matrix including the effect of axial load
$[k_m]$ = beam stiffness matrix excluding the effect of axial load
$[k_{\mathrm{Gm}}]$ = element geometric stiffness matrix as defined above

The presence of the axial load in the beam reduces its lateral stiffness. Knowing the effective global stiffness matrix of the structure, the second-order effect can be included in dynamic analysis by adding the geometric stiffness matrix to the system. This can be expressed in the following form:

$$[M]\{\ddot{U}\}+[C]\{\dot{U}\}+[K_{\mathrm{eff}}]\{U\}=-[M]\{R\}\ddot{u}_g$$

or

$$[M]\{\ddot{U}\}+[C]\{\dot{U}\}+[[K]-[K_G]]\{U\}=-[M]\{R\}\ddot{u}_g$$

The equation above is the general dynamic equation that includes material and geometric nonlinearity of the structure. The solution of this equation must be solved incrementally using numerical methods as shown in the next chapter.

3.9.6 Mass Matrix in 2D

The development of the mass matrix in 1D, as given in previous sections, may be extended to 2D problems. The consistent mass matrix in 2D may be developed with the same procedures implemented in 1D analysis. As an example, the consistent mass matrix of a 2D triangle plane stress element shown in Fig. 3-51 may be developed using the three linear shape functions of N_1, N_2, and N_3 for the three nodes 1, 2, and 3. The shape function N_1 is used if either $\delta_1 = 1$ or $\delta_2 = 1$. An example of shape function, N_1 when $\delta_1 = 1$, is also shown in Fig. 3-51.

In 2D analysis, the approximate field displacement, $\bar{u}$, consists of two components in two dimensions, x and y, as $\bar{u}_1$ and $\bar{u}_2$.

$$\bar{U} = \bar{u}_1 + \bar{u}_2$$

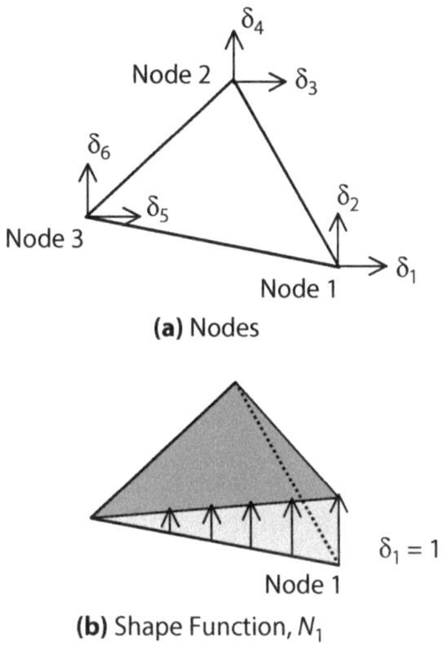

FIGURE 3-51 Linear shape functions of triangle element.

where

$$\bar{u}_1 = N_1\delta_1 + N_2\delta_3 + N_3\delta_5$$
$$\bar{u}_2 = N_1\delta_2 + N_2\delta_4 + N_3\delta_6$$

This may be expressed in matrix form as

$$\{\bar{U}\} = [N]\{\delta\}$$

where

$$[N]_{2\times 6} = \begin{bmatrix} N_1 & 0 & N_2 & 0 & N_3 & 0 \\ 0 & N_1 & 0 & N_2 & 0 & N_3 \end{bmatrix}$$

The elements of the mass matrix are evaluated by the following expression:

$$m_{ij} = \int m\{[N]^T[N]\}dx$$

3.9.7 Application of Consistent Mass Matrix

Application of the consistent mass matrix arises when a more accurate dynamic solution is needed, or when the use of the lumped mass approximation becomes infeasible. For example, if the frame shown in Fig. 3-52(a) must be analyzed taking into account the effect of mass distribution over its members, the use of the consistent mass matrix becomes the logical choice for analysis.

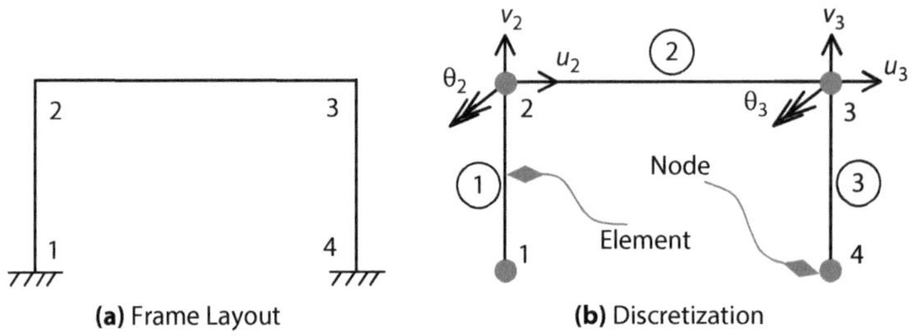

FIGURE 3-52 Discretization of frame structure.

The frame shown in Fig. 3-52(a) consists of three beam elements connected by four joints and has fixed supports at joints 1 and 4. The frame may be modeled with three beam elements as marked in circles, and with four nodes as shown in Fig. 3-52(b). Since the frame has fixed supports at joints 1 and 4, it would only have degrees of freedom at joints 2 and 3. The degrees of freedom are given as two displacements and one rotation at each of nodes 2 and 3, as shown in Fig. 3-52(b). Accordingly, the displacement vector will have six components, u_2, v_2, θ_2, u_3, v_3, and θ_3. In matrix notations, the displacement vector may be expressed as follows:

$$\{U\} = \{u_2\ v_2\ \theta_2\ u_3\ v_3\ \theta_3\}^T$$

Consequently, the corresponding dynamic equation takes the following form:

$$[M]\{\ddot{U}\} + [C]\{\dot{U}\} + [K]\{U\} = -[M]\{R\}\ddot{u}_g$$

where $[M] = 6 \times 6$ consistent mass matrix (nondiagonal)
$[C] = 6 \times 6$ damping matrix (nondiagonal)
$[K] = 6 \times 6$ stiffness matrix (nondiagonal)

Unlike the lumped mass formulation, the consistent mass matrix formulation includes the rotational dynamic degrees of freedom of the system. The rotational structural degrees of freedom exist in the stiffness matrix in both cases. They explicitly exist in the case of the consistent mass matrix solution in terms of θ_2 and θ_3. However, they implicitly exist in the case of the lumped mass solution as they are kinematically condensed and incorporated in the displacement degrees of freedom as shown in previous sections.

The solution of the equation of motion above depends on the orthogonality of the damping matrix $[C]$. The modal analysis procedures outlined in previous sections require orthogonality to be valid. It was proven that orthogonality exists for both mass and stiffness matrices $[M]$ and $[K]$. In order to apply modal analysis to the equation above, the matrix, $[C]$ also has to be orthogonal. Therefore, the damping matrix can be considered orthogonal if it can be expressed as a linear combination of the two matrices, $[M]$ and $[K]$. In mathematical form, this may be expressed as follows:

$$[C] = \alpha[M] + \beta[K]$$

where α and β are constants.

If the damping matrix $[C]$ is orthogonal, the solution of the equation of motion follows exactly the same procedures given in previous sections. Otherwise, the matrices will be mathematically coupled, and recurrence to numerical methods becomes necessary as will be shown in the next chapter.

3.10 Incoherence

If all the supports of the structure are excited by the same acceleration signal that reaches all supports at the same instance, the excitation is said to be coherent. Therefore, incoherence is a term used to describe the excitation of the structure by the application of different excitation signals at different support locations. In building structures, incoherence may not be of concern insofar as the foundations are usually tied together and located in compact region. Thus, it is usually assumed that the same earthquake signal will reach all supports at the same time with the same acceleration profile.

However, incoherence may need to be considered in bridges, especially in long bridges where different earthquake signals may reach their abutments and piers.

Incoherence may be included in the analysis using the earthquake directional vector $\{R\}$, as defined in the MDOF presentation. The earthquake directional vector is defined and specified for building structures as the resulting displacement of all DOFs due to a unit rigid body motion in the direction of the earthquake. This definition is valid for buildings given before, since coherence was implied in the analysis. The more general definition of the earthquake directional vector $\{R\}$, in case of incoherence, will be the resulting displacement of all DOFs due to a unit displacement of the excited support. This definition implies that the deflected shape of the structure due to unit displacement of each support is known in order to formulate such problems.

The next section will illustrate this by examining for incoherent one span beam with different cases of supporting conditions.

Case (1): Simple support with discrete mass

The simply supported beam with two lumped masses, m_1 and m_2, as shown in Fig. 3-53 is considered a 2DOF system with DOFs at the mass locations. If this beam is subjected to coherent excitation, that is, both supports A and B are excited by the same earthquake at the same instant, the earthquake directional vector, $\{R\}$, will be the

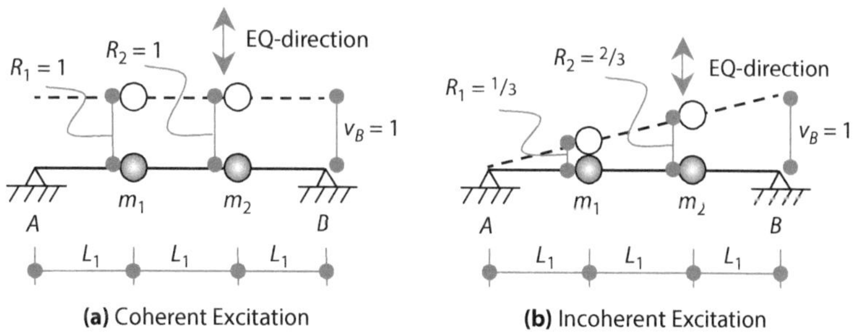

FIGURE 3-53 Simple support with discrete mass.

displacement of masses m_1 and m_2 due to rigid body motion in the direction of the earthquake as shown in Fig. 3-53(a). Accordingly, the resulting $\{R\}$ will be

$$\{R\}_{\text{coh}} = \begin{Bmatrix} 1 \\ 1 \end{Bmatrix}$$

On the other hand, if the beam is subjected to incoherent excitation, for example, if only support B is excited by the earthquake as shown in Fig. 3-53(b), the resulting $\{R\}$ will be the displacement of the two masses due to unit movement of support B without any movement of support A. Because the beam is simply supported, the deflected shape due to this motion is a straight line, as shown in Fig. 3-53(b). Because the masses are equally spaced across the beam, $\{R\}$ will be

$$\{R\}_{\text{inc}} = \begin{Bmatrix} 1/3 \\ 2/3 \end{Bmatrix}$$

The general equation of motion derived earlier is still applicable with vector $\{R\}$ as derived above. Therefore, the equation of motion for a discrete system with this incoherent excitation is still given as

$$[M]\{\ddot{U}\}+[C]\{\dot{U}\}+[K]\{U\}=-[M]\{R\}\ddot{u}_g$$

where $\{R\} = \{1/3\ 2/3\}^T$ and all other components are the same as defined in the MDOF section.

Case (2): Simple support with uniform mass

The simply supported beam with uniformly distributed mass, m, as shown in Fig. 3-54 is considered a continuous system with infinite DOFs. As a result, it must be treated at the differential level. If this beam is subjected to coherent excitation, that is, both supports A and B are excited by the same earthquake at the same instant, the earthquake directional vector $r(x)$ will be a continuous function and is given as defined before. $r(x)$ will be the displacement of the mass density, m, due to rigid body motion in the direction of the earthquake as shown in Fig. 3-54(a). Accordingly, the resulting $r(x)$ will be

$$r_{\text{coh}}(x) = 1$$

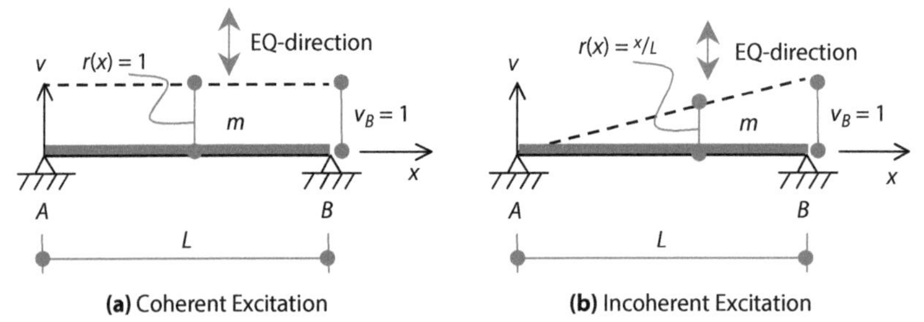

FIGURE 3-54 Simple support with distributed mass.

On the other hand, if the beam is subjected to incoherent excitation, for example, if only support *B* is excited by the earthquake as shown in Fig. 3-54(b), the resulting $r(x)$ will be the displacement of the mass density due to unit movement of support *B* without any movement of support *A*. Because the beam is simply supported, the deflected shape due to this motion is a straight line as shown in Fig. 3-54(b). Thus,

$$r_{\text{inc}}(x) = \frac{x}{L}$$

The general equation of motion derived earlier still applies with vector $r(x)$ as derived above. Therefore, the equation of motion for a continuous system with this incoherent excitation is still given as

$$m\ddot{v} + EI\,v^{\text{iv}} = -mr(x)\ddot{v}_g$$

where $r(x) = x/L$, and all other components are the same as defined in the MDOF section.

Case (3): Fixed support with discrete mass

The fixed-end beam with the lumped masses, m_1 and m_2, shown in Fig. 3-55 is considered a 2DOF system with DOFs at the mass locations. If this beam is subjected to coherent excitation, that is, both supports *A* and *B* are excited by the same earthquake at the same instant, the earthquake directional vector, $\{R\}$, will be the displacement of masses, m_1 and m_2, due to rigid body motion in the direction of the earthquake as shown in Fig. 3-55(a). Accordingly, the resulting $\{R\}$ will be

$$\{R\}_{\text{coh}} = \begin{Bmatrix} 1 \\ 1 \end{Bmatrix}$$

On the other hand, if the beam is subjected to incoherent excitation, for example, if only support *B* is excited by the earthquake as shown in Fig. 3-55(b), the resulting $\{R\}$ will be the displacement of the two masses due to unit movement of support *B* without any movement of support *A*. Because the beam is fixed at both supports, the deflected shape due to this motion is a not a straight line as in case (1). Instead, it takes the shape

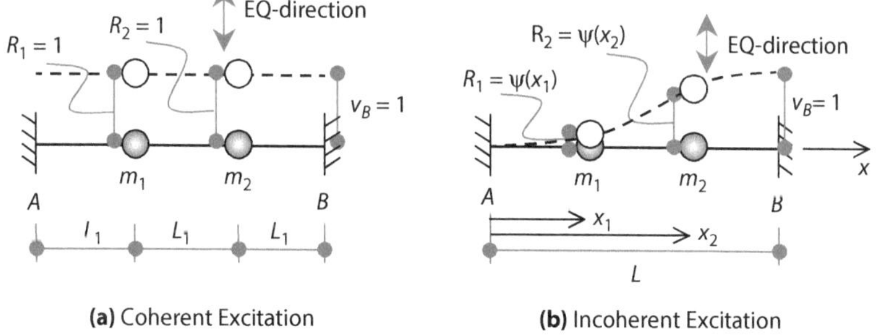

FIGURE 3-55 Fixed beam with discrete mass.

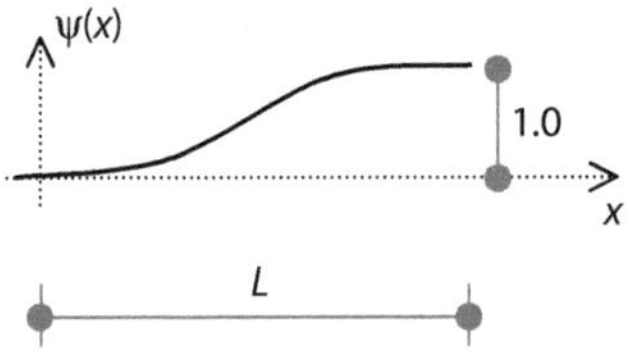

FIGURE 3-56 Shape function.

of the shape function of this displacement. The cubic deflected shape developed for this displacement as derived in the finite element section may be used. This shape, $\psi(x)$, is shown in Fig. 3-56 and given as

$$\psi(x) = +3\left(\frac{x}{L}\right)^2 - 2\left(\frac{x}{L}\right)^3$$

Therefore, the $\{R\}$ values are given as

$$R_1 = \psi\left(\frac{L}{3}\right) = 3\left(\frac{1}{3}\right)^2 - 2\left(\frac{1}{3}\right)^3 = 0.259$$

$$R_2 = \psi\left(\frac{2L}{3}\right) = 3\left(\frac{2}{3}\right)^2 - 2\left(\frac{2}{3}\right)^3 = 0.741$$

Accordingly, the $\{R\}$ vector will be

$$\{R\}_{\text{inc}} = \begin{Bmatrix} 0.259 \\ 0.741 \end{Bmatrix}$$

As stated before, the equation of motion for a discrete system with this incoherent excitation is still given as

$$[M]\{\ddot{U}\} + [C]\{\dot{U}\} + [K]\{U\} = -[M]\{R\}\ddot{u}_g$$

where $\{R\} = \{0.259\ 0.741\}^T$ and all other components are the same as defined in the MDOF section.

Case (4): Fixed support with uniform mass

The fixed-end beam with uniformly distributed mass, m, as shown in Fig. 3-57 is considered a continuous system with infinite DOFs. Thus, it must be treated at the differential level. If this beam is subjected to coherent excitation, that is, both supports A and B are excited by the same earthquake at the same instant, the earthquake directional vector, $r(x)$ will be a continuous function and is given as defined before. $r(x)$, will be the displacement of the mass density, m, due to rigid body motion in the direction of the earthquake as shown in Fig. 3-57(a). Accordingly, the resulting $r(x)$ will be

$$r_{\text{coh}}(x) = 1$$

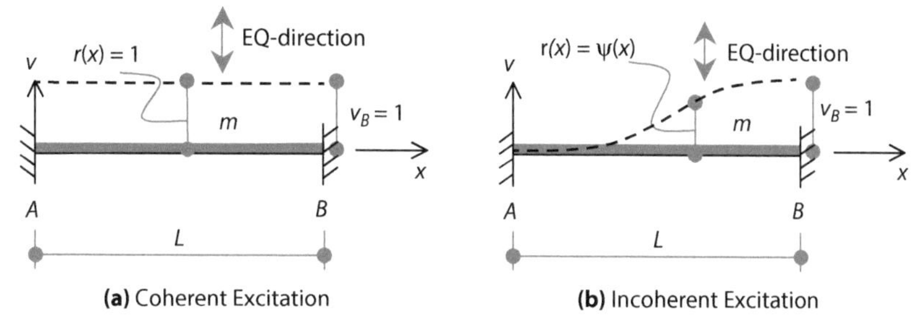

FIGURE 3-57 Fixed beam with distributed mass.

On the other hand, if the beam is subjected to incoherent excitation, for example, if only support B is excited by the earthquake as shown in Fig. 3-57(b), the resulting $r(x)$ will be the displacement of the mass density due to unit movement of support B without any movement of support A. Because the beam is fixed at both supports, the deflected shape due to this motion is not a straight line as in case (1). Instead, it takes the shape of the shape function of this displacement. The same cubic deflected shape used in case (3) may be used as

$$r_{\text{inc}}(x) = \psi(x)$$

where

$$\psi(x) = +3\left(\frac{x}{L}\right)^2 - 2\left(\frac{x}{L}\right)^3$$

The general equation of motion derived earlier still applies with vector $r(x)$ as derived above. Therefore, the equation of motion for a continuous system with this incoherent excitation is still given as

$$m\ddot{v} + EIv^{\text{iv}} = -mr(x)\ddot{v}_g$$

where $r(x) = \psi(x)$ and all other components are the same as defined in the MDOF section.

Example 3-11 A simply supported bridge with span, L, and uniform mass, m, is excited by an earthquake in the direction shown in Example 3-11, Fig. 1. Evaluate the participation factor of the first mode in the following two cases:

1. Both supports are excited by the earthquake at the same time with the same intensity (coherent excitation).
2. Only support B is excited by the earthquake with the same intensity given in (1) (incoherent excitation):

Solution

The participation factor of the continuous system is given for the first mode by the following expression:

$$\Gamma = \frac{£}{m^*} = \frac{\int \phi m r dx}{\int \phi m \phi dx}$$

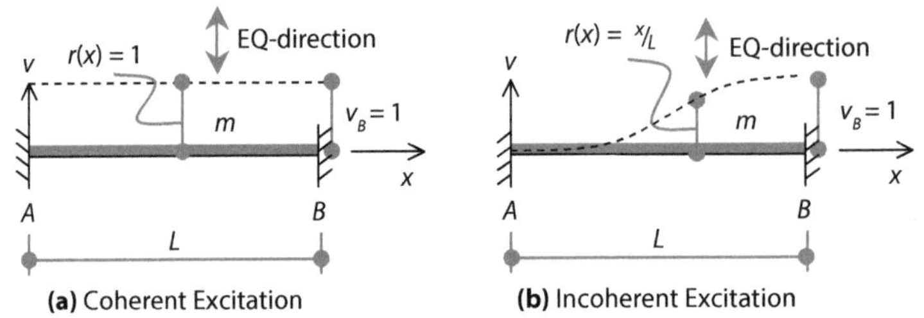

EXAMPLE 3-11, FIGURE 1 Plan of bridge.

where ϕ is the first mode shape of a simple beam, which is derived in Section 3.7, simple flexure beam. ϕ is given as

$$\phi = \sin\left(\frac{\pi x}{L}\right)$$

The denominator m^* is common to both cases and will be evaluated at this stage as follows:

$$m^* = \int \phi m \phi dx = m \int \phi^2 dx = m \int \sin^2\left(\frac{\pi x}{L}\right) dx = \frac{mL}{2}$$

Part (1)
Coherent excitation, Example 3-11, Fig. 1(a)
Since $r(x) = 1$, the integration of (£) is evaluated as follows:

$$£ = \int \phi m r(x) dx = m \int \phi dx = m \int \sin\left(\frac{\pi x}{L}\right) dx = \frac{2mL}{\pi}$$

Therefore,

$$\Gamma = \frac{£}{m^*} = \frac{\left(\frac{2mL}{\pi}\right)}{\left(\frac{mL}{2}\right)} = \frac{4}{\pi}$$

Part (2)
Incoherent excitation, Example 3-11, Fig. 1(b)
Since $r(x) = \frac{x}{L}$, the integration of £ is evaluated as follows:

$$£ = \int \phi m r(x) dx = m \int \phi \left(\frac{x}{L}\right) dx$$

$$= m \int \left(\frac{x}{L}\right) \sin\left(\frac{\pi x}{L}\right) dx = \frac{mL}{\pi}$$

Therefore,

$$\Gamma = \frac{£}{m^*} = \frac{\left(\frac{mL}{\pi}\right)}{\left(\frac{mL}{2}\right)} = \frac{2}{\pi}$$

The participation factor that results from the excitation of one support is only one-half its value when both supports are excited. This means that, for this example the displacements and the earthquake-induced forces in the incoherence case will be half the values of the coherence case. Remember that the displacement and induced forces are directly proportional to the modal participation factor as follows:

$$v_{\max} = \phi \Gamma S_d$$
$$f_{s,\max} = m\phi \Gamma S_a$$

3.11 Problems

3-1 A stiff beam is supported by two columns and a viscous damper as shown in Prob. 3-1, Fig. 1. The frame is fixed at A and hinged at B. The total mass is concentrated in the beam. The mass, the flexure stiffness of each column, and the coefficient of damping are given as follows:

$$m = 500 \text{ kN}\cdot\text{s}^2/\text{m}$$
$$EI = 30 \times 10^3 \text{ kN}\cdot\text{m}^2$$
$$c = 200 \text{ kN}\cdot\text{s/m}$$

If the structure is excited by the El Centro earthquake:

1. Determine the maximum displacement and base shear.
2. Determine the distribution of the base shear to the columns.

3-2 A structure is idealized as a 3DOF system as shown in Prob. 3-2, Fig. 1. The structure is subjected to an earthquake that has the direction shown in the figure. The velocity response spectrum of the earthquake is given in Prob. 3-2, Fig. 2. Assuming that the damping ratio, $\zeta = 2\%$, and the cross section of the frame is constant with, $EI = 100 \times 10^3$ kN·m².

It is required to do the following:

1. The frequencies and their mode shapes
2. The modal displacements of each mass due to first mode only
3. The modal base shears due to first mode only
4. The modal base moments due to first mode only

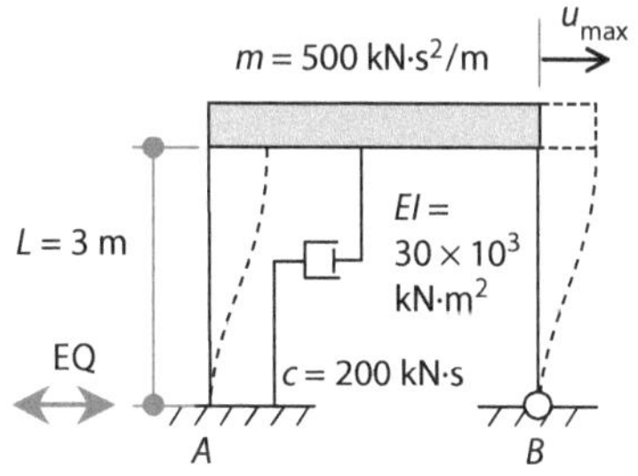

PROBLEM 3-1, FIGURE 1

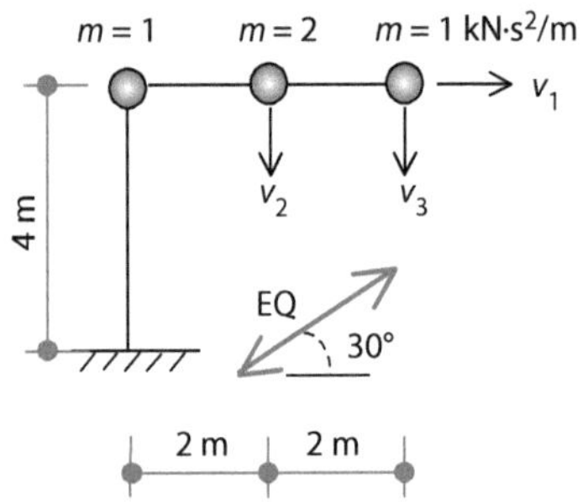

PROBLEM 3-2, FIGURE 1 Structure layout.

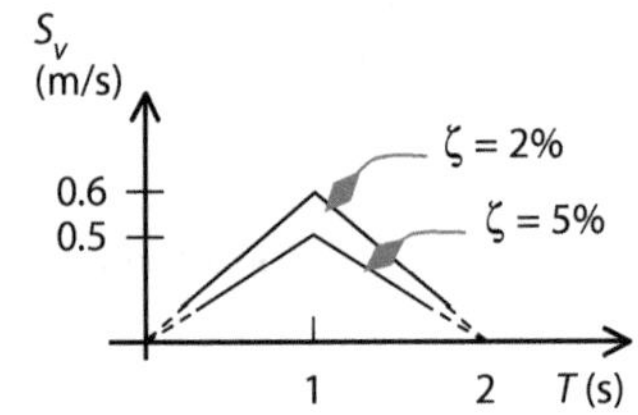

PROBLEM 3-2, FIGURE 2 EQ response spectrum.

3-3 The building shown in Prob. 3-3, Fig. 1, has the mass matrix and vibration properties as shown. The structure is excited by a horizontal earthquake with acceleration response spectrum values for the three modes as given in the S_a column shown.

$$m = 30\begin{bmatrix} 1 & 0 & 0 \\ 0 & 1 & 0 \\ 0 & 0 & 1 \end{bmatrix} \text{kN·S}^2/\text{m}, \qquad S_a = \begin{bmatrix} 2.94 \\ 1.57 \\ 3.93 \end{bmatrix} \text{m/s}^2$$

$$\omega = \begin{bmatrix} 3.88 \\ 9.15 \\ 15.31 \end{bmatrix} \text{rad/s} \qquad \phi = \begin{bmatrix} 1.000 & 1.000 & 1.000 \\ 0.548 & -1.522 & -6.260 \\ 0.198 & -0.872 & 12.10 \end{bmatrix}$$

1. For each mode of vibration, calculate the maximum displacement, overturning moment and shear force at each story level.
2. By the square root of sum of squares (SRSS) method, determine approximate total maximums for each of the response quantities of Item (1).

3-4 A structure is idealized as a 3DOF system as shown in Prob. 3-4, Fig. 1. The structure is subjected to an earthquake having the direction shown in the same figure with a velocity response spectrum as shown in the Prob. 3-4, Fig. 2. The damping ratio is given as $\zeta = 2\%$.

The mass of the structure is lumped into three equal masses, m = 1 kN·s²/m, located as shown in Prob. 3-4, Fig. 1.

The following are required:

1. Find the frequencies and their mode shapes.
2. Find the SRSS displacements of each mass.

$$m = 30\begin{bmatrix}1 & 0 & 0\\ 0 & 1 & 0\\ 0 & 0 & 1\end{bmatrix}\ \text{kN·s}^2/\text{m},\qquad S_a = \begin{bmatrix}2.94\\ 1.57\\ 3.93\end{bmatrix}\ \text{m/s}^2$$

$$\omega = \begin{bmatrix}3.88\\ 9.15\\ 15.31\end{bmatrix}\ \text{rad/s},\qquad \phi = \begin{bmatrix}1.000 & 1.000 & 1.000\\ 0.548 & -1.522 & -6.260\\ 0.198 & -0.872 & 12.10\end{bmatrix}$$

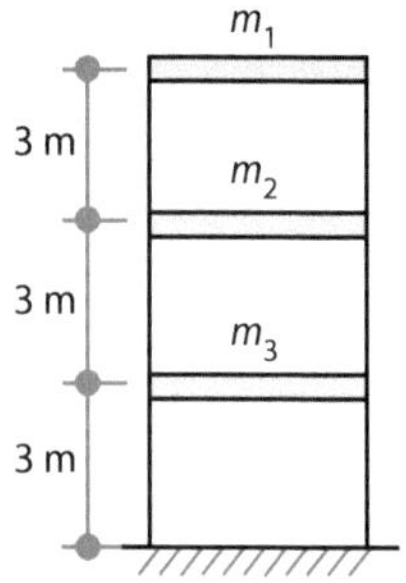

PROBLEM 3-3, FIGURE 1

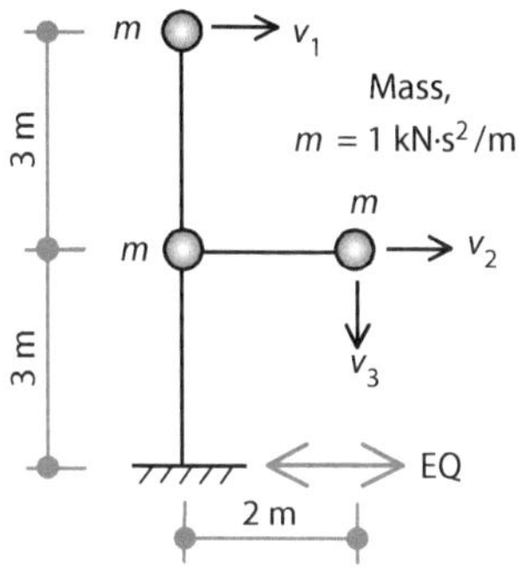

PROBLEM 3-4, FIGURE 1 Structure layout.

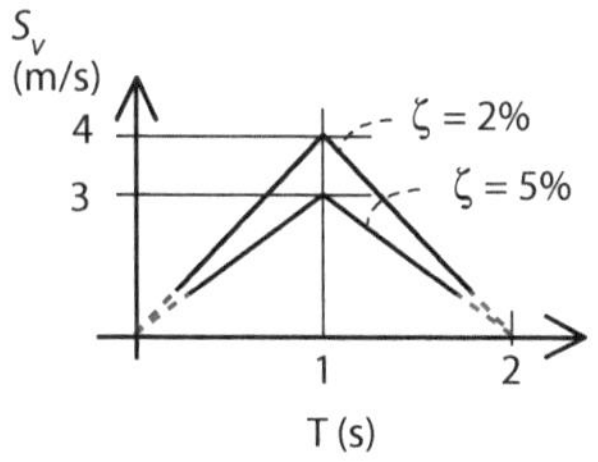

PROBLEM 3-4, FIGURE 2 EQ response spectrum.

3. Find the SRSS base shear.
4. Find the SRSS base moment.

3-5 A 150 mm concrete slab is supported by hollow circular shaft, which is located as shown in Prob. 3-5, Fig. 1. The slab is considered rigid, the shaft is weightless and totally fixed to the slab. The clear height of the shaft is 5 m. The concrete mass is given as $\rho_c = 2.5$ kN·s²/m/m³.

1. Express mass matrix and mode shapes in terms of the slab centroid coordinates u, v, θ_u, θ_v, and θ_z as shown in Prob. 3-5, Fig. 2.
2. If the El Centro earthquake acts in the direction of coordinate, v, determine the maximum dynamic displacement at the top of the shaft in the first mode of vibration. Damping ratio is taken as $\zeta = 2\%$.

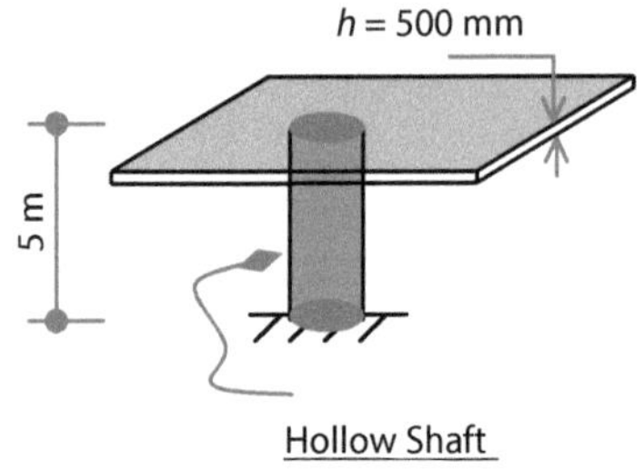

Problem 3-5, Figure 1 Isometric.

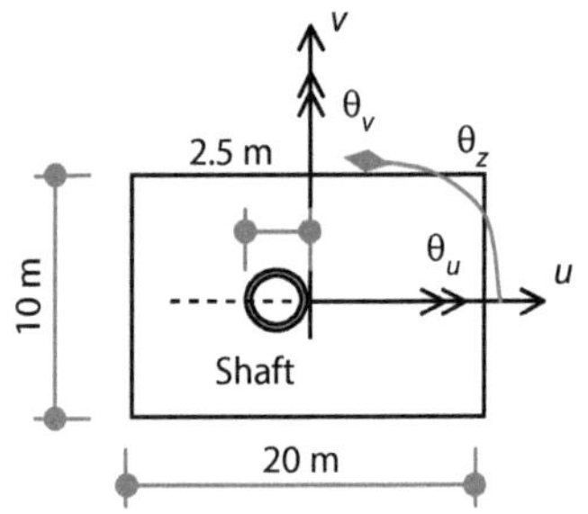

Problem 3-5, Figure 2 Plan.

3-6 Part (A): The frame shown in Prob. 3-6, Fig. 1, has the uniform mass and flexural stiffness shown in the figure. If the structure is excited by the El Centro earthquake in the direction, *Z-Z*, as given in the figure, use the consistent mass matrix to

1. Find the first two frequencies and mode shapes.
2. Find the SRSS displacement and rotation of the tip of cantilever, B, due to the first two modes.
3. Find the SRSS base shear and moment due to the first two modes.

Part (B): If the frame shown in Prob. 3-6, Fig. 1, is idealized as 2DOFs as shown in Prob. 3-6, Fig. 2, and is subjected to the same excitation, repeat items (1), (2), and (3) given in Part (A). Compare results from both parts.

3-7 Problem 3-7, Fig. 1(a) shows the building that is solved in the book (Example 3-9) using shear building concept.

It is required to use the concept of generalized SDOF system to find the fundamental frequency of the building. Consider the mass to be lumped at the floor level; and use linear mode shape, that is,

$$\psi(x) = \frac{x}{L}$$

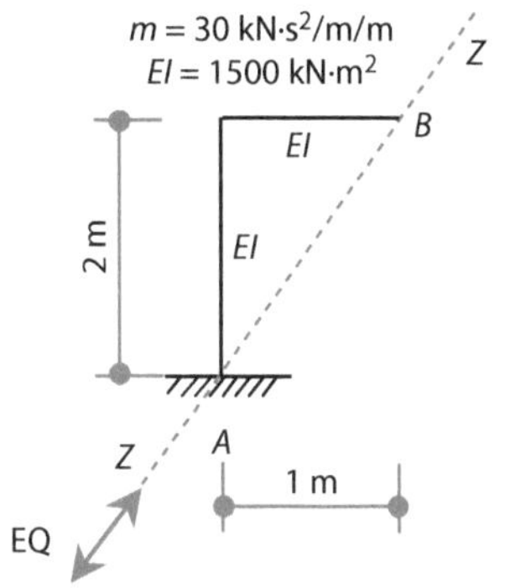

Problem 3-6, Figure 1

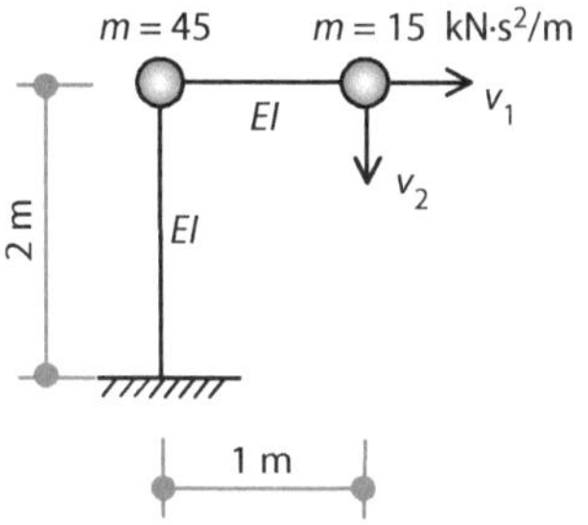

Problem 3-6, Figure 2

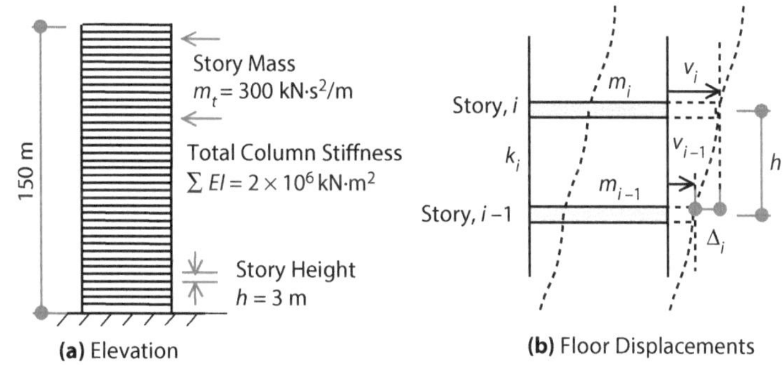

Problem 3-7, Figure 1 Building layout.

Note that the beams are assumed to be rigid, and hence, the stiffness of each floor k_i is taken as,

$$k_i = \frac{\Sigma 12EI}{h^3}$$

Compare your results with shear beam solution.

Hint: The internal energy stored in each floor may be evaluated as the story shear, V_i, going through story virtual relative displacement, Δ_i, as shown in Prob. 3-6, Fig. 1(b), that is,

$$W_{\text{int}} = \sum V_i \Delta_{\text{iv}} = \sum k\Delta_i \Delta_{\text{iv}}$$

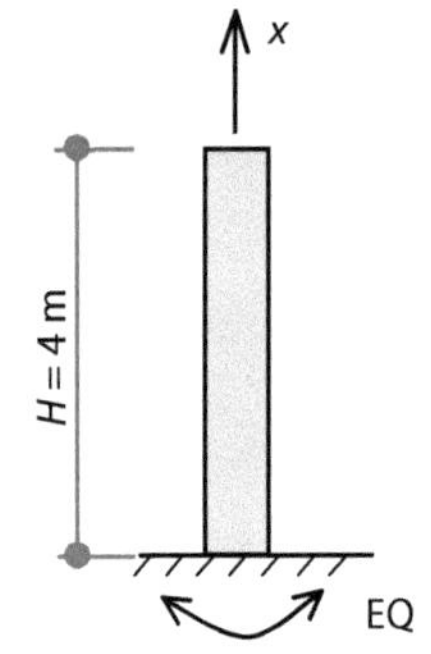

Problem 3-8, Figure 1

where v_i = total displacement of the story under consideration
v_{i-1} = total displacement of the story below

and

$\Delta_i = v_i - v_{i-1}$
Δ_{iv} = virtual relative displacement

3-8 The vertical cantilever wall shown in Prob. 3-8, Fig. 1, has the following properties

Uniform mass: $m = 1$ kN·s²/m²
Uniform inertia: $EI = 0.5 \times 10^6$ kN·m²

The deflected shape of this structure may be expressed as

$$\psi(x) = 1 - \cos\left(\frac{\pi x}{2H}\right)$$

The structure is excited by a rotational earthquake component having constant response spectral velocity,

$$S_v = 0.5 \text{ rad/s}$$

If the damping ratio, ζ, is 5 percent, find the maximum top displacement, base shear and base moment.

3-9 A simply supported bridge with span, L, and uniform mass, m, is excited by an earthquake in the direction shown in Prob. 3-9, Fig. 1. Evaluate the participation factor of the first mode in the following two cases:

1. Both supports are excited by the earthquake at the same time with the same intensity (coherent excitation).
2. Only support B is excited by the earthquake with the same intensity given in (1) (incoherent excitation).

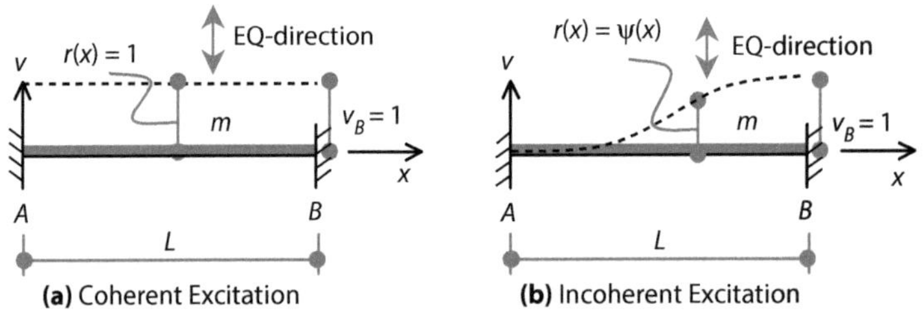

Problem 3-9, Figure 1 Fixed beam with distributed mass.

CHAPTER 4

Nonlinear and Inelastic Dynamic Analysis

4.1 Introduction

As noted in Chap. 1, it is important to differentiate between nonlinear elastic behavior and inelastic behavior. Nonlinear elastic behavior is characterized by a unique relationship between load and displacement, and between stress and strain. Such elastic behavior does not result in energy dissipation in the system as in the case of cable structures. However, inelastic behavior is not characterized by such uniqueness; instead, inelastic behavior depends mainly on the loading history. Such inelastic behavior results in hysteresis loops, which are the source of energy dissipation in the system, as in the case of material yielding. Because closed-form solutions of nonlinear and inelastic systems do not exist in general, reverting to incremental analysis is necessary.

Earthquake-resistant structures are typically designed with ultimate resistance from two to eight times less than their elastic strength demand (ESD). This reduction in ESD is possible owing to the inelastic response of structures as long as they also exhibit some desirable properties. Such properties include frequency shift, ductility, and energy dissipation capacity of the structure, provided that the structure also has self-centering capacity.

For this reason, understanding the inelastic behavior of structures is essential. Inelastic behavior is mainly a structural property that depends on the shape and size of the structure hysteresis model. The hysteresis model is defined as the relationship between the structure displacement, Δ, and its restoring force, F, under cyclic loading as shown in Fig. 4-1. The restoring force, F, is history dependent, that is, the force-displacement relationship is not unique and can only be evaluated according to loading history of the system.

Figure 4-2 shows several examples of hysteresis models used in practice. A series of straight lines define the first three types, namely, the elastoplastic, the bilinear, and the stiffness degrading models; the fourth type, the smooth stiffness degrading model, is defined with curved lines. Most of these models exhibit kinematic hardening,

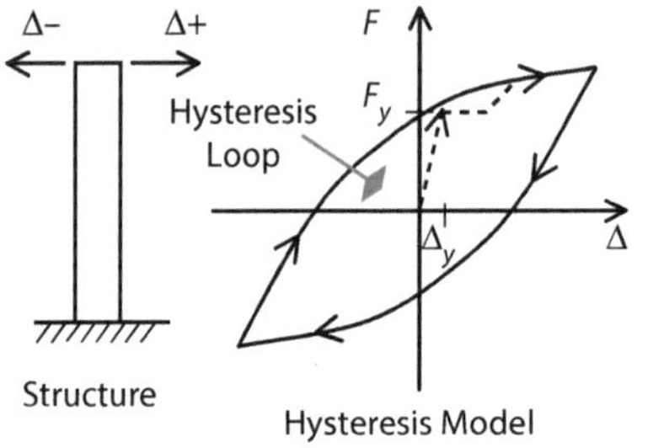

FIGURE 4-1 Cyclic behavior.

that is, the yield surface shifts, but does not expand as the maximum displacement changes.

For example, in Fig. 4-2(b), the line o-a-b is defined as the backbone curve with yielding point at location a. The unloading branch, b-c-d, may be characterized as twofold magnification of the backbone curve, that is, line b-c is twice line o-a. This behavior is observed in most models except in the stiffness degrading model shown in Fig. 4-2(c), which is specifically developed for reinforced concrete structures. It is worthwhile to also mention that the smooth stiffness degrading model shown in Fig. 4-2(d) is more representative in describing the energy dissipation in small cycle vibration. For example, cycling between points g-h in Fig. 4-2(d) results in small loops (energy dissipation), whereas such cycling in all other models does not result in loops (thus, no energy dissipation has occurred).

In the inelastic response, the structure stiffness, k, is not constant. Therefore, the classical solution of the elastic systems presented in the previous chapter will not apply. Clearly, all of the techniques given in the elastic systems, such as superposition, cannot be applied to this system. The following sections will address how to handle inelastic systems.

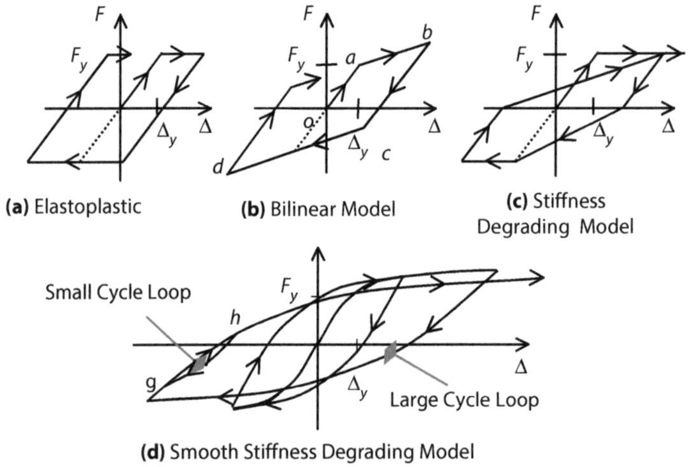

FIGURE 4-2 Examples of hysteresis models.

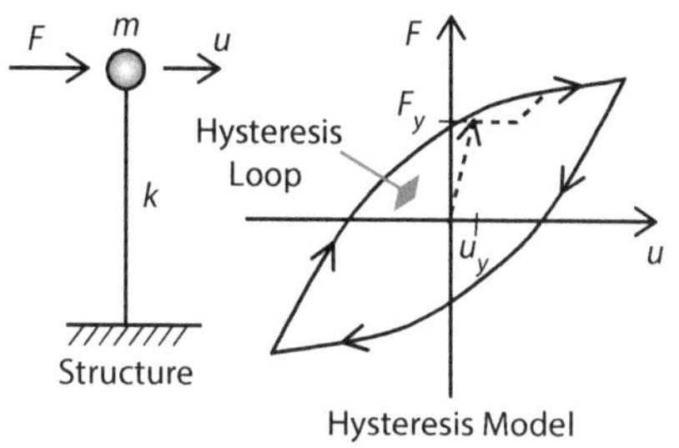

Figure 4-3 SDOF system.

4.2 Single Degree of Freedom System

An inelastic single degree of freedom (SDOF) system consists of concentrated mass, m, damping, c, and restoring force, $F = f(u)$, as shown in Fig. 4-3. This system is characterized by the inelasticity of its restoring force (by the relationship between displacement, u, and restoring force, F). The equilibrium equation developed in the previous chapter for elastic SDOF system is still valid

$$m\ddot{u} + c\dot{u} + ku = -m\ddot{u}_g$$

Because k is not constant but is a function of displacement, u, the differential equation above cannot be solved in the usual manner. Instead, numerical methods may be used as a powerful technique for operating directly on the differential equation. The following sections review some of the popular numerical methods that are frequently used in dynamic analysis.

4.3 Numerical Methods

Numerical methods are used as the general solution to the general problems of nonlinear inelastic systems. Numerical methods can also be used to solve elastic systems.

For convenience, the equation of motion may be written in the following form:

$$m\ddot{u} + c\dot{u} + f_i = p_i$$

where f_i = restoring force (hysteresis model)
p_i = external applied load $= -m\ddot{u}_g$

The damping may also exhibit nonlinearity similar to stiffness.

Three popular methods used in dynamic analysis are considered to be solutions to an initial value problem:

1. Central differences method (a branch of the general finite differences method)
2. Newmark-β method
3. Wilson-θ method

4.3.1 Central Differences Method

The central differences method is an approximation of derivatives over small intervals of the function. Because the velocity is the tangent of the displacement function, it can

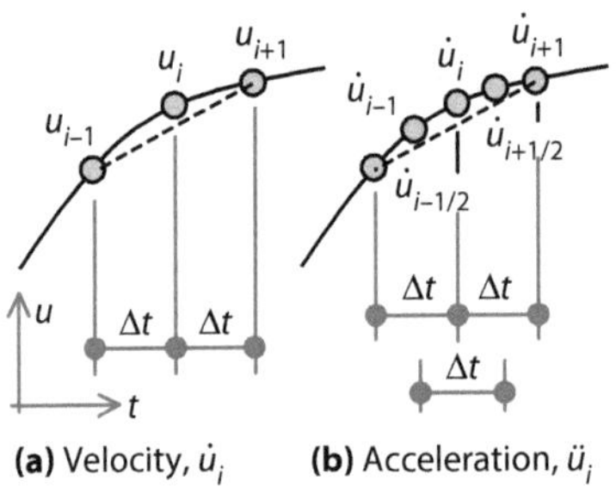

FIGURE 4-4 Central differences operators.

be approximated as the slope of the cord between the two displacements of u_{i-1} and u_{i+1}, over a finite interval of time, $2\Delta t$, as shown in Fig. 4-4(a). Accordingly, the velocity, $\dot{u}_i$, at time step, i, and time interval, Δt, is approximated as

$$\dot{u}_i = \frac{u_{i+1} - u_{i-1}}{2\Delta t} \tag{4-1}$$

Similarly, the acceleration can be approximated as the difference between two velocities at time steps $i - 1$ and $i + 1$, over time intervals, $2\Delta t$, as shown in Fig. 4-4(b). To increase the accuracy of acceleration calculations, intermediate steps at $i \pm ½$ may be introduced as shown in Fig. 4-4(b). As a result, the acceleration may be given as

$$\ddot{u}_i = \frac{\dot{u}_{i+1/2} - \dot{u}_{i-1/2}}{\Delta t} = \frac{1}{\Delta t}\left\{\frac{u_{i+1} - u_i}{\Delta t} - \frac{u_i - u_{i-1}}{\Delta t}\right\}$$

$$\ddot{u}_i = \frac{1}{\Delta t^2}\left[u_{i+1} - 2u_i + u_{i-1}\right] \tag{4-2}$$

Inserting the two operators of velocity and acceleration, from Eqs. (4-1) and (4-2), into the equation of motion yields

$$m\ddot{u} + c\dot{u} + f_i = p_i$$

$$m\left\{\frac{1}{\Delta t^2}\left[u_{i+1} - 2u_i + u_{i-1}\right]\right\} + c\left\{\frac{u_{i+1} - u_{i-1}}{2\Delta t}\right\} + f_i = p_i$$

$$u_{i+1} = \left[m + \frac{\Delta t}{2}c\right]^{-1}\left\{\Delta t^2(p_i - f_i) + \left[\frac{\Delta t}{2}c - m\right]u_{i-1} + 2mu_i\right\}$$

The equation above gives the central differences method in its basic form. The basic form is given in contrast to the central differences method in its summed form, which is usually implemented to reduce the roundoff errors in computers. The summed form of the central differences method takes the following form.

If the quantity z is defined as

$$z_i = \frac{u_i - u_{i-1}}{\Delta t}$$

The central differences equation takes the following form:

$$z_{i+1} = \left[\frac{m}{\Delta t} + \frac{c}{2}\right]^{-1} \left\{ p_i - f_i + z_i \left[\frac{m}{\Delta t} - \frac{c}{2}\right] \right\}$$

then,

$$u_{i+1} = u_i + \Delta t z_{i+1}$$

The central differences method is not a self-starting method. For given initial conditions, a Taylor series is usually used to start the algorithm as follows:

$$u_1 = u_0 + \Delta t \dot{u}_0 + \frac{\Delta t^2}{2} \ddot{u}_0$$

The selected time interval (Δt) should be small enough to capture the variation and details of the hysteresis model and the earthquake record. In addition, Δt should be small enough to ensure convergence (stability of algorithm). A time interval from one-fifth to one-tenth of the fundamental period is usually good enough to satisfy these conditions

$$\Delta t \leq \left(\frac{1}{5} \rightarrow \frac{1}{10}\right) T$$

4.3.2 Newmark-β Methods

Unlike the central differences method, Newmark-β methods lack mathematical evidence. However, they are popular methods in dynamic analysis proposed by Newmark. For this purpose, Newmark proposed the following system of equations:

$$m\ddot{u}_{i+1} + c\dot{u}_{i+1} + ku_{i+1} = p_{i+1} \tag{4-3}$$

$$\dot{u}_{i+1} = \dot{u}_i + \frac{\Delta t}{2}\ddot{u}_i + \frac{\Delta t}{2}\ddot{u}_{i+1} \tag{4-4}$$

$$u_{i+1} = u_i + \Delta t \dot{u}_i + \left(\frac{1}{2} - \beta\right)\Delta t^2 \ddot{u}_i + \beta \Delta t^2 \ddot{u}_{i+1} \tag{4-5}$$

Knowing u_i, $\dot{u}_i$, and $\ddot{u}_i$ at step i, the three unknowns u_{i+1}, $\dot{u}_{i+1}$, and $\ddot{u}_{i+1}$ at step $i+1$, can be found by simultaneously solving Eqs. (4-3), (4-4), and (4-5) above.

The parameter β may be chosen anywhere between $0 < \beta < 1/4$. The variation of the parameter β offers a group of submethods known by the following names:

$\beta = 1/4$ (constant acceleration method)

$\beta = 1/6$ (linear acceleration method)

$\beta = 1/8$ (method exists, but is unnamed)

$\beta = 1/12$ (Fox and Goodman method)

$\beta = 0$ (explicit Newmark method)

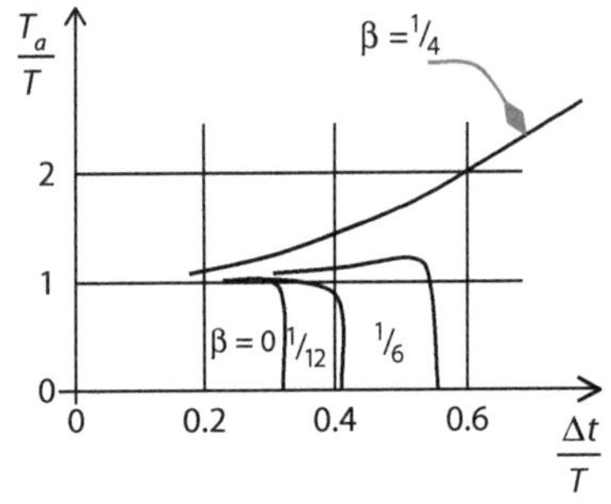

FIGURE 4-5 Relationship between time interval and apparent period.

The stability of algorithm of the methods above depends on the selection of the time interval, Δt. Figure 4-5 shows the relationship between the time interval and the apparent period, T_a, normalized to the structural period, T. Because the constant acceleration method ($\beta = 1/4$) does not collapse, as shown in this figure, this is known as the unconditionally stable method. The numerical stability of the other methods depends mainly on the value of the time interval, Δt. The linear acceleration method, $\beta = 1/6$, is numerically stable for values of $\Delta t < \approx 0.55T$.

4.3.3 Wilson-θ Method

The Wilson-θ method is simply an extension of the linear acceleration method in its differential form. It is applied by following the same procedures of the linear acceleration method, $\beta = 1/6$, using the time increment of τ instead of Δt where $\tau = \theta \Delta t$.

This method is unconditionally stable if $\theta > 1.37$.

Example 4-1 The structure in Example 4-1, Fig. 1, has an SDOF system with properties as shown in the figure. If the structure is subjected to the forcing function in Example 4-1, Fig. 1(c), calculate the first few displacements using the central differences method and the Newmark-β method. Take the time increment as $\Delta t = 0.1$ s.

Solution

1. Central differences method
   Because this method is not self-starting, a Taylor series is required to start the algorithm.

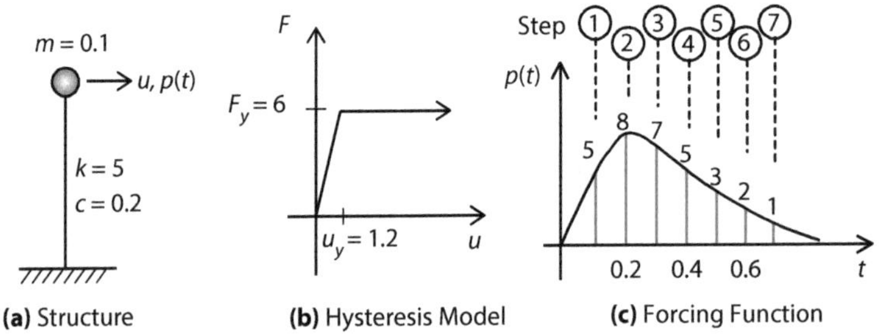

EXAMPLE 4-1, FIGURE 1 SDOF system (units are consistent).

Step 1

Because initial conditions are all zero, the displacement u_1 at step 1 will be zero and the restoring force f_1 will also be zero.

$$u_1 = u_0 + \Delta t \dot{u}_0 + \frac{\Delta t^2}{2}\ddot{u}_0 = 0, \quad \rightarrow \quad f_1 = 0$$

Step 2

$$u_0 = 0, \quad u_1 = 0, \quad f_1 = 0, \quad p_1 = 5$$

$$u_2 = \left[m + \frac{\Delta t}{2}c\right]^{-1}\left\{\Delta t^2(p_1 - f_1) + \left[\frac{\Delta t}{2}c - m\right]u_0 + 2mu_1\right\}$$

$$u_2 = \left[0.1 + \frac{0.1}{2}0.2\right]^{-1}\left\{(0.1)^2(5-0) + \left[\frac{0.1}{2}0.2 - 0.1\right]0 + 2(0.1)(0)\right\}$$

$$= 0.455 < u_y$$

Similarly, the next step of displacements is found by calculating steps 3 and 4 as follows:

Step 3

$$u_3 = \left[0.1 + \frac{0.1}{2}0.2\right]^{-1}\left\{(0.1)^2(8 - 5\times0.455) + \left[\frac{0.1}{2}0.2 - 0.1\right]0 + 2(0.1)(0.455)\right\}$$

$$= 1.347 > u_y$$

Step 4

$$u_4 = \left[0.1 + \frac{0.1}{2}0.2\right]^{-1}\left\{(0.1)^2(7-6) + \left[\frac{0.1}{2}0.2 - 0.1\right](0.455) + 2(0.1)(1.347)\right\}$$

$$= 2.176 > u_y$$

These systematic procedures can be continued up to any required elapsed time, u_5, u_6, . . . etc.

2. Newmark-β method

For the given initial conditions of zero displacement, zero velocity, and zero acceleration, that is, at step 0, the displacement, velocity, and acceleration at step 1 are found by solving the following three equations:

$$m\ddot{u}_1 + c\dot{u}_1 + ku_1 = p_1 \tag{4-6}$$

$$\dot{u}_1 = \dot{u}_0 + \frac{\Delta t}{2}\ddot{u}_0 + \frac{\Delta t}{2}\ddot{u}_1 \tag{4-7}$$

$$u_1 = u_0 + \Delta t \dot{u}_0 + \frac{\Delta t^2}{2}\ddot{u}_0 \tag{4-8}$$

Substituting the initial conditions,

$$0.1\ddot{u}_1 + 0.2\dot{u}_1 + 5(0) = 5 \;\rightarrow\; 0.1\ddot{u}_1 + 0.2\dot{u}_1 = 5 \tag{4-9}$$

$$\dot{u}_1 = 0 + \frac{0.1}{2}(0) + \frac{0.1}{2}\ddot{u}_1 \;\rightarrow\; \dot{u}_1 = 0.05\ddot{u}_1 \tag{4-10}$$

$$u_1 = 0 + 0.1(0) + \frac{(0.1)^2}{2}(0) \;\rightarrow\; u_1 = 0 \tag{4-11}$$

Equation (4-11) yields, $u_1 = 0$.

Substitution of Eq. (4-10) into Eq. (4-9) yields

$$0.1\ddot{u}_1 + 0.2(0.05\ddot{u}_1) = 5 \;\rightarrow\; \ddot{u}_1 = 45.45 \tag{4-12}$$

Substitution of Eq. (4-12) into Eq. (4-10) yields

$$\dot{u}_1 = 0.05(45.45) \rightarrow \dot{u}_1 = 2.27 \tag{4-13}$$

The displacement, velocity, and acceleration in step 1 are found as given by Eqs. (4-11), (4-13), and (4-12), respectively. To find the displacement, velocity, and acceleration in step 2, repeat the procedures above by solving the three equations. For example, Eq. (4-8) yields

$$u_2 = u_1 + \Delta t \dot{u}_1 + \frac{\Delta t^2}{2}\ddot{u}_1 \tag{4-14}$$

$$u_2 = 0 + 0.1(2.27) + \frac{\Delta t^2}{2}(45.45) = 0.439 \tag{4-15}$$

The displacement in step 1 is given as 0.439 instead of 0.455 as found by central differences method. Of course, the results should be closer and will be almost identical if the time increment, Δt, is small enough (e.g., 0.001 s).

Example 4-2

1. The SDOF structure in Example 4-2, Fig. 1, is excited by an earthquake that has the accelerogram shown in Example 4-2, Fig. 2. If the structure is elastic with an infinite elastic strength, find the maximum force, F_e, and the maximum displacement, u_e, reached during this excitation, assuming the structure has no damping, $\zeta = 0$. Plot displacement and restoring force histories. NOTE: Take $\Delta t = 0.2$ s and displacement history up to 6 s for Secs. (1) and (2).
2. The structure in Sec. (1) is designed to carry a lateral static load equal to 50 kN at its ultimate state with elastoplastic hysteretic behavior as shown in Example 4-2, Fig. 3. Find the maximum force, F_{max}, and maximum displacement, u_{max}, under the excitation of the same earthquake shown in Example 4-2, Fig. 2. Plot displacement and restoring force history.
3. Make a graphical comparison between the resulting maximum forces and maximum displacements resulting from Secs. (1) and (2). In other words, make a comparison between the elastic response and the inelastic response by plotting these values on a figure with u as abscissa and F as ordinate.

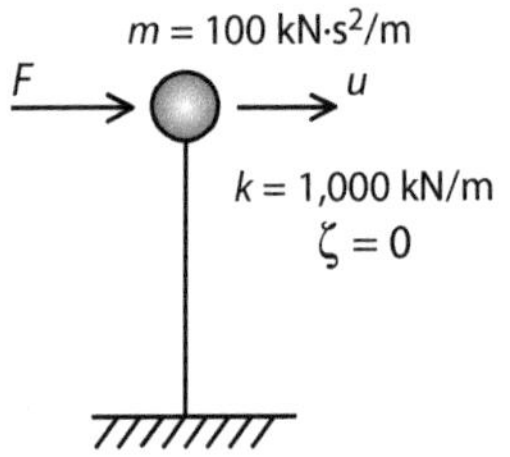

Example 4-2, Figure 1 Structural system.

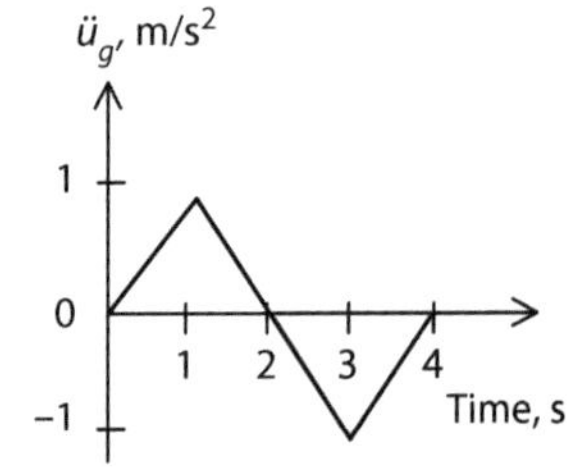

Example 4-2, Figure 2 Accelerogram.

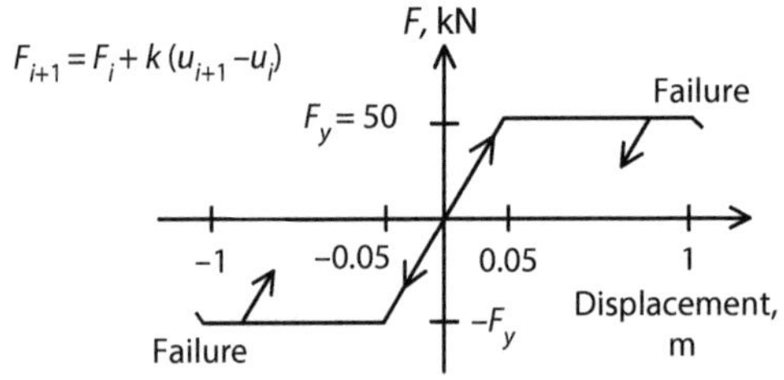

Example 4-2, Figure 3 Hysteresis model.

Solution The solution can be obtained by integrating the equation of motion using the central differences method with procedures identical to those laid out in Example 4-1. The results are summarized in Example 4-2, Table 1 and are described in the following sections.

1. The solution to this section is the elastic case where the stiffness is constant and equal to 1,000 kN/m. Step-by-step integration yields the displacement history from step 0, with zero initial values, and proceeds to step 31 at elapsed time equals to 6.2 s. The resulting

		Elastic Response		Inelastic Response	
Step No.	**Time, s**	u_e, m	F_e, kN	u_{ine}, m	F_{ine}, kN
0	0.00	0.000	0.0	0.000	0.0
1	0.20	0.000	0.0	0.000	0.0
2	0.40	–0.008	–8.0	–0.008	–8.0
3	0.60	–0.029	–28.8	–0.029	–28.8
4	0.80	–0.062	–62.1	–0.062	–50.0
5	1.00	–0.103	–102.5	–0.107	–50.0
6	1.20	–0.142	–142.0	–0.173	–50.0
7	1.40	**–0.157**	**–156.6**	–0.250	–50.0
8	1.60	–0.133	–132.6	–0.331	–50.0
9	1.80	–0.072	–71.6	–0.408	–50.0
10	2.00	0.010	10.1	–0.474	–50.0
11	2.20	0.088	87.7	–0.519	–50.0
12	2.40	0.138	138.3	**–0.536**	**–50.0**
13	2.60	**0.150**	**149.5**	–0.518	–31.3
14	2.80	0.125	124.9	–0.462	24.0
15	3.00	0.082	82.4	–0.385	50.0
16	3.20	0.047	46.9	–0.287	50.0
17	3.40	0.025	24.6	–0.177	50.0
18	3.60	0.017	16.5	–0.064	50.0
19	3.80	0.018	17.8	0.046	50.0
20	4.00	0.020	20.0	0.144	50.0
21	4.20	0.014	14.1	0.221	50.0
22	4.40	0.003	2.7	0.279	50.0
23	4.60	–0.010	–9.9	0.316	50.0
24	4.80	–0.018	–18.5	**0.334**	**50.0**
25	5.00	–0.020	–19.7	0.332	47.7
26	5.20	–0.013	–13.0	0.310	26.2
27	5.40	–0.001	–1.1	0.278	–5.7
28	5.60	0.011	11.2	0.249	–35.3
29	5.80	0.019	19.0	0.233	–50.0
30	6.00	0.019	19.3	0.238	–45.5
31	6.20	0.012	11.8	0.260	–22.8

Example 4-2, Table 1 Structural Response

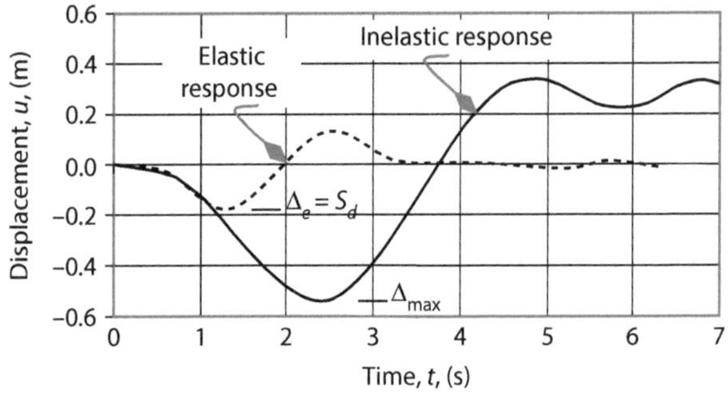

EXAMPLE 4-2, FIGURE 4 Displacement history.

displacement at each step is given in meters as shown in Example 4-2, Table 1. Note that the absolute maximum elastic force, 156.6 kN, and absolute maximum elastic displacement, 0.157 m, have been reached at step 7 at elapsed time equals to 1.4 s. A graphical plot of the displacement history is also shown in Example 4-2, Fig. 4.

2. The solution to this section is the inelastic case where the stiffness is not constant and must be tracked according to the previous displacement reached as given by the hysteresis model in Example 4-2, Fig. 3. The step-by-step integration yields the displacement history from step 0, with zero initial values, and proceeds to step 31 at elapsed time equals to 6.2 s. The resulting displacement at each step is given in meters as shown in Example 4-2, Table 1. Note that the absolute maximum inelastic displacement, 0.536 m, has been reached at maximum strength, 50 kN, and at step 12 at elapsed time equal to 2.4 s. A graphical plot of the displacement history is shown in Example 4-2, Fig. 4.

3. A graphical relationship is plotted between the restoring force and the displacement as shown in Example 4-2, Fig. 5. The elastic response is plotted in dashed lines, whereas the inelastic response is plotted with solid lines.

Example 4-2, Fig. 4, shows that the displacement in the inelastic case is much larger than the displacement in the elastic case. This result is also obvious in Example 4-2, Fig. 5. However, the force in the elastic case is larger than the force in the inelastic case. As expected, the force in the inelastic case cannot exceed the structure strength (the yield level of the structure).

This example highlights an extremely important conclusion mentioned in earlier chapters with respect to inelastic response of structures to earthquake excitations. If the structure is designed with strength $F_y = 50$ kN less than the ESD, $F_e = 156.6$ kN, then the structure can survive the excitation as long as the maximum displacement does not reach the ultimate displacement capacity of the structure, $\Delta_u = 1$ m.

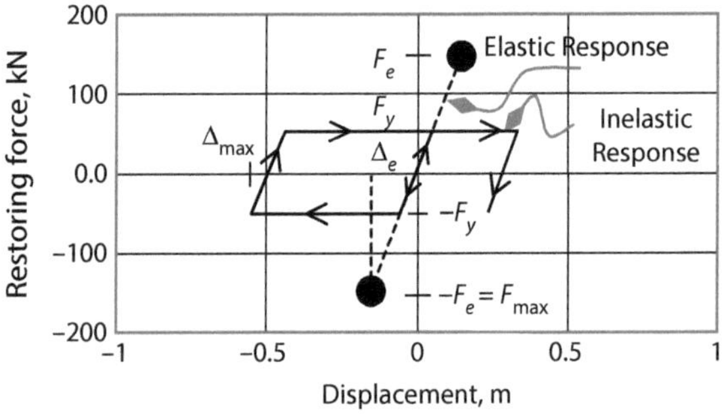

EXAMPLE 4-2, FIGURE 5 Hysteretic behavior.

4.4 Multiple Degrees of Freedom System

Similar to an SDOF system, an inelastic multiple degrees of freedom (MDOF) system can only be solved by numerical methods. The numerical methods explained in the previous sections for an SDOF system are equally applicable for an MDOF system.

The formulas given for an SDOF are identical, namely, central differences, Newmark-β and Wilson-θ methods, are identical for MDOF system. For application, we simply replace the single quantities, *m*, *c*, *k*, *u*, *f*, and *p* by their matrix counterparts such that

$$\begin{aligned} m &\rightarrow [M]_{N\times N} \\ c &\rightarrow [C]_{N\times N} \\ k &\rightarrow [K]_{N\times N} \\ u &\rightarrow \{U\}_{N\times 1} \\ f &\rightarrow \{F\}_{N\times 1} \\ p &\rightarrow \{P\}_{N\times 1} \end{aligned}$$

For example, if the MDOF equation of motion is given as

$$[M]\{\ddot{U}\}+[C]\{\dot{U}\}+[K]\{U\}=-[M]\{R\}\ddot{u}_g$$

then the central differences method for the next step displacement at step $i+1$ is given as a function of the step i as follows:

$$\{U_{i+1}\}=\left[[M]+\frac{\Delta t}{2}[C]\right]^{-1}\left\{\Delta t^2\{\{P_i\}-\{F_i\}\}+\left[\frac{\Delta t}{2}[C]-[M]\right]\{U_{i-1}\}+2[M]\{U_i\}\right\}$$

For large structures, computation time and storage requirements also become very large in terms of computer resources, especially when time history is needed. Special problems arise if the true hysteretic behavior is needed. Therefore, explicit inelastic dynamic analysis is usually performed either in research areas or for very complex and special structures such as nuclear power plants. Of course, this problem becomes worse if finite element analysis is used.

Because of these shortcomings, inelastic dynamic analysis is not used in practice. Instead, research and codes have focused on indirectly incorporating inelastic dynamic analysis into elastic analysis. An example of this would be using parameters such as force reduction factor, *R*, and global ductility demand, μ_d, with some relationships between them. This will be discussed in the following chapters.

Example 4-3

A frame with a diagonal viscous damper is shown in Example 4-3, Fig. 1. The frame members have uniform mass distribution and constant cross section. The frame is intended to behave elastically under the excitation of its supports by a horizontal component of an earthquake profile along the line *A*-*B* as shown in Example 4-3, Fig. 2. The frame members have the following properties:

$$m = 100\ \text{kN·s}^2/\text{m/m}$$

$$EA = 6\times 10^6\ \text{kN}$$

$$EI = 0.3\times 10^6\ \text{kN·m}^2$$

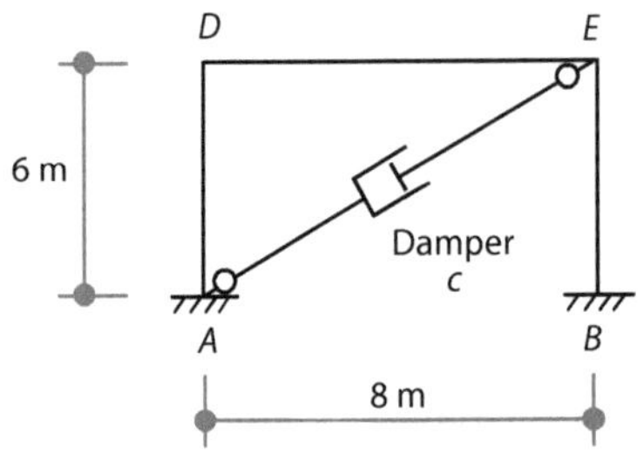

Example 4-3, Figure 1 Frame Layout.

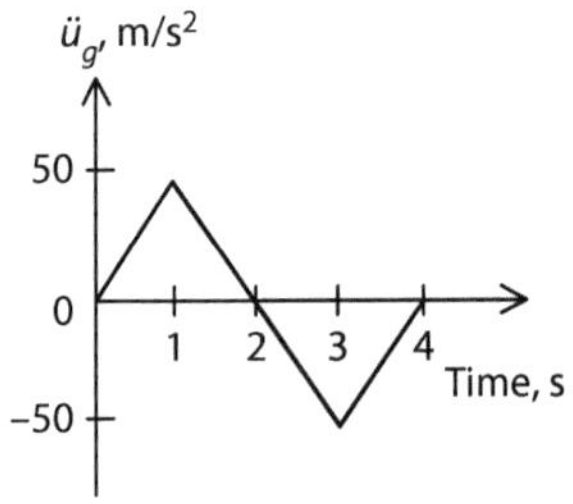

Example 4-3, Figure 2 Accelerogram.

The damper has the following coefficient of viscous damping:

$$c = 220 \text{ kN·s/m}$$

Use the central differences method and the consistent mass matrix to find the horizontal displacement history.

For calculation purposes, take the time interval $\Delta t = 0.1$ s.

Solution

Because the consistent mass matrix is needed, the frame may be discretized into three finite elements and four nodes as shown in Example 4-3, Fig. 3. Because the supports are fixed at nodes 1 and 4, the only left degrees of freedom are the displacements at joints 2 and 3. In plane frames, each node would have three degrees of freedom, two displacements and one rotation. Consequently, the frame would have a total of six degrees of freedom, two displacements, and one rotation at joint 2 and also at joint 3. These degrees of freedom are designated d_1 through d_6 as shown in Example 4-3, Fig. 3.

The solution of this problem requires the construction of the relevant matrices in the general equation of motion as stated before

$$[M]\{\ddot{U}\}+[C]\{\dot{U}\}+[K]\{U\}=-[M]\{R\}\ddot{u}_g$$

The relevant matrices in the equation of motion above are the global mass matrix, $[M]$, the global damping matrix, $[C]$, the global stiffness matrix, $[K]$, and the earthquake directional vector, $\{R\}$, that is, associated with the global displacement vector, $\{U\}$. Construction of these matrices may be accomplished by any convenient standard assembly process given in matrix structural analysis context. In this example, sample elements will be constructed by inspection for demonstration purposes.

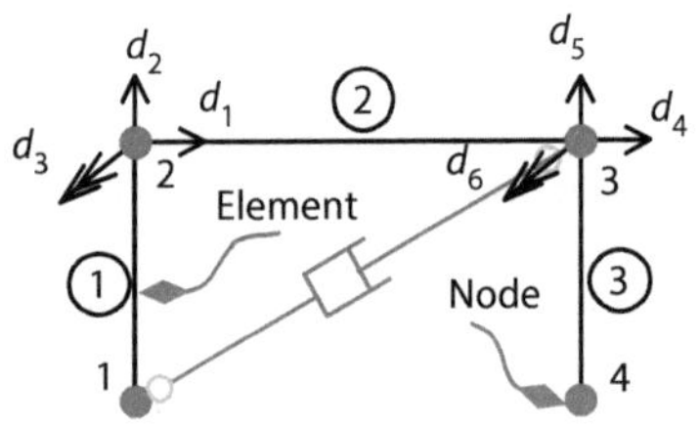

Example 4-3, Figure 3 Discretization of frame.

Section 3.9 provides finite element formulation and relevant element matrices. For convenience, the stiffness and mass element matrices are repeated below.

1. Beam element stiffness matrix

$$[k_m]=\begin{bmatrix} \frac{12EI}{L^3} & \frac{6EI}{L^2} & \frac{-12EI}{L^3} & \frac{6EI}{L^2} \\ \frac{6EI}{L^2} & \frac{4EI}{L} & \frac{-6EI}{L^2} & \frac{2EI}{L} \\ \frac{-12EI}{L^3} & \frac{-6EI}{L^2} & \frac{12EI}{L^3} & \frac{-6EI}{L^2} \\ \frac{6EI}{L^2} & \frac{2EI}{L} & \frac{-6EI}{L^2} & \frac{4EI}{L} \end{bmatrix}$$

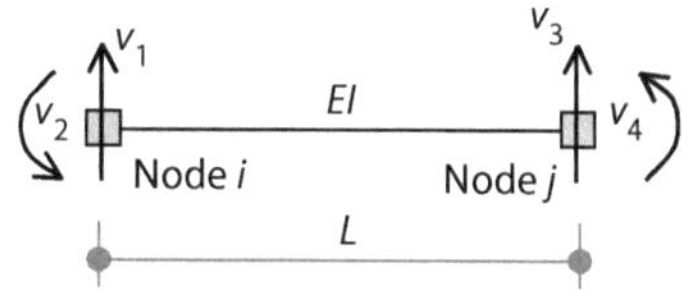

2. Truss element stiffness matrix

$$[k_m]=\frac{EA}{L}\begin{bmatrix} 1 & -1 \\ -1 & 1 \end{bmatrix}$$

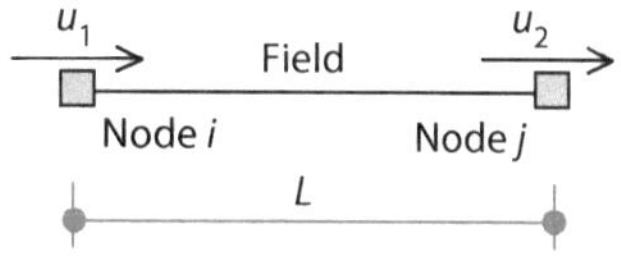

3. Beam element mass matrix

$$[m_m]=\frac{\bar{m}L}{420}\begin{bmatrix} 156 & 22L & 54 & -13L \\ 22L & 4L^2 & 13L & -3L^2 \\ 54 & 13L & 156 & -22L \\ -13L & -3L^2 & -22L & 4L^2 \end{bmatrix}$$

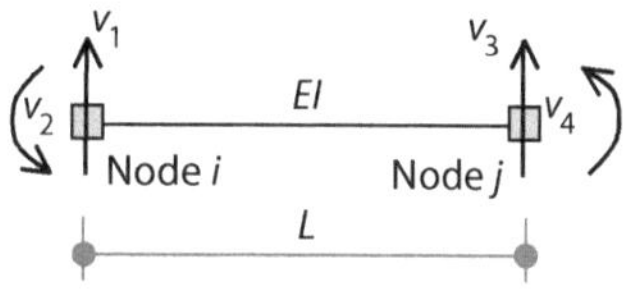

4. Truss element stiffness matrix

$$[m_m]=\frac{\bar{m}L}{420}\begin{bmatrix} 140 & 70 \\ 70 & 140 \end{bmatrix}$$

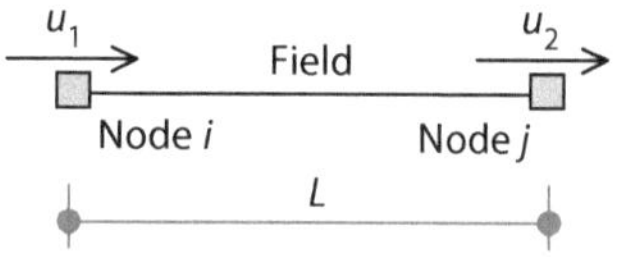

5. The damping matrix can be deduced for a linear damper by noticing that the force system resembles truss behavior. Therefore, by replacing the stiffness of the truss by the damping coefficient, *C*, the element coefficient of damping matrix becomes

$$[c_m]=C\begin{bmatrix} 1 & 1 \\ -1 & 1 \end{bmatrix}$$

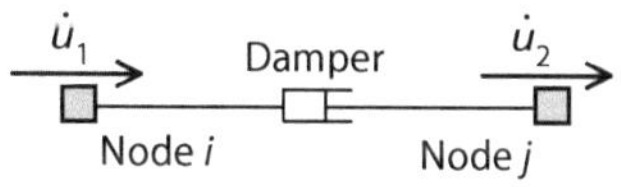

Note that the force develops in the damper due to nodal velocities not nodal displacements.

1. Displacement and loading vectors $\{U\}$ and $\{R\}$
   The displacement vector will contain the number of rows equal to the number of degrees of freedom, or six rows. Thus, $\{U\}$ is given as

$$\{U\}=\begin{Bmatrix} d_1 \\ d_2 \\ d_3 \\ d_4 \\ d_5 \\ d_6 \end{Bmatrix}$$

   Because the earthquake is horizontal and will excite all supports simultaneously, the loading vector will take values according to the displacement of the DOFs as a result of rigid body displacement in the direction of the earthquake. By inspection, the $\{R\}$ vector associated with the displacement vector takes the following values:

$$\{R\}=\begin{Bmatrix} 1 \\ 0 \\ 0 \\ 1 \\ 0 \\ 0 \end{Bmatrix}$$

   where d_3 and d_6 are rotational degrees of freedom.
2. Stiffness matrix
   The global stiffness matrix is constructed by the assembly process. The assembly process may be described as the summation of all nodal forces (or moments) at each node as a result of the contribution of all member reactions at that node. In order to visualize such a contribution, the frame may be separated into its discretized components, or into three beams as shown in Example 4-3, Fig. 4. Consequently, the matrix elements are evaluated by inspection as follows.
   The global matrix element K_{11} is given as the forces that develop in direction 1 due to unit displacement in direction 1. Thus, the contribution of all elements meeting at the node will

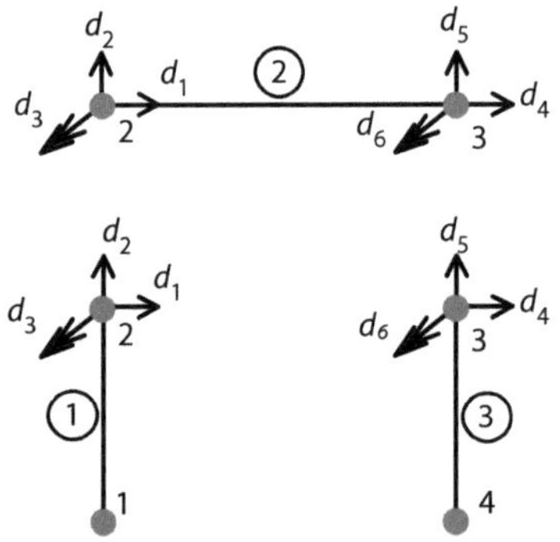

Example 4-3, Figure 4 Assembly process.

be a shearing force from element 1 and an axial force from element 2 due to this unit displacement in direction 1

$$\begin{aligned}K_{11} &= k^1_{m11} + k^2_{m11}\\ &= \frac{12EI}{L^3} + \frac{EA}{L}\\ &= \frac{12(0.3\times10^6)}{6^3} + \frac{6\times10^6}{8}\\ &= 767 \times 10^3 \text{ kN/m}\end{aligned}$$

Similarly, K_{21} is given as the forces that develop in direction 2 due to unit displacement in direction 1. Thus, the contribution of all elements meeting at the node will be an axial force from element 1 and a shearing force from element 2 due to this unit displacement in direction 1

$$K_{21} = 0 + 0 = 0$$

Similarly, K_{31} is given as the moment in direction 3 due to unit displacement in direction 1. Thus, the contribution of all elements meeting at the node will be a moment from element 1 and a moment from element 2 due to unit displacement in direction 1

$$\begin{aligned}K_{31} &= k^1_{m21} + 0\\ &= \frac{6EI}{L^2} + 0 = \frac{6(0.3\times10^6)}{6^2} + 0\\ &= 50 \times 10^3 \text{ kN/m}\end{aligned}$$

The same reasoning used to obtain answers for K_{11}, K_{21}, and K_{31} above can be applied to the rest of the matrix elements to yield the global stiffness matrix $[K]$ as follows:

$$[K]=10^3\begin{bmatrix} 767 & 0 & 50 & -750 & 0 & 0\\ 0 & 1007 & 28 & 0 & -7 & 28\\ 50 & 28 & 350 & 0 & -28 & 75\\ -750 & 0 & 0 & 767 & 0 & 50\\ 0 & -7 & -28 & 0 & 1007 & -28\\ 0 & 28 & 75 & 50 & -28 & 350 \end{bmatrix}$$

Note that $[K]$ above has constant elements because the behavior is elastic. If inelastic analysis were performed, these elements will be a function of the displacement history as shown under the section of SDOF.

3. Mass matrix
   The global mass matrix is constructed by the assembly process with the same reasoning applied to produce the stiffness matrix in (2). For example, the global matrix element M_{11} is given as the forces that develop in direction 1 due to unit acceleration in direction 1. Thus, the contribution of all elements meeting at the node will be a shearing force from element 1 and an axial force from element 2 due to this unit acceleration in direction 1

$$\begin{aligned}M_{11} &= m^1_{m11} + m^2_{m11}\\ &= mL_1\left(\frac{156}{420}\right) + mL_2\left(\frac{140}{420}\right)\\ &= 100(6)\left(\frac{156}{420}\right) + 100(8)\left(\frac{140}{420}\right)\\ &= 490 \text{ kN·s}^2\text{/m}\end{aligned}$$

Similarly, the next two elements, M_{21} and M_{31}, are evaluated as follows:

$$M_{21} = 0 + 0 = 0$$

and

$$M_{31} = m^1_{m21} + 0$$
$$= mL_1\left(\frac{22L}{420}\right) = 100(6)\left(\frac{22(6)}{420}\right)$$
$$= 189 \text{ kN·s}^2/\text{m}$$

The same reasoning used to obtain answers in M_{11}, M_{21}, and M_{31}, above can be applied to the rest of the matrix elements to yield the global mass matrix, [M], as follows:

$$[M] = \begin{bmatrix} 490 & 0 & 189 & 133 & 0 & 0 \\ 0 & 497 & 335 & 0 & 103 & -198 \\ 189 & 335 & 693 & 0 & 198 & -366 \\ 133 & 0 & 0 & 490 & 0 & 189 \\ 0 & 103 & 198 & 0 & 497 & -335 \\ 0 & -198 & -366 & 189 & -335 & 693 \end{bmatrix} \text{kN·s}^2/m$$

4. Damping matrix

The global damping matrix may be constructed by examining the state of force displacement at the nodes of concern. By referring to Example 4-3, Fig. 1, it can be seen that the damping force will only develop due to velocity of node E (note the activation of the DOFs, d_4 and d_5, shown in Example 4-3, Fig. 3). The damping matrix can be developed with reasoning similar to that used to develop both mass and stiffness matrices.

The global matrix element, C_{11}, is given as the force that develops in direction 1 due to unit velocity in direction 1. Thus, the contribution of all elements meeting at the node will be an axial force from the damper in the direction of the damper. The component of this force is then evaluated by resolving this force in direction 1.

Example 4-3, Fig. 5, shows the forces that develop in the structure due to unit velocity in direction, d_4. Insofar as the force in the damper is along its axis and is also proportional to the coefficient of damping, C, then

$$F_D = C(0.8\dot{d}_4) = 0.8C$$

Resolving this force in directions 4 and 5 yields

$$F_4 = F_D \cos\theta = 0.8\,C\,(0.8)$$
$$= 0.64\,C = 0.64\,(220)$$
$$= 141 \text{ kN}$$

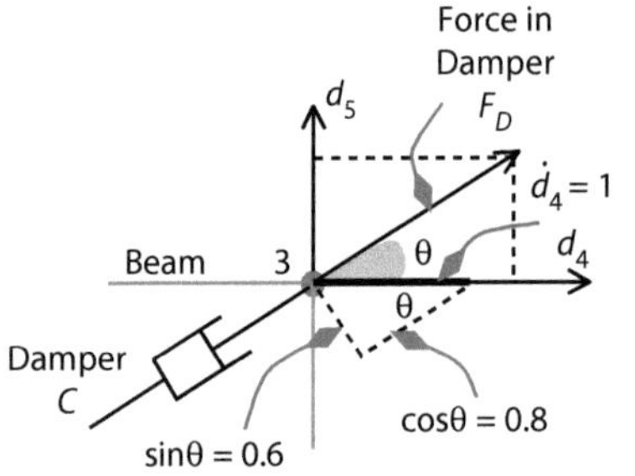

EXAMPLE 4-3, FIGURE 5 Damper force at node 3.

$$F_5 = F_D \sin\theta = 0.8\, C\,(0.6)$$
$$= 0.48\, C = 0.48\,(220)$$
$$= 106 \text{ kN}$$

Similarly, it can be shown that F_4 and F_5, due to unit velocity in direction 5, will be 106 and 141 kN, respectively. By noting that the damper will not be affected by the other degrees of freedom, the damping matrix can be given as

$$[C] = \begin{bmatrix} 0 & 0 & 0 & 0 & 0 & 0 \\ 0 & 0 & 0 & 0 & 0 & 0 \\ 0 & 0 & 0 & 0 & 0 & 0 \\ 0 & 0 & 0 & 141 & 106 & 0 \\ 0 & 0 & 0 & 106 & 141 & 0 \\ 0 & 0 & 0 & 0 & 0 & 0 \end{bmatrix} \text{kN}$$

Note also that the formal development of the damping matrix is usually accomplished through the assembly process using tensor transformation.

5. Central differences method
   In order to solve the given dynamic equation, namely

$$[M]\{\ddot{U}\} + [C]\{\dot{U}\} + [K]\{U\} = -[M]\{R\}\ddot{u}_g$$

we must apply the following central differences operator:

$$\{U_{i+1}\} = \left[[M] + \frac{\Delta t}{2}[C]\right]^{-1} \left\{\Delta t^2 \{\{P_i\} - \{F_i\}\} + \left[\frac{\Delta t}{2}[C] - [M]\right]\{U_{i-1}\} + 2[M]\{U_i\}\right\}$$

Because this method is not self-starting, Taylor series is used to start the algorithm.

Step 1

Because initial conditions are all zeros, the displacement vectors, $\{U_1\}$, at step 1 will be zero, and the restoring force vector, $\{F_1\}$, will also be zero

$$\Delta t = 0.1, \qquad t_1 = 0 \text{ s}, \qquad \ddot{u}_g = 5 \text{ m/s}^2$$

$$\{U_1\} = \{U_0\} + \Delta t\,\{\dot{U}_0\} + \frac{\Delta t^2}{2}\{\ddot{U}_0\} = \{0\}$$

$\{F_1\}$ is calculated as: $\{F_1\} = [K]\{U_1\} = [K]\{0\} = \{0\}$

$\{P_1\}$ is calculated as: $\{P_1\} = -[M]\{R\}\ddot{u}_g = -[M]\{R\}(5) = \begin{Bmatrix} -.31150E+04 \\ +.00000E+00 \\ -.94500E+03 \\ -.31150E+04 \\ +.00000E+00 \\ -.94500E+03 \end{Bmatrix} \text{kN}$

Step 2

$$\Delta t = 0.1, \qquad t_2 = 0.2 \text{ s}, \qquad \ddot{u}_g = 10 \text{ m/s}^2$$
$$\{U_0\} = \{0\}, \qquad \{U_1\} = \{0\},$$
$$\{F_1\} = \{0\}, \qquad \{P_1\} = \text{as in Step 1}$$

Therefore,

$$\{U_{i+1}\}=\left[[M]+\frac{\Delta t}{2}[C]\right]^{-1}\left\{\Delta t^2\{\{P_i\}-\{F_i\}\}+\left[\frac{\Delta t}{2}[C]-[M]\right]\{U_{i-1}\}+2[M]\{U_i\}\right\}$$

$$\{U_2\}=\left[[M]+\frac{\Delta t}{2}[C]\right]^{-1}\left\{\Delta t^2\{\{P_1\}-\{F_1\}\}+\left[\frac{\Delta t}{2}[C]-[M]\right]\{U_0\}+2[M]\{U_1\}\right\}$$

$$\{U_2\}=\left[[M]+\frac{0.1}{2}[C]\right]^{-1}\left\{(0.1)^2\{\{P_1\}-\{0\}\}+\left[\frac{0.1}{2}[C]-[M]\right]\{0\}+2[M]\{0\}\right\}$$

$$\{U_2\}=\begin{Bmatrix} -.50170E-01 \\ -.58114E-04 \\ -.80461E-04 \\ -.49258E-01 \\ +.57090E-03 \\ +.12614E-04 \end{Bmatrix} \text{m}$$

$\{F_2\}$ is calculated as: $$\{F_2\}=[K]\{U_2\}=\begin{Bmatrix} -.15409E+04 \\ -.64417E+02 \\ -.25533E+04 \\ -.15276E+03 \\ +.57720E+03 \\ -.24821E+04 \end{Bmatrix} \text{kN}$$

$\{P_2\}$ is calculated as: $$\{P_2\}=-[M]\{R\}\ddot{u}_g=-[M]\{R\}(10)=\begin{Bmatrix} -.62300E+04 \\ +.00000E+00 \\ -.18900E+03 \\ -.62300E+04 \\ +.00000E+00 \\ -.18900E+03 \end{Bmatrix} \text{kN}$$

Step 3

$\Delta t = 0.1,\quad t_2 = 0.3$ s, $\quad \ddot{u}_g = 15$ m/s^2

$\{U_1\} = \{0\},\quad \{U_2\}$ = as in step 2,

$\{F_2\}$ = as in step 2, $\quad \{P_2\}$ = as in step 2

Therefore,

$$\{U_{i+1}\}=\left[[M]+\frac{\Delta t}{2}[C]\right]^{-1}\left\{\Delta t^2\{\{P_i\}-\{F_i\}\}+\left[\frac{\Delta t}{2}[C]-[M]\right]\{U_{i-1}\}+2[M]\{U_i\}\right\}$$

$$\{U_3\}=\left[[M]+\frac{\Delta t}{2}[C]\right]^{-1}\left\{\Delta t^2\{\{P_2\}-\{F_2\}\}+\left[\frac{\Delta t}{2}[C]-[M]\right]\{U_1\}+2[M]\{U_2\}\right\}$$

$$\{U_3\}=\left[[M]+\frac{0.1}{2}[C]\right]^{-1}\left\{(0.1)^2\{\{P_2\}-\{F_2\}\}+\left[\frac{0.1}{2}[C]-[M]\right]\{0\}+2[M]\{U_2\}\right\}$$

$$\{U_3\}=\begin{Bmatrix} -.55685E-01 \\ -.28052E-01 \\ +.94445E-01 \\ -.94257E-01 \\ +.24491E-01 \\ +.98861E-01 \end{Bmatrix} \text{m}$$

These systematic procedures can be continued up to any required elapsed time: $\{U_4\}$, $(U_5\}$, etc. Note that each step would also yield the restoring force and, if required, a software program may be developed to carry these procedures and calculate the internal forces as well.

4.5 Equivalent Linearization

The energy dissipation capacity explained in previous sections is also termed hysteretic damping. The hysteretic damping and softening effect can be explained and quantified using the concept of equivalent linearization, which may be explained by consideration of the response of linear damped system in free vibration. If a cycle is considered with an average amplitude, A, and frequency, ω, one can write

$$u = A\sin\omega t \tag{4-16}$$

$$\dot{u} = \omega A\cos\omega t \tag{4-17}$$

Recall that the spring force, f_s, is proportional to the displacement

$$f_s = ku \tag{4-18}$$

which may be plotted as shown in Fig. 4-6.

Since the damping force, f_D, is proportional to velocity, then

$$f_D = c\dot{u} \tag{4-19}$$

Substitution of velocity from Eq. (4-17) into Eq. (4-19) yields

$$f_D = c\omega A\cos\omega t$$

$$= c\omega A\sqrt{1-\sin^2\omega t}$$

Replacing $(\sin^2\omega t)$ by $(u/A)^2$ from Eq. (4-16) yields

$$f_D = c\omega A\sqrt{1-\left(\frac{u}{A}\right)^2}$$

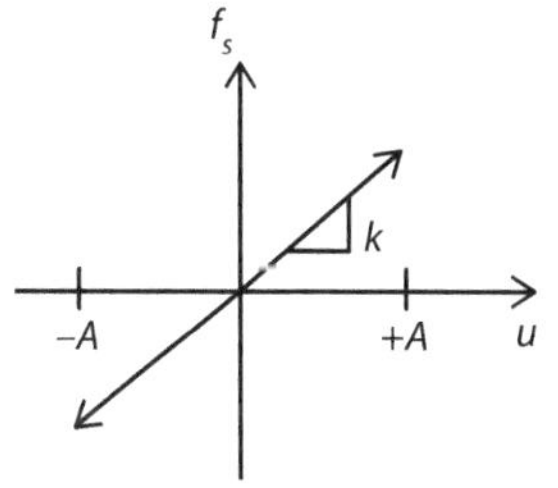

FIGURE 4-6 Spring force-displacement relationship.

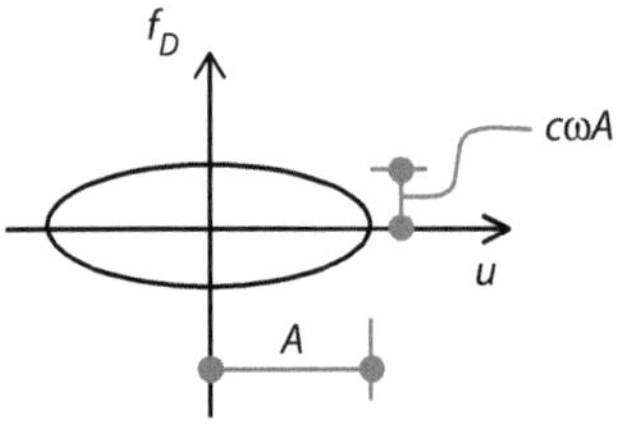

FIGURE 4-7 Damping force-displacement relationship.

or

$$f_D^2 = c^2\omega^2 A^2\left\{1-\left(\frac{u}{A}\right)^2\right\}$$

or

$$\left(\frac{f_D}{c\omega A}\right)^2 + \left(\frac{u}{A}\right)^2 = 1.0$$

The equation above is an equation of ellipse, which is plotted graphically in Fig. 4-7. It can be noted from Fig. 4-7 that the area of the ellipse represents a hysteretic loop, which is also the energy dissipated in one cycle of vibration due to viscous damping.

If Figs. 4-6 and 4-7 are graphically superposed, the resulting graph resembles a nonlinear system as shown in Fig. 4-8. In other words, the relationship between the summation of the elastic force and the viscous force, $f_s + f_D$, and between the displacement, u, forms a hysteresis loop similar to nonlinear systems.

Consequently, the relationships above may be used to evaluate the hysteretic damping (the energy dissipated by a nonlinear system in terms of the damping ratio of a viscous system).

If the energy dissipated by a nonlinear system is approximated by an elliptical shape, the energy dissipated will be the area of the ellipse, which is given as π times its major and minor radii a and b

$$\text{Area} = \pi ab = \pi(A)(c\omega A) = \pi A^2(c)(\omega)$$

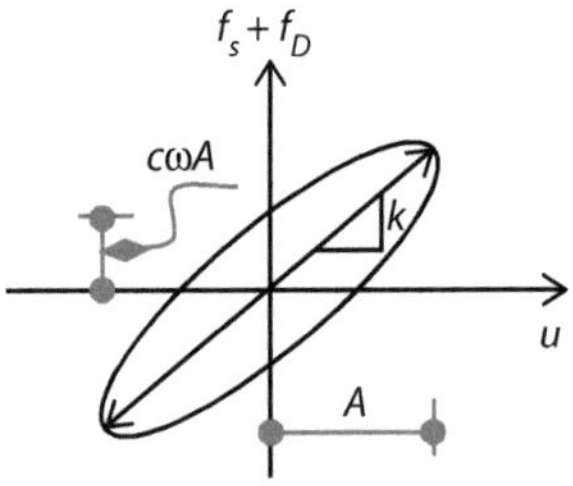

FIGURE 4-8 Spring and damping force-displacement relationship.

When c and ω, $(c = 2\zeta\sqrt{km})$ and $(\omega = \sqrt{k/m})$, are substituted in the equation above, the area becomes

$$\text{Area} = \pi A^2 (2\zeta\sqrt{km})\left(\sqrt{\frac{k}{m}}\right)$$

$$= 2\pi\zeta A^2(k) = 2\pi\zeta(kA)A$$

Therefore, the area of the ellipse, or the energy dissipated (ED) becomes

$$\text{ED} = 2\pi\zeta(f_{s,\max})A$$

The area of the strain energy, $U_{e,\max}$, (the area of the two elastic energy triangles shown in Fig. 4-9) equals to

$$U_{e,\max} = \frac{1}{2} f_{s,\max} A + \frac{1}{2} f_{s,\min} A$$

$$= f_{s,\max} A$$

If $f_{s,\max} A$ is substituted into the energy dissipation expression given above, then

$$\text{ED} = 2\pi\zeta(U_{e,\max})$$

Rearranging the expression above, the equivalent viscous damping of the hysteresis loop may be given with reference to Fig. 4-10 as

$$\zeta_{\text{eq}} = \frac{1}{2\pi} \frac{\text{ED}}{U_{e,\max}}$$

$$\zeta_{\text{eq}} = \frac{1}{2\pi} \frac{A_{\text{Hysteresis Loop}}}{A_{\text{trianges (OAB+OCD)}}}$$

Note that ζ is proportional to energy dissipation.

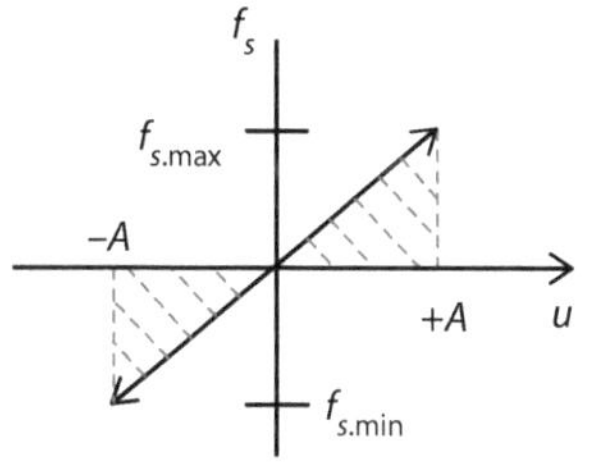

FIGURE 4-9 Elastic energy.

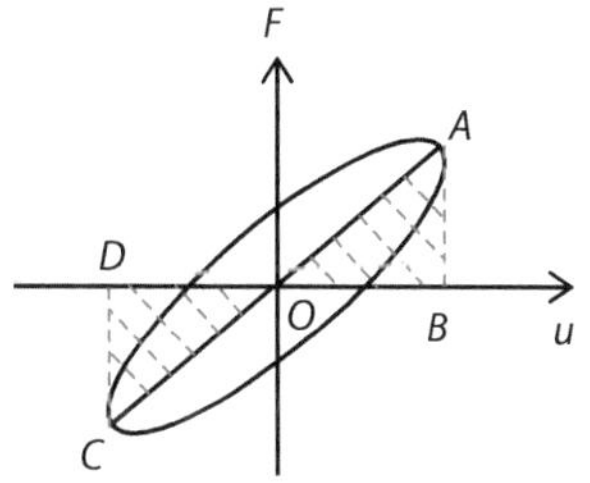

FIGURE 4-10 Equivalent viscous damping.

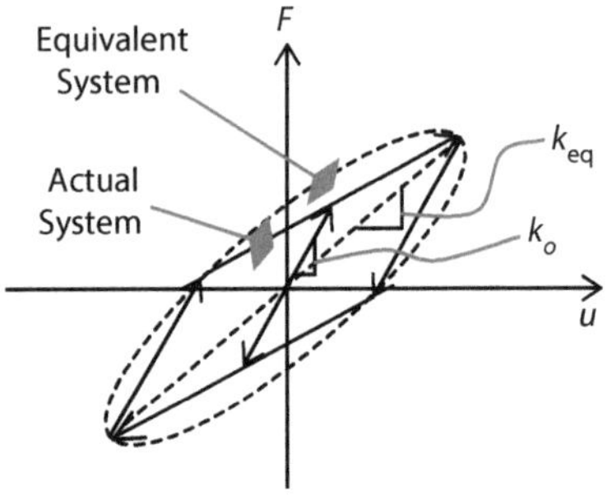

FIGURE 4-11 Equivalent system in reference to actual system.

The development above implies that it would be possible to approximate an inelastic system with a hysteretic loop similar to the system shown in Fig. 4-11 by using an equivalent linear system with equivalent frequency, ω_{eq}, equivalent stiffness, k_{eq}, and equivalent viscous damping, ζ_{eq}, such that

$$k_{eq} = \frac{F_{u,\max}}{u_{\max}}$$

$$\omega_{eq} = \sqrt{\frac{k_{eq}}{m}}$$

$$\zeta_{eq} = \frac{1}{2\pi}\frac{ED}{U_{e,\max}} = \frac{c_{eq}}{c_{cr,eq}}$$

This system is also described by an equivalent equation of motion

$$m\ddot{u} + c_{eq}\dot{u} + k_{eq}u = -m\ddot{u}_g$$

or,

$$\ddot{u} + 2\zeta_{eq}\,\omega_{eq}\,\dot{u} + \omega_{eq}\,u = -\ddot{u}_g$$

The system in this case will have an initial elastic stiffness, k_o, and an initial frequency, ω_o, given as

$$\omega_o = \sqrt{\frac{k_o}{m}}$$

Example 4-4 An SDOF structure is shown in Example 4-4, Fig. 1. The structure is excited by an earthquake that has the response spectrum shown in Example 4-4, Fig. 2(b). The figure also shows the structural properties and its response parameters in terms of the hysteresis loop shown in Example 4-4, Fig. 2(a). The system viscous damping is $\zeta_v = 0$. If the system viscous damping is ignored, (i.e., $\zeta_v = 0$), do the following:

1. Find the elastic strength demand by finding the maximum elastic response in terms of maximum elastic force and maximum elastic displacement due to this excitation.
2. Use the equivalent linearization concept to find the maximum actual displacement due to this excitation.

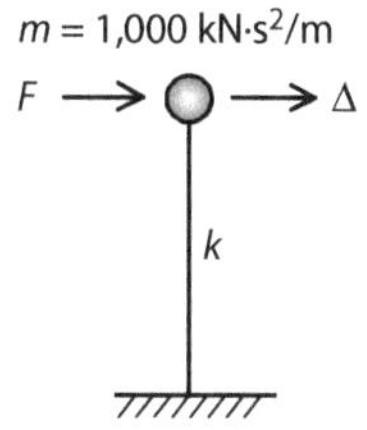

Example 4-4, Figure 1 SDOF structure.

Solution

1. Elastic response

$$k_o = \frac{F_y}{\Delta_y} = \frac{1{,}600}{0.02} = 80{,}000 \text{ kN/m}$$

$$\omega_o = \sqrt{\frac{k_o}{m}} = \sqrt{\frac{80{,}000}{1{,}000}} = 8.94 \text{ rad/s}$$

$$T = \frac{2\pi}{\omega_o} = \frac{2\pi}{8.92} = 0.7 \text{ s}$$

Using response spectrum with $T = 0.7$ s and $\zeta = 0$, spectral velocity is given in Example 4-4, Fig. 1(c), as $S_v = 0.56$ m/s. As a result,

$$S_d = \frac{S_v}{\omega} = \frac{0.56}{8.94} = 0.062 \text{ m}$$

$$S_a = \omega S_v = 8.94(0.56) = 5.006 \text{ m/s}^2$$

$$\Delta_e = S_d = 0.062 \text{ m}$$

$$F_e = mS_a = 1{,}000(5.006) = 5{,}006 \text{ kN}$$

2. Inelastic response

In the case of inelastic response, the maximum displacement will not be known at this stage. Therefore, the solution is obtained by iterative procedures through assuming initial displacement and finding the resulting force until the force converges to the yield force.

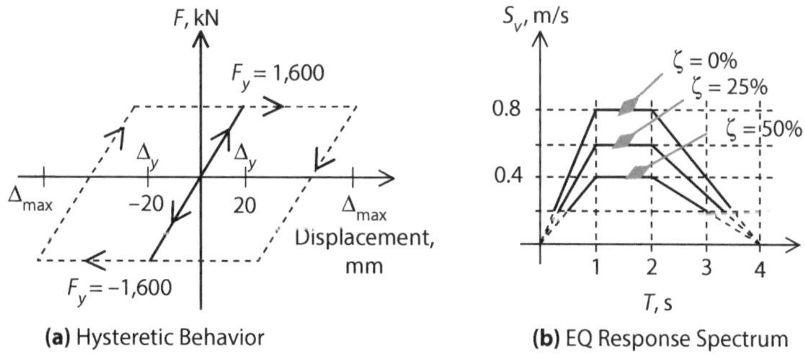

Example 4-4, Figure 2

Trial 1

Starting with $\Delta_{max} = \Delta_e = 0.062$ m

$$k_{eq} = \frac{F_y}{\Delta_{max}} = \frac{1{,}600}{0.062} = 25{,}806 \text{ kN/m}$$

$$\omega_{eq} = \sqrt{\frac{k_{eq}}{m}} = \sqrt{\frac{25{,}806}{1{,}000}} = 5.080 \text{ rad/s}$$

$$T_{eq} = \frac{2\pi}{\omega_{eq}} = \frac{2\pi}{5.080} = 1.24 \text{ s}$$

Equivalent damping

$$\zeta_{eq} = \frac{1}{2\pi}\frac{ED}{U_{e,max}}$$

where

$$ED = \text{area of loop} = 2\,F_y(2\Delta_{max} - 2\Delta_y)$$

$$= 2(1{,}600)\{2(0.062) - 2(0.02)\} = 268.8 \text{ kN·m}$$

$$U_{e,max} = F_{max}\Delta_{max} = 1{,}600(0.062) = 99.2 \text{ kN·m}$$

hence,

$$\zeta_{eq} = \frac{1}{2\pi}\frac{268.8}{99.2} = 0.43 = 43\%$$

with $T_{eq} = 1.24$ s and $\zeta_{eq} = 43\%$, spectral velocity is given in Example 4-4, Fig. 1(c), as $S_v \approx 0.45$ m/s. As a result,

$$S_d = \frac{S_v}{\omega_{eq}} = \frac{0.45}{5.08} = 0.089 \text{ m}$$

$$S_a = \omega_{eq}\,S_v = 5.08(0.45) = 2.286 \text{ m/s}^2$$

$$\Delta_{max} = S_d = 0.089 \text{ m}$$

$$F_{max} = mS_a = 1{,}000(2.286) = 2{,}286 \text{ kN}$$

Trial 2

Use $\Delta_{max} = 0.1$ m

$$k_{eq} = \frac{F_y}{\Delta_{max}} = \frac{1{,}600}{0.10} = 16{,}000 \text{ kN/m}$$

$$\omega_{eq} = \sqrt{\frac{k_{eq}}{m}} = \sqrt{\frac{16{,}000}{1{,}000}} = 4 \text{ rad/s}$$

$$T_{eq} = \frac{2\pi}{\omega_{eq}} = \frac{2\pi}{4} = 1.57 \text{ s}$$

Equivalent damping

$$\zeta_{eq} = \frac{1}{2\pi}\frac{ED}{U_{e,max}}$$

where

$$\text{ED} = \text{area of loop} = 2F_y(2\Delta_{max} - 2\Delta_y)$$
$$= 2(1{,}600)\{2(0.1) - 2(0.02)\} = 512 \text{ kN·m}$$
$$U_{e,max} = F_{max}\,\Delta_{max} = 1{,}600(0.1) = 160 \text{ kN·m}$$

hence,

$$\zeta_{eq} = \frac{1}{2\pi}\frac{512}{160} = 0.51 = 51\%$$

with $T_{eq} = 1.57$ s and $\zeta_{eq} = 51\%$, spectral velocity is given in Example 4-4, Fig. 1(c), as $S_v \approx 0.4$ m/s. As a result,

$$S_d = \frac{S_v}{\omega_{eq}} = \frac{0.45}{4} = 0.1 \text{ m}$$
$$S_a = \omega_{eq}\,S_v = 4\,(0.4) = 1.6 \text{ m/s}^2$$
$$\Delta_{max} = S_d = 0.1 \text{ m}$$
$$F_{max} = mS_a = 1{,}000\,(1.6) = 1{,}600 \text{ kN}$$

Because the resulting force is 1,600 kN, which coincides with the yield force of the structure, the correct solution is reached, and the final result is

$$\Delta_{max} = 0.1 \text{ m}$$

4.6 Problems

4-1 A frame with a diagonal viscous damper is shown in Prob. 4-1, Fig. 1. The frame members have uniform mass distribution and constant cross section. The frame is intended to behave inelastically under the excitation of its supports by a horizontal component of an earthquake profile along line A-B as shown in Prob. 4-1, Fig. 2. The frame members have the following properties:

$$m = 100 \text{ kN·s}^2\text{/m/m}$$
$$E = 25 \times 10^6 \text{ kN/m}^2$$
$$A = 0.24 \text{ m}^2$$
$$I = 0.0128 \text{ m}^4$$
$$M_y = 1{,}500 \text{ kN·m}$$

The damper has the following coefficient of viscous damping:

$$c = 220 \text{ kN·s/m}$$

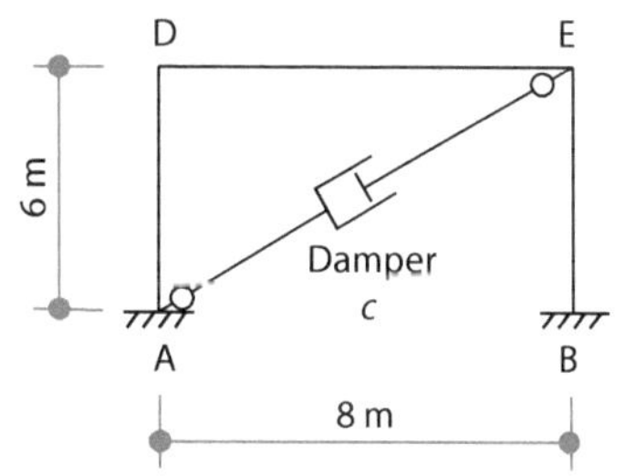

Problem 4-1, Figure 1 Frame layout.

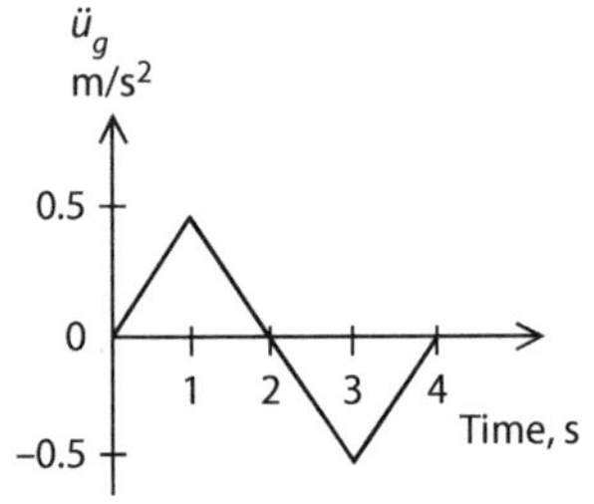

Problem 4-1, Figure 2 Accelerogram.

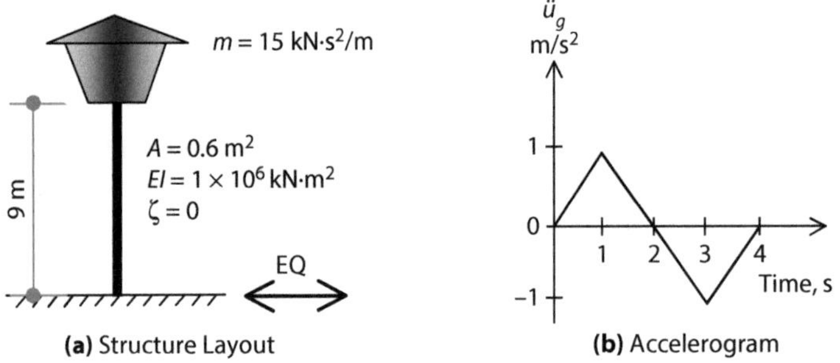

Problem 4-2, Figure 1

Use the central differences method and the consistent mass matrix to find the horizontal displacement history. For calculation purposes, take the time interval, $\Delta t = 0.1$ s.

4-2 An elevated tank is supported by a single column as shown in Prob. 4-2, Fig. 1(a). The structure is subjected to an earthquake as shown in Prob. 4-2, Fig. 1(b). Find the displacement history of the horizontal displacement of the tank in the two following cases:

1. Ignore the effect of the tank mass on the stiffness of the column (use first-order analysis).
2. Consider the effect of the tank mass on the stiffness of the column (use second-order analysis). The geometric stiffness matrix is given in Sec. 3.9.2.

CHAPTER 5

Behavior of Structures under Seismic Excitation

5.1 Introduction

Previous chapters have shown that structures can be designed and constructed with much less strength than their elastic strength demand in order to reduce the extremely high cost associated with elastic response. This reduction imposes additional structural requirements to the traditional structural requirements of ductility, energy dissipation, and self-centering capacity. Such requirements can be evaluated and quantified using explicit inelastic dynamic analysis.

Ideally, one would use simplified rules and procedures when doing an explicit inelastic dynamic analysis on account of the complex, time-consuming nature of this analysis, and the need for practical and relatively quick solutions to design problems in practice.

Such simplifications do exist in earthquake engineering in terms of seismic structural properties and force-reduction parameters. Figure 5-1 illustrates this concept by showing the inelastic response of a structure to earthquake excitation. If the structure is designed to remain elastic during excitation, the structure will experience an induced elastic force, V_e, and inelastic displacement Δ_e. However, experience from previous actual earthquakes and inelastic dynamic analysis shows that the structure can be designed with strength less than the elastic demand, V_y, resulting in maximum inelastic displacement, Δ_M. Because Δ_M is larger than the yield displacement, Δ_y, there will be a trade off between large force and large displacement.

For these reasons, let us consider the following necessary quantities and expressions before proceeding with this concept.

5.1.1 Force-Reduction Factor, *R*

The force-reduction factor is defined as the ratio of the elastic strength demand to the actual yield level of the structure. With reference to Fig. 5-2, *R* may be expressed as

$$R = \frac{F_e}{F_y}$$

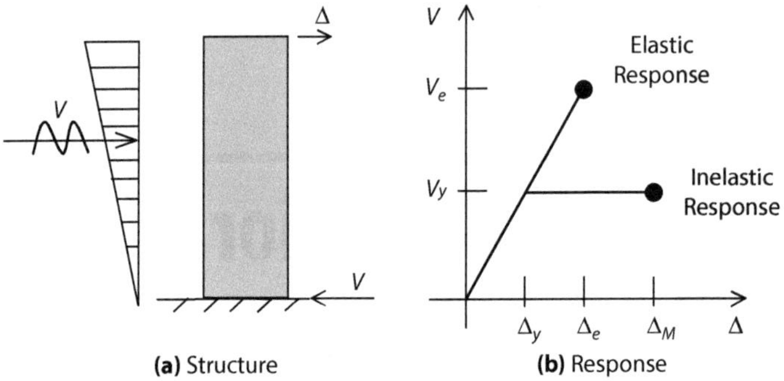

Figure 5-1 Elastic versus inelastic response of structures.

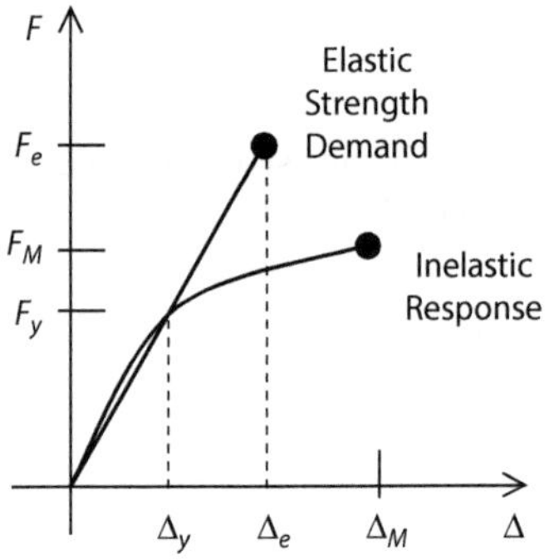

Figure 5-2 Force-reduction factor.

It should be noted that R may also be expressed in terms of displacements as

$$R = \frac{\Delta_e}{\Delta_y}$$

where F_e and Δ_e are the maximum force and maximum displacement due to the linear elastic response of the structure during the entire time of excitation by the earthquake.

F_y and Δ_y are the force and displacement at first significant yield as shown in Fig. 5-2. Because the inelastic response of actual structures is not well defined by straight line segments, as in the ideal case, the first significant yield is not necessarily the formation of the first plastic hinge in the structure. Figure 5-3 provides various definitions for the first significant yield, namely, the first yielding point of the structure as shown in Fig. 5-3(a), the intersection of the initial tangent and the collapse load as shown in Fig. 5-3(b), or equal energy absorption as shown in Fig. 5-3(c).

5.1.2 Ductility

Ductility is the capacity of the structure to undergo large deformations without any significant reduction in strength. Figure 5-4 shows how the ductility of structures may be described as high, low, or even brittle. Ductility is usually expressed as a ductility ratio of the maximum deformation under consideration to the actual yield deformation.

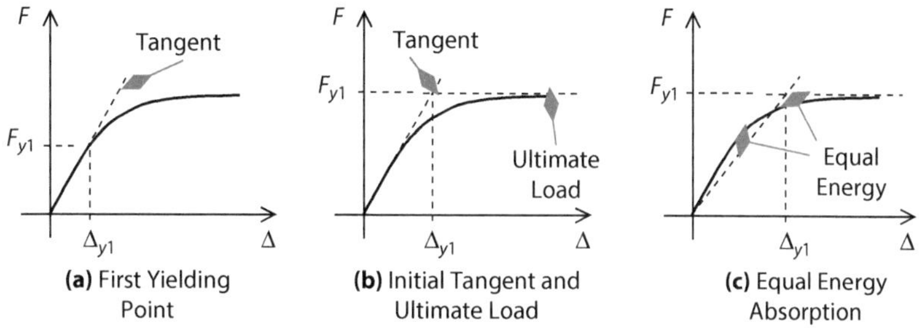

FIGURE 5-3 Definition of first significant yield.

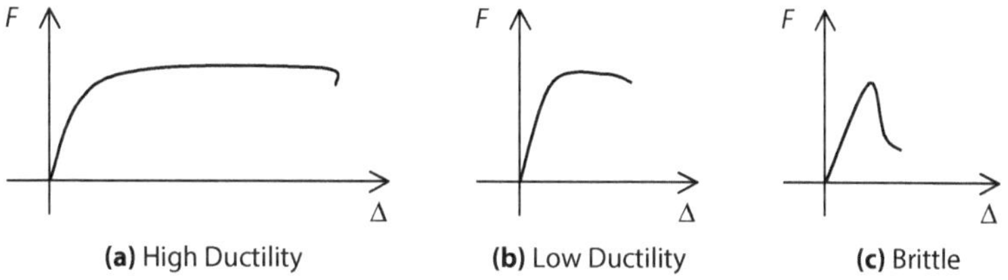

FIGURE 5-4 Types of ductilities.

Deformation under consideration may include top displacement, plastic hinge rotation, curvature, and cyclic inelastic displacement. Ductility ratios for different structural elements are defined below.

For buildings, ductility is expressed in terms of the displacement at the top of the building, as shown in Fig. 5-5, and is defined as the top displacement ductility ratio, μ_Δ

$$\mu_\Delta = \frac{\Delta}{\Delta_y}$$

For beam elements, the ductility may be expressed in terms of rotation at specified locations along the axis of the beam. This is usually expressed as a rotational ductility ratio, μ_θ

$$\mu_\theta = \frac{\theta}{\theta_y}$$

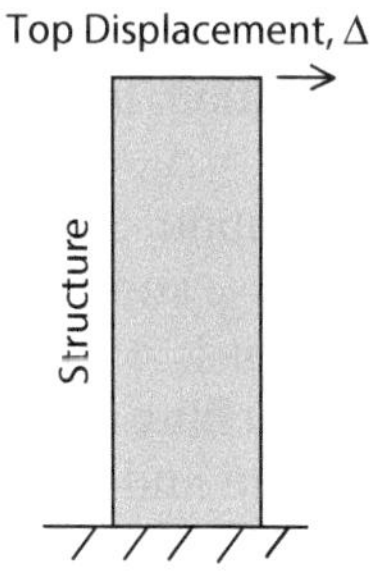

FIGURE 5-5 Top displacement ductility.

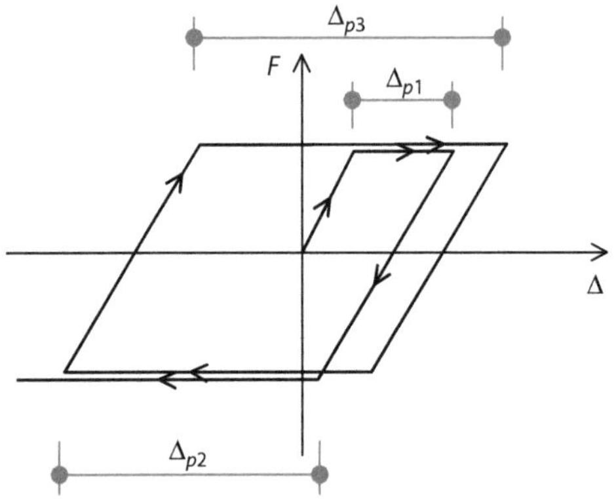

FIGURE 5-6 Cyclic ductility.

The ductility of beam elements may also be expressed in terms of the curvature, κ, of sections along the axis of the beam. This is usually given as a curvature ductility ratio, μ_κ

$$\mu_\kappa = \frac{\kappa}{\kappa_y}$$

For cyclic behavior, the ductility may be expressed as an accumulation of plastic deformations. For example, the cyclic ductility ratio, μ_c, for the model in Fig. 5-6 is expressed as

$$\mu_c = \frac{\Delta_y + \Sigma|\Delta_{\mathrm{pi}}|}{\Delta_y}$$

For each ductility ratio defined above, two sets of ductility ratios are specified, namely, ductility capacity and ductility demand. Ductility capacity is the maximum usable ductility of the structure (the top displacement ductility capacity, $\mu_{\Delta c}$) and is expressed as

$$\mu_{\Delta c} = \frac{\Delta_u}{\Delta_y}$$

where Δ_u is the maximum usable top displacement of the structure.

Similarly, there is rotational ductility capacity, $\mu_{\theta c}$; curvature ductility capacity, $\mu_{\kappa c}$; and so forth.

The second set of ductility ratios, ductility demand, is the maximum ductility that is reached during earthquake excitation. In other words, ductility demand is the top displacement ductility demand ($\mu_{\Delta d} = \Delta_M/\Delta_y$), where Δ_M is the maximum displacement reached during earthquake excitation. Similarly, there is rotational ductility demand, $\mu_{\theta d}$; curvature ductility demand, $\mu_{\kappa d}$; and so forth.

Because the top displacement is a global quantity, it may be referred to as the global ductility; hence, global ductility demand, μ_d, and global ductility capacity, μ_c, are defined. Conversely, curvature and rotation are considered local ductility. Thus, they are referred to as local ductility; hence, local ductility demand, μ_{ld}, and local ductility capacity, μ_{lc}, are defined.

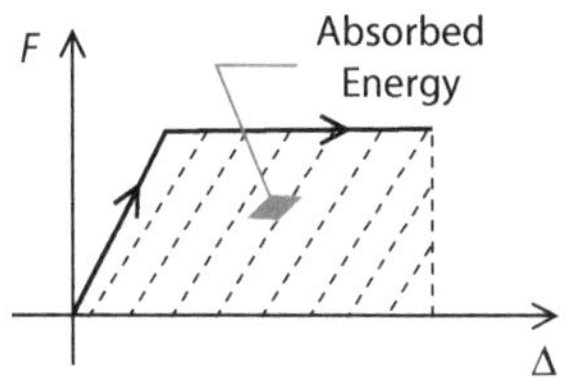

FIGURE 5-7 Energy absorption.

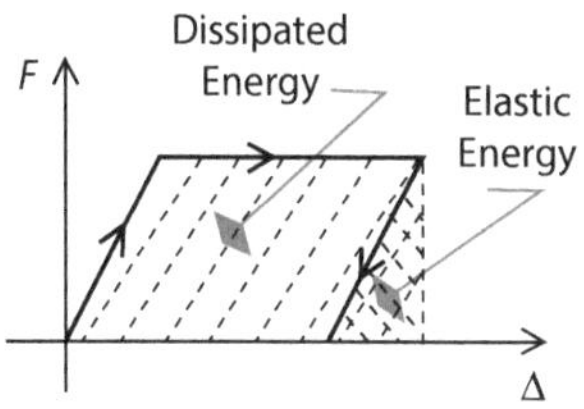

FIGURE 5-8 Energy dissipation.

The global ductility ratio is generally a function of the local ductility ratio, as will be shown later in this chapter.

5.1.3 Energy Dissipation Capacity

The energy dissipation capacity of the structure is its capacity to dissipate portions of the absorbed energy. In this context, it is important to distinguish between energy absorption and energy dissipation. Energy absorption is the total energy imparted to the structure during excitation as shown in Fig. 5-7. Energy dissipation is the energy dissipated in the structure in any form. For example, the plastic deformations shown in Fig. 5-8 dissipate energy in the form of heat. A portion of the absorbed energy is not dissipated in the structure. In Fig. 5-8, the elastic energy is converted back into kinetic energy, causing the structure to swing in the other direction during vibration.

The energy dissipation in these figures is characterized by the size of the hysteresis loop, and the loop size is proportional to hysteretic damping. In addition, the energy dissipation capacity of the structure is different from its ductility. The energy dissipation may be characterized as low or high, regardless of the ductility of the structure. This difference is illustrated in Fig. 5-9, which shows systems of low and high energy dissipation capacities for the same ductility ratio of the structure. Elastic systems exhibit zero energy dissipation without formation of hysteresis loops as shown in Fig. 5-9(a).

5.1.4 Self-Centering Capacity

Self-centering capacity is the capacity of an inelastic structure to return to its original position when subjected to vibration in the inelastic range as shown in Fig. 5-10(a). If the structure does not return to its original position, the system does not have self-centering capacity, as shown in Fig. 5-10(b). Instead, the system exhibits permanent plastic deformation known as plastic drift. Analysis and experience have shown that structures with

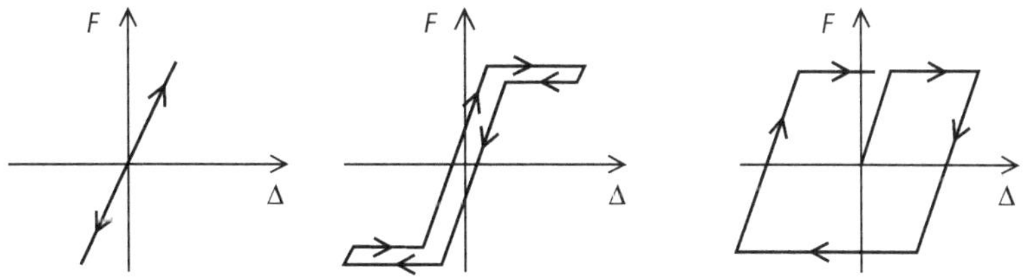

(a) Elastic System **(b)** Low Energy Dissipation **(c)** High Energy Dissipation

FIGURE 5-9 Measures of energy dissipations.

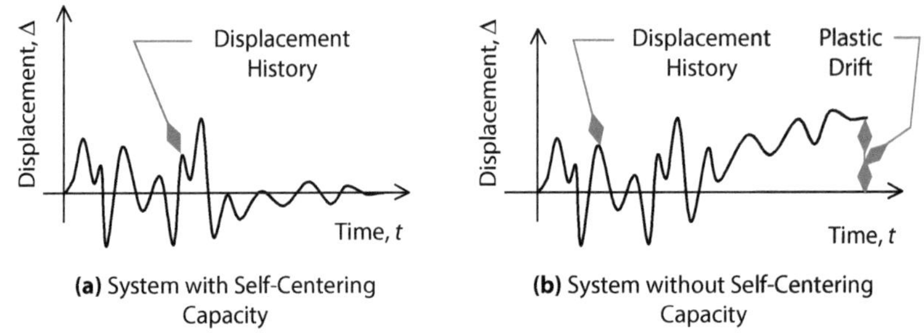

Figure 5-10 Self-centering phenomenon.

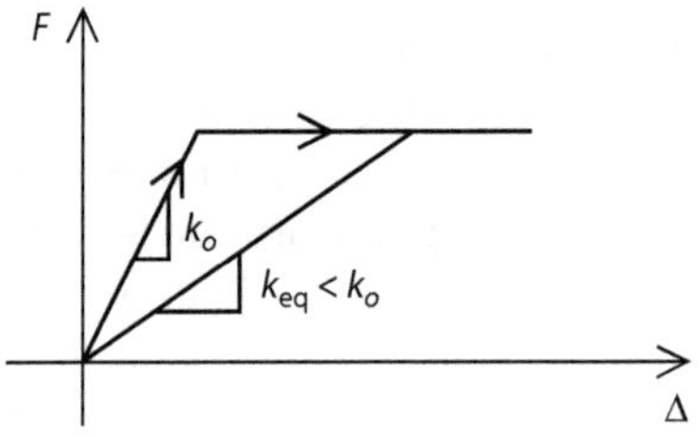

Figure 5-11 Frequency shift.

post-yielding stiffness around 10 percent or higher than their initial elastic stiffness exhibit self-centering capacity.

5.1.5 Frequency Shift

Frequency shift is a property of inelastic systems that is related to the change of stiffness due to yielding of the structure. Because the frequency is proportional to the square root of the stiffness ($\omega_{eq} = \sqrt{k_{eq}/m}$), any yielding of the system causes changes in the stiffness, as shown in Fig. 5-11. This results in a change in the frequency and in the period. Remember that attracted forces from earthquakes are a function of the structural period (as discussed in the section on response spectrum analysis). Therefore, the force and displacement will interact with this frequency shift during excitation.

General Note

In earthquake engineering, ductility is a directly used quantity in design. For example, the force-reduction factor is usually expressed in terms of ductility demand (as will be shown in the following sections). The other properties, such as energy dissipation, self-centering capacity, and frequency shift, are only implicitly considered in the relationships of force reduction even though they have a large effect on such relations.

5.2 Relationship between Force Reduction and Ductility Demand

Because of the extreme randomness and uncertainty of earthquakes, the relationship between the force-reduction factor, R, and the global ductility demand, μ_d, can only be established through statistical or probabilistic means. Experience and inelastic

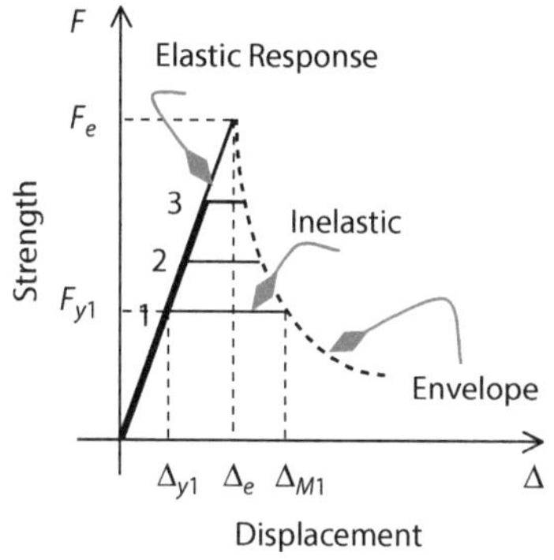

FIGURE 5-12 Structural response.

dynamic analysis indicate that the ductility demand is a function of the amount of force reduction. As a general trend, ductility demand increases as the force reduction increases as shown in Fig. 5-12. This figure shows that a reduction of the actual strength of the structure from elastic strength demand, F_e, to level, F_{y1}, will result in the maximum inelastic displacement of Δ_{M1}. Each level of reduction in the figure, that is, to level 2 or 3, for example, requires ductility demand associated with that level as marked by the envelope line. The envelope also indicates that large reduction in the strength requires larger ductility demand from the structure.

Newmark has proposed two approximate rule of thumb relationships between R and μ_d, based on the structural period. For long period structures, the maximum elastic and maximum inelastic displacements remain within the same range. For short period structures, the maximum elastic energy and the maximum inelastic energy absorbed by the structure remain within the same range. Accordingly, Newmark has proposed the following two famous relationships:

1. Equal displacement criterion (EDC) for long period structures
2. Equal energy criterion (EEC) for short period structures

Nassar and Krawinkler propose a more general relationship. These criteria will be explained in the following sections.

5.2.1 Equal Displacement Criterion

As mentioned earlier, in equal displacement criterion (EDC), the maximum elastic displacement and the maximum inelastic displacements remain the same during the excitation. This can be seen in Fig. 5-13. Because of similarity in triangles, O-F_y-Δ_y and O-F_e-Δ_M, the following relationships are deduced.
Since

$$\Delta_e = \Lambda_M$$

and

$$R = \frac{F_e}{F_y} = \frac{\Delta_e}{\Delta_y} \tag{5-1}$$

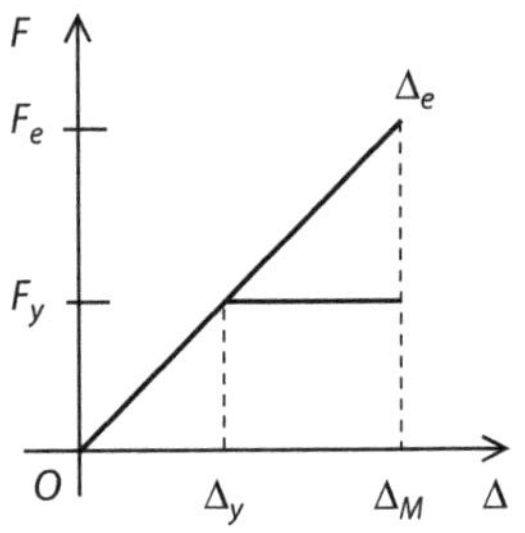

FIGURE 5-13 Equal displacement criterion.

and

$$\mu_d = \frac{\Delta_M}{\Delta_y} = \frac{\Delta_e}{\Delta_y} \tag{5-2}$$

by equating Eqs. (5-1) and (5-2), the relationship between R and μ_d becomes

$$R = \mu_d$$

This relation is only valid for long period structures. The long period term is only a relative quantity associated with the dominant period of the earthquake.

5.2.2 Equal Energy Criterion

As mentioned earlier, in equal energy criterion (EEC), the maximum elastic absorbed energy and the maximum inelastic absorbed energy remain the same during the excitation. This can be seen in Fig. 5-14 due to equality of areas, O-F_e-Δ_e and O-F_y-F_y-Δ_M. For example, A_1 and A_2 yields
since

$$A_1 = A_2$$

and

$$R = \frac{F_e}{F_y} = \frac{\Delta_e}{\Delta_y}$$

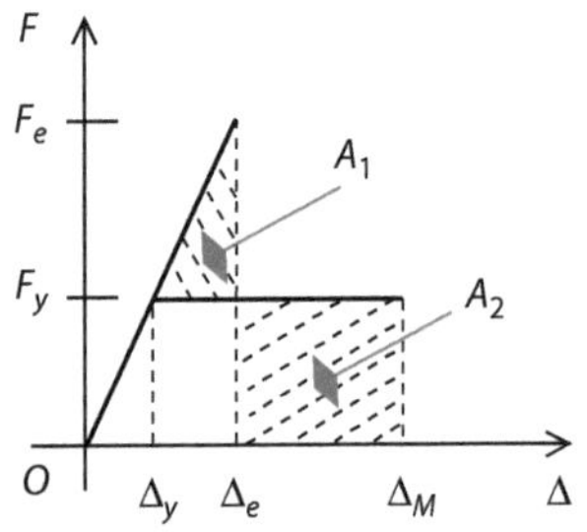

FIGURE 5-14 Equal energy criterion.

and

$$\mu_d = \frac{\Delta_M}{\Delta_y}$$

by equating area A_1 and A_2, the relationship between R and μ_d can be shown to be

$$R = \sqrt{2\mu_d - 1}$$

This relationship is only valid for short period structures. The short period is only a relative quantity associated with the dominant period of the earthquake.

5.2.3 General Relationship between R and μ_d

All relationships between the force-reduction factor and the global ductility demand are based on statistical measures. This section illustrates the need for statistical evaluation by presenting and discussing the inelastic response of a single degree of freedom system to nine earthquake records. The parametric study is performed using inelastic dynamic analysis with variation of the period and the yielding level of a bilinear hysteresis model. The parameter variation includes six periods and six levels of yielding of the hysteresis model for each earthquake record that results in 324 pairs of R and μ_d values. The bilinear model is selected with 10 percent post-yielding stiffness.

Figure 5-15 shows the results of this parametric study. The response is extremely scattered, and ductility demand increases tremendously for short periods. This structural behavior is considered the norm in the response of structures to earthquakes due to the extreme randomness in earthquakes, as noted in previous chapters.

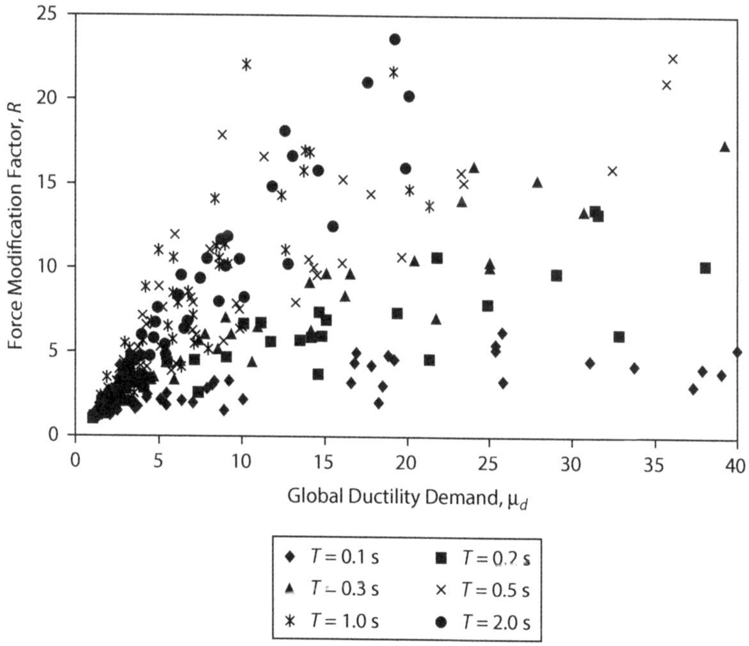

Figure 5-15 Response to nine actual earthquakes.

For these reasons, Nassar and Krawinkler have conducted a statistical evaluation of the response of different hysteresis models to 16 actual earthquake records. Nassar and Krawinkler's findings were consistent with the general trend. They concluded that the relationship between R and μ_d can only be evaluated by statistical means and proposed the following expression to describe this relationship:

$$R = [C(\mu_d - 1) + 1]^{1/C}$$

where C is given as

$$C = \frac{T^a}{1+T^a} + \frac{b}{T}$$

For a bilinear model with 10 percent post-yielding stiffness, Nassar and Krawinkler used nonlinear regression analysis to produce values of $a = 0.8$ and $b = 0.29$. Figure 5-16 uses these values to show a plot of the parameter C versus the period T. This figure also shows results from the response of the elastoplastic model ($p = 0$).

Both the Newmark criteria of equal displacement and equal energy are embedded in the above relationship as follows.
When

$$C = 1 \rightarrow \quad R = \mu_d$$

and when

$$C = 2 \rightarrow \quad R = \sqrt{2\mu_d - 1}$$

The parameter C can also be viewed as an indication for ductility demand. For example, if $C = 1$, the relationship tends to the EDC where ductility demand is equal to the force-reduction value. However, if $C = 2$, the relationship tends to the EEC where ductility demand exceeds the force-reduction value. Therefore, higher values of C indicate higher ductility demand which increases by shortening the structural period as clearly shown in Fig. 5-16.

In conclusion, the structural response to earthquakes is extremely random. The data points are extremely scattered. For these reasons, there is always a need

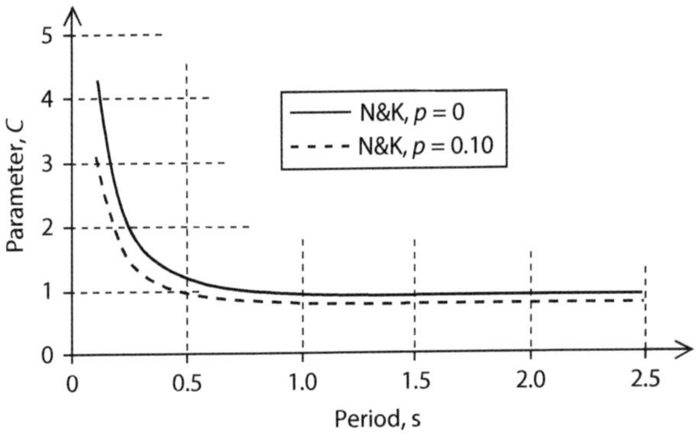

FIGURE 5-16 Relationship between parameter C and structural period.

for statistical evaluation of the structural demand. In addition, seismic codes are generally cautious when they address this issue (the high uncertainty with the structural demand). As a result, the assignment of R and μ_d in seismic codes is based in part on judgment, experiments, and experience from previous earthquakes and from practice. The codes always emphasize that the given requirements are minimum requirements.

In general, the EDC and the EEC are still the most popular in practice. The EEC may be viewed as an upper bound criterion because it requires more ductility demand than the EDC.

Example 5-1 Example 5-1, Fig. 1, shows a short period building that is excited by an earthquake. Inelastic dynamic analysis results in the hysteretic behavior shown in this figure, which also shows one *asymmetric* hysteresis loop at the maximum response.

1. Calculate the hysteretic damping in terms of equivalent viscous damping.
2. Calculate the global ductility demand for this building.
3. Calculate the elastic design force and the maximum elastic displacement as if this building is required to be constructed as an essential facility to remain elastic during the earthquake given in the figure.

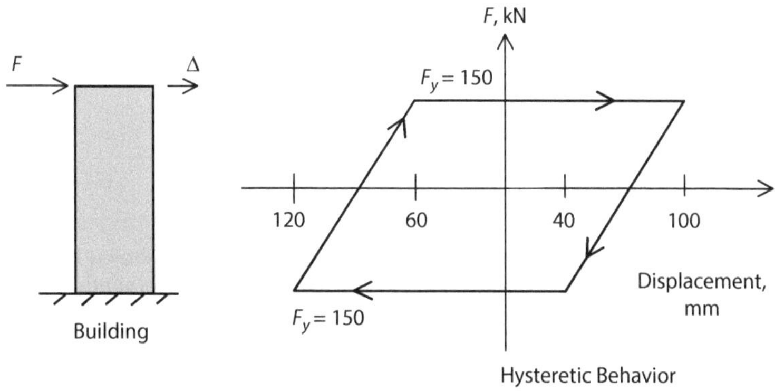

EXAMPLE 5-1, FIGURE 1

Solution

Part (1)

Equivalent viscous damping (see Chap. 4)

$$\zeta_{eq} = \frac{1}{2\pi}\frac{ED}{U_{e,\max}}$$

where $ED = (\text{area of loop}) = (100+60)(150)(2) = 48{,}000\ \text{kN}\cdot\text{mm}$

$$U_{e,\max} = F_{\max}\,\Delta_{\max}$$

$$= 0.5\{(150)(100)+(150)(120)\} = 16{,}500\ \text{kN}\cdot\text{mm}$$

$$\zeta_{eq} = \frac{1}{2\pi}\frac{48{,}000}{16{,}500} = 0.46 = 46\%$$

Part (2)
Global ductility demand

Initial stiffness: $$k_o = \frac{F_y}{\Delta_e} = \frac{2F_y}{2\Delta_e}$$

$$= \frac{2(150)}{100-40} = 5 \text{ kN/mm}$$

Yield displacement: $$\Delta_y = \frac{F_y}{k_o} = \frac{150}{5} = 30 \text{ mm}$$

Maximum displacement: Δ_M (from Example 5-1, Fig. 1) = 120 mm

Therefore, ductility demand: $$\mu_d = \frac{\Delta_M}{\Delta_y} = \frac{120}{30} = 4$$

Part (3)
Elastic strength demand
Because the building is of the short period type, use EEC for $R - \mu_d$ relationship

$$R = \sqrt{2\mu_d - 1} = \sqrt{2(4)-1} = 2.65$$

Since

$$R = \frac{F_e}{F_y} \rightarrow \text{or,} \qquad 2.65 = \frac{F_e}{150}$$

required elastic force: $F_e = 397$ kN

and elastic displacement: $$\Delta_e = \frac{F_e}{k} = \frac{397}{5} = 79.4 \text{ mm}$$

Example 5-2 The smooth stiffness-degrading (SSD) hysteresis shown in Example 5-2, Fig. 1, is subjected to an excitation based on the record of an earthquake in southern California (EQ15.stn). Example 5-2, Table 1 below summarizes the results from inelastic dynamic analysis in consistent units. Use the following steps to analyze these results.

1. Calculate the force-reduction factor, R, and the global ductility demand, μ_d, for each case in the table.

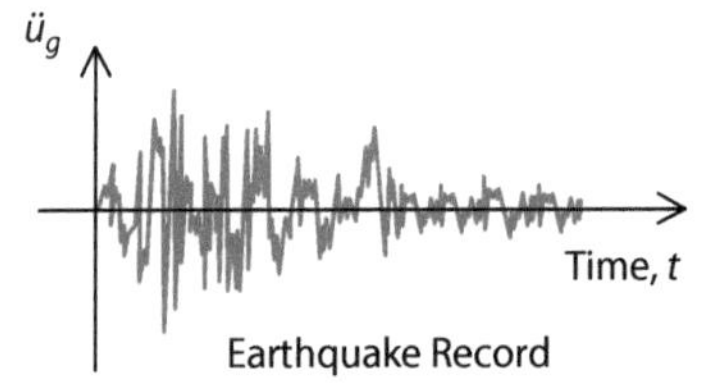

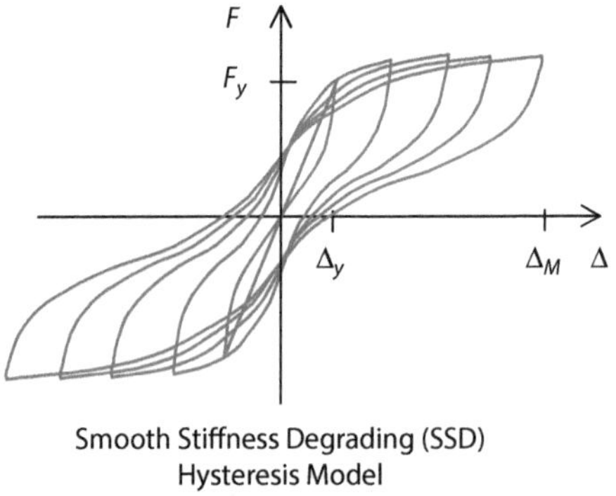

Example 5-2, Figure 1

Case No.	T (s)	F_y	Δ_y	F_e	Δ_M	Case No.	T (s)	F_y	Δ_y	F_e	Δ_M
1	0.1	800	0.66	800	0.66	19	0.5	30	0.68	30	0.68
2	0.1	800	0.66	1030	0.58	20	0.5	30	0.68	60	1.36
3	0.1	800	0.66	1260	1.02	21	0.5	30	0.68	90	1.95
4	0.1	800	0.66	1490	1.36	22	0.5	30	0.68	120	2.49
5	0.1	800	0.66	1720	1.61	23	0.5	30	0.68	150	2.90
6	0.1	800	0.66	1950	3.42	24	0.5	30	0.68	180	2.96
7	0.2	180	0.64	180	0.64	25	1	25	0.64	25	0.64
8	0.2	180	0.64	252	0.68	26	1	25	0.64	56	1.39
9	0.2	180	0.64	324	1.10	27	1	25	0.64	87	1.85
10	0.2	180	0.64	396	1.74	28	1	25	0.64	118	2.65
11	0.2	180	0.64	468	2.22	29	1	25	0.64	149	2.50
12	0.2	180	0.64	540	3.85	30	1	25	0.64	180	3.11
13	0.3	90	0.71	90	0.71	31	2	20	0.69	20	0.69
14	0.3	90	0.71	124	1.01	32	2	20	0.69	32	1.05
15	0.3	90	0.71	158	1.32	33	2	20	0.69	44	1.41
16	0.3	90	0.71	192	2.42	34	2	20	0.69	56	1.59
17	0.3	90	0.71	226	2.98	35	2	20	0.69	68	2.23
18	0.3	90	0.71	260	3.53	36	2	20	0.69	90	3.34

Example 5-2, Table 1 Inelastic Dynamic Analysis Results

2. Plot all data points obtained from (1) as discrete points, using μ_d as an abscissa and R as an ordinate.
3. Use the chart from (2) to plot the Newmark rule of thumb relationships between R and μ_d (namely, the equal displacement criterion, EDC, and equal energy criterion, EEC, on the same chart).

Solution

Part (1)

Consider Case 6 in Example 5-2, Table 1.

Force-reduction factor:
$$R = \frac{F_e}{F_y} = \frac{1,950}{800} = 2.44$$

Global ductility demand:
$$\mu_d \frac{\Delta_M}{\Delta_y} = \frac{3.42}{0.66} = 5.18$$

The rest of the values are calculated in a similar way and are provided in Example 5-2, Table 2.

Parts (2) and (3)

Example 5-2, Fig. 2, is a graphical presentation of the results in Example 5-2, Table 2. The figure shows that the results are generally random and scattered. Note the wide spread of results at large ductility demand. In addition, the response of all short period cases is close to the EEC. This implies higher ductility demand for structures with short periods than for those with long periods.

Case No.	T (s)	F_y	F_e	Δ_y	Δ_M	μ_d	R
1	0.1	800	800	0.66	0.66	1.00	1.00
2	0.1	800	1030	0.66	0.58	0.88	1.29
3	0.1	800	1260	0.66	1.02	1.55	1.58
4	0.1	800	1490	0.66	1.36	2.05	1.86
5	0.1	800	1720	0.66	1.61	2.44	2.15
6	0.1	800	1950	0.66	3.42	5.18	2.44
7	0.2	180	180	0.64	0.64	1.00	1.00
8	0.2	180	252	0.64	0.68	1.07	1.40
9	0.2	180	324	0.64	1.10	1.73	1.80
10	0.2	180	396	0.64	1.74	2.74	2.20
11	0.2	180	468	0.64	2.22	3.49	2.60
12	0.2	180	540	0.64	3.85	6.06	3.00
13	0.3	90	90	0.71	0.71	1.00	1.00
14	0.3	90	124	0.71	1.01	1.44	1.38
15	0.3	90	158	0.71	1.32	1.87	1.76
16	0.3	90	192	0.71	2.42	3.42	2.13
17	0.3	90	226	0.71	2.98	4.22	2.51
18	0.3	90	260	0.71	3.53	5.00	2.89
19	0.5	30	30	0.68	0.68	1.00	1.00
20	0.5	30	60	0.68	1.36	2.00	2.00
21	0.5	30	90	0.68	1.95	2.88	3.00
22	0.5	30	120	0.68	2.49	3.67	4.00
23	0.5	30	150	0.68	2.90	4.27	5.00
24	0.5	30	180	0.68	2.96	4.35	6.00
25	1	25	25	0.64	0.64	1.00	1.00
26	1	25	56	0.64	1.39	2.17	2.24
27	1	25	87	0.64	1.85	2.89	3.48
28	1	25	118	0.64	2.65	4.14	4.72
29	1	25	149	0.64	2.50	3.90	5.96
30	1	25	180	0.64	3.11	4.86	7.20
31	2	20	20	0.69	0.69	1.00	1.00
32	2	20	32	0.69	1.05	1.52	1.60
33	2	20	44	0.69	1.41	2.05	2.20
34	2	20	56	0.69	1.59	2.31	2.80
35	2	20	68	0.69	2.23	3.24	3.40
36	2	20	80	0.69	3.34	4.86	4.00

Example 5-2, Table 2 Calculated R and μ_d Values

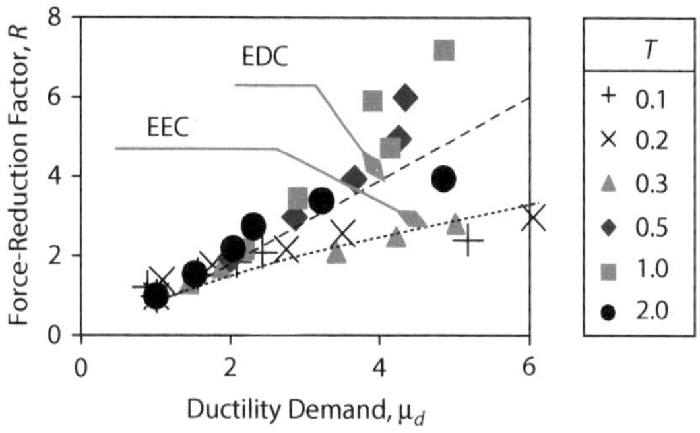

EXAMPLE 5-2, FIGURE 2

5.3 Relationship between Global Ductility and Local Ductility

As discussed in previous sections, reduction of the elastic strength demand by an *R*-value imposes ductility demand on the structure, μ_d. For safe design, the structure must possess enough ductility capacity to equal or exceed the imposed ductility demand. In terms of global ductility, this condition is given as

$$\mu_c \geq \mu_d \qquad \text{(as safe design condition)}$$

The global ductility capacity, μ_c, is a function of local ductility capacity, μ_c, which in turn is a function of local section properties.

In structural engineering, the deformations of frame structures are generally dominated by flexure. Therefore, ductility is usually expressed in terms of flexure deformation (curvature, κ)

$$\mu_c = \mu_\kappa = \frac{\kappa_u}{\kappa_y}$$

Because shear behavior is brittle, shear failure must be avoided in seismic design. Instead, the ductility is provided through flexure behavior. Global and local ductility are related by the laws of the theory of structures. By knowing forces and moments, the rotations and curvatures can be found, and deflection can be evaluated.

For example, Fig. 5-17 shows a vertical elastic steel column that is subjected to a horizontal force, *H*. The moment and curvature diagrams for this system are drawn in Fig. 5-17. For a linear elastic system, the curvature will be directly proportional to the moment ($\kappa = M/EI$). Knowing curvature, the top displacement deflection can be evaluated, for example, by integrating the curvature over the height of the column. Structural theory methods such as conjugate beam, virtual work, and others will also be valid.

As noted earlier, seismic design requires the calculation of ductility ratios in the inelastic stage. As a result, deformations are required in the inelastic range. For example, if the steel column in Fig. 5-17 is loaded beyond its elastic limits, the top displacement at first yield and at ultimate state can be found using the same principle as in the elastic case, that is, by integrating the curvature at first yield and then at ultimate state.

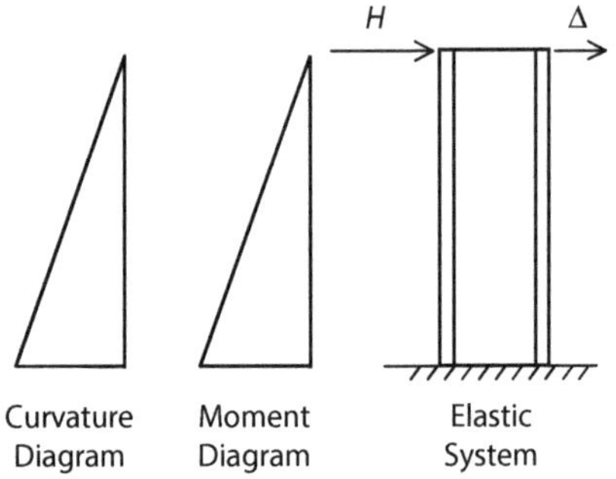

FIGURE 5-17 Linear elastic system.

The difference from the elastic case will be in the relationship between the moment and the curvature, which is nonlinear.

Figure 5-18(a) illustrates the column section properties in terms of the M–κ relationship. The moment at first yield is drawn in Fig. 5-18(b). The corresponding curvature distribution will also be linear insofar as the yield state is not exceeded. Thus, the behavior will be identical to the linear elastic case. The displacement at first yield is calculated by integration of the curvature given in Fig. 5-18(b).

The moment at ultimate state is shown in Fig. 5-18(c). This will also be linear due to the given loading condition. However, the curvature will not be linear as in the first yield case. The curvature, in this case, must be found from the moment-curvature relationship given in Fig. 5-18(a), which by inspection, yields the curvature distribution shown in Fig. 5-18(c). Finally, the top displacement at ultimate state is simply calculated by integration of the curvature distribution given in Fig. 5-18(c).

An empirical expression is developed for the length of the plastic hinge region for reinforced concrete beams. For Fig. 5-19, this expression is

$$\ell_p = 0.5\,d + 0.05\,Z$$

where d = effective depth of the section
Z = distance from critical section to the point of contraflexure
Z = H, for cantilevers

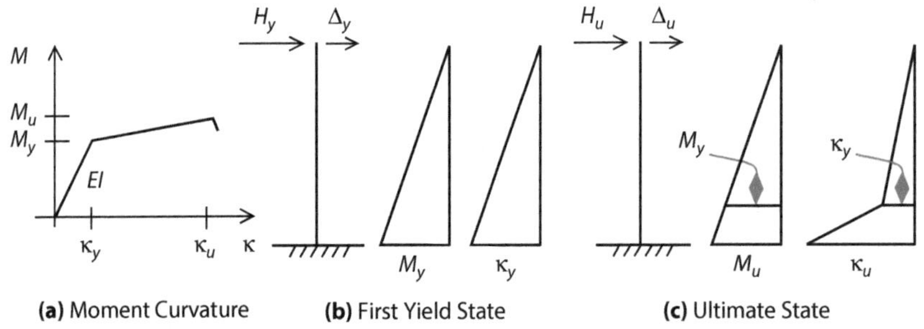

FIGURE 5-18 Inelastic system.

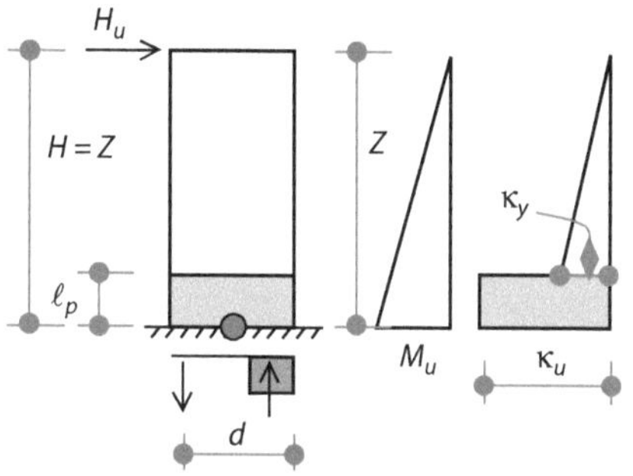

FIGURE 5-19 Equivalent plastic hinge.

5.4 Local Ductility Capacity

For ultimate curvature, local ductility is evaluated by the laws of structural mechanics using equilibrium, compatibility (strain compatibility), and constitutive relationships (material properties). Note that local ductility is highly affected by axial loads. In general, axial loads reduce the ductility of the cross section.

An understanding of how material properties affect ductility capacity is essential because such properties depend on loading schemes and responding modes. For example, the response of concrete in the flexure mode is different from its response in the shear mode. To evaluate material properties, the loading schemes may be classified in terms of state of stresses and in terms of loading histories. Although the state of stress affects concrete behavior as a material, loading history affects both concrete and steel behavior at the material level.

In general, the behavior of the material and the corresponding evaluation of ductility capacity may be classified according to loading schemes as monotonic and cyclic behavior. The evaluation of ductility capacity for the monotonic case and the cyclic case will be addressed in separate sections below.

The evaluation of cyclic ductility capacity is more complicated than monotonic ductility capacity. However, the vibrations caused by earthquakes induce large cyclic displacements in the structures. Thus, cyclic ductility capacity is the more realistic evaluation to consider when addressing the reliable safety of a structure.

5.5 Evaluation of Monotonic Local Ductility Capacity

The evaluation of monotonic local ductility capacity depends on the monotonic material behavior of both concrete and steel. Such behavior is described in the following sections.

5.5.1 Monotonic Behavior of Concrete

Plain concrete is known to be a brittle material. However, the ductility and strength of concrete under monotonic compressive loading can be improved by lateral confinement. Because concrete behaves according to the Coulomb failure criterion, the lateral confinement of concrete creates a triaxial state of stresses that improves its strength. The confinement of concrete improves the monotonic as well as cyclic behavior of concrete.

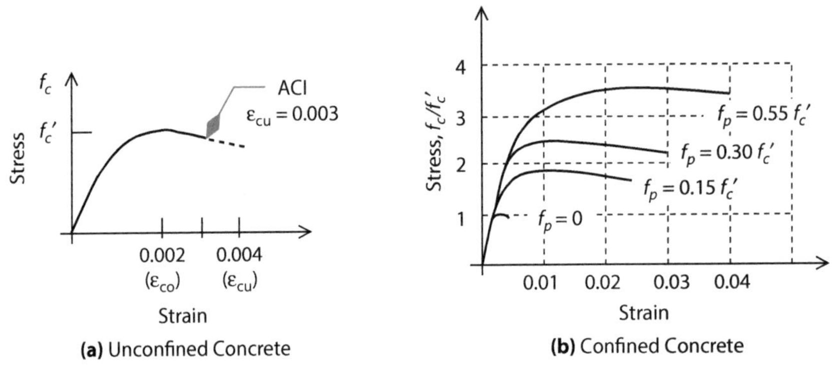

FIGURE 5-20 Monotonic behavior of concrete.

The improvement in concrete quality due to confinement is tremendous. Figure 5-20(b) indicates that the strength due to confinement may be increased to three times its unconfined strength, whereas its confined strain capacity can be as high as 10 times its unconfined strain capacity.

The compressive strength capacity of confined concrete may be expressed in terms of confined lateral pressure

$$f_{con} = f_c' + 4.1 f_p$$

where f_c' = unconfined compressive strength of concrete
f_{con} = confined compressive strength of concrete
f_p = lateral confining pressure

Steel and concrete perform in different ways in terms of material failure. Both materials fail eventually in shear. However, concrete material depends on cohesion and internal friction to resist external forces, whereas steel material depends only on cohesion to resist such forces. In other words, because steel as a material does not have an internal angle of friction, its behavior is analogous to cohesion soil (pure clay). In contrast, the behavior of concrete as a material is analogous to cohesion-friction soils (a clay-sand mix).

Consequently, the failure of steel is independent of any confinement lateral pressure. As shown in Fig. 5-21(a), the shear strength of steel remains constant regardless of any external lateral pressure. The steel strength bounded by Coulomb envelope as shown in Fig. 5-21(a) may be expressed mathematically as follows:

$$\tau_u = C$$

where τ_u = ultimate shear strength capacity of steel
C = internal cohesion

In contrast, the failure of concrete depends on the external lateral pressure. Figure 5-21(b) shows how the shear strength of concrete increases by increasing lateral pressure.

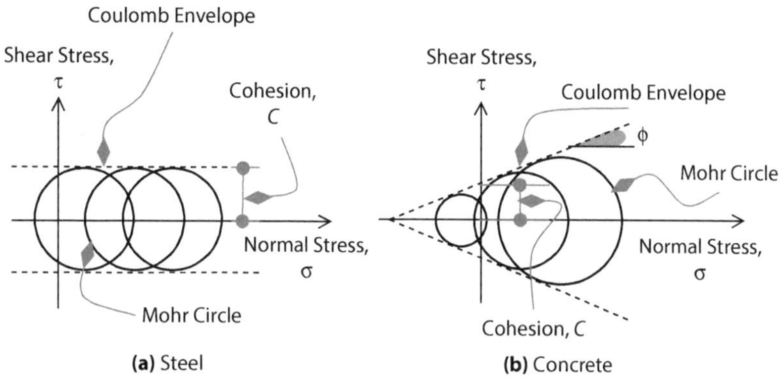

FIGURE 5-21 Coulomb failure criteria.

The concrete strength bounded by the Coulomb envelope shown in Fig. 5-21(b) may be expressed mathematically as follows:

$$\tau_u = C + \sigma_n \tan \phi$$

where τ_u = ultimate shear strength capacity of concrete
C = internal cohesion
σ_n = normal stress on failure plane
ϕ = internal angle of friction

5.5.2 Monotonic Behavior of Steel

Unlike concrete, the properties of steel as a material are well established and well defined. Figure 5-22(a) shows the stress-strain diagram of steel as obtained from a monotonic test on a steel coupon that is free of residual stresses. The steel behavior is linear up to its yielding stress, followed by a well-defined yielding plateau, and continues into a strain-hardening region, which marks the increase of stress up to its ultimate capacity. The steel shown in the figure exhibits a high level of ductility.

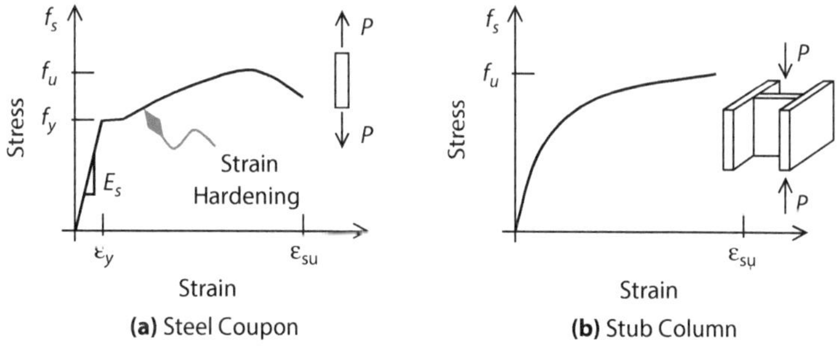

FIGURE 5-22 Monotonic behavior of steel.

However, in structural members, the behavior will not be linear as in the coupon test case. The residual stresses that usually result from production processes change the properties of the steel. Figure 5-22(b) shows the shape of the stress-strain curve of steel that would occur during the stub column test. As can be observed, the steel loses its linearity and shows nonlinear behavior similar to concrete material. This behavior is extremely important in the study of stability in steel structures. In fact, inelastic stability, where the effect of residual stresses is considered in the stability of steel structures, is reflected in the column curves in the Load and Resistance Factor Design (LRFD) manual of the American Institute of Steel Construction (AISC) code.

5.5.3 Idealized Strain Compatibility Analysis

Local ductility capacity may be evaluated in the simplest ideal case when the material is linear elastic and perfectly plastic, sections are homogeneous, and the loading is monotonic. Note that steel material behaves linearly if it is free of residual stresses, whereas concrete is considered linear as long as its stress does not exceed $0.45 f_c'$. In this case, the curvatures at first yield and at ultimate state are evaluated as follows.

Curvature at First Yield

Curvature at first yield is evaluated when the maximum strain in the section reaches yield stress. Figure 5-23 shows the state of stresses and strains at first yield for steel and for reinforced concrete sections. Figure 5-23(a) shows the state of strain and stress for a steel section subjected to bending moment and axial load. Because of the presence of the axial load, the neutral axis will not be at the midpoint of the section. Thus, one side of the section will yield before the other. The curvature in this case is evaluated as the yield strain divided by the distance to the neutral axis

$$\kappa_y = \frac{\varepsilon_y}{x_1}$$

Figure 5-23(b) shows the state of strain and stress for a reinforced concrete section subjected to bending moment with or without axial load. Because reinforced concrete sections are designed as under-reinforced sections, the steel will yield long before concrete crushing occurs. If the stress in concrete is within half its crushing strength, f_c', the curvature at first yield is evaluated in reference to the strain in the steel

$$\kappa_y = \frac{\varepsilon_y}{x_2}$$

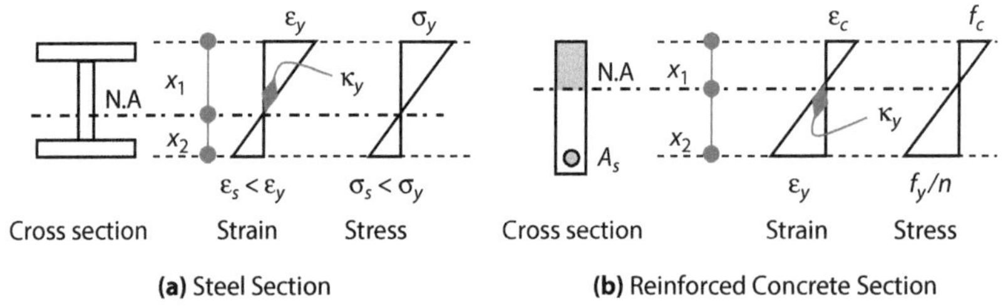

FIGURE 5-23 Curvature at first yield.

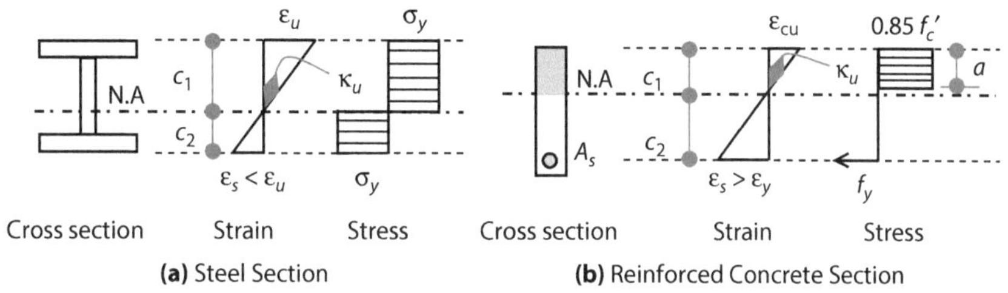

FIGURE 5-24 Curvature at ultimate state.

Curvature at Ultimate State

Curvature at ultimate state is evaluated when the maximum strain in the section reaches its capacity. Figure 5-24 shows the state of stresses and strains at ultimate state for steel and for reinforced concrete sections. Figure 5-24(a) shows the state of strain and stress for a steel section subjected to bending moment and axial load. The curvature in this case is evaluated as the ultimate strain divided by the distance to the neutral axis

$$\kappa_u = \frac{\varepsilon_u}{c_1}$$

Figure 5-24(b) shows the state of strain and stress for a reinforced concrete section subjected to bending moment with or without axial load. Note that the distribution of concrete stresses at ultimate state is nonlinear. However, the equivalent rectangular block given by the American Concrete Institute (ACI) code may be considered an alternative. Consequently, the curvature at ultimate state is evaluated in reference to the strain in the concrete

$$\kappa_u = \frac{\varepsilon_{cu}}{c_1}$$

Unfortunately, the earthquake loading effect is not monotonic. Vibrations during an earthquake induce large cyclic displacements in the structure. Because cyclic behavior is completely different than monotonic behavior, cyclic behavior should be used when studying the behavior of structures. Cyclic behavior of structures is addressed in the next section.

Example 5-3 The steel column in Example 5-3, Fig. 1, is subjected to a horizontal monotonic loading, F, and a constant axial load, $P = 0$.

1. Calculate the moment and curvature at first yield, M_y, κ_y.
2. Calculate the moment and curvature at ultimate state, M_u, κ_u.
3. Calculate the local ductility (curvature) ratio, μ_{lc}.
4. Calculate the top displacement ductility ratio, μ_c.

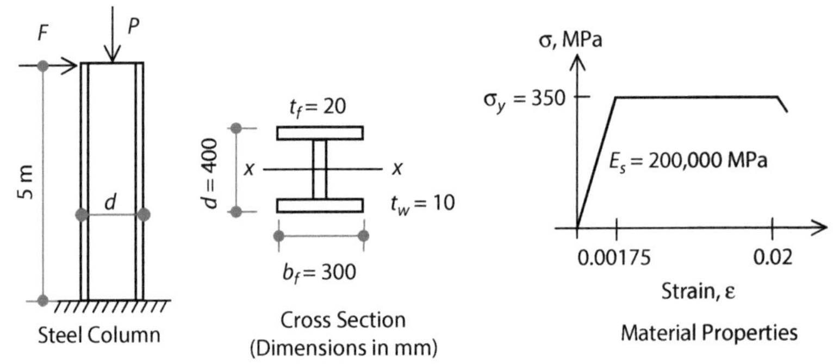

EXAMPLE 5-3, FIGURE 1

Solution For first yield, refer to Example 5-3, Fig. 2.

Moment of inertia:
$$I = \frac{300(400)^3}{12} - \frac{290(360)^3}{12}$$
$$= 473 \times 10^6 \text{ mm}^4$$

Elastic section modulus:
$$S_x \frac{I}{x} = \frac{473 \times 10^6}{200}$$
$$= 2{,}362 \times 10^3 \text{ mm}^3$$

Moment:
$$M_y = S_x \sigma_y = 2{,}362 \times 10^3 (350)$$
$$= 827 \times 10^6 \text{ N·mm}$$

Curvature:
$$\kappa_y = \frac{\varepsilon_y}{x_1} = \frac{0.00175}{200}$$
$$= 9 \times 10^{-6} \text{ rad/mm}$$

Yield displacement: Refer to Example 5-3, Fig. 3.
By using the conjugate beam method, the moment of the curvature distribution about the top yields

$$\Delta_y = \frac{\kappa_y L^2}{3} = \frac{9 \times 10^{-3}(5)^2}{3} = 0.075 \text{ m}$$

For ultimate state, refer to Example 5-3, Fig. 4.
Neutral axis = area bisector

$$c_1 = c_2$$

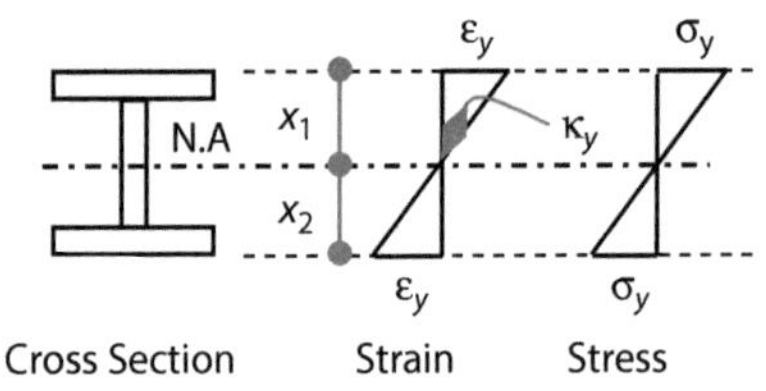

EXAMPLE 5-3, FIGURE 2

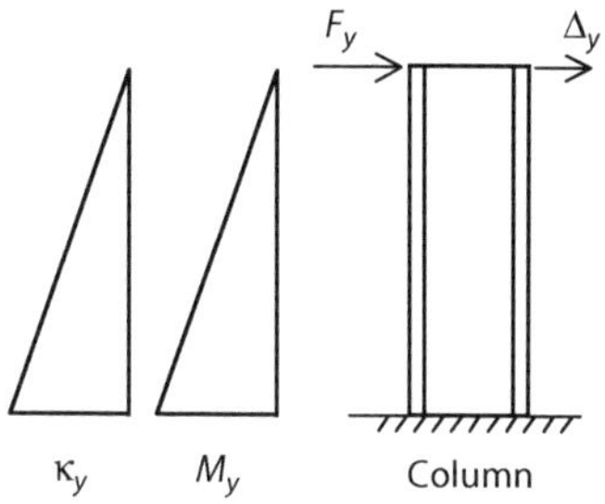

Example 5-3, figure 3 Yield displacement.

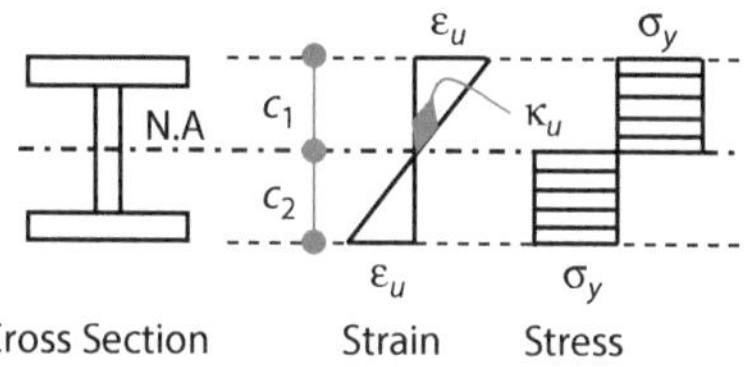

Example 5-3, figure 4

Plastic section modulus: $$Z_x = \left\{300(20)(190) + \frac{10(180)^2}{2}\right\}(2)$$
$$= 2{,}604 \times 10^3 \text{ mm}^3$$

Moment: Ultimate moment is equal to plastic moment. Thus,

$$M_u = M_p = Z_x \sigma_y$$
$$= 2{,}604 \times 10^3 (350) = 911 \times 10^6 \text{ N·mm}$$

Curvature: $$\kappa_u = \frac{\varepsilon_u}{c_1} = \frac{0.02}{200} = 100 \times 10^{-6} \text{ rad/mm}$$

For ultimate displacement, refer to Example 5-3, Fig. 5.

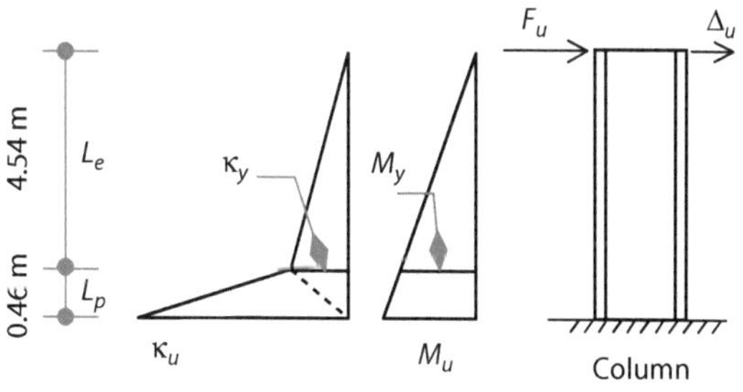

Example 5-3, figure 5 Ultimate displacement.

Length of plastic zone, L_p, is obtained by proportionality of moments

$$\frac{M_y}{M_u} = \frac{L_e}{L_e + L_p}$$

$$\frac{827}{911} = \frac{L_e}{5}$$

hence,

$$L_e = 4.54 \text{ m}$$
$$L_p = 0.46 \text{ m}$$

By using the conjugate beam method and dividing the plastic zone into two triangles by the dashed line as shown in Example 5-3, Fig. 5, the moment of the curvature distribution about the top yields

$$\Delta_u = \frac{\kappa_y L_e^2}{3} + \frac{1}{2}\kappa_y L_p\left(L_e + \frac{L_p}{3}\right) + \frac{1}{2}\kappa_u (L_p)\left(L_e + \frac{2}{3}L_p\right)$$

$$= \frac{9\times10^{-3}(4.54)^2}{3} + \frac{1}{2}(9\times10^{-3})(0.46)\left(4.54 + \frac{0.46}{3}\right)$$

$$+\frac{1}{2}(100\times10^{-3})(0.46)\left\{4.54 + \frac{2}{3}(0.46)\right\}$$

$$= 0.183 \text{ m}$$

Ductility ratios

Local ductility: $\mu_{lc} = \mu_\kappa = \frac{\kappa_u}{\kappa_y} = \frac{100}{9} = 11.1$

Global ductility: $\mu_c = \mu_1 = \frac{\Delta_u}{\Delta_y} = \frac{183}{75} = 2.4$

Global and local ductility ratios are not the same. In general, local ductility demand is much higher than global ductility demand.

Example 5-4 Consider the column in Example 5-3 to be subjected to an axial load of $P = 1{,}000$ kN.

1. Calculate moment and curvature at first yield, M_{ya}, k_{ya}.
2. Calculate moment and curvature at ultimate state, M_{ua}, κ_{ua}.
3. Calculate local ductility (curvature) ratio, μ_{lc}.
4. Calculate top displacement ductility ratio, μ_c.

Solution

For first yield, refer to Example 5-4, Fig. 1.
Neutral axis, distance x_1
Cross-sectional area: $A = 300(20)(2) + 360(10)$
$= 15{,}600 \text{ mm}^2$

Moment: Stress at top flange is yield stress

$$\sigma_y = -\frac{P}{A} - \frac{M_{ya}}{S_x}$$

$$-350 = -\frac{1{,}000\times10^3}{15{,}600} - \frac{M_{ya}}{2{,}362\times10^3}$$

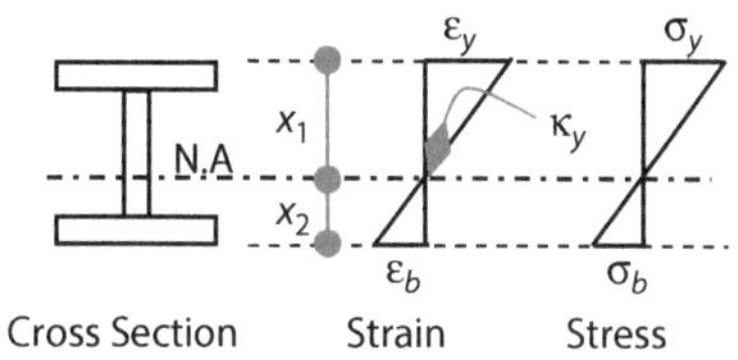

Example 5-4, figure 1

Thus,

$$M_{ya} = 675 \times 10^6 \text{ N·mm}$$

Stress at bottom flange: $\sigma_b = -\dfrac{P}{A} + \dfrac{M_{ya}}{S_x}$

$$= -\frac{1{,}000 \times 10^3}{15{,}600} + \frac{675 \times 10^6}{2{,}362 \times 10^3} = 222 \text{ N/mm}^2 (\text{MPa})$$

Find x_1 by proportionality to extreme fiber stresses: $\dfrac{x_1}{x_1 + x_2} = \dfrac{\sigma_y}{\sigma_y + \sigma_b}$

$$\frac{x_1}{400} = \frac{350}{350 + 222}$$

Thus,

$$x_1 = 245 \text{ mm}$$

Curvature: $\kappa_{ya} = \dfrac{\varepsilon_y}{x_1} = \dfrac{0.00175}{245}$

$$= 7 \times 10^{-6} \text{ rad/mm}$$

For yield displacement, refer to Example 5-4, Fig. 2. Use the conjugate beam method to determine the moment of curvature about the top

$$\Delta_y = \frac{\kappa_y L^2}{3}$$

$$= \frac{7 \times 10^{-3} (5)^2}{3} = 0.058 \text{ m}$$

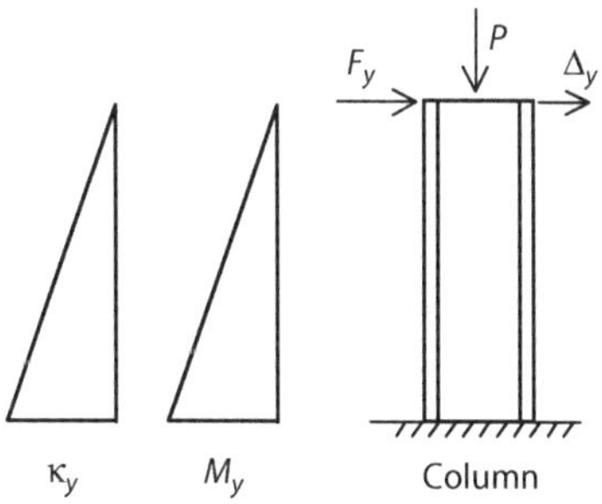

Example 5-4, figure 2 Yield displacement.

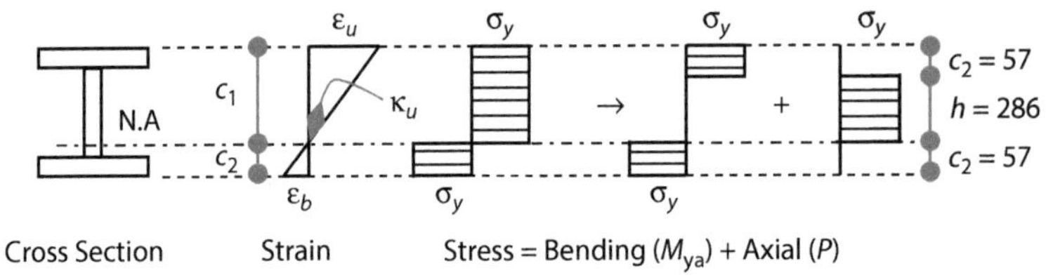

Example 5-4, Figure 3

For the ultimate state, refer to Example 5-4, Fig. 3.
The neutral axis can be found by considering the central symmetrical shape for axial load

$$P = \sigma_y t_w h$$

$$1{,}000 \times 10^3 = 350(10)(h) \rightarrow h = 286 \text{ mm}$$

$$c_2 = \left(\frac{(c_1 + c_2) - h}{2}\right) = \frac{400 - 286}{2} = 57 \text{ mm}$$

$$c_1 = 400 - 57 = 343 \text{ mm}$$

The plastic section modulus in the presence of the axial load, Z_{xa}, equals the plastic section modulus minus the central core

$$Z_{xa} = Z_x - \text{first moment area of central core}$$

$$= 2{,}604 \times 10^3 - \frac{10(286)^2}{4} = 2{,}400 \times 10^3 \text{ mm}^3$$

Moment: Ultimate moment = Plastic moment

$$M_{ua} = M_{pa} = Z_{xa}\,\sigma_y$$

$$= 2{,}400 \times 10^3(350) = 840 \times 10^6 \text{ N·mm}$$

Curvature: $\kappa_{ua} = \dfrac{\varepsilon_u}{c_1} = \dfrac{0.02}{343} = 58 \times 10^{-6} \text{ rad/mm}$

Ultimate displacement: Refer to Example 5-4, Fig. 4.
The length of plastic zone, L_p, is obtained by proportionality of moments

$$\frac{M_{ya}}{M_{ua}} = \frac{L_e}{L_e + L_p}$$

$$\frac{675}{840} = \frac{L_e}{5}$$

Thus,

$$L_e = 4.02 \text{ m}$$

$$L_p = 0.98 \text{ m}$$

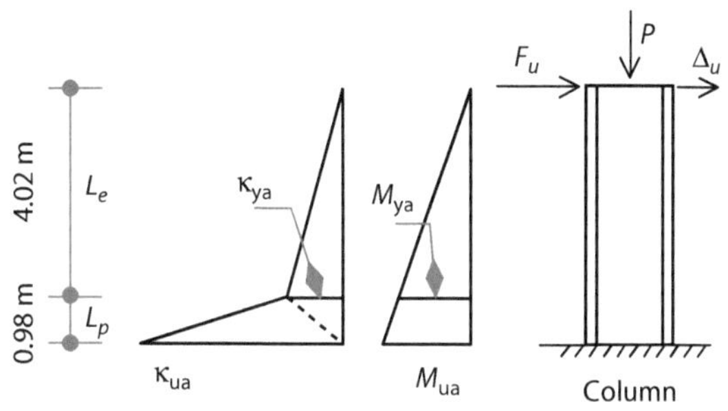

EXAMPLE 5-4, FIGURE 4 Ultimate displacement.

Determine the moment of curvature about the top by using the conjugate beam method and dividing the plastic region into two triangles by the dashed line

$$\Delta_u = \frac{\kappa_{ya} L_e^2}{3} + \frac{1}{2}\kappa_{ya} L_p \left(L_e + \frac{L_p}{3} \right) + \frac{1}{2}\kappa_{ua}(L_p)\left(L_e + \frac{2}{3} L_p \right)$$

$$= \frac{7\times10^{-3}(4.02)^2}{3} + \frac{1}{2}(7\times10^{-3})(0.98)\left(4.02 + \frac{0.98}{3} \right)$$

$$+ \frac{1}{2}(58\times10^{-3})(0.98)\left\{ 4.02 + \frac{2}{3}(0.98) \right\}$$

$$= 0.185 \text{ m}$$

Ductility ratios

Local ductility: $\mu_{1c} = \mu_\kappa = \dfrac{\kappa_{ua}}{\kappa_{ya}} = \dfrac{58}{7} = 8.3$

Global ductility: $\mu_c = \mu_\Delta = \dfrac{\Delta_u}{\Delta_y} = \dfrac{185}{58} = 3.2$

Remember that the global and local ductility ratios are not the same. In general, local ductility demand is much higher than global ductility demand.

Example 5-5 The shear wall in Example 5-5, Fig. 1, is subjected to a horizontal monotonic loading, F, and a constant axial load, $P = 5{,}000$ kN.

1. Evaluate the yield curvature, κ_y, assuming linear elastic concrete behavior.
2. Evaluate the ultimate curvature, κ_u, using the ACI rectangular stress block distribution.
3. Evaluate the curvature ductility factor, μ_k.

Solution

Concrete modulus of elasticity: $E_c = 4{,}700\sqrt{f_c'} = 4{,}700\sqrt{30} = 26{,}000$ MPa

ACI parameter, β_1: $\beta_1 = 0.85$

Modular ratio: $n = \dfrac{E_s}{E_c} = \dfrac{200}{26} = 7.7$

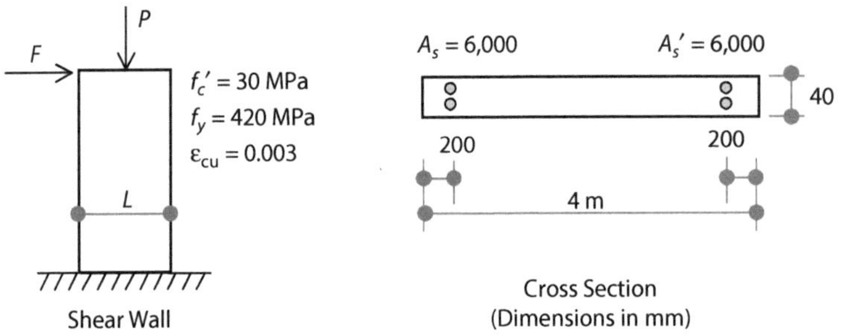

EXAMPLE 5-5, FIGURE 1

For first yield, refer to Example 5-5, Fig. 2.
Similarity of triangles (0-1-2 and 4-1-3)

$$\frac{x_1}{d} = \frac{\varepsilon_c}{\varepsilon_c + \varepsilon_y}$$

Multiply top and bottom by E_c

$$\frac{x_1}{d} = \frac{f_c}{f_c + \frac{f_y}{n}}$$

$$= \frac{f_c}{f_c + \frac{420}{7.7}} \tag{5-3}$$

Also similarity of triangles within area 0-1-2

$$\frac{\varepsilon_s'}{\varepsilon_c} = \frac{x_1 - d'}{x_1} = 1 - \frac{d'}{x_1}$$

Multiply top and bottom by E_c

$$\frac{f_s'}{f_c} = n\left(1 - \frac{d'}{d}\frac{d}{x_1}\right)$$

$$= 7.7 - 0.405\frac{d}{x_1}$$

Substitute from Eq. (5-3) into the equation above

$$\frac{f_s'}{f_c} = 7.3 f_c - 22.11 \tag{5-4}$$

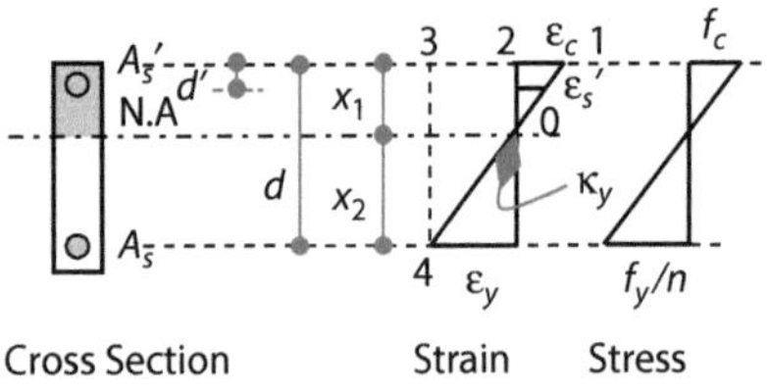

EXAMPLE 5-5, FIGURE 2

Equilibrium: $P = C_c + C_s - T$

$$5\times10^6 = \frac{1}{2} f_c x_1 b + A_s' f_s' - A_s f_y$$

$$5\times10^6 = \frac{1}{2} f_c' x_1 (400) + (6{,}000) fs - (6{,}000)(420)$$

Substitute x_1 from Eq. (5-3) and f_s' from Eq. (5-4) in the equation above, resulting in

$$803.8 f_c^2 - 5{,}263.4 f_c - 417{,}452.6 = 0$$

Solving the quadratic equation above in f_c results in

$$f_c = 26.3 \text{ MPa}$$

from Eq. (5-3): $x_1 = \dfrac{3{,}800(26.3)}{26.3 + 420/7.7} = 1{,}236 \text{ mm}$

from Eq. (5-4): $f_s' = 7.3\,(26.3) - 22.11 = 170 \text{ MPa}$

Curvature: $\kappa_y = \dfrac{e_y}{d - x_1} = \dfrac{0.0021}{3{,}800 - 1{,}236}$

$= 0.82 \times 10^{-6}$ rad/mm

For ultimate state, refer to Example 5-5, Fig. 3.

Similarity of triangles: $\dfrac{\varepsilon_s'}{\varepsilon_c} = \dfrac{c_1 - d'}{c_1}$

$$\frac{\varepsilon_s'}{0.003} = \frac{c_1 - 200}{c_1}$$

Multiply top and bottom by $E_s = 200{,}000$

$$f_s' = 600\frac{c_1 - 200}{c_1} \tag{5-5}$$

Equilibrium: $P = C_c + C_s - T$

$$5\times 10^6 = 0.85 f_c' a\, b + A_s' f_s' - A_s f_y$$

$$5\times 10^6 = 0.85(30)(0.85\, c_1)(400) + (6{,}000) f_s' - (6{,}000)(420)$$

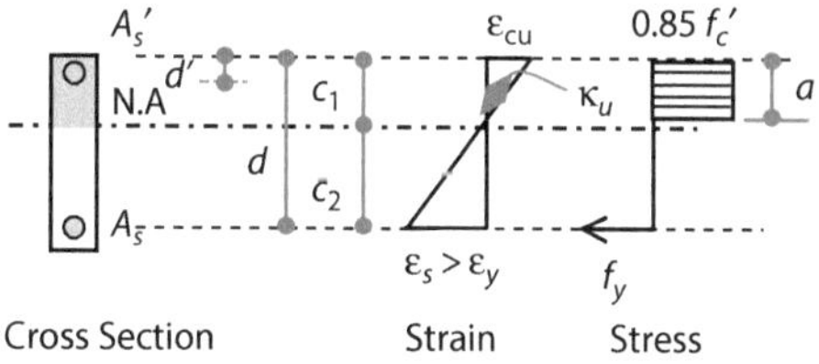

EXAMPLE 5-5, FIGURE 3

Substituting f_s' from Eq. (5-5) results in

$$8.67\,c_1^2 - 3{,}920\,c_1 - 720{,}000 = 0$$

Solving the quadratic equation above in c_1 results in

$$c_1 = 592 \text{ mm}$$

Curvature: $$\kappa_u = \frac{\varepsilon_{cu}}{c_1} = \frac{0.003}{592} = 5.07 \times 10^{-6} \text{ rad/mm}$$

Ductility ratio: $$m_\kappa = \frac{\kappa_u}{\kappa_y} = \frac{5.07}{0.82} = 6.18$$

Example 5-6 The steel frame in Example 5-6, Fig. 1, is subjected to a horizontal load, *F*. The columns have the cross section and material properties given in Example 5-3. Use those results (M_y and M_u) and ignore the effect of axial load on the curvature and moment to determine the following:

1. Construct the load displacement curve, *F*-Δ relationship.
2. Calculate ductility capacity, μ_c.

Solution To enhance understanding, reread Example 5-3 before reading this example.

At first yield

From Example 5-3: $M_y = 827$ kN·m

$$k_y = 9 \times 10^{-6} \text{ rad/mm}$$

The static solution of this frame indicates that the horizontal shears are equal at the supports as shown in Example 5-6, Fig. 2, hence,

$$M_y = 5\left(\frac{F_y}{2}\right)$$

$$827 = 5\left(\frac{F_y}{2}\right)$$

Therefore,

$$F_y = 331 \text{ kN}$$

Yield displacement may be found using the virtual work concept. Example 5-6, Fig. 3(b), shows the bending moment that occurs when a unit force is applied in the direction of the required displacement. Example 5-6, Fig. 3(a), shows the curvature distribution at first yield due to yielding force, F_y.

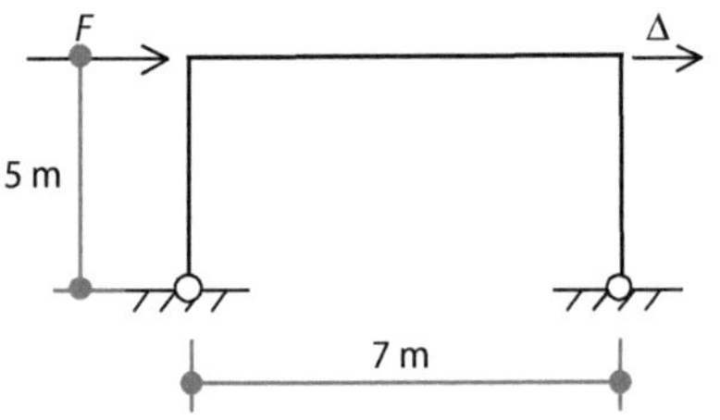

Example 5-6, Figure 1

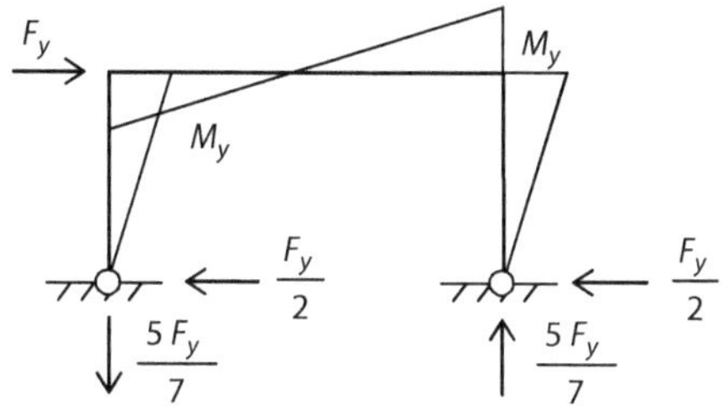

Example 5-6, Figure 2 Static solution.

Using the virtual work equation,

$$\Delta_y = \int M_1 \, \kappa ds = \text{integration of two diagrams shown in Example 5-6, Fig. 3}$$

For this purpose, the known expression for integration of two trapezoids may be used

$$\text{Integration} = \frac{S}{6}(2u_1v_1 + 2u_2v_2 + u_1v_2 + u_2v_1) = \text{integration of two triangles} + \text{trapezoid}$$
$$= \frac{5}{6}\{(2)(0.009)(2.5)\}(2) + \frac{7}{6}\{(2)(0.009)(2.5)(2) - (0.009)(2.5)(2)\}$$

hence,

$$\Delta_y = 0.128 \text{ m}$$

At ultimate state

From Example 5-3: $M_u = 911$ kN·m

$\kappa_u = 100 \times 10^{-6}$ rad/mm

The static solution of this frame also indicates that the horizontal shears are equal at the supports as shown in Example 5-6, Fig. 4. Thus,

$$M_u = 5\left(\frac{F_u}{2}\right)$$
$$911 = 5\left(\frac{F_u}{2}\right)$$

hence,

$$F_u = 364 \text{ kN}$$

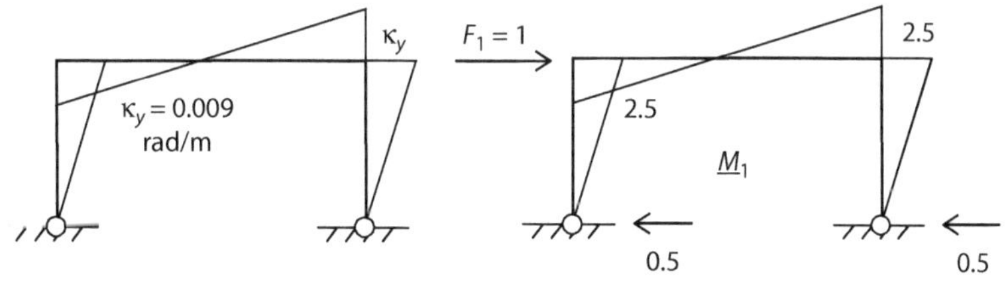

(a) Elastic Curvature Distribution **(b)** Moment Diagram Due to Unit Force

Example 5-6, Figure 3 Virtual work formulation for yield displacement (kn·M) units.

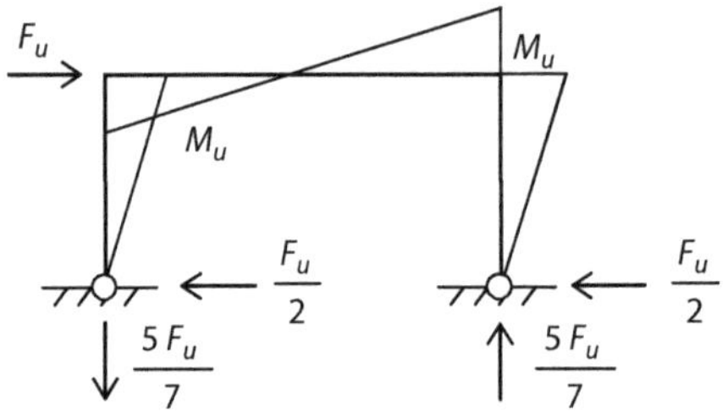

Example 5-6, Figure 4 Static solution.

Ultimate displacement may also be found using the virtual work concept. Example 5-6, Fig. 5(b) shows the bending moment that occurs when unit force is applied in the direction of the required displacement. Example 5-6, Fig. 5(a) provides the curvature distribution at ultimate state due to ultimate force, F_u. The procedures for finding the yield curvature are identical to the procedures in Example 5-3, which are obtained by the proportionality of the moments along the member.

Using virtual work equation,

$$\Delta_u = \int M_1 \kappa ds = \text{integration of diagrams shown in Example 5-6, Fig. 5}$$

$$Columns = \text{integration of two columns}$$

$$= \frac{4.54}{6}\{(2)(0.009)(2.27)\}(2)$$

$$+\frac{4.54}{6}\{(2)(0.009)(2.27)+(2)(0.1)(2.5)+(0.009)(2.5)+(0.1)(2.27)\}(2)$$

$$= 0.183$$

$$Beam = \text{integration of one beam, consider twice half beam}$$

$$= \frac{3.18}{6}\{(2)(0.009)(2.27)\}(2)$$

$$+\frac{0.32}{6}\{(2)(0.009)(2.27)+(2)(0.1)(2.5)+(0.009)(2.5)+(0.1)(2.27)\}(2)$$

$$= 0.084$$

$$\Delta_u = 0.183 + 0.084 = 0.267 \text{ m}$$

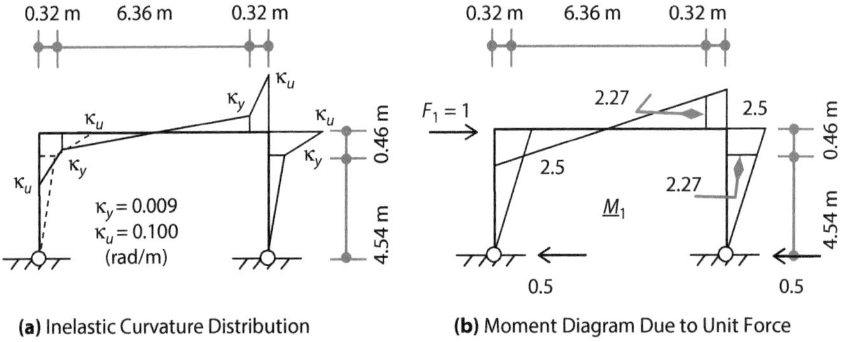

Example 5-6, Figure 5 Virtual work formulation for ultimate displacement moments are given in kN·M.

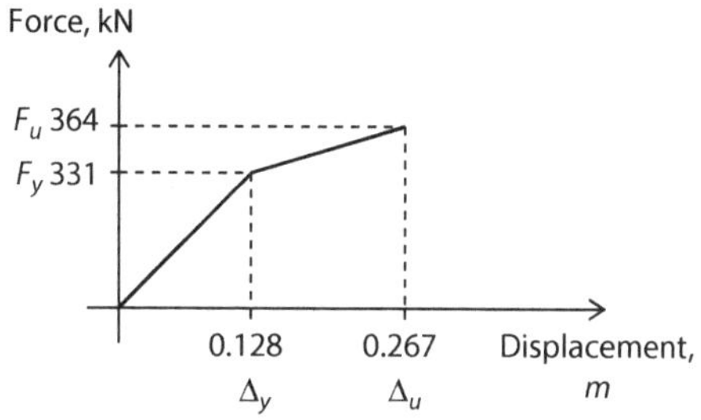

Example 5-6, Figure 6 Force-displacement relationship.

Knowing the forces and displacements at yielding and at ultimate makes it possible to construct the force displacement diagram shown in Example 5-6, Fig. 6.

Ductility capacity: $\mu_c = \dfrac{\Delta_u}{\Delta_y} = \dfrac{0.267}{0.128}$

$= 2.09$

5.5.4 General Strain Compatibility Analysis

The general strain compatibility analysis is usually needed when a more precise evaluation of ductility capacity is required. It would also be required if the actual material properties are not idealized where the actual ultimate section capacity needs to be evaluated. For example, the actual stress strain of concrete is nonlinear and can best be approximated by parabola. Previous sections have shown that the stress-strain curve of steel is also nonlinear in the range of strain hardening.

The general strain compatibility analysis is usually performed by iterative procedures to establish force equilibrium in the section by variation of the state of strain and, in turn, to determine the state of stress in the section. Such a procedure is best illustrated by the following example.

Example 5-7 An 800 × 800 mm box section is reinforced with eight bars, ϕ 20, located as shown in Example 5-7, Fig. 1(a). The cross-sectional area of each bar is given as 314 mm². The material properties are given as follows:

Steel: Idealized elastoplastic curve with yield stress, f_y = 420 MPa, and modulus of elasticity, E_s = 200 GPa.

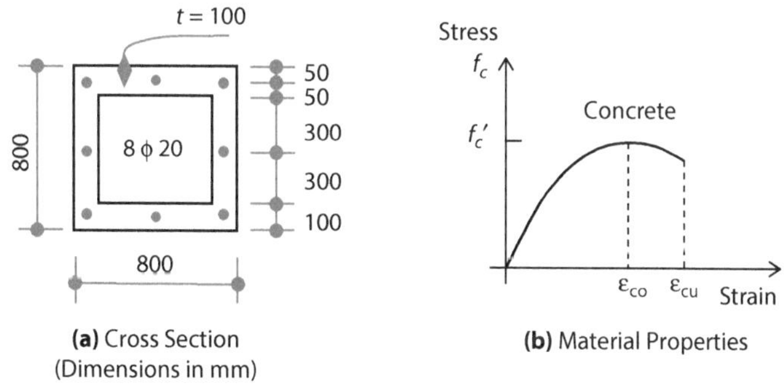

Example 5-7, Figure 1

Concrete: Concrete stress strain is approximated as parabola as shown in Example 5-7, Fig. 1(b). This may be expressed as follows:

$$\frac{f_c}{f_c'}=2\left(\frac{\varepsilon}{\varepsilon_{co}}\right)-\left(\frac{\varepsilon}{\varepsilon_{co}}\right)^2$$

where $f_c'=30$ MPa
$\varepsilon_{co}=0.002$
$\varepsilon_{cu}=0.003$

Consider this cross section to be subjected to an axial load, $P_n=900$ kN.

1. Evaluate curvature at first yield, κ_y.
2. Evaluate curvature at ultimate state, κ_u.

Solution

1. *At first yield*

At first yield, the state of strain and stress will be as shown in Example 5-7, Fig. 2. The strain in the steel will be at its yield level, whereas the strain in the concrete is unknown. A starting point value of the neutral axis, x_1, is first assumed in order to find the strain and stress in the concrete. Knowing x_1 makes it possible to establish the strain in the concrete and all steel levels.

Trial 1

Let $x_1=250$ mm.
Thus,

$$x_2=(d_4-x_1)=750-250=500\text{ mm}$$

Strain in steel at level d_4 will be at yielding, $f_s=f_y=420$ MPa.
Thus,

$$\varepsilon_y=\frac{f_y}{E_s}=\frac{420}{200{,}000}=+0.0021$$

Therefore, the curvature at this stage, κ, is given as

$$\kappa=\frac{\varepsilon_y}{x_2}=\frac{0.0021}{500}=4.2\times10^{-6}\text{ rad/mm}$$

Strains

Strain in the extreme compression fibers of concrete (compression)

$$\varepsilon_{cm}=\kappa x_1=-4.2\times10^{-6}(250)=-0.001050$$

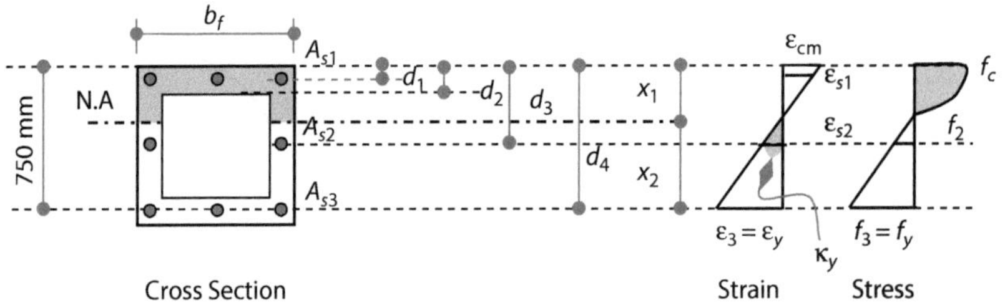

Example 5-7, Figure 2

Strain in the upper edge of the hollow core (compression)

$$\varepsilon_2 = \kappa(x_1 - d_2) = -4.2 \times 10^{-6}(250 - 100) = -0.000630$$

Strain in steel at level d_1 (A_{s1} compression)

$$\varepsilon_1 = \kappa(x_1 - d_1) = -4.2 \times 10^{-6}(250 - 50) = -0.000840$$

Strain in steel level d_3 (A_{s2} tension)

$$\varepsilon_2 = \kappa(d_3 - x_1) = +4.2 \times 10^{-6}(400 - 250) = +0.000630$$

Stresses

Stress in extreme compression fibers of concrete (compression)

$$f_{\text{cm}} = f'_c\left\{2\left(\frac{\varepsilon_m}{\varepsilon_{\text{co}}}\right) - \left(\frac{\varepsilon_m}{\varepsilon_{\text{co}}}\right)^2\right\}$$

$$f_{\text{cm}} = -30\left\{2\left(\frac{0.001050}{0.002}\right) - \left(\frac{0.001050}{0.002}\right)^2\right\} = -23.231 \text{ MPa}$$

Stress in the upper edge of the hollow core (compression)

$$f_{c2} = f'_c\left\{2\left(\frac{\varepsilon_2}{\varepsilon_{\text{co}}}\right) - \left(\frac{\varepsilon_2}{\varepsilon_{\text{co}}}\right)^2\right\}$$

$$f_{c2} = -30\left\{2\left(\frac{0.000630}{0.002}\right) - \left(\frac{0.000630}{0.002}\right)^2\right\} = -15.923 \text{ MPa}$$

Stress in steel at level d_1 (A_{s1} compression)

$$f_{s1} = E_s\varepsilon_1 = 200{,}000(-0.000840) = -168.000 \text{ MPa}$$

Stress in steel at level d_3 (A_{s2} tension)

$$f_{s2} = E_s\varepsilon_3 = 200{,}000(0.000630) = +126.000 \text{ MPa}$$

Forces

The force in the concrete will be equal to the area of the stress diagram. Because the neutral axis passes through the hollow core, the concrete force may be calculated as a hollow core force, F_{c2} [at depth = 100 mm (3.94 in)], subtracted from a rectangular section force, F_{c1} [at depth = 250 mm (9.84 in)].

The area of the stress diagram may be evaluated by the integration of the parabola that describes the stress as given above. For ease of calculations, the average stress under the stress diagram, f_{cav}, may be calculated by dividing the integrated function by its length.

Since

$$f_c = f'_c\left\{2\left(\frac{\varepsilon}{\varepsilon_{\text{co}}}\right) - \left(\frac{\varepsilon}{\varepsilon_{\text{co}}}\right)^2\right\}$$

then

$$f_{\text{cav}} = \left[\int f_c' \left\{ 2\left(\frac{\varepsilon}{\varepsilon_{\text{co}}}\right) - \left(\frac{\varepsilon}{\varepsilon_{\text{co}}}\right)^2 \right\} d\varepsilon \right] \div \varepsilon$$

or

$$f_{\text{cav}} = f_c' \left\{ \left(\frac{\varepsilon}{\varepsilon_{\text{co}}}\right) - \left(\frac{1}{3}\right)\left(\frac{\varepsilon}{\varepsilon_{\text{co}}}\right)^2 \right\}$$

Therefore

$$f_{\text{cav},1} = 30\left\{ \frac{0.001050}{0.002} - \left(\frac{1}{3}\right)\left(\frac{0.001050}{0.002}\right)^2 \right\} = 12.994 \text{ MPa}$$

$$f_{\text{cav},2} = 30\left\{ \frac{0.000630}{0.002} - \left(\frac{1}{3}\right)\left(\frac{0.000630}{0.002}\right)^2 \right\} = 8.458 \text{ MPa}$$

$$F_{c1} = f_{\text{cav},1}\, x_1 b_f = (-12.994)(250)(800) = -2{,}598{,}800 \text{ N}$$
$$F_{c2} = -f_{\text{cav},2}(x_1 - d_2)(b_f - 2t)$$
$$= (8.458)(250 - 100)(600) = 761{,}220 \text{ N}$$
$$F_{s1} = A_{s1} f_{s1} = 3(314)(-168.000) = -158{,}256 \text{ N}$$
$$F_{s2} = A_{s2} f_{s2} = 2(314)(+126.000) = +79{,}128 \text{ N}$$
$$F_{s3} = A_{s3} f_y = 3(314)(420) = +395{,}640 \text{ N}$$

Because all internal forces in the section are calculated, equilibrium requires that the summation of these forces be equal to the external force of 900 kN. If compression is taken as positive, the summation of force in kN is calculated as

$$\Sigma F_i = F_{c1} - F_{c2} + F_{s1} - F_{s2} - F_{s3}$$
$$= 2{,}599 - 761 + 158 - 79 - 396$$
$$= 1{,}521 \text{ kN} > P_n = 900 \text{ kN}$$

When the summation of internal forces is larger than the external applied load, the resulting compressive internal force must be reduced to reach 900 kN. Therefore, the second trial must proceed with a smaller value of concrete strain, for example, a smaller value of x_1, such as 150 mm. This iterative procedure is continued until equilibrium is reached. If iterative trials continue, the final equilibrium state will be reached at $x_1 = 200.5$ mm, which also yields the following final results:

$$x_1 = 200.5 \text{ mm}$$
$$x_2 = 549.5 \text{ mm}$$
$$\kappa_y = 3.82 \times 10^{-6} \text{ rad/mm}$$
$$\varepsilon_{\text{cm}} = 0.000766,\ \varepsilon_2 = 0.000384,\ \varepsilon_{s1} = 0.000575,\ \varepsilon_{s2} = 0.000762$$
$$f_{\text{cm}} = 18.584,\ f_{c2} = 10.416,\ f_{s1} = 115,\ f_{s2} = 152,\ f_{s3} = 420$$
$$F_{\text{cm}} = 1{,}608,\ F_{c2} = -325,\ F_{s1} = 108,\ F_{s2} = -95,\ F_{s3} = -396$$
$$\Sigma F_i = 900 \text{ kN}$$

The moment at first yield, M_y, can be readily calculated from the final trial by simply summing the moments of internal forces about the gross centroid of the section

$$M_y = \Sigma F_i\, r_i$$

where r_i is the distance from the section's gross centroid to the point of application of force i.

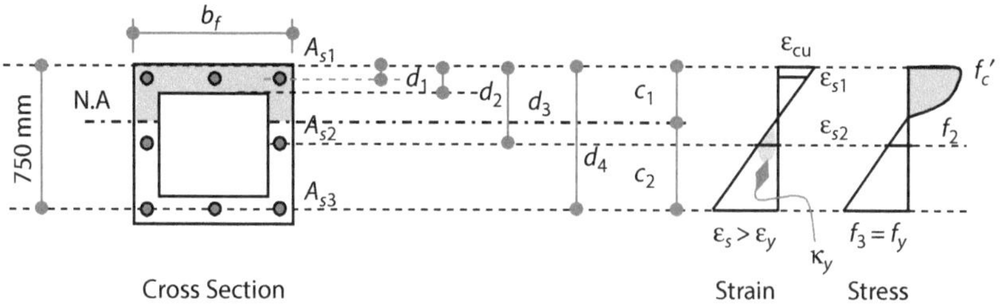

EXAMPLE 5-7, FIGURE 3

2. *At ultimate state*

The state of strain and stress at ultimate state is as shown in Example 5-7, Fig. 3. The strain in the concrete will be at its ultimate level, ε_{cu}, whereas the strain in the steel is unknown. The first step to finding the strain and stress in the concrete is to assume a starting point value of the neutral axis, c_1. However, the neutral axis depth at ultimate state will be smaller than its depth at yield state. Knowing c_1 makes it possible to establish the strain in the concrete and in all steel levels.

Trial 1:
Let $c_1 = 100$ mm, thus,

$$c_2 = (d_4 - c_1) = 750 - 100 = 650 \text{ mm}$$

The curvature at this stage, κ, is given as

$$k = \frac{\varepsilon_{cu}}{c_1} = \frac{0.003}{100} = 30 \times 10^{-6} \text{ rad/mm}$$

Strains
Strain in the extreme compression fibers of concrete (compression)

$$\varepsilon_{cm} = \kappa c_1 = -30 \times 10^{-6}(100) = -0.003$$

Strain in the upper edge of the hollow core (compression)

$$\varepsilon_2 = \kappa(c_1 - d_2) = -300^{-6}(100 - 100) = 0$$

Strain in steel at level $d_1(A_{s1}$ compression)

$$\varepsilon_1 = \kappa(c_1 - d_1) = -30 \times 10^{-6}(100 - 50) = -0.001500$$

Strain in steel level $d_3(A_{s2}$ tension)

$$\varepsilon_2 = \kappa(d_3 - c_1) = +30 \times 10^{-6}(400 - 100) = +0.009$$

Strain in steel level $d_4(A_{s3}$ tension)

$$\varepsilon_2 = \kappa c_2 = +30 \times 10^{-6}(650) = +0.0195$$

Stresses
Stress in extreme compression fibers of concrete (compression)

$$f_{cm} = f'_c\left\{2\left(\frac{\varepsilon_m}{\varepsilon_{co}}\right) - \left(\frac{\varepsilon_m}{\varepsilon_{co}}\right)^2\right\}$$

$$f_{cm} = -30\left\{2\left(\frac{0.003}{0.002}\right) - \left(\frac{0.003}{0.002}\right)^2\right\} = -22.500 \text{ MPa}$$

Stress in the upper edge of the hollow core (compression)

$$f_{c2} = f'_c\left\{2\left(\frac{\varepsilon_2}{\varepsilon_{co}}\right) - \left(\frac{\varepsilon_2}{\varepsilon_{co}}\right)^2\right\} = 0 \text{ MPa}$$

Stress in steel at level d_1(A_{s1} compression)

$$f_{s1} = E_s\varepsilon_1 = 200{,}000(-0.001500) = -300 \text{ MPa}$$

Stress in steel at level d_3 (A_{s2} tension)

$$f_{s2} = E_s\varepsilon_3 = 200{,}000(0.009) = +1{,}800 \text{ MPa}$$

Since the calculated

$$f_{s2} > 420 \text{ MPa}$$

then

$$f_{s2} = 420 \text{ MPa}$$

Stress in steel at level d_4(A_{s3} tension)

$$f_{s3} = E_s\varepsilon_3 = 200{,}000(0.0195) = +3{,}900 \text{ MPa}$$

Since the calculated

$$f_{s2} > 420 \text{ MPa}$$

then

$$f_{s3} = 420 \text{ MPa}$$

Forces

The force in the concrete will be equal to the area of the stress diagram. Because the neutral axis passes through the upper edge of the hollow core, the concrete force will only be, F_{c1} (at depth = 100 mm).

Similar to the yield stage, the average stress under the stress diagram, f_{cav}, may be calculated as follows:

$$f_{cav} = f'_c\left\{\left(\frac{\varepsilon}{\varepsilon_{co}}\right) - \left(\frac{1}{3}\right)\left(\frac{\varepsilon}{\varepsilon_{co}}\right)^2\right\}$$

Thus,

$$f_{cav,1} = 30\left\{\frac{0.003}{0.002} - \left(\frac{1}{3}\right)\left(\frac{0.003}{0.002}\right)^2\right\} = 22.500 \text{ MPa}$$

$$F_{c1} = f_{cav,1}\, c_1 b_f = (-22.500)(100)(800) = -1{,}800{,}000 \text{ N}$$
$$F_{s1} = A_{s1} f_{s1} = 3(314)(-300) = -282{,}600 \text{ N}$$
$$F_{s2} = A_{s2} f_{s2} = 2(314)(+420) = +263{,}760 \text{ N}$$
$$F_{s3} = A_{s3} f_y = 3(314)(+420) = +395{,}640 \text{ N}$$

Because all internal forces in the section are calculated, equilibrium requires that the summation of these forces be equal to the external force of 900 kN. If compression is taken as positive, the summation of force in kN is calculated as

$$\Sigma F_i = F_{c1} - F_{c2} + F_{s1} - F_{s2} - F_{s3}$$
$$= 1{,}800 - 0 + 283 - 264 - 396$$
$$= 1{,}423 \text{ kN} > P_n = 900 \text{ kN}$$

Because the summation of internal forces is larger than the external applied load, the resulting compression internal force must be reduced to reach 900 kN. Therefore, the second trial must proceed with a smaller value of concrete strain, a smaller value of c_1, such as 50 mm. Such an iterative procedure would continue until equilibrium is reached. If iterative trials continue, the final equilibrium state will be reached at $c_1 = 81.5$ mm, which also yields the following final results:

$$c_1 = 79.9 \text{ mm}$$

$$c_2 = 674.1 \text{ mm}$$

$$k_u = 39.52 \times 10^{-6} \text{ rad/mm}$$

$$\varepsilon_{cm} = 0.003,\ \varepsilon_{s1} = 0.001024,\ \varepsilon_{s2} = 0.012810,\ \varepsilon_{s3} = 0.026644$$

$$f_{cm} = 22.500,\ f_{c2} = 0.,\ f_{s1} = 205,\ f_{s2} = 420,\ f_{s3} = 420$$

$$F_{cm} = 1{,}366,\ F_{c2} = 0.,\ F_{s1} = 193,\ F_{s2} = -264,\ F_{s3} = -396$$

$$\Sigma F_i = 900 \text{ kN}$$

Similar to the case of yield state, the moment at ultimate state, M_u, can be readily calculated from the final trial by simply summing the moments of internal forces about the gross centroid of the section

$$M_u = \Sigma F_i r_i$$

where r_i is the distance from the section's gross centroid to the point of application of force i.

5.6 Evaluation of Cyclic Local Ductility Capacity

As noted earlier, the evaluation of cyclic local ductility capacity depends on the cyclic material behavior of both concrete and steel. Such behavior is described in the following sections.

5.6.1 Cyclic Behavior of Concrete

Similar to monotonic behavior, ductility and strength of concrete under cyclic loading can also be improved by lateral confinement. Figure 5-25 shows the effect of confinement on the cyclic behavior of concrete. This effect is similar to the case of monotonic loading in terms of the great improvement in strength and ductility. However, cyclic

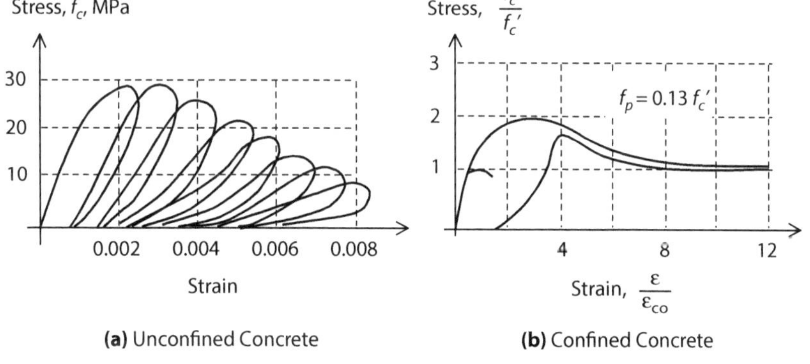

FIGURE 5-25 Experimental results from cyclic tests on concrete cylinders.

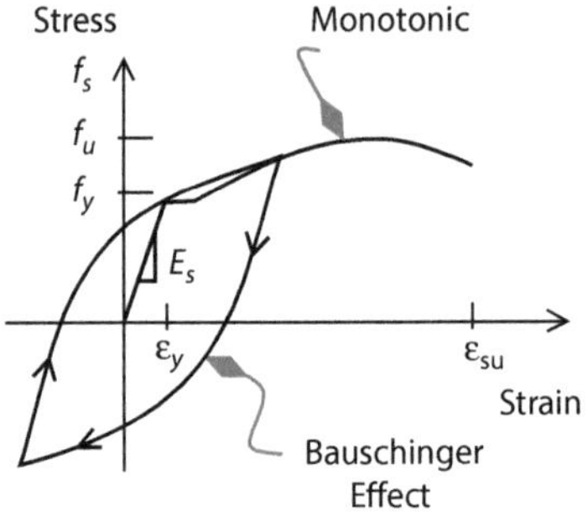

FIGURE 5-26 Cyclic behavior of steel.

loading results in a reduction in stiffness upon reloading, causing a phenomenon known as stiffness degradation of the material.

Stiffness degradation is usually associated with cyclic behavior of both confined and unconfined concrete. However, the previous experimental results show that the ultimate deformation remains the same. In addition, reloading results in permanent deformation in concrete which may be characterized as plastic offset.

5.6.2 Cyclic Behavior of Steel

Cyclic tests on steel coupons also indicate that the linearity of steel is lost once the yield stress is exceeded. Figure 5-26 shows the cyclic behavior of steel as a material after yielding has occurred. The figure shows that pure linear behavior disappears in the unloading stage, but is more defined in the reloading region. This loss of linearity is known as the Bauschinger effect.

Figure 5-27 shows experimental results from cyclic tests on steel. Figure 5-27(a) shows results of specimens under the large compressive strains expected to occur in steel structures. For specific behavior of reinforcing steel, Fig. 5-27(b) shows test results of specimens under small compressive strains. The steel undergoes small strain values due to the relative rigidity of the surrounding concrete. Both results show that the

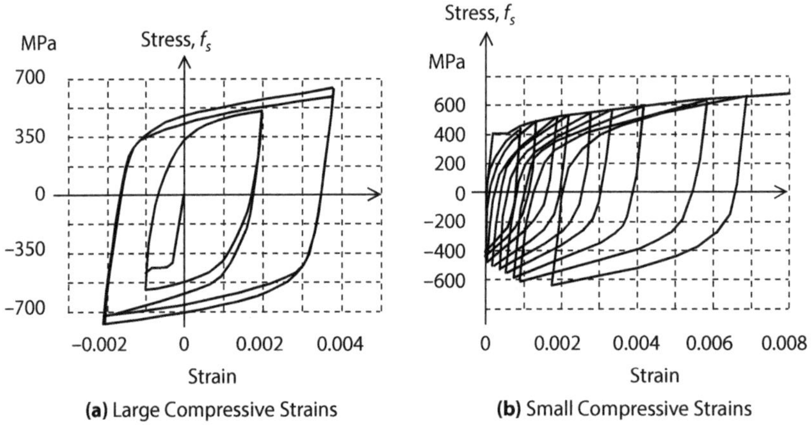

FIGURE 5-27 Experimental results from cyclic tests on steel.

behavior of steel as a material is characterized by nonlinear behavior and expansion of steel cycles with increasing maximum stress amplitudes.

5.6.3 Cyclic Strain Compatibility Analysis

Previous sections have shown that a material cyclic properties are different than its monotonic properties. This difference has to be considered in the evaluation of cyclic strain compatibility analysis. The obvious difference is the change of the shape and the size of the stress-strain curves that result from cyclic effect. This effect may be incorporated by describing such shapes by proper functions (e.g., by approximation with straight line segments).

The modified Ramberg-Osgood function is useful for expressing cyclic steel behavior for the purpose of determining strain compatibility. This function was developed to describe the stress-strain curves of prestressed strands and bars. The modified Ramberg-Osgood function is expressed as follows:

$$\sigma = \sigma_o + E_s \varepsilon \left\{ A + \frac{1-A}{[1+(B\varepsilon)^C]^{(1/C)}} \right\}$$

where C = tuning parameter obtained from cyclic experimental tests
E_s = modulus of elasticity of steel (200 GPa)
ε = strain of steel at any instant
σ = stress of steel for a given strain value
σ_o, A, B = as defined in Fig. 5-28

The other important difference between cyclic strain compatibility analysis and monotonic strain compatibility analysis as presented earlier is the origin of the stress-strain cycle at the start of calculations. In monotonic analysis, the state of strain (and stress) at rest will be zero, thus, the strain at any instant is measured from a reference of zero as shown previously. For cyclic behavior, strain compatibility analysis does not start from rest where stresses and strains are zero. In cyclic loading, a previous load cycle induces compressive strains in the concrete and in the steel.

Therefore, the origin of the strain and stress at the start of the analysis has to be found according to conditions of compatibility and equilibrium. Such a requirement creates additional unknown needs to be found (such as the value of strain and, in turn, the value of stress at the start of the iterative cycle of analysis). The additional unknown in

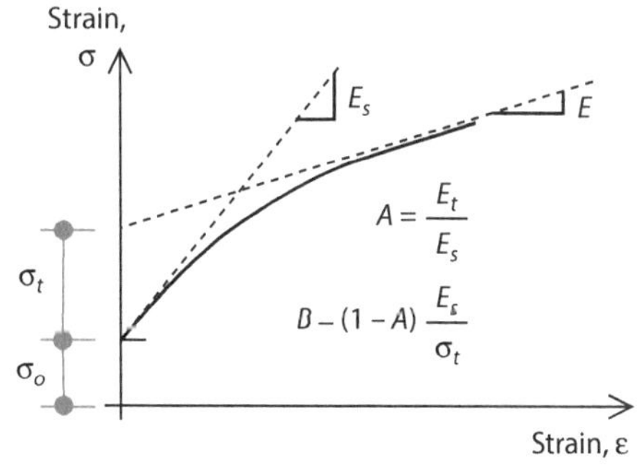

FIGURE 5-28 Modified Ramberg-Osgood function.

this analysis may be found by introducing two compound cycles of iterations in the analysis, the first cycle will be to find the state of equilibrium, whereas the second cycle will be to find the strains and stresses at the starting point of analysis.

This procedure is similar to the case of monotonic analysis and may be used to establish the cyclic moment-curvature curves of the section. More details on the method, development and its implementation may be found in the author dissertation included in the bibliography at the end of the book.

5.7 Precast Concrete Structures

For many years, precast concrete has been regarded as a nonseismic system and been prevented from being used as a seismic system. The main reason for this perception was the lack of confidence in the performance of the connections during earthquakes. It has been felt that precast elements are not tied well enough at the connection to withstand severe shaking. It was not until 2002, when the ACI code began to permit precast systems to be used as earthquake-resistant structural systems.

Recent developments of connection hardware have proven by analysis and experimentation that connections built with such hardware can actually withstand earthquake shaking and can hold precast elements tied together during an earthquake. The literature is rich with information on connection details that can successfully be used for seismic resistance. For example, the connection manuals of the Prestressed Concrete Institute (PCI) offer a wide variety of such detailing.

Strong connections can be constructed with a variety of systems and techniques. For example, spliced sleeve, angles, and plates anchored to concrete by steel bars or studs can be used for nonprestressed concrete. In contrast, wet joints can be cast in place to provide strong connections as well. In prestressed members, continuity can be provided by couplers and other anchorage hardware provided by the manufacturers.

Figure 5-29 shows several examples of such connections. Figure 5-29(a) shows an example of a strong connection for a precast shear wall, which includes a splice sleeve

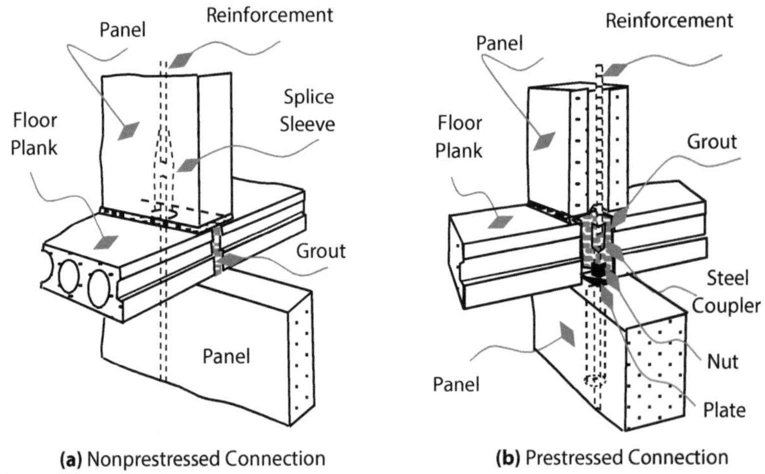

FIGURE 5-29 Precast concrete connections.

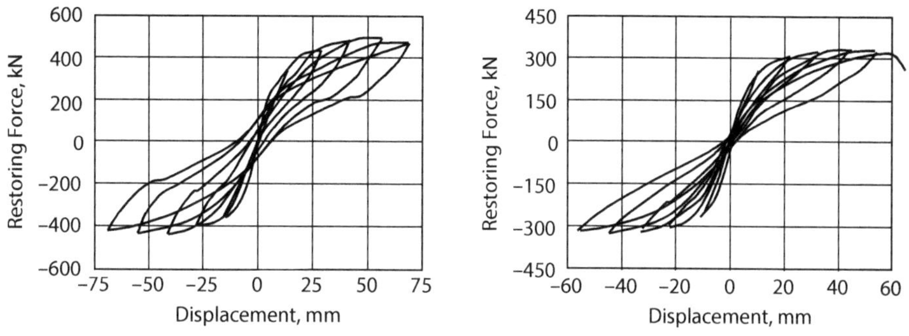

(a) Nonprestressed Wall (b) Prestressed Wall

FIGURE 5-30 Cyclic loading tests of precast shear walls.

across the connections to provide continuity of the vertical main reinforcement. Figure 5-29(b) shows a steel coupler that maintains continuity of the bars across prestressed precast concrete connection.

Experimental tests of the connections shown in Fig. 5-29 as part of shear wall systems show that such connections have a performance similar to a cast-in-place system if they are designed properly. Figure 5-30 shows the resulting hysteretic behavior of the tested walls that use these connections, illustrating that precast systems can develop both ductility and energy dissipation capacity. The ductility levels achieved in these tests reach up to a ductility ratio of 5. These values are comparable to those provided by cast-in-place systems.

Note that the hysteresis loops in Fig. 5-30 are generally narrow and the size of the prestressed wall exhibits narrower loops than its nonprestressed counterpart. This narrow appearance of the loops is not necessarily a precast property. It can be attributed to the level of the axial load in the member, as will be explained in Sec. 5.11.

The design of precast concrete systems follows in principle the capacity design philosophy by directing yielding in the members while ensuring that the connections remain within their elastic range. To guarantee such behavior, the connections are usually designed for a strength that is at least 50 percent higher than the maximum forces that can develop in the joining members.

The behavior of precast elements depends mainly on the presence of prestressing. In general, prestressed members tend to exhibit narrower hysteretic behavior, indicating a reduced capacity for energy dissipation. This behavior is due to the fact that prestressing induces a large amount of strain in the steel before the application of the external forces. In addition, prestressing induces large internal axial loads in the system, which tends to reduce the size of the hysteresis loops. This will be shown in a later section.

5.8 Effect of Structure Configuration on Ductility

In general, behavior of structures depends on the material itself, the construction detailing, and the structural configuration. For example, steel as a material is very good in terms of ductility and energy dissipation. However, it has to be properly designed and carefully detailed in order to achieve such material characteristics. Because of its high strength-to-weight ratio, steel is generally used in construction in relatively thin components. From a stability point of view, such construction is vulnerable under loading. Steel elements

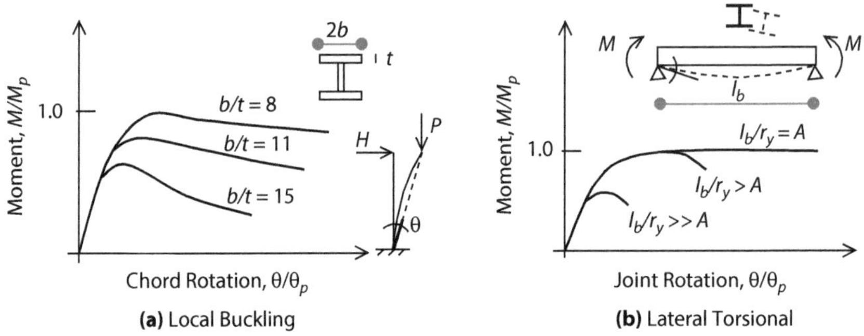

FIGURE 5-31 Effect of structural configuration on ductility.

and members must be properly sized and detailed to prevent premature local and global buckling before attaining their full material ductile capacity. Figure 5-31 shows examples of the effect of the size and bracing system on the local and global ductility.

5.9 Second-Order Effect on Ductility

Ductility is affected by the presence of axial loads both at the section level and at the system level. Previous sections explained the effect of the axial load on section ductility and demonstrated that axial loads reduce section ductility. At the system level, axial loads also reduce ductility through the phenomenon of the P-Δ effect. The behavior of frames under heavy axial loads shown in Fig. 5-32 indicates that the second-order effect would result in total loss of the system ductility as the axial load approaches the axial plastic capacity of the columns.

5.10 Undesirable Hysteretic Behavior

Because hysteretic behavior is an indicator of a structure ductility and energy dissipation, it can be viewed as an indicator of good (or bad) behavior of a structure. Poor hysteretic behavior is an indicator of bad structural behavior under seismic excitation. Undesirable hysteretic behavior can result from both material deterioration and unfavorable structural configuration that may be summarized as follows.

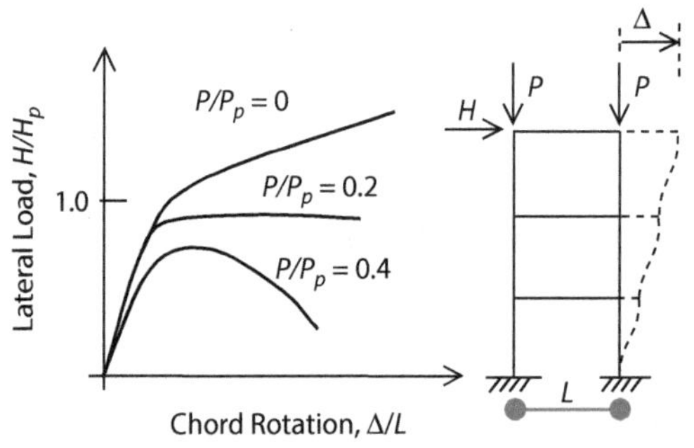

FIGURE 5-32 P-Δ effect on ductility.

Undesirable Hysteretic Behavior Due to Material Deterioration

1. Stiffness degradation
   Stiffness degradation is a typical behavior of well-detailed reinforced concrete structures. Such degradation is shown in Fig. 5-33(a).
2. Pinching
   Pinching is typical of shear-controlled or bond-controlled behavior in reinforced concrete members. Pinching is usually associated with severe stiffness degradation as shown in Fig. 5-33(b).
3. Slip-lock behavior
   Slip-lock behavior is characterized as extreme pinching degradation. Slip lock usually creates shock waves through the structure similar to brake and passenger reactions in cars. This type of behavior may be described by the hysteresis loops shown in Fig. 5-33(c).
4. Strength degradation
   Strength degradation is another type of material deterioration that may result from cyclic loading effect in structures. Such behavior is shown in Fig. 5-33(d).

Undesirable Hysteretic Behavior Due to Unfavorable Structural Configuration

1. Unsymmetrical strength
   Unsymmetrical strength can result in accumulation of plastic drift in one direction that affects the stability of vibration. Such behavior is shown in Fig. 5-34(a).
2. P-Δ effect
   The P-Δ effect can result in overall negative stiffness. The stability of the structure under large loads becomes a problem that can lead to premature catastrophic failures. Such behavior is shown in Fig. 5-34(b).

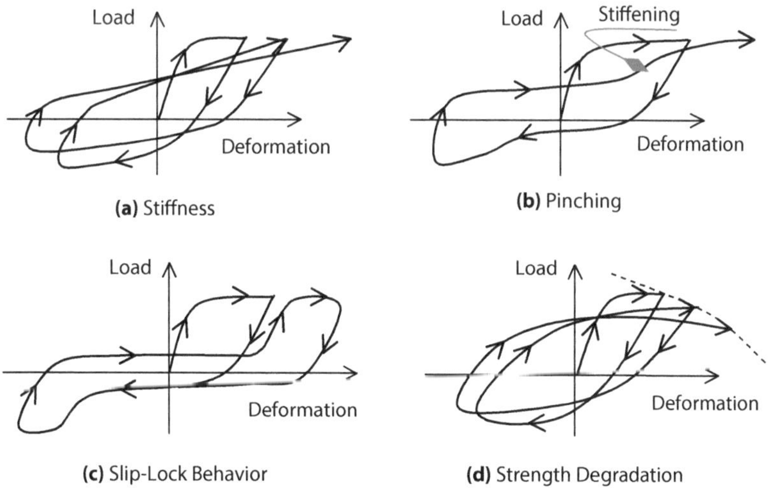

(a) Stiffness
(b) Pinching
(c) Slip-Lock Behavior
(d) Strength Degradation

Figure 5-33 Material degradation effect on hysteretic behavior.

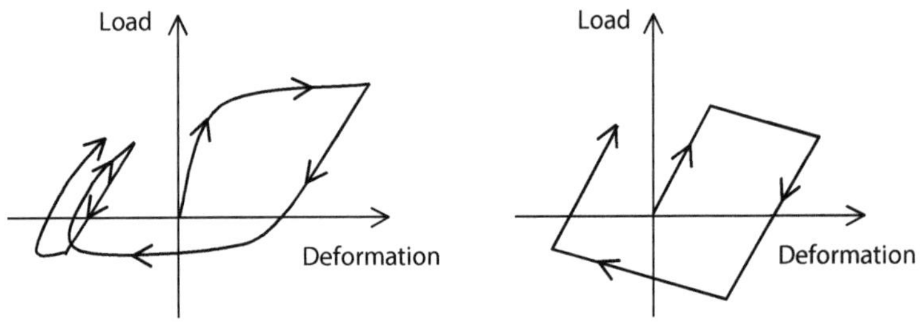

(a) Unsymmetrical Strength **(b)** *P*-Δ Effect

Figure 5-34 Structural configuration effect on hysteretic behavior.

In general, stiffness degradation is unavoidable in reinforced concrete structures. However, analysis shows that such structures behave better than elastoplastic models. With the exception of stiffness degradation, the undesirable hysteretic properties defined above will affect the validity of the force-reduction factor-ductility demand relationships given in previous sections. Therefore, the following characteristics should be avoided in design:

1. Low energy dissipation (small loops).
2. Severe stiffness degradation (pinching of hysteresis loops).
3. Strength degradation.
4. Yielding in only one direction.

5.11 Effect of Axial Load on Hysteretic Behavior

Experimental tests demonstrate that axial loads tend to reduce the energy dissipation of structures (as shown in the reduced size of the resulting hysteresis loops). Because the presence of axial loads in the structure increases the amount of energy imparted to the structure due to the work done by the external loads, the apparent decrease in the size of the hysteresis loops is due to the increase in the energy input to the structure. This state results in a decrease in the relative (not the absolute) energy dissipated in the structure.

Hysteretic behavior of steel material is known to follow the kinematic hardening laws of plasticity, which postulates that the yielding surface shifts around according to the applied loading history without any expansion of the yielding surface. Hysteresis loops of steel material can also be described by Masing type behavior. Masing loops are characterized as a twofold magnification of the skeleton curve (the virgin monotonic load-deformation curve at first loading). Such behavior is portrayed in Fig. 5-35.

Reinforced concrete sections follow Masing type behavior because their behavior under pure bending moment is basically controlled by reinforcing steel. Figure 5-35 shows how such behavior exhibits fatness of the hysteresis loops and, as a result, indicates good energy dissipation capacity, a desired property in earthquake-resistant structures.

Even though experimental results show that the size of hysteresis loops tends to decrease by increasing axial loads, the amount of energy dissipated by structures remains the same. The apparent reduction in energy dissipation capacity is attributed to the increase in the energy input to the structure rather than to the decrease in the

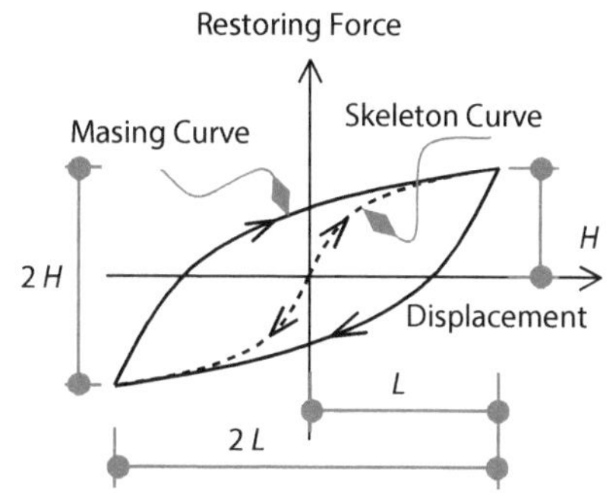

FIGURE 5-35 Masing type hysteresis model.

absolute energy dissipation capacity of the material. Furthermore, it will be shown later that the apparent reduction in the size of the loops is directly related to the level of axial load value in the section.

5.11.1 Rigid Bar Idealization

The effect of axial load on hysteretic behavior may be explained by examining an idealized structure. The structure is represented by a simplified rigid bar model supported by springs as shown in Fig. 5-36. This idealization may be viewed as a model for a shear wall. The rigid bar represents the concrete portion, which is supported by springs as its reinforcement. The bar is also supported by rigid pipes. The interface between the bar and pipe marks the critical section of the wall at the base as also shown in the same figure. The model dimensions are given as width, L, and height, H. The springs are identified by their stiffness, k_s, and their yield strength, T_y. The model is subjected to a constant axial load, N, and to a varying lateral force, Q. To understand the behavior of this model, the effect of axial load and the effect of springs on its behavior will be considered separated. Three cases will be examined as follows: Case 1 considers the model without springs and subject to forces, Q and N, only. Case 2 considers the model with springs and subject to force, Q, only (i.e., in the absence of N). Case 3 will be examined by combining Cases 1 and 2 together.

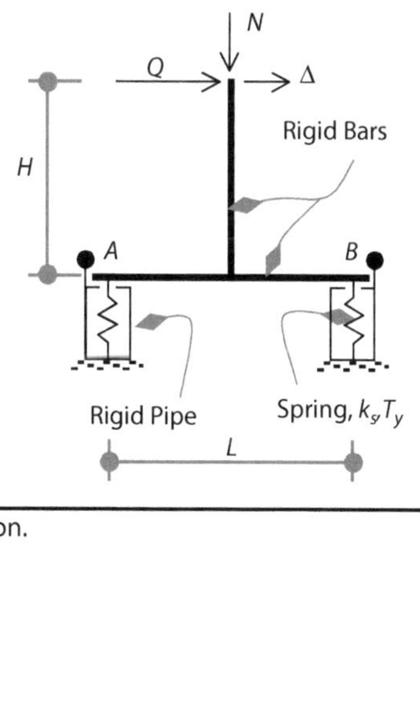

FIGURE 5-36 Rigid bar idealization.

Case 1: Rigid Bar under Axial Load and without Springs

In the absence of springs, the rigid bar exhibits rocking type behavior. The behavior of the bar in this case may be examined with reference to Fig. 5-37. If the bar is subject to lateral force, Q, as shown in Fig. 5-37(a), the bar remains in position ($\Delta = 0$) as long as the force, Q, is less than the quantity, Q_{yN}, which may be viewed as the yielding force of the bar only in the presence of axial load.

The force, Q_{yN}, may be evaluated by considering the bar equilibrium in the state shown in Fig. 5-37(a). Taking moments of all forces about point B yields the following:

$$Q_{yN} H = \frac{N}{L/2}$$

Thus,

$$Q_{yN} = \frac{N}{2} \cdot \frac{L}{H}$$

The equation above indicates that as far as $Q < Q_{yN}$, the bar remains in position. On the other hand, if $Q = Q_{yN}$, or slightly exceeded, the bar will move horizontally as shown in Fig. 5-37 (b). Upon removal of the force, Q, the bar returns to its original position. This behavior is depicted graphically by the hysteretic behavior shown in Fig. 5-37(b). Such behavior is characterized by ductile behavior without any energy dissipation capacity.

Case 2: Rigid Bar with Springs and without Axial Load

In the absence of the axial load, the rigid bar behaves in a way similar to sections under pure bending moment. The behavior of the bar in this case may be examined with reference to Fig. 5-38. If the bar is subject to lateral force Q, as shown in Fig. 5-38 (a), the top of the bar displaces horizontally proportional to the amount of the force Q, as long as Q is less than the quantity Q_{yT}, which may also be viewed as the yielding force of the bar only in the presence of springs. The force Q_{yT}, may be evaluated by consideration of the

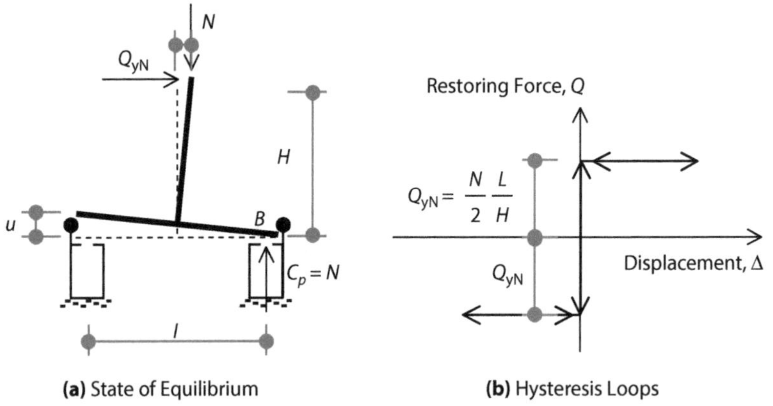

FIGURE 5-37 Case 1: rigid bar under axial load without springs.

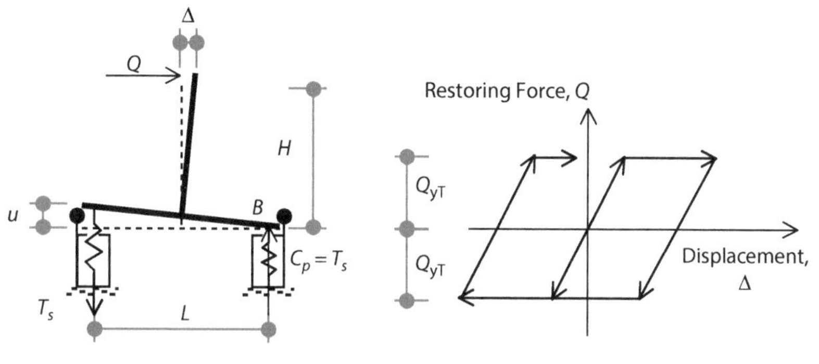

(a) State of Equilibrium **(b)** Hysteresis Loops

FIGURE 5-38 Case 2: springs without axial load.

bar equilibrium in the state shown in Fig. 5-38(a). Taking moments of all forces about point B yields the following:

$$Q_{yT}H = \frac{T_y}{L}$$

Thus,

$$Q_{yT} = T_y \frac{L}{H}$$

The equation above indicates that as long as $Q < Q_{yT}$, the bar behaves elastically. However, if $Q = Q_{yT}$, or slightly exceeded, the bar will yield with a force in the left spring equal to T_y as shown in Fig. 5-38(b). Upon removal of the force Q, the bar returns to a position that depends on the amount of plastic deformation induced in the spring. This behavior is demonstrated by the hysteretic behavior shown in Fig. 5-38(b). Such behavior is characterized by ductile behavior that results in energy dissipation proportional to the size of the hysteresis loops.

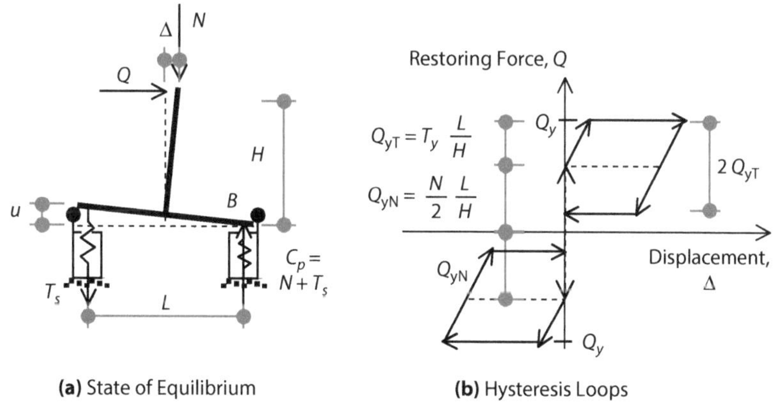

(a) State of Equilibrium **(b)** Hysteresis Loops

FIGURE 5-39 Case 3: axial load with springs.

Case 3: Rigid Bar with Springs and under Axial Load

If Cases 1 and 2 are combined, the behavior of the bar in Case 3 may be examined with reference to Fig. 5-39. If the bar is subject to lateral force Q, as shown in Fig. 5-39(a), the bar remains in position ($\Delta = 0$) as long as the force Q, is less than the quantity Q_{yN}, as defined in Case 1. If Q exceeds Q_{yN}, the load-deformation curve will follow a linear elastic path as shown in Fig. 5-39(b), and similar to the path in Case 2. The bar yielding force Q_y may be evaluated by considering the bar equilibrium in the state shown in Fig. 5-39(a). Taking moments of all forces about point B yields the following:

$$Q_y H = \frac{N}{L/2} + \frac{T_y}{L}$$

Thus,

$$Q_y = \frac{N}{2} \cdot \frac{L}{H} + T_y \frac{L}{H}$$

or,

$$Q_y = \left(\frac{N}{2} + T_y \right) \frac{L}{H}$$

Upon removal of the force Q, the bar displaces back in the negative direction along the linear elastic path first and then horizontally as indicated in Fig. 5-39(b) to the position of zero displacement. A comparison of Figs. 5-37 to 5-39 indicates that the energy dissipation of the system is constant in the three cases.

This is represented by the area of the hysteresis loop marked with depth $2Q_{yT}$, on each side of the displacement axis. However, the energy absorbed by the system is simply the summation of the energy absorbed in Cases 1 and 2. Consequently, the ratio of the energy dissipated to the energy absorbed in the system of Case 2 will be relatively larger than the ratio in Case 3, even though the amount of energy dissipated in both cases is the same. This reduction in the relative energy dissipation is what makes hysteresis loops of actual structures appear narrower in the presence of axial loads.

Furthermore, Fig. 5-39(b) indicates that the upper hysteresis loop in the positive load-displacement region remains in the first quarter of the figure as long as the yielding force of the spring T_y, is less ($N/2$). Otherwise, the two loops on both sides of the displacement axis will interfere with each other, rendering this analysis invalid. In this context, a bearing wall may be defined as a wall in which its hysteresis loops are separated by the displacement axis. It can be deduced from Fig. 5-38(b) that this definition may be expressed in terms of, N and T_y, by defining the bearing wall as a wall that is subject to an axial load (N) larger than the yielding force of the total flexural reinforcement ($2T_y$) in the section.

5.11.2 Energy Dissipation Factor (α_N)

This section introduces an energy dissipation factor to quantify the reduction in the energy dissipation capacity of the system that is due to the presence of axial load and also to construct the unloading branch of the hysteresis loops. Figure 5-40 shows the rigid bar model from Case 3 combined with Masing type loops as presented in previous sections.

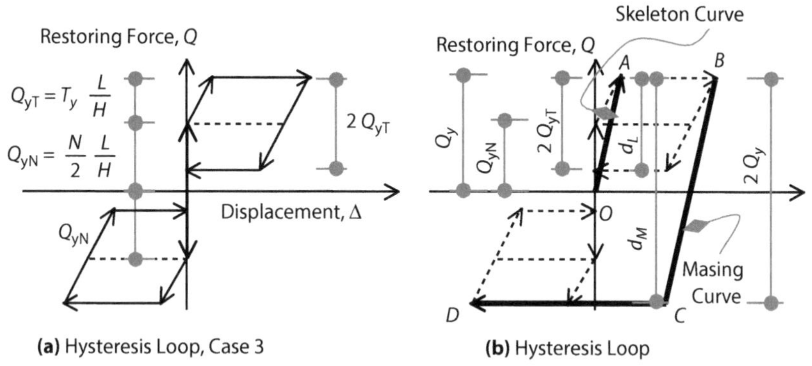

FIGURE 5-40 Definition of energy dissipation factor, α_N.

Recall that this simplified rigid bar is assumed to model concrete as a rigid portion with reinforcement as elastic springs. In real structures, concrete exhibits some elastic stiffness. Therefore, if the concrete flexibility is considered, the loading branch from rest to yielding point of the skeleton curve of the Masing loop may be replaced by an equivalent elastic segment, *O-A*, as shown in Fig. 5-40(b). When the skeleton curve is defined, which is segment *O-A-B* in Fig. 5-40(b), the Masing loop can now be drawn as twofold magnification of the skeleton curve as segment *B-C-D*.

The energy dissipation factor, α_N, is now introduced as the ratio of the depth of the actual loop, marked with distance, d_L, in Fig. 5-40(b), to the depth of the Masing loop, marked with distance d_M in Fig. 5-40(b). In reference to Fig. 5-40 (a) and (b), α_N can be expressed mathematically as

$$\alpha_N = \frac{d_L}{d_M} = \frac{2Q_{yT}}{2Q_y} = \frac{2T_y(L/H)}{2(T_y + N/2)(L/H)}$$

or

$$\alpha_N = \frac{T_y}{T_y + N/2}$$

According to the definition of bearing walls given previously, the maximum value of α_N will be reached when $T_y = N/2$, which results in $\alpha_N = 0.5$. This result indicates that the maximum reduction in energy dissipation capacity in bearing walls is 50 percent.

The other important application of this energy dissipation factor is the prediction of the unloading branch of actual hysteresis loops in the presence of axial loads. The unloading branch of reinforced concrete structures, and also of steel structures, under pure bending moment without axial loads can be constructed from the skeleton curve using Masing criterion. This makes it possible to construct the unloading branch of reinforced concrete and steel structures in the presence of axial loads using the energy dissipation factor, α_N. Accordingly, if the skeleton curve is constructed for a wall (or for a beam, for example), the unloading branch can be simply calculated by multiplying the Masing unloading branch by the energy dissipation factor, α_N.

This result has a special useful application in inelastic dynamic analysis, because it tremendously reduces the effort to model the unloading branches of hysteresis models. In this case, it would only be necessary to construct the loading branch (skeleton curve) of the model. The unloading branch will simply be an outcome as a function of the loading branch, the Masing loop, and the energy dissipation factor α_N. Full development of this factor and the effect of prestressing on the hysteretic behavior of shear walls may be found in Armouti (1993).

The application of energy dissipation factor in actual structures may be demonstrated through its application to the experimental results shown in Fig. 5-30(a). The calculations of the unloading branch of the hysteresis loops may be outlined, with reference to Fig. 5-41, as follows:

1. Define the skeleton curve (the virgin monotonic curve shown as curve *O-A*) in Fig. 5-41.
2. From the end of the skeleton curve at point, *A*, draw a Masing loop that is a twofold magnification of the skeleton curve (evaluate the quantity d_M and draw the curve *A-B* through depth d_M as shown in Fig. 5-41).
3. Calculate the energy dissipation factor, a_N, as defined above

$$\alpha_N = \frac{T_y}{T_y + N/2}$$

4. Evaluate the reduced loop depth d_L as defined earlier ($d_L = \alpha_N d_M$).
5. Draw the actual (reduced) loop (draw the curve, *A-B*, through depth d_L as shown in Fig. 5-41).

The results of applying the outlined procedures above to the wall shown in Fig. 5-30(a) are shown in Fig. 5-42.

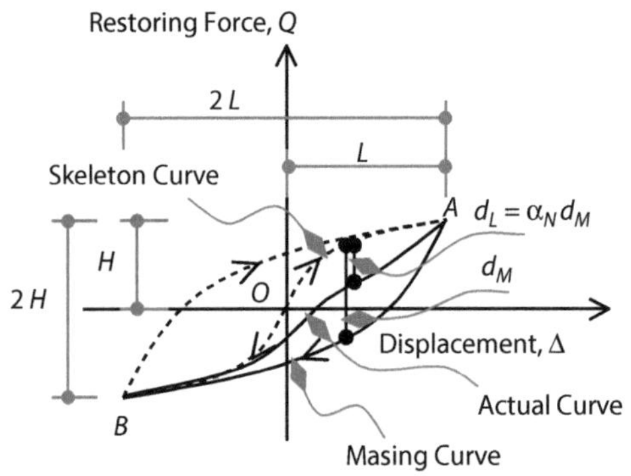

Figure 5-41 Construction of actual loop from α_N and Masing loop.

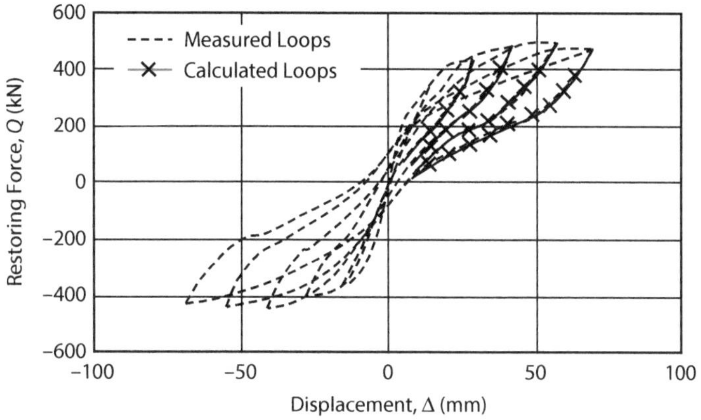

FIGURE 5-42 Application of energy dissipation factor to actual loops.

5.12 Design Considerations

Because structures are designed with strength less than their elastic strength demand, and because ductility is required to achieve such reduction, design must ensure that ductility can be achieved without premature and brittle failures. To achieve this, flexural ductility must be guaranteed by preventing brittle failures such as shear, connection, bond, stability, and foundation failures.

Examples of ductility assurance may be given as in the case of concrete confinement by closely spaced stirrups (hoops) and by prevention of local and global buckling of steel structures. In addition, brittle components, such as shear elements, connections, and foundations, must be designed with forces that exceed the maximum probable developed forces in the ductile elements.

Such requirements apply to shear calculations as shown in Fig. 5-43. If the actual plastic moments in the beam are given as M_p as shown in Fig. 5-43(a), the shear force is calculated to exceed the maximum probable shear in the section by considering the plastic moments are M_{pr} as shown in Fig. 5-43(b). As a result,

$$M_{pr} = f(1.25 f_y \rightarrow 1.40 f_y)$$

where M_p is $f(f_y)$ only.

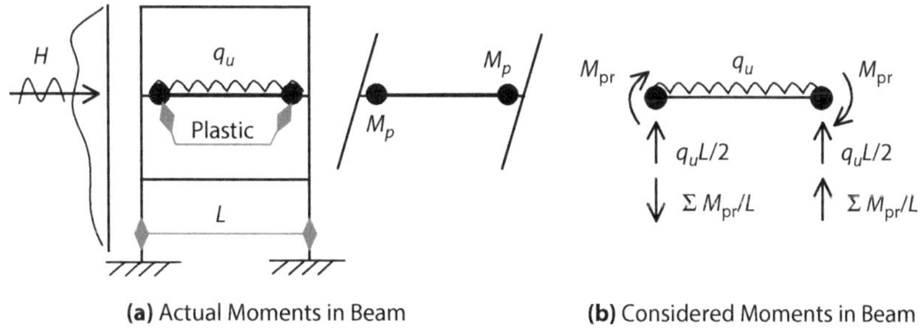

FIGURE 5-43 Shear design requirements.

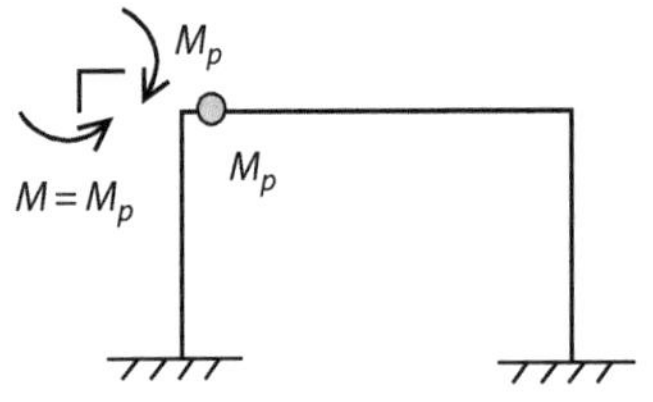

FIGURE 5-44 Connection requirements.

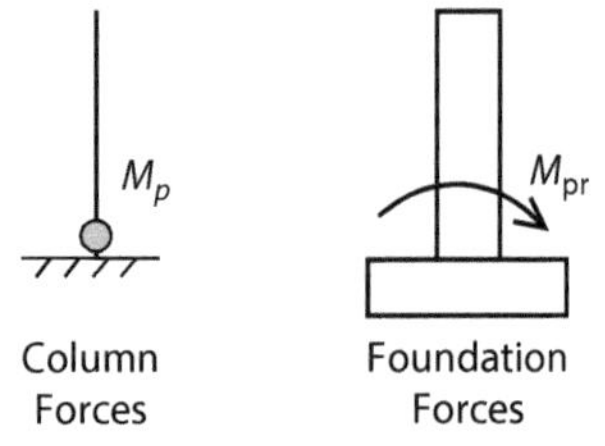

FIGURE 5-45 Foundation requirements.

The design shearing force, V_u, shown in Fig. 5-43(b) is given as

$$V_u = V_u(q_u) + V_u(M_{\text{pr}})$$
$$= q_u \frac{L}{2} + \frac{\Sigma M_{\text{pr}}}{L}$$

The same concept applies to connections. For example, if the beam in Fig. 5-44 has a plastic moment capacity, M_p, the connection cannot develop more than M_p in the column. Of course, this can be seen by equilibrium consideration of the joint as a free body diagram. In this case, the connection should be designed for, M_{pr}. Thus, the beam yields first, leaving the connection in the elastic range (stronger than the beam).

The same concept also applies to foundations, that is, foundation design is based on, M_{pr}, rather than on, M_p. Figure 5-45 illustrates such requirements.

M_{pr} may actually develop in the beam due to strain hardening as noted in the section on cyclic behavior of material. This additional strength beyond yielding level is known as overstrength. To summarize this lengthy consideration in simple words, brittle elements must be kept elastic during earthquake excitation.

The short column phenomenon also lies in this category of ductility assurance. Because the shear in the column is equal to the summation of the end moments divided by the height of the column, the shear force will be inversely proportional to its length. Consequently, very short columns develop very high shears at the development of end plastic moments. This results in premature brittle failures prior to the development of the end plastic moments. Such phenomena must be avoided.

Keep in mind that inelastic displacements, Δ_M, are also relatively large. Because this may result in a high, P-Δ effect, all members including nonseismic elements should be designed accordingly.

5.13 Capacity Design

The capacity design concept has been developed and used for many years in New Zealand. Although this concept is not explicitly stated in seismic codes developed in the United States by the International Code Council (ICC), it is implicitly a natural consequence of the force-reduction factor. As a result, it is implicitly embedded in many systems, including the special truss moment frames given in Chap. 9.

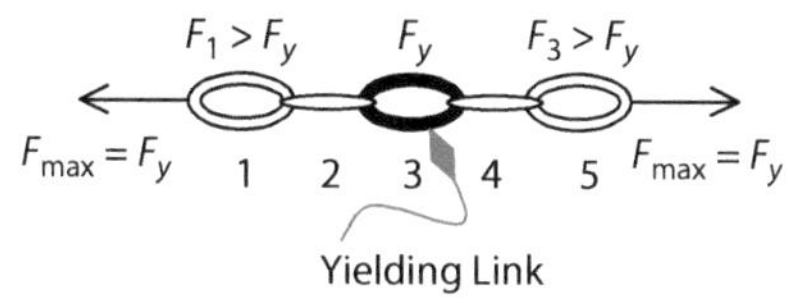

FIGURE 5-46 Strength of ductile chain.

Capacity design can be understood by considering the well-known ductile chain shown in Fig. 5-46. This chain has five links numbered 1 to 5. The yielding force capacity of links 1, 2, 4, and 5 is larger than the yield force capacity of link 3. It can be seen from equilibrium consideration of each link, as a free body diagram, that the maximum force that the chain can take is limited to the maximum capacity of its weakest link (link 3). Because the strength of the chain is limited by its weakest link, the other links will never be subjected to a force larger than the force in link 3.

Consequently, if link 3 is provided with ductile material, the chain can sustain the maximum force and experience yielding, whether the other links are ductile or not, as long as the strength of the links is larger than the yielding of the ductile link (link 3).

Once link 3 has yielded, the chain will develop a mechanism. Furthermore, the weak link in the chain may be any of the other four links. This means that the failure mode of the structure (the chain) can be controlled by the designer as desired.

Previous sections discuss the extreme randomness and uncertainty in the prediction of earthquake characteristics. In addition, because the structural response to earthquakes is extremely scattered, the most sophisticated structural analysis will only produce crude force predictions. Because of these high uncertainties, it would be prudent to control the structural behavior rather than try to predict extremely random events. This concept may be implemented through capacity design.

This simple chain concept can be applied to more complex structures. To do so, a designer first selects a proper mechanism and then identifies the plastic links and hinges along with their roles in the chain concept. This selection is analogous to selecting a weak link to be ductile in the chain described above. The other elements of the structure are then designed to remain elastic by distributing internal equilibrium through the rest of the structure, starting from the selected plastic regions. Consequently, the structure behavior will be guided by the designer selection of events, not by earthquake characteristics. The following example further illustrates this simple concept.

Example 5-8 A simple three-member steel truss, *A-B-C*, is shown in Example 5-8, Fig. 1. The truss is located in a seismic zone where a truss must be designed to resist a seismic force, Q_E, equal to 600 kN.

1. Use capacity design to size the truss members if it is only feasible to detail the bottom chord as a ductile member. The material is given as steel, $\sigma_y = 350$ MPa.

Solution Because the bottom chord is feasible to be ductile, this member will be selected as the weak link in the structure. The bottom chord will be designed to yield, whereas the rest of the members will be designed to remain elastic.

To understand the force distribution in the truss, the truss is analyzed for unit force at joint C. The resulting internal forces are shown in Example 5-8, Fig. 2.

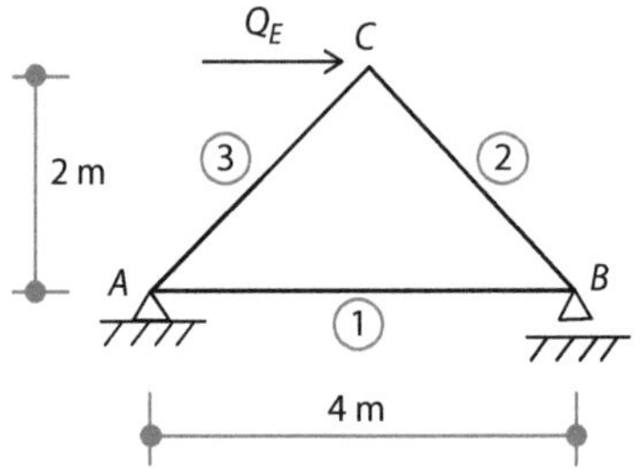

Example 5-8, Figure 1 Truss configuration.

It can be realized from the equilibrium of joint B that the internal forces, F, in the members are given as

$$\Sigma F_x = 0$$

$$F_1 = F_2 \cos 45 = 0.707\ F_2$$

Because equilibrium cannot be violated, the result above indicates that the force in one of these two members will be limited to the maximum of the other. Therefore, if member 1 is designed to yield with, $F_1 = F_{1y}$, then the force, F_2, will be limited to

$$F_2 = 1.414\ F_1 = 1.414\ F_{1y}$$

The force F_2 cannot exceed, 1.414 F_{1y}, regardless of the actual strength of member 2. This is an important conclusion because if member 2 is designed for strength larger than, 1.414 F_{1y}, the member does not have to be ductile insofar as it will never reach its maximum strength, regardless of the intensity of the earthquake.

The design force of member 1 can be obtained from Example 5-8, Fig. 2.

Since

$$F_1 = 0.5\ Q_E$$

hence,

$$F_1 = 0.5\ Q_E = 0.5(600) = 300\ \text{kN}$$

Cross-sectional area: $$A_1 = \frac{F_1}{\sigma_y} = \frac{300{,}000}{350} = 857\ \text{mm}^2$$

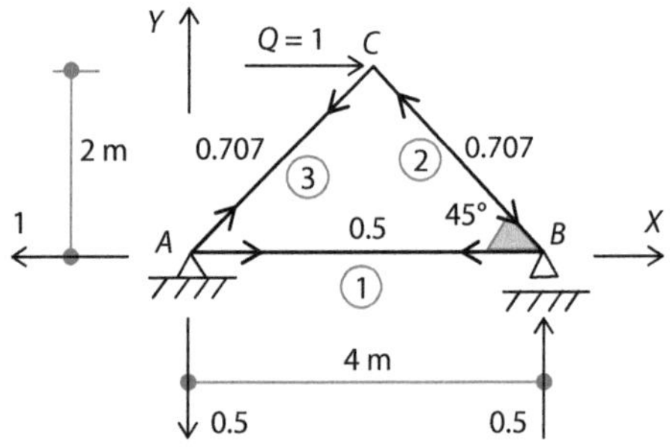

Example 5-8, Figure 2 Member forces.

To keep the other two members elastic, they must be designed with a force that exceeds their expected equilibrium forces. Thus, by spreading internal equilibrium from joint, B, the member forces are found as follows.

At joint B,

$$\Sigma F_x = 0$$

$$F_2 = \frac{F_1}{\cos 45} = 1.414\,F_1 = 1.414(300) = 424 \text{ kN}$$

The design force of member 2 must be increased above 424 kN in order to remain elastic (e.g., by 15 percent). Thus, the area is given as follows.

Cross-sectional area: $$A_2 = 1.15\frac{F_2}{\sigma_y} = 1.15\frac{424{,}000}{350} = 1{,}393 \text{ mm}^2$$

Member 3 will be treated in a similar way in order to find the same force limitation from equilibrium.

Cross-sectional area: $$A_3 = 1.15\frac{F_2}{\sigma_y} = 1.15\frac{424{,}000}{350} = 1{,}393 \text{ mm}^2$$

In this example, design can be accomplished with a controlled failure pattern. Such a controlled failure scheme needs fewer detailing requirements for ductility because the mechanism is selected beforehand and is directed toward member 1. Of course, the design may take another direction by directing the mechanism toward any other member and then proceeding with design as shown above. To make this design feasible, only member 1 has to be detailed for ductility (members 2 and 3 do not).

5.14 Pushover Analysis

Pushover analysis is a relatively modern technique used to construct the global load-deformation curve, known as the pushover curve, of the structure up to its failure. Pushover analysis includes nonlinear static analysis procedures that utilize monotonic inelastic properties of the material to construct the pushover curve. Such procedures were explained in previous sections to evaluate the local and global ductility capacities. This analysis is mainly intended for use in performance-based seismic design, which basically defines performance in terms of displacement of the pushover curve. Levels of performance are usually defined as shown in Fig. 5-47 through immediate occupancy level (IO), life safety level (LS), and collapse prevention level (CP).

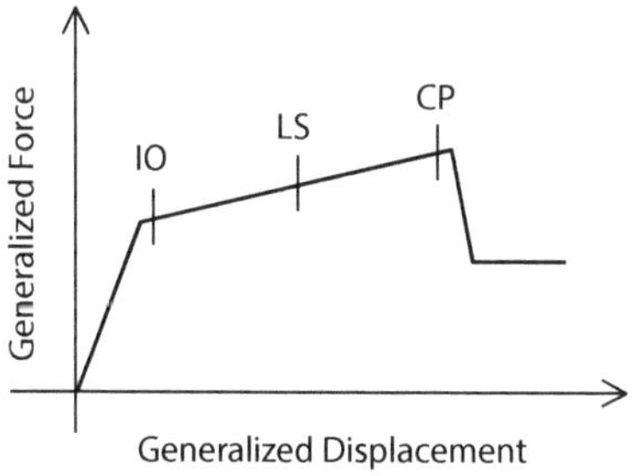

Figure 5-47 Generalized pushover curve.

More details of the analysis procedures and the acceptance criteria of these levels are found in the Federal Emergency Management Agency (FEMA) publication, *NEHRP Guidelines for the Seismic Rehabilitation of Buildings*, FEMA 273.

The author believes that such performance-based levels cannot be guaranteed to work as intended by the designer because of the inherent high levels of randomness, scatter, and uncertainty in the level of earthquake intensity characteristics. Such high levels of uncertainties render the level of response of structures highly unpredictable. In addition, utilization of monotonic material properties may not reflect the actual behavior of structures under earthquake excitation. Therefore, it may be worth the effort to examine pushover analysis under cyclic properties.

5.15 Recommended versus Undesirable Structural Systems

In general, seismic design prefers simplicity in structural systems and structural forms. Structures that exhibit good ductility, energy dissipation, and self-centering capacity are recommended. The following systems are recommended:

1. Systems with simplicity in plans. Structures with square and circular shapes are preferred.
2. Systems with compactness in shape. Avoid structures with long extended wings.
3. Systems with symmetry and large torsional resistance.
4. Systems with vertical uniformity and continuity. Avoid structures with sudden changes in mass and stiffness.

Examples of various systems are shown in Figs. 5-48 through 5-50. Figure 5-48(a) shows structural walls installed inside the building with small torsional rigidity, whereas Fig. 5-49(a) shows better arrangement of the structural walls that provide large torsional resisting capacity. Figure 5-48(b), for example, shows a structure with long extended wings. This kind of structure will usually perform poorly in earthquakes because the arrangement could result in severe torsional moments at the junction regions. A better arrangement would be to separate the wings from the main structure as shown in Fig. 5-49(b).

Another undesirable system is to have plastic hinges develop in columns, which results in the story mechanism shown in Fig. 5-48(c). This system is also known as the weak column-strong beam system. Such a system puts severe ductility demand on the columns and creates severe stability problems for the columns.

A better solution would be to enforce the plastic hinges to develop in the beams to create what is known as strong column-weak beam system as shown in Fig. 5-49(c). This system is also known as overall mechanism. Such a system is capable of dissipating energy through plastic deformation in the beams while preserving the integrity of the columns in terms of stability and low ductility demands.

The soft story is another undesirable system that should be avoided. Such a system is shown in Fig. 5-48(d). Because of the extreme sudden change between the columns on the ground floor and the rest of the building, the building would behave as a rigid body supported by the columns below. This puts very high ductility demands on the columns that cannot be achieved and also creates a severe stability problem for these

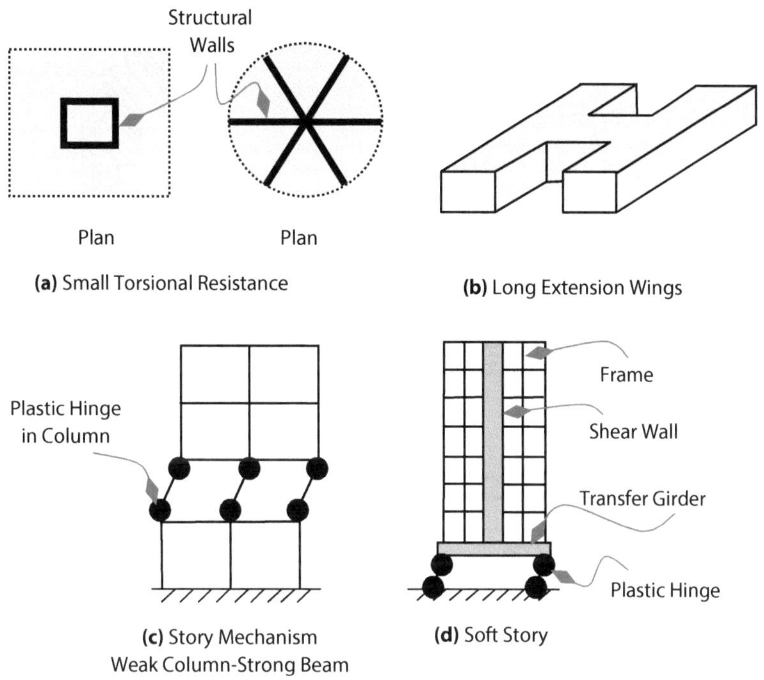

Figure 5-48 Undesirable seismic systems.

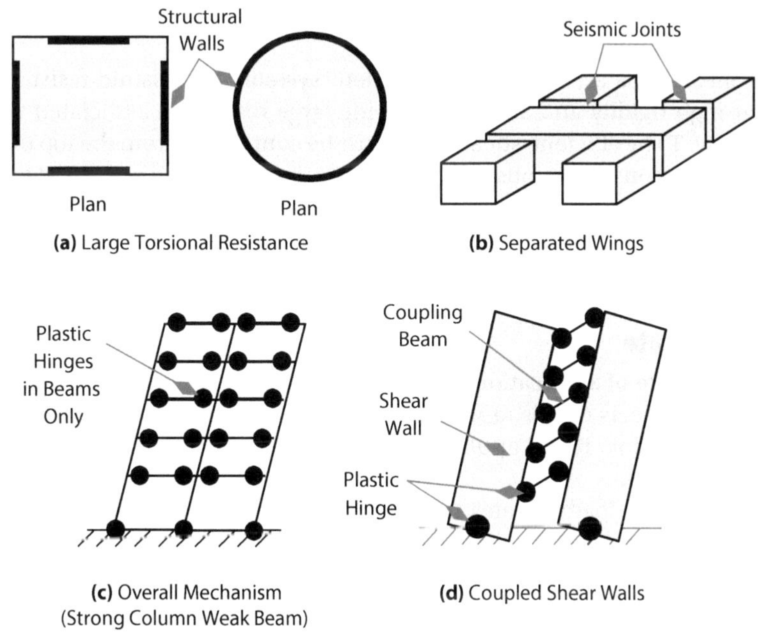

Figure 5-49 Recommended seismic systems.

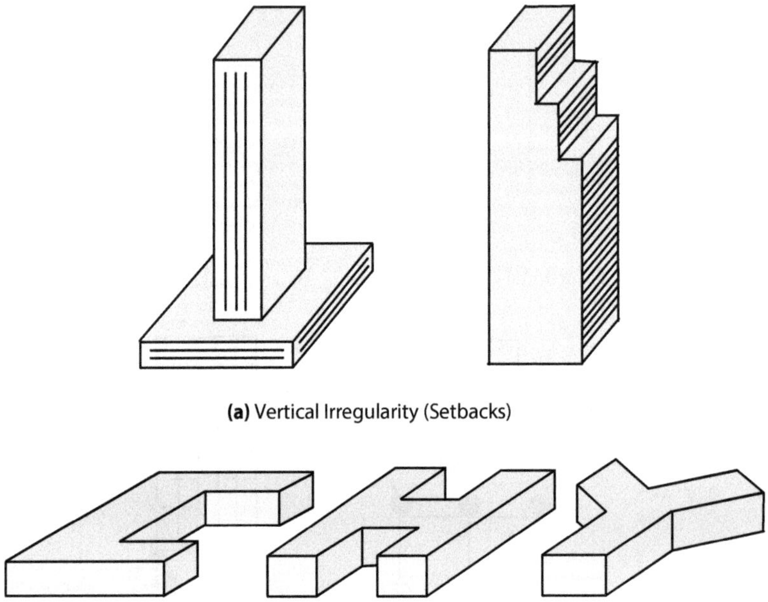

(a) Vertical Irregularity (Setbacks)

(b) Complicated Plan Shapes

Figure 5-50 Undesirable seismic irregularities.

columns. Many failures have occurred in previous earthquakes that are due to this soft story phenomenon.

Shear walls are considered excellent systems for seismic resistance because of their high rigidity and ability to provide large resistance associated with reasonable ductility. To be efficient, shear walls must be continuous from the top of the building to the foundations. They must not, in any case, be terminated at the first floor. Shear walls may be designed to stand alone as independent members, or they can be coupled with beams as shown in Fig. 5-49(d).

5.16 Strain Rate

The time rate of application of external loads affects the properties of material. The dynamic aspects of this rapid application is extensively addressed in various chapters of this book. Rapid application also affects the material properties of the structural elements.

In general, loading rates may be classified into three categories, namely, static, quasi-static, and dynamic rate. Static loading is merely a theoretical case insofar as it assumes that the load is applied to the structure during an infinite period of time. The practical slow application of loads that is used in laboratories results in a negligible dynamic effect known as quasi-static loading. Rapid loading induces dynamic effect in the structure and is known as dynamic loading.

From a practical point of view, dynamic loading may or may not affect material properties. Very rapid application of loads induces relatively high strain rates that can change the material properties such as stiffness. A ship striking a bridge is an example of a high rate of application of loads. In contrast, a relatively less rapid application of loads may induce a negligible effect on the material properties.

Earthquake loading is considered to be the second type of dynamic loading in which the effect of strain rates on the material properties may be neglected. In this case, the analysis may be performed using the material properties obtained from the quasi-static response.

5.17 Problems

5-1 An elevated tank is supported by a weightless, reinforced concrete single column with the following properties:

$$f'_c = 30 \text{ MPa} \qquad \varepsilon_{cu} = 0.003$$

$$f_y = 300 \text{ MPa} \qquad A_{s,tot} = 1\%\, A_g$$

The structure is subjected to an earthquake as shown in Prob. 5-1, Fig. 1(a), and having a constant spectral pseudo-velocity of $S_v = 0.7$ m/s.

For simplicity, consider an idealized moment-curvature relationship as shown in Prob. 5-1, Fig. 1(c), and also consider that compression steel is yielding with tension steel.

The structure's global ductility capacity, μ_c, is 5. Check the structural safety against both EDC and EEC during the earthquake excitation.

NOTE: Consider the effect of the tank weight on the capacity and ductility of the column.

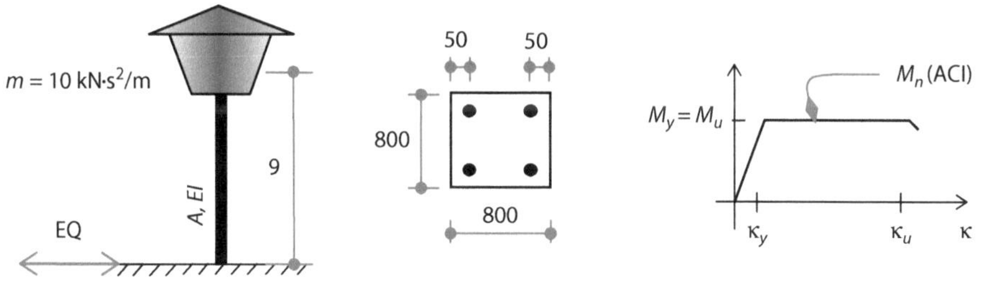

(a) Structure Layout

(b) Section (Dimensions in mm)

(c) Idealized M-κ diagram

PROBLEM 5-1, FIGURE 1

5-2 A long period shear wall is shown in Prob. 5-2, Fig. 1(a). The elastic response to an earthquake excitation is shown in Prob. 5-3, Fig. 1(b). The wall has been designed using the force-reduction factor concept. Structural analysis shows that the wall moment-curvature relationship is as shown in Prob. 5-2, Fig. 2.

1. Calculate the force-reduction factor and global ductility demand for this wall.
2. Determine whether this wall can survive this earthquake without collapse.

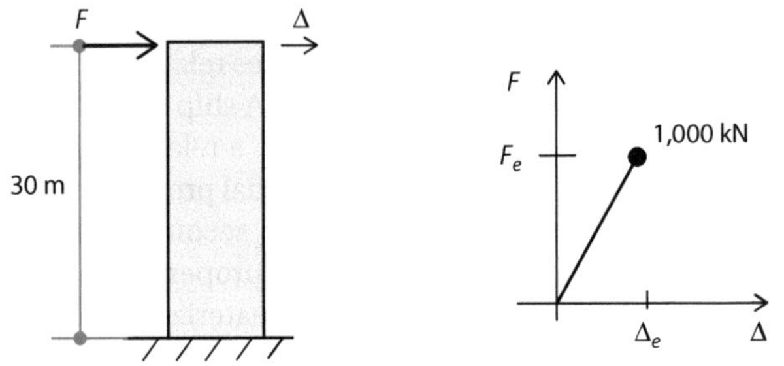

(a) Shear Wall

(b) Maximum Response Results from Elastic Dynamic Analysis

Problem 5-2, Figure 1

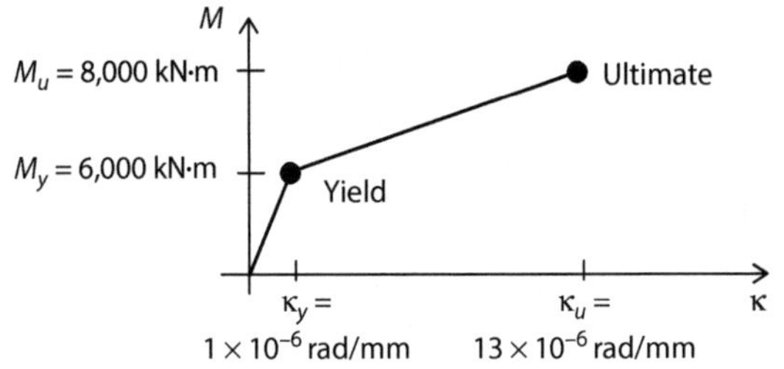

Problem 5-2, Figure 2 Moment-curvature relationship.

5-3 A frame with two fixed supports at its base has a total mass, m, and dimensions as shown in Prob. 5-3, Fig. 1. The frame has an infinite stiff beam, with steel columns having the cross section shown in Prob. 5-3, Fig. 2. The column stress-strain curve is also given in Prob. 5-3, Fig. 3.

The frame is excited by an EQ having the response spectrum shown in Prob. 5-3, Fig. 4, and acting in the direction shown in Prob. 5-3, Fig. 1.

Assuming $\zeta = 2\%$, do the following:

1. Calculate the elastic force and elastic displacement due to this EQ (e.g., ESD).
2. Find the force-reduction factor, R.

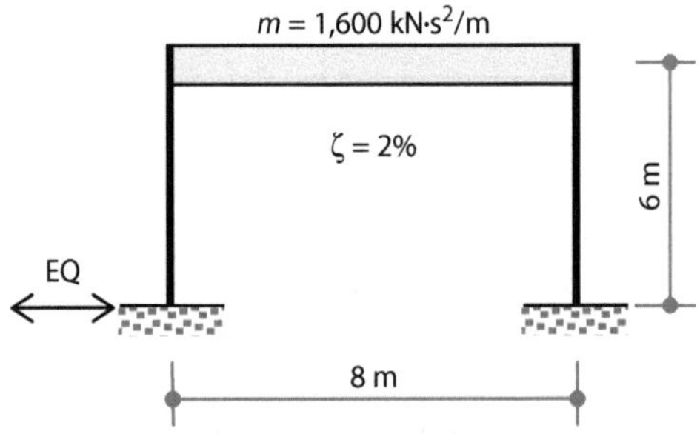

Problem 5-3, Figure 1 Steel frame.

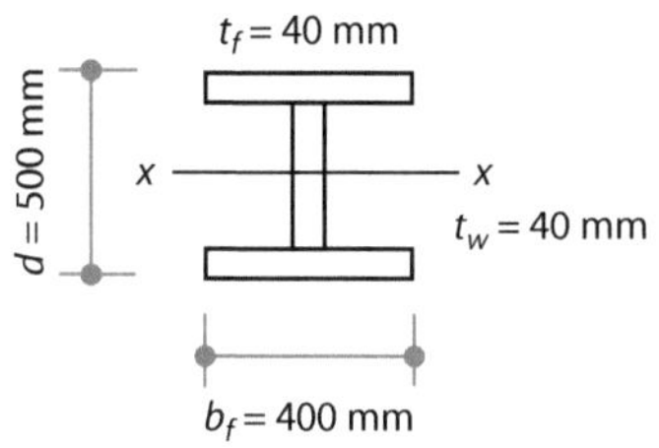

PROBLEM 5-3, FIGURE 2 Column cross section.

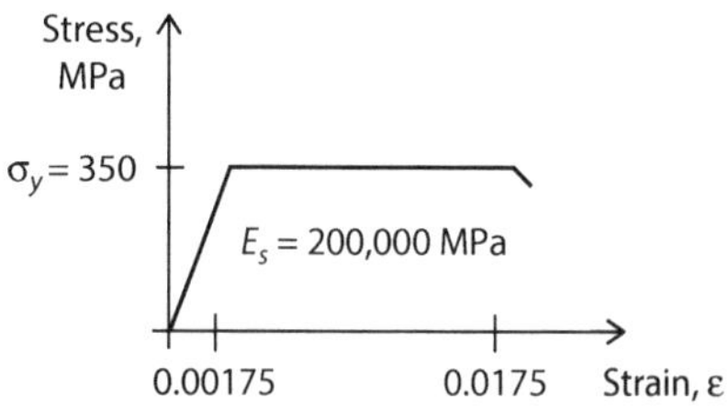

PROBLEM 5-3, FIGURE 3 Material properties.

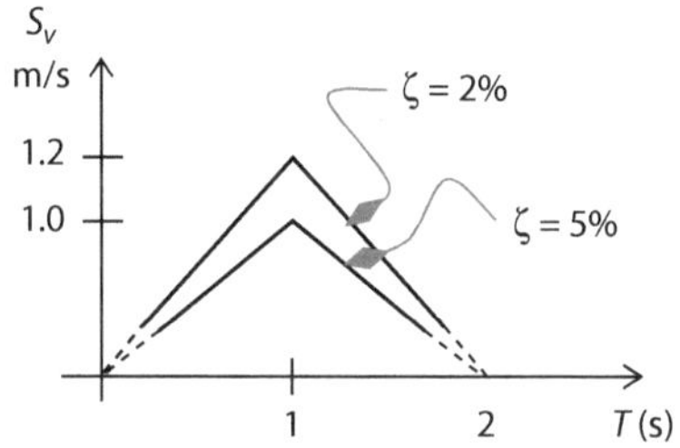

PROBLEM 5-3, FIGURE 4 EQ response spectrum.

3. Find the global ductility demand, μ_d.
4. Check the frame safety in terms of the global ductility capacity, μ_c.

5-4 A rigid concrete slab is supported by four circular columns as shown in Prob. 5-4, Fig. 1. The columns are fixed at both ends and located as shown. The columns are 240 mm in diameter with effective depth, d, of 180 mm. The effective height of columns is 4 m.

The mass and stiffness matrices are given in terms of the centroid coordinates u, v, and θ as follows:

$$m = 2\begin{bmatrix} 60 & 0 & 0 \\ 0 & 60 & 0 \\ 0 & 0 & 680 \end{bmatrix} (\text{kN}\cdot\text{m}\cdot\text{s}), \quad k = 410\begin{bmatrix} 4 & 0 & 0 \\ 0 & 4 & -8 \\ 0 & -8 & 88 \end{bmatrix} (\text{kN}\cdot\text{m})$$

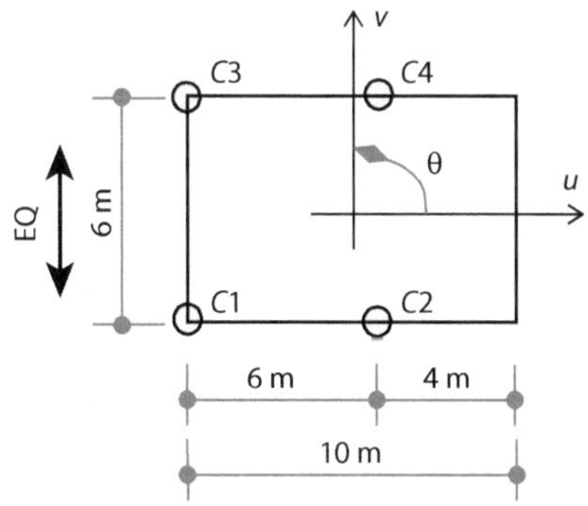

PROBLEM 5-4, FIGURE 1 Structural layout.

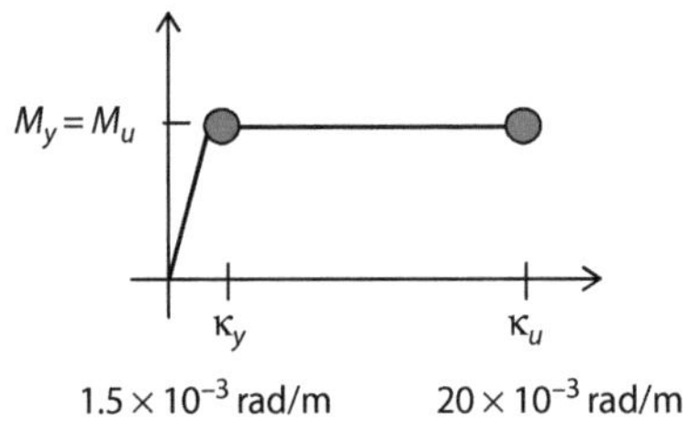

Problem 5-4, Figure 2 Moment-curvature.

The moment, M, and curvature, κ, relationship is given as shown in Prob. 5-4, Fig. 2.

If the earthquake response spectrum is given as, $S_a = \frac{1}{T} \leq 1\ \mathrm{m/s^2}$, do the following:

1. Find the fundamental frequency and its mode shape for this structure.
2. For only column 3, find the top displacement and top force due to the first mode.
3. Check the safety of column 3 against the EEC and the criterion in IBC Code.

5-5 The reinforced concrete I-section shown in Prob. 5-5, Fig. 1, is reinforced with 7 ϕ 25 bars located as shown. The cross-sectional area of each bar is given as 491 mm².

If this cross section is subjected to factored axial load, P_u = 900 kN, then calculate the ultimate curvature capacity of the cross section.

Material properties: $f_c' = 30$ MPa

$\varepsilon_{cu} = 0.003$

$f_y = 420$ MPa

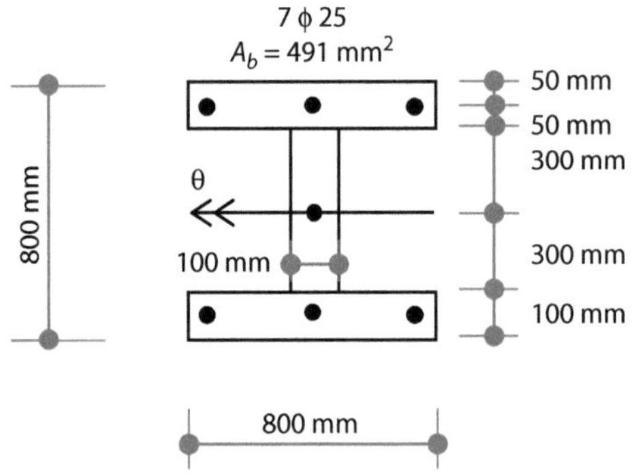

Problem 5-5, Figure 1

CHAPTER 6

Design of Earthquake-Resistant Buildings (IBC)

6.1 Introduction

For many years, the Uniform Building Code (UBC) has been recognized internationally as a model code, especially for seismic provisions. In fact, many countries have adopted UBC provisions as the foundation of their local building codes. After the last version of the UBC was issued in 1997, the UBC was merged with two other building codes to form the International Building Code (IBC). The first version of the IBC was published in 2000 as a unified building code in the United States.

Because the IBC is a merger of the UBC and two other building codes, it continues the same philosophy of all three codes. In fact, most of the seismic codes in the United States are based on the provisions of the National Earthquake Hazard Reduction Program (NEHRP), as administered by the Federal Emergency Management Agency (FEMA). As a result, the IBC addresses the same main issues as the last version of the UBC, with some changes in the format of force calculations and seismic maps.

This chapter presents the IBC as the current model code for seismic provisions and design. Because the IBC also refers to ASCE 7, Minimum Design Loads for Buildings and Other Structures, published by the American Society of Civil Engineers, this chapter will also discuss ASCE 7 when necessary.

Because of the extreme uncertainty in earthquakes and their effect on structures, building codes cannot clearly and precisely define the required seismic parameters, such as the response modification factor, R, and the global ductility demand, μ_d. Therefore, in addition to theory, codes rely on experience and engineering judgment to define levels of strength and ductility of structures in different seismic regions. As a result, codes do not leave much room for the engineer to decide on the proper selection of R and μ_d combinations. Instead, codes limit the use of structural systems in seismic regions to those that have performed well in previous earthquakes and have been extensively tested in research laboratories.

For example, the IBC identifies and permits structural systems such as, but not limited to, ordinary, intermediate, and special moment frames (SMFs). Ordinary shear walls, special shear walls, concentrically braced frames, and eccentrically braced frames are also identified and permitted. In addition, the IBC allows combination of frames and shear walls or braced frames, which are identified as dual systems.

The codes explicitly state that any system not listed in their tables shall not be permitted unless the system is proven by theory and experiment to be capable of resisting earthquakes. As may be expected, this is a very difficult and involved task for engineers to complete in practice.

For each identified system, the codes usually assign fixed values of R and μ_d without any flexibility for the engineer to examine or define such values. Instead, the codes describe specific conditions and level of detailing for each identified system, implying that the systems achieve the intended behavior. Furthermore, codes do not guarantee intended performance. Instead, they explicitly state that such requirements are minimum requirements that need to be included in the system. This statement clearly reflects the uncertainty in structural performance and in earthquake characteristics.

In general, codes define seismic hazard through the use of seismic zoning maps, which are based on probabilistic measures. The strength level of the structure is then provided according to these maps.

6.2 Definition of Structural Components

The basic structural components of an earthquake-resistant system are shown in Fig. 6-1. The main structural system usually consists of frames, shear walls, or a combination of both (a dual system). These elements are connected at the floor level by diaphragms.

If the earthquake excitation is as shown in Fig. 6-1, the seismic resisting elements are given as the marked frames and the shear walls shown as active in this direction. These elements are forced to act as one unit because they are connected through diaphragms. The diaphragms can be rigid or flexible. Diaphragms may be idealized as rigid according to ASCE 7 if they consist of concrete slabs or concrete-filled metal decks with span-to-depth ratios of 3 or less that have no horizontal irregularities. In addition, the American Concrete Institute (ACI) code requires such concrete decks to be at least 50 mm thick to qualify as rigid diaphragms. In rigid diaphragms, forces are distributed to frames and walls according to their relative rigidity. Flexible diaphragms, however, require special modeling of diaphragms because they deform in plane and interact with other structural elements. Flexible diaphragms may be modeled as shell elements in finite element analysis in conjunction with overall modeling of the structure. To determine the design forces in lateral force resisting systems, ASCE 7 allows the treatment of flexible diaphragms as simple beams between vertical supporting elements.

Rigid diaphragms are designed to transfer the in-plane forces generated by an earthquake to the lateral force supporting elements, such as frames or shear walls. A diaphragm may be analyzed as an I-beam in the transverse direction with its thickness as its web and the surrounding exterior beams as its flanges. These are the marked diaphragm chords in Fig. 6-1, which may also be termed boundary elements. The

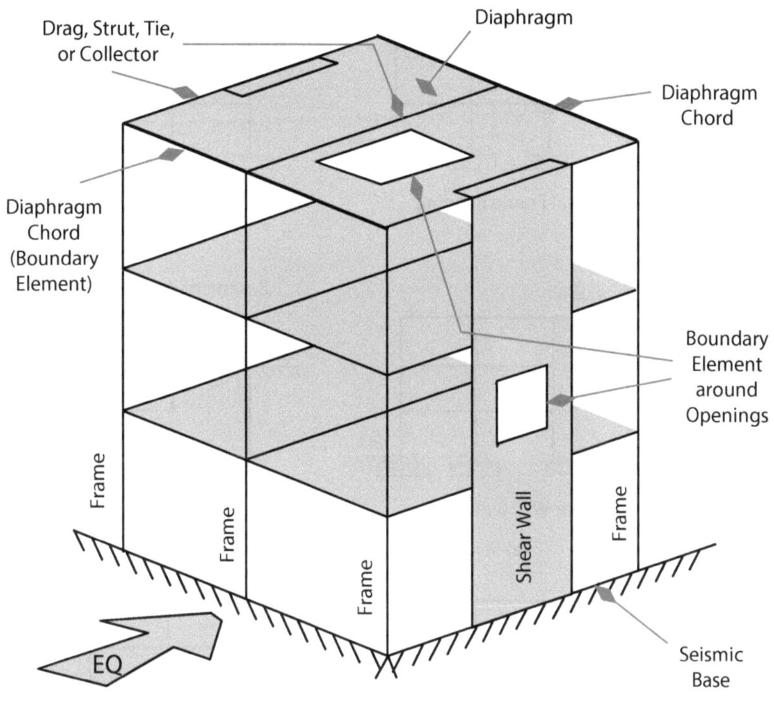

FIGURE 6-1 Basic components of earthquake-resistant building.

junctions between the diaphragms and the lateral force supporting elements must be designed to transfer the diaphragm forces to these elements. Such components are known as drags, struts, or collectors because they collect forces from diaphragms and transfer them to supporting elements.

6.2.1 Seismic Base

The location of seismic base is one of the most important factors that affect seismic analysis, especially in the presence of basement walls, as it defines the level where the earthquake effect is imparted to the building. The seismic base depends mainly on the relative stiffness of the lateral resisting structural system components around the ground level. This relative stiffness is affected by many factors, such as, but not limited to, the location of the grade level, surrounding soil conditions, depth of basement location, and grade slopes around the building.

ASCE 7 standards outline various cases to define the seismic base for buildings. Some of these cases are presented for illustration as shown in Fig. 6-2. Figure 6-2(a) shows typical cases where the seismic base is considered at the top of foundation level.

Figure 6-2(b) shows special cases where the seismic base is determined according to the relative stiffness of the structure below the grade level, and also on the stiffness of the surrounding soil. For example, if the soil is too soft that it will compress during earthquake excitation, the seismic base should be taken at the foundation level. If on the other hand, the basement walls or foundation walls exist, the seismic base may be

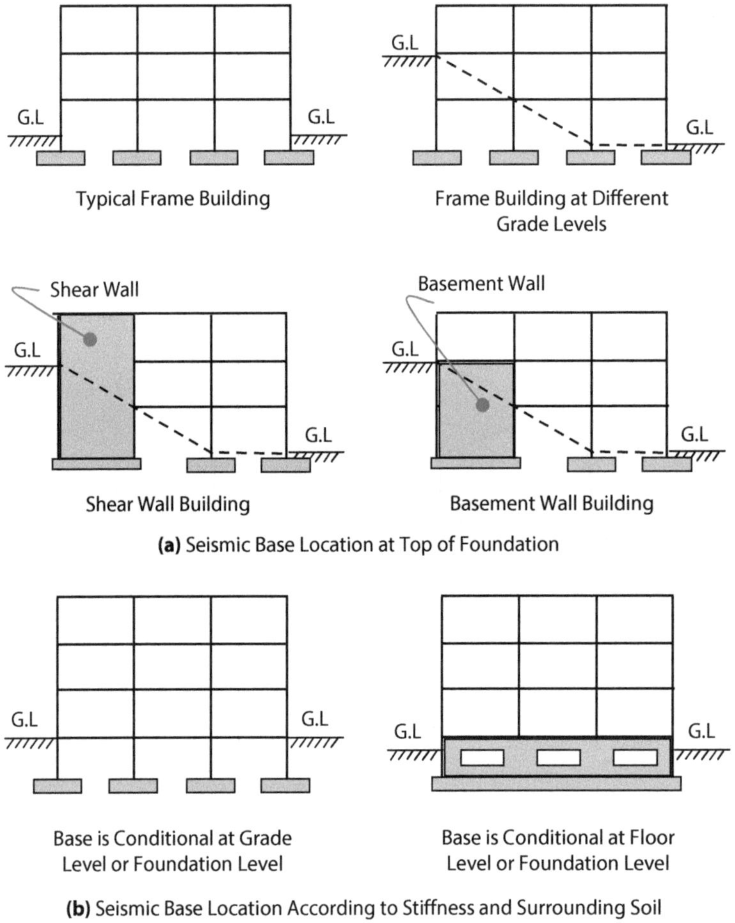

Figure 6-2 Seismic base location.

considered at the floor level above the grade. In this case, the conditions laid out for the two-stage analysis procedures, as outlined later in this chapter, may be considered for analysis.

6.3 Seismic Design Category

The IBC classifies a building into a Seismic Design Category (SDC) according to the building occupancy categorization and its region of seismic hazard. The risk categories are classified into four groups as I, II, III, and IV. Each category is assigned an importance factor, I_e, according to its functionality as given in Table 6-1. The SDCs are classified into the six categories—A, B, C, D, E, and F, as given in Table 6-2. For convenience, Tables 6-1 and 6-2 are also provided in App. 6-1 at the end of this chapter.

S_{DS} and S_{D1}, given in Table 6-2, are the design spectral acceleration at short period and at a 1-s period, respectively. These are defined in the next section.

Risk Category		Seismic Importance Factor, I_e
I	Low hazard to humans, such as agricultural facilities	1
II	Other than I, III, and IV	1
III	Hazardous to humans because of large number of occupants, such as schools and public areas	1.25
IV	Essential facilities either important for security or for major services such as rescue operations, health services, transportation, and communications	1.5

Table 6-1 Importance Factor

S_{DS}	S_{D1}	Risk Category	
		I, II, or III	IV
$S_{DS} < 0.167$	$S_{D1} < 0.067$	A	A
$0.167 \leq S_{DS} < 0.33$	$0.067 \leq S_{D1} < 0.133$	B	C
$0.33 \leq S_{DS} < 0.50$	$0.133 \leq S_{D1} < 0.20$	C	D
$0.50 \leq S_{DS}$	$0.20 \leq S_{D1}$	D	D
$S_1 \geq 0.75$		E	F

Table 6-2 Seismic Design Category (SDC)

6.4 Zoning Classification

The IBC divides the United States into contour lines that plot the spectral acceleration for a maximum considered earthquake (MCE) at bedrock. The MCE is based on a 2 percent probability of exceedance in a 50-year period. This probability implies a return period of 2,500 years. Note that the spectral acceleration in these maps is not the design spectral acceleration. As will be shown later, the design spectral acceleration is taken as two-thirds of the spectral acceleration obtained from the maximum considered earthquake indicated in these maps.

For each region, the IBC assigns two spectral acceleration values for the maximum considered earthquake; one value for spectral acceleration at short period, S_s, and another value for spectral acceleration at a 1-s period, S_1. These spectral accelerations given in seismic hazard maps (listed in percent of g) are measured at bedrock. The effects of the site characteristics are considered by using factors F_a and F_v for short period and at a 1-s period, respectively. Values of F_a and F_v are listed in Tables 6-A3 and 6-A4 at the end of this chapter. As shown in those two tables, sites are classified into six classes as A, B, C, D, E, and F implying that there are six distinct response spectra in this classification.

Accordingly, the maximum considered earthquake spectral acceleration at the base level modified for each site is defined at short period as S_{MS} and at 1-s period as S_{M1}. These quantities are evaluated as follows:

$$S_{MS} = F_a S_s$$
$$S_{M1} = F_v S_1$$

where F_a and F_v = soil factor for short period and at 1-s period, respectively, as listed in Tables 6-A3 and 6-A4 at the end of this chapter

S_s and S_1 = spectral acceleration values listed in seismic maps for short period and at 1-s period, respectively

6.5 Response Spectra

The response spectra in IBC are defined for six site classes as A, B, C, D, E, and F. The design spectral acceleration for each site class is given as S_{DS} and S_{D1} for short period and at 1-s period, respectively. The response spectra parameters, S_{DS} and S_{D1}, are taken as two-thirds of the spectral acceleration for the maximum considered earthquake S_{MS} and S_{M1}, that is,

$$S_{DS} = \frac{2}{3} S_{MS}$$

$$S_{D1} = \frac{2}{3} S_{M1}$$

where S_{MS} and S_{M1} are defined by the expressions above in the previous section.

Once S_{DS} and S_{D1} are defined for the site, the response spectrum can be constructed as shown in Fig. 6-3, which is expressed mathematically as follows.

Controlling periods

$$T_s = \frac{S_{D1}}{S_{DS}}$$

$$T_o = 0.2\,T_s$$

T_L = long-period transition period assigned to regions by the code

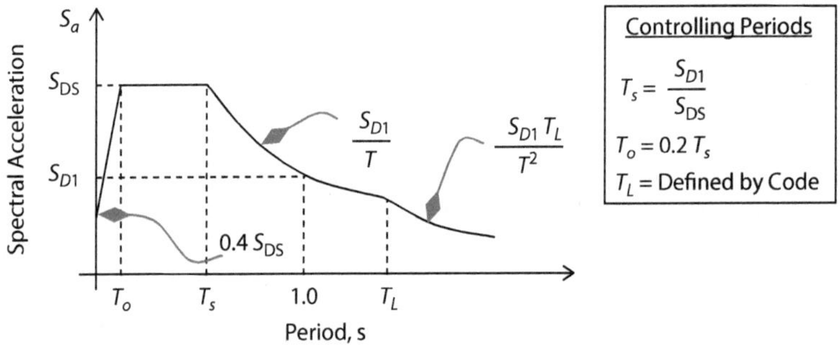

FIGURE 6-3 IBC response spectra.

For $T \leq T_o$, the response spectrum S_a is given as

$$S_a = S_{DS}\left(0.4 + 0.6\frac{T}{T_o}\right)$$

For T between T_o and T_s, the response spectrum S_a is given as

$$S_a = S_{DS}$$

For $T \geq T_s$, and less than T_L, the response spectrum S_a is given as

$$S_a = \frac{S_{D1}}{T}$$

For $T > T_L$, the response spectrum S_a is given as

$$S_a = \frac{S_{D1}\,T_L}{T^2}$$

The response spectrum value starts at $S_a = 0.4\ S_{DS}$, at $T = 0$ in the absence of any other information, the code allows peak ground acceleration (PGA) to be taken as $0.4\ S_{DS}$.

The six classifications of soil classes follow, in principle, the response spectra curves defined by Seed et al., and explained in Fig. 3-13. The spectra curves reflect the fact that the response spectra are simply a function of the site characteristics. At this point, IBC soil classification may be summarized in broad terms as follows:

1. Soil Type A: Hard rock
2. Soil Type B: Rock
3. Soil Type C: Very dense soil and soft rock
4. Soil Type D: Stiff soil profile
5. Soil Type E: Soft soil profile
6. Soil Type F: Needs special evaluation

6.6 Design Requirements of Seismic Design Categories

The design forces and requirements of structures are given according to their SDC. The effective seismic weight to be considered for seismic force calculations is given as:

1. Total dead load above the base
2. 25 percent of live load in storage areas and warehouses
3. Actual partition weight; but shall not be taken less than 0.48 kN/m^2
4. 20 percent of the snow load if intensity of the snow exceeds 1.44 kN/m^2
5. Weight of landscaping and other material at roof gardens and similar areas
6. Permanent equipment

Seismic Design Category A

The seismic forces in the structure may be calculated directly for each floor as follows:

$$F_x = 0.01\ w_x$$

where F_x = seismic force applied to level x
w_x = share of story from the effective seismic weight as defined earlier

The force is allowed to be applied separately in each of two orthogonal directions of the structure. The orthogonality effect can be neglected in this category. In addition, the special seismic load combination that includes seismic force, E_m, does not need to be considered in this category. (Seismic force will be explained later.)

Connections in this category must be designed for horizontal forces equal to 5 percent of the vertical dead and live loads applied to these connections.

Seismic Design Category B and C

The seismic forces in the structure may be calculated using any of the four methods given in the next section as applicable. Other design requirements will be given as they develop throughout the next sections.

Seismic Design Category D, E, and F

The seismic forces in the structure may be calculated using any of the four methods given in the next section as applicable. Other design requirements will be given as they develop throughout the next sections.

6.7 Earthquake-Induced Forces

The IBC does not explicitly give details about force calculations. However, the code does require that all force calculations be made according to ASCE 7 standards. In its standard, ASCE offers four methods of analysis to evaluate earthquake-induced forces in structures as follows:

1. Simplified lateral force analysis procedure
2. Equivalent lateral force procedure
3. Modal response spectrum analysis
4. Time-history analysis

Because all methods expect the structure to go into its inelastic range, the structure must be designed accordingly. In methods 1, 2, and 3, inelastic response is incorporated in the structure through the concept of response modification factor, R, in association with inelastic displacement, δ, as explained in previous chapters.

In method 4, elastic dynamic analysis may be performed, resulting in elastic earthquake-induced forces. In this case, the concept of $R - \mu_d$ is also utilized to find design forces and displacements as done in methods 1, 2, and 3. However, if method 4 is used and utilizes the structures actual inelastic property, then R and δ relations will not apply. Instead, an actual hysteresis model based on experimental data must be used. In this case, the resulting forces and displacements will be the final design quantities as explained in previous chapters.

The IBC also uses an importance factor, I_e, to reflect the importance of the structure as given in Table 6-A1. In summary, I_e is given a value of 1.25 for hazardous facilities whose failure may result in a large number of casualties. I_e is given a value of 1.5 for essential facilities that are expected to remain functional during and after earthquakes.

The IBC also classifies structures into building and nonbuilding types. In general, the nonbuilding classification includes building-like structures such as water tanks, towers, and vessels. This classification does not include bridges, which are addressed in bridge codes such as the ones published by the American Association of State Highway and Transportation Officials (AASHTO). The IBC assigns R-values for each group of the types above. Selected systems are given in Tables 6-A6 and 6-A7.

The IBC also assigns ductility ratios for each system individually, which is designated as displacement amplification factor, C_d. Values of C_d are also listed in Tables 6-A6 and 6-A7.

According to the code, only structural systems that are listed in those tables can be used as seismic-resistant systems. If a system is not listed, it can be used only if its seismic structural properties can be verified by an approved cyclic test and adequate analysis that prove its integrity. As may be expected, this is a very difficult task to complete.

The IBC defines the first significant yield as the design level similar to the definition in provisions issued by the NEHRP. Accordingly, the design level V_s is given as shown in Fig. 6-4. This strength level is reduced from the elastic strength demand by the R-factor. The code also defines a system overstrength factor, Ω_o, which brings the structure to its fracture level, V_m. This level includes any strain-hardening effect of the material that may come early in the structure because of cyclic effect.

Accordingly, the code format for maximum force, E_m, that may develop in any member is given as

$$E_m = \Omega_o Q_E + 0.2\, S_{DS} D$$

where Q_E is the force resulting from the horizontal component of the design of earthquake (base shear). This will be explained later in this chapter. The second term includes the effect of the vertical component of the earthquake, which is given proportional to the dead load, D.

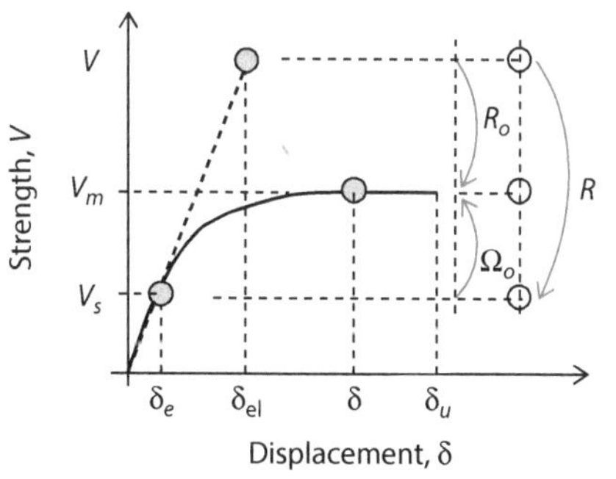

FIGURE 6-4 IBC definition of strength and displacements.

The weight of the structure included in the force calculations that represent the structure mass will be as defined earlier. The sum of these weights will be referred to as the effective seismic weight, W.

Because the application of these four methods of analysis depends on structural regularity, the regularity of structures will be discussed before an in-depth consideration of the analysis methods.

6.7.1 Regularity of Structures

In general terms, the response of regular structures to dynamic forces is dominated by the first mode of vibration. Therefore, dynamic analysis will be greatly simplified and may be performed with an equivalent static procedure that uses a force distribution close to the first mode of vibration. In contrast, the response of irregular structures to dynamic forces cannot easily be predicted. Therefore, explicit dynamic analysis is required to obtain the equivalent forces and their distribution in the structure. The IBC allows an equivalent lateral static procedure for regular structures and some irregular structures, among other conditions (as will be seen later). Irregularity is defined in ASCE Table 12.3-1 for types of horizontal irregularity, and in Table 12.3-2 for types of vertical irregularity.

In addition, the code prohibits the irregularities outlined in the following sections to exist in SDC D through F as follows:

SDC D: Vertical irregularity Type 5b shall not be permitted.

SDC E or F: Horizontal irregularity Type 1b shall not be permitted. Vertical irregularity Type 1b, 5a, or 5b shall not be permitted.

Types of horizontal and vertical irregularities are summarized as follows.

Horizontal Types of Irregularity

In a plan, irregularity may exist because of displacements, geometry, and discontinuity. These irregularities may be summarized as follows:

Type 1a. Torsional irregularity in rigid diaphragms: When the maximum displacement of one corner of the diaphragm exceeds 1.2 times the average displacement of both corners in each direction.

Type 1b. Extreme torsional irregularity in rigid diaphragms: When the maximum displacement of one corner of the diaphragm exceeds 1.4 times the average displacement of both corners in each direction.

Type 2. Reentrant corners: When the projection of both ends of a reentrant exceeds 15 percent of the structure dimension.

Type 3. Diaphragm discontinuity: Diaphragm discontinuity is considered when opening in the diaphragms exceeds 50 percent of its area. A 50 percent difference in a diaphragm's stiffness between adjacent stories is also considered irregular.

Type 4. Out-of-plane offsets in the vertical direction.

Type 5. Nonparallel systems: When lateral-load-resisting elements are not parallel or not symmetric about the major orthogonal axes of the structure.

Vertical Types of Irregularity

In the vertical direction, irregularity may exist due to changes in stiffness, weight, geometry, and discontinuity. These irregularities may be summarized as follows:

Type 1a. Stiffness irregularity—soft story: A story is considered soft if its stiffness is less than 70 percent of the story above or less than 80 percent of the average stiffness of the three stories above.

Type 1b. Stiffness irregularity—extreme soft story: A story is considered extremely soft if its stiffness is less than 60 percent of the story above or less than 70 percent of the average stiffness of the three stories above.

Type 2. Weight (mass) irregularity: Weight irregularity exists if there is a 150 percent difference in the mass of adjacent floors. A lighter roof is excluded from this definition.

Type 3. Vertical geometric irregularity: Vertical geometric irregularity exists if there is a 130 percent difference in the horizontal dimension of the lateral force system between adjacent floors.

Type 4. In-plane discontinuity in a vertical lateral force resisting element: In-plane discontinuity exists if there is an in-plane offset of the element in the vertical direction equal to the length of the element.

Type 5a. Discontinuity in lateral strength-weak story: A story is considered weak if its strength is less than 80 percent of the story above.

Type 5b. Discontinuity in lateral strength-extremely weak story: A story is considered extremely weak if its strength is less than 65 percent of the story above.

6.7.2 Simplified Lateral Force Analysis Procedure

This method may be applied to simple, three-story buildings with bearing wall or building frame systems, subject to strength and geometry limitations. Some of these limitations may be highlighted as follows:

1. SDC is determined according to S_{DS} value.
2. Structures shall only be in Risk Categories I and II.
3. Site class shall be a class other than *E* or *F*.
4. Structures will have at least two lines of lateral resisting systems in each major direction, provided that at least one line of resistance lies on each side of the center of mass.
5. Story strength shall not be less than 80 percent of the story above.

If the limitations above, among others, are satisfied, then the total base shear, *V*, may be calculated using the following expression:

$$V = \frac{FS_{DS}}{R} W$$

where $F = 1.0$ for one-story buildings
$= 1.1$ for two-story buildings
$= 1.2$ for three-story buildings
R = response modification factor as per Table 6-A6
S_{DS} = design spectral acceleration at short period as defined earlier. In calculating S_{DS}, S_S need not be taken as larger than 1.5
W = the effective seismic weight defined earlier

Because the period of low-rise buildings is very short, such buildings will fall in the range of periods less than T_s, as given in Fig. 6-3. For such structures, the code implies that the force may conservatively be calculated at the maximum values in the response spectrum range, S_{DS}. However, the code takes one more precaution by increasing the force by the factor, F, as seen in the final expression earlier.

Vertical Distribution of Base Shear

In this simplified method, floor force is directly proportional to floor effective seismic weight

$$F_x = \frac{w_x}{W} V$$

where F_x is the floor force, and w_x is the floor effective seismic weight as shown in Fig. 6-5.

Substituting the base shear value, V, in the equation above, the floor forces may be expressed in the following form:

$$F_x = \frac{FS_{DS}}{R} w_x$$

Unless otherwise calculated by more refined methods, the story drift, Δ, in this method may be taken as 1 percent of the story height. The story drift, Δ, given in this section is the relative displacement between stories.

Example 6-1 A two-story residential building with story height equal to 3 m is shown in Example 6-1, Fig. 1. The building will be constructed in a region with the mapped acceleration coefficient at short period, S_s, equal to 1.23. The site is classified as Soil Type C.

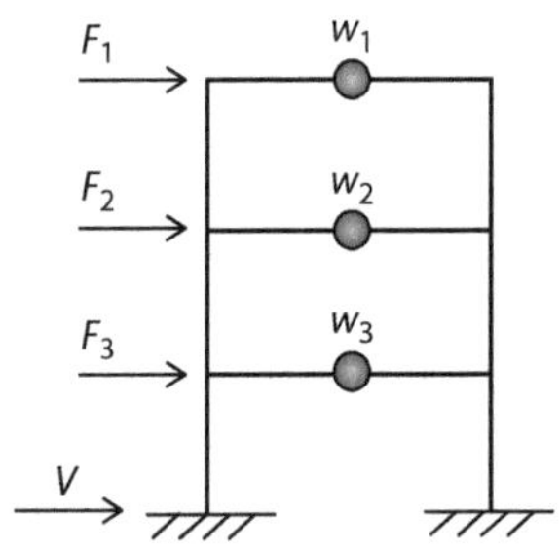

Figure 6-5 Floor forces.

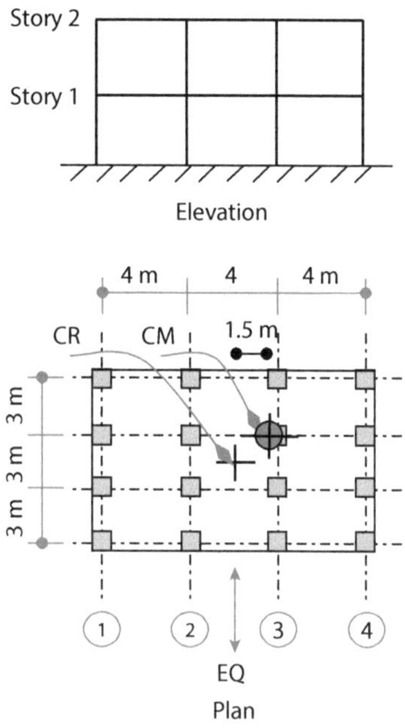

EXAMPLE 6-1, FIGURE 1 Two-story building.

The structural system of the building consists of four reinforced concrete moment frames in each direction, which are spaced as shown in the plan. The interior frames have double the stiffness of the exterior frames. The center of mass, CM, of each floor is offset by 1.5 m from the center of rigidity in both directions.

The first and the second floor carry a total effective seismic weight of intensity equal to q_D = 10 kN/m^2 and 7 kN/m^2, respectively.

One must find the design seismic force for each of the four frames marked 1 through 4 on the plan if the earthquake direction is along these frames (in the short direction as shown in the figure).

Solution The simplified lateral force analysis procedure may be used because the building is residential, placing it in Risk Category I, and the building has only two stories. Refer to App. 6-1 for relevant IBC tables.

1. Seismic coefficients, F_a and S_{DS}
   For the given S_S value and Soil Type C, the site coefficient F_a is 1.0 (from Table 6-A3). Accordingly, the seismic coefficient S_{DS} is calculated as follows:

$$S_{MS} = F_a S_s = 1.0\,(1.23) = 1.23$$

$$S_{DS} = \frac{2}{3}\,S_{MS} = \frac{2}{3}\,(1.23) = 0.82$$

2. Risk and SDC
   A residential building is in Risk Category I. Knowing that S_{DS} = 0.82 and by referring to Table 6-A2, for Risk Category I, it can be determined that the building is in SDC D.

3. Response modification factor, R
   Table 6-A6 states that ordinary and intermediate moment frames are not permitted for SDC D. Nevertheless, SMF are permitted. Therefore, if an SMF system is selected for the building, the response modification factor, R, is obtained for SMF as $R = 8$.

4. Effective seismic weight, W

Area of each floor:	$A = 9\,(12) = 108\ \text{m}^2$
Effective seismic weight of first floor:	$w_1 = q_D A = 10\,(108) = 1{,}080\ \text{kN}$
Effective seismic weight of second floor:	$w_2 = q_D A = 7\,(108) = 756\ \text{kN}$

Total effective seismic weight of the building:

$$W = w_1 + w_2 = 1{,}080 + 756 = 1{,}836\ \text{kN}$$

5. Base shear, V

The building is a two-story building with factor $F = 1.1$. Thus,

$$V = \frac{FS_{DS}}{R}W$$

$$V = \frac{1.1(0.82)}{8}(1{,}836) = 0.112(1{,}836) = 206\ \text{kN}$$

6. Vertical force distribution (floor forces), F_x

$$F_x = \frac{w_x}{W}V$$

$$F_1 = \frac{1{,}080}{1{,}836}(207) = 121\ \text{kN}$$

$$F_2 = \frac{756}{1{,}836}(207) = 85\ \text{kN}$$

These forces are shown in Example 6-1, Fig. 2.

7. Horizontal distribution of forces due to F_x excluding effect of eccentricity
The frame forces due to a force applied at the center of rigidity is proportional to frame stiffness because all frames displace the same amount, Δ, as shown in Example 6-1, Fig. 3. The relative stiffness of the four frames is stated in the example above as $k_1 = k$, $k_2 = 2k$, $k_3 = 2k$, and $k_4 = k$ for frames 1 through 4, respectively. Thus, the share of each frame from the floor force becomes

$$Q_i = \frac{F}{\Sigma k_j}k_i = \frac{F}{(k+2k)(2)}k_i = \frac{F}{6k}k_i$$

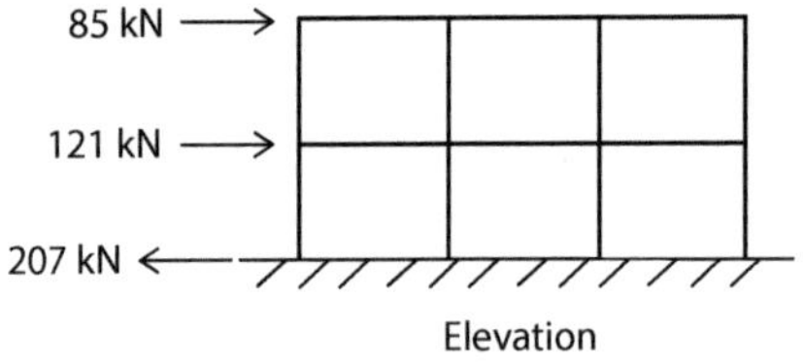

Example 6-1, Figure 2 Floor forces.

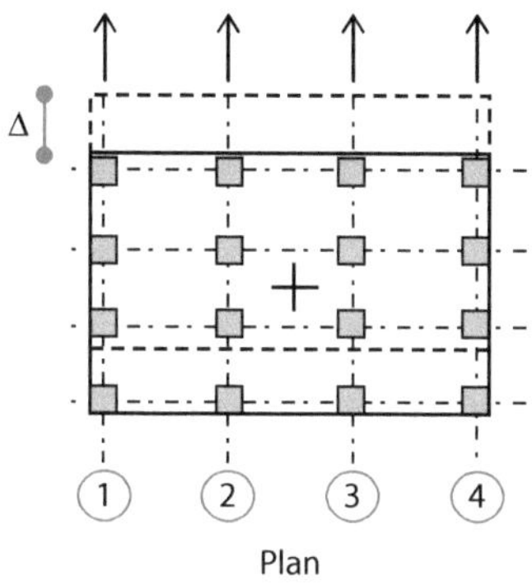

EXAMPLE 6-1, FIGURE 3 Equal displacement.

Because the first floor force is $F_1 = 121$ kN, then

$$Q_1 = Q_4 = \left(\frac{121}{6k}\right)(k) = 20.17 \text{ kN}$$

$$Q_2 = Q_3 = \left(\frac{121}{6k}\right)(2k) = 40.33 \text{ kN}$$

Because the second floor force is $F_2 = 85$ kN, then

$$Q_1 = Q_4 = \left(\frac{85}{6k}\right)(k) = 14.17 \text{ kN}$$

$$Q_2 = Q_3 = \left(\frac{85}{6k}\right)(2k) = 28.33 \text{ kN}$$

8. Horizontal force distribution due to eccentricity of force, F_x
   Because the center of mass does not coincide with the center of rigidity, this eccentricity will generate a torsional moment M_{tor}. The force distribution due this moment is proportional to rotational stiffness of each frame because all frames have the same rotational angle as shown in Example 6-1, Fig. 4. Because the moment is acting counterclockwise, the forces in frames

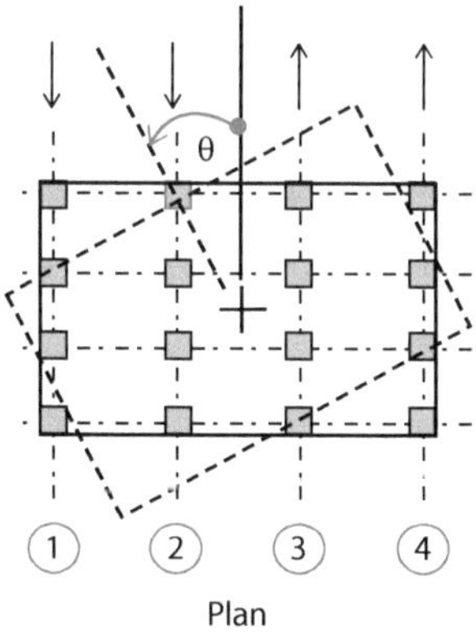

EXAMPLE 6-1, FIGURE 4 Equal rotation.

3 and 4 will be opposite in direction to the forces in frames 1 and 2 as shown in Example 6-1, Fig. 4. Thus, the share of each frame from the floor force becomes

$$Q_i = \frac{M_{\text{tor}}}{\sum k_j r_j^2} k_i r_i$$

$$= \frac{M_{\text{tor}}}{\{2k(2)^2 + k(6)^2\}(2)} k_i r_i = \frac{M_{\text{tor}}}{88k} k_i r_i$$

where torsional moment M_{tor} is equal to *Fe*.

Because the first floor force is $F_1 = 121$ kN and eccentricity is $e = 1.5$ m, then

$$M_{\text{tor},1} = 121(1.5) = 182 \text{ kN·m}$$

$$Q_1 = Q_4 = \left(\frac{182}{88k}\right)(k)(6) = 12.41 \text{ kN}$$

$$Q_2 = Q_3 = \left(\frac{182}{88k}\right)(2k)(2) = 8.27 \text{ kN}$$

Because the second floor force is $F_2 = 85$ kN and eccentricity is $e = 1.5$ m, then

$$M_{\text{tor},2} = 85\,(1.5) = 128 \text{ kN·m}$$

$$Q_1 = Q_4 = \left(\frac{128}{88k}\right)(k)(6) = 8.73 \text{ kN}$$

$$Q_2 = Q_3 = \left(\frac{128}{88k}\right)(2k)(2) = 5.82 \text{ kN}$$

The final forces applied to the frames are obtained by the summation of both force and moment components as follows.

First floor:

$$Q_1 = 20.17 - 12.41 = 7.76 \text{ kN}$$
$$Q_2 = 40.33 - 8.27 = 32.06 \text{ kN}$$
$$Q_3 = 40.33 + 8.27 = 48.60 \text{ kN}$$
$$Q_4 = 20.17 + 12.41 = 32.58 \text{ kN}$$

Second floor:

$$Q_1 = 14.17 - 8.73 = 5.44 \text{ kN}$$
$$Q_2 = 28.33 - 5.82 = 22.51 \text{ kN}$$
$$Q_3 = 28.33 + 5.82 = 34.15 \text{ kN}$$
$$Q_4 = 14.17 + 8.73 = 22.90 \text{ kN}$$

The frame forces at the first floor level are shown in Example 6-1, Fig. 5, while the forces in frame 3 are shown in Example 6-1, Fig. 6.

It should be pointed out that the torsional moment, in principle, should be distributed to all frames in both directions. However, it will be conservative to consider the torsional moment to be resisted by the system in the direction under consideration as chosen in this example.

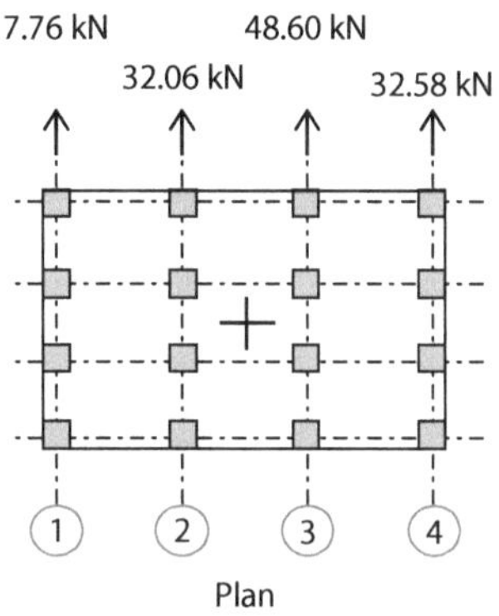

Example 6-1, Figure 5 Frame forces at the first floor.

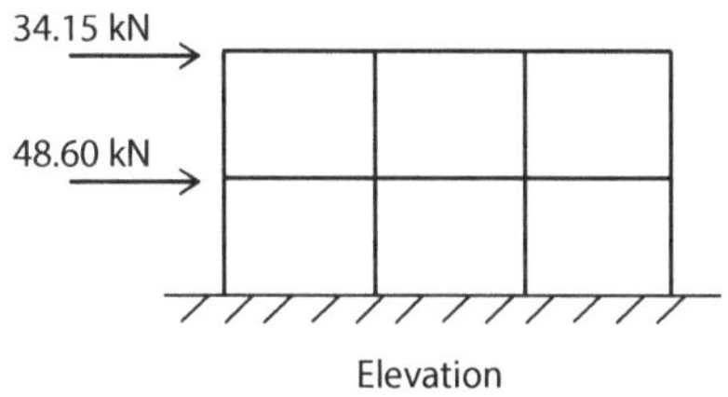

Example 6-1, Figure 6 Forces in frame 3.

6.7.3 Equivalent Lateral Force Procedure

The IBC allows the use of this method as presented by ASCE 7. It may be applied to the following cases:

a. SDC B and C: All structures.

b. SDC D, E, and F:
   1. Up to two-story buildings in Risk Category I or II.
   2. All light-frame structures.
   3. Regular structures less than 48 m in height.
   4. Regular structures exceeding 48 m with period $T < 3.5\ T_s$, where $T_s = S_{D1}/S_{DS}$, which is the controlling period in the response spectra of Fig. 6-3.
   5. Irregular structures with height less than 48 m, and having only plan irregularity 2, 3, 4, or 5, or vertical irregularity 4 or 5, as defined in previous sections.

The total base shear, V, is obtained from the response spectra shown in Fig. 6-3, which may be expressed as follows:

$$V = C_s W$$

where

$$C_S = \frac{S_{DS}I_e}{R} = \begin{pmatrix} \le \dfrac{S_{D1}\,I_e}{RT} & \text{for } T \le T_L \\ \le \dfrac{S_{D1}\,T_L\,I_e}{RT^2} & \text{for } T > T_L \\ \ge 0.044 S_{DS} I_e \ge 0.01 & \\ \ge \dfrac{0.5\,S_1\,I_e}{R} & \text{for } S_1 \ge 0.6 \end{pmatrix}$$

where C_S = seismic response coefficient
I_e = importance factor as defined earlier
R = response modification factor as given in Table 6-A6
S_{D1} = design spectral acceleration at 1-s period
S_{DS} = design spectral acceleration at short period
T = the fundamental period of the structure
W = the effective seismic weight as defined earlier

An approximate fundamental period may be calculated using the following expression:

$$T_a = C_t(h_n)^x$$

where C_t = coefficient from Table 6-3
x = coefficient from Table 6-3
h_n = total height of the building above the base in meters

Alternatively, for concrete and steel moment frames in buildings up to 12 stories with a minimum story height of 3 m, the approximate period may be estimated as a function of the number of stories, N, as

$$T_a = 0.1\,N$$

The fundamental period of the structure, T, may also be calculated by established methods in structural dynamics as given in Chap. 3. However, if the period, T, is calculated by rigorous analysis, the period shall not exceed the following value:

$$T \le C_u T_a$$

Structure Type	C_t	x
Steel moment frames free to deflect under seismic forces	0.0724	0.8
Reinforced concrete moment frames free to deflect under seismic forces	0.0466	0.9
Eccentrically braced steel frames	0.0731	0.75
All other structural systems	0.0488	0.75

Table 6-3 Approximate Period Calculation Coefficients

S_{D1}	≥ 0.4	0.3	0.2	0.15	≤ 0.1
C_u	1.4	1.4	1.5	1.6	1.7

TABLE 6-4 Coefficient for Upper Limit on Calculated Period

where C_u = coefficient for upper limit on calculated period, which is dependant on the design spectral response acceleration at 1 s; S_{D1}, as given in Table 6-4
T_a = approximate fundamental period as calculated earlier

Vertical Distribution of Base Shear

In this method, the code assumes a power distribution of forces as shown in Fig. 6-6. This distribution, of course, assumes that a structure responds in its first mode of vibration only. The floor force, F_x, is given by the following expression:

$$F_x = C_{vx} V$$

$$C_{vx} = \frac{w_x h_x^k}{\sum w_i h_i^k}$$

where w_x and h_x are the weight and height of the story under consideration, whereas w_i and h_i are the weight and height of all stories, including the story under consideration. The power k is taken as follows:

$k = 1$ for T ≤ 0.5 s
$k = 2$ for T ≥ 2.5 s
k = by interpolation for periods in between

The values of k indicate that for long-period structures, the value of k increases the degree of curvature, resulting in the transfer of more force to the upper stories as illustrated in Fig. 6-6. This action is meant to include the effect of higher modes for long-period structures. Because the value of $k = 2$ is on the conservative side, k may be taken as 2 instead of interpolation for the period range between 0.5 and 2.5 s.

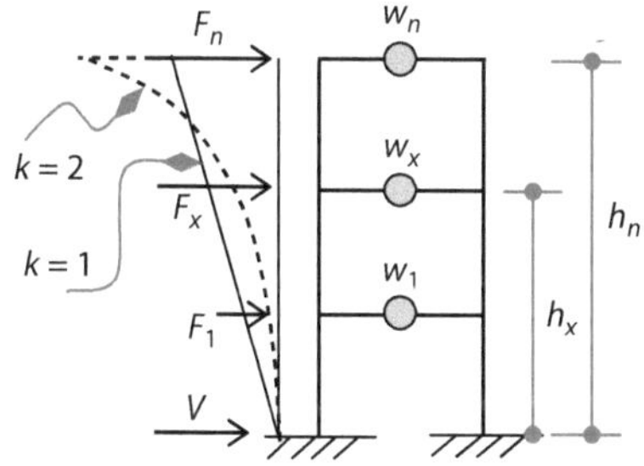

FIGURE 6-6 Floor forces.

Example 6-2 An eight-story building with story height equal to 3 m is shown in Example 6-2, Fig. 1. The building will be constructed in a region with mapped acceleration coefficient at short period and at 1-s period equal to $S_s = 0.50$ and $S_1 = 0.16$, respectively. The long-period transition period, $T_L = 8$ s. The site is classified as Soil Type C.

The structural system of the building consists of four reinforced concrete moment frames spaced in each direction as shown in the plan. The interior frames have double the stiffness of the exterior frames. The center of mass, CM, of each floor is offset by 1.5 m from the center of rigidity in both directions.

The first seven floors carry a total effective seismic weight of intensity equal to $q_D = 10$ kN/m^2 each. The eighth floor carries a total effective seismic weight of intensity equal to 7 kN/m^2.

One must find the design seismic force for each of the four frames marked, 1 through 4, on the plan if the earthquake direction is along these frames (in the short direction as also shown in the same figure).

Solution Because the building is regular, the equivalent lateral force procedure may be used if it satisfies the period limitations outlined earlier. This condition is illustrated in step (3) below.

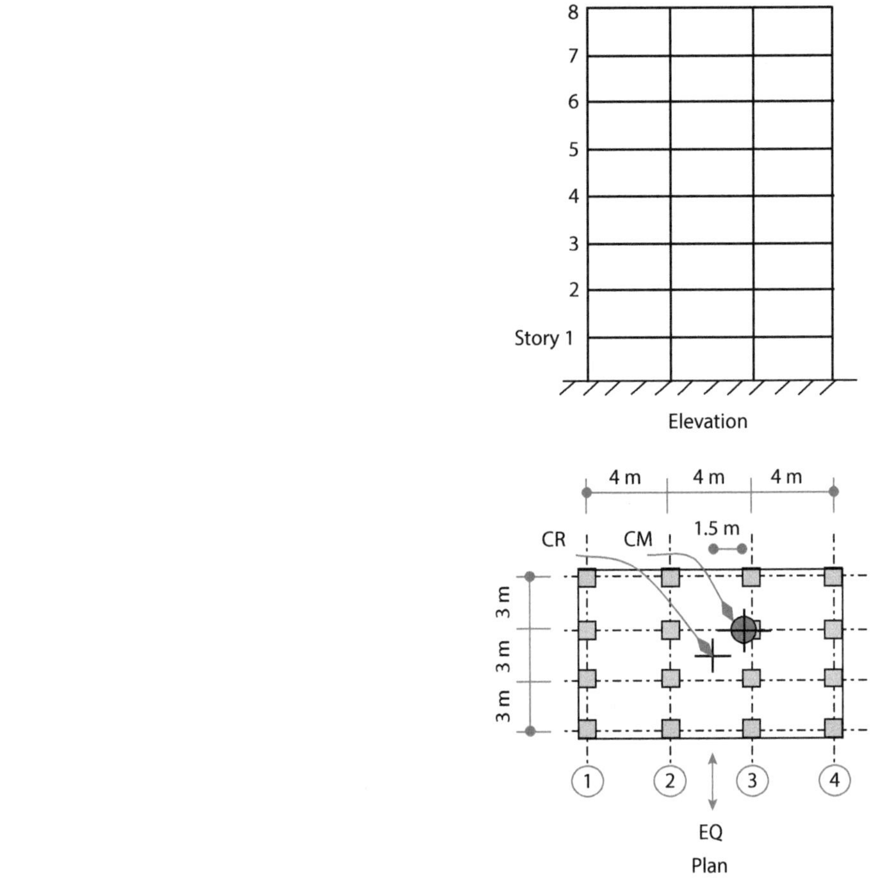

Example 6-2, Figure 1 Eight-story building.

Refer to App. 6-1 for the relevant IBC tables.

1. Seismic coefficients F_a, F_v, S_{DS}, and S_{D1}
   The seismic coefficient F_a and F_v are obtained from IBC tables included as Tables 6-A3 and 6-A4. For $S_s = 0.50$, $F_a = 1.2$ and $S_1 = 0.16$, $F_v = 1.65$.
   Accordingly, the seismic coefficients S_{DS} and S_{D1} are calculated as follows:

$$S_{MS} = F_a S_s = 1.2\,(0.50) = 0.60$$
$$S_{DS} = \frac{2}{3}\,S_{MS} = \frac{2}{3}\,(0.60) = 0.40$$
$$S_{M1} = F_v S_1 = 1.65\,(0.16) = 0.264$$
$$S_{D1} = \frac{2}{3}\,S_{M1} = \frac{2}{3}\,(0.264) = 0.176$$

2. Risk, SDC, and Importance, I_e
   A residential building is in Risk Category I. When referred to Table 6-A2 for information on that category and on an S_{DS} of 0.40 and an S_{D1} of 0.176, it can be determined that the building is in SDC C.
   The importance factor, I_e, is 1.
3. Period, T, and response modification factor, R
   Determine the response spectrum-controlling period, T_s, and the limiting period, T_{limit}, as follows:

$$T_s = \frac{S_{D1}}{S_{DS}} = \frac{0.176}{0.4} = 0.44\text{ s}$$
$$T_{\text{limit}} = 3.5\ T_s = 3.5\ (0.44) = 1.54\text{ s}$$

   The approximate period, T_a, is calculated as

$$T_a = C_t(h_n)^x$$

   where building height $h_n = 8$ (3 m) = 24 m, $C_t = 0.0466$, and $x = 0.9$ for reinforced moment frames. Consequently, the period is given as

$$T_a = C_t\,(h_n)^x = 0.0466\,(24)^{0.9} = 0.81\text{ s}$$

   Because, $T_a = 0.81$ s $< T_{\text{limit}} = 1.54$ s and the structure is regular, the equivalent lateral force procedure may be used.
   Table 6-A6 indicates that ordinary moment frames are not allowed for SDC C, but intermediate and SMFs are permitted. Therefore, if an intermediate moment frame is selected for the building, the response modification factor, R, is obtained for IMF as $R = 5$.
4. Effective seismic weight, W

Area of each floor:	$A = 9(12) = 108\text{ m}^2$
Effective seismic weight of each of the first seven floors:	$w_i = q_D A = 10(108) = 1{,}080\text{ kN}$
Effective seismic weight of the eighth floor:	$w_8 = q_D A = 7(108) = 756\text{ kN}$

   Total effective seismic weight of the building:

$$W = \sum w_i = 1{,}080(7) + 756 = 8{,}316\text{ kN}$$

5. Base shear, V
   Because $T < T_L$ and $S_1 < 0.6$, the coefficient, C_s, is given as follows:

$$C_s = \frac{S_{DS} I_e}{R} = \left(\begin{array}{l} \le \dfrac{S_{D1} I_e}{RT} \\ \ge 0.044 S_{DS} I_e \ge 0.01 \end{array} \right)$$

$$C_s = \frac{0.40(1)}{5} = \begin{pmatrix} \leq \dfrac{0.176(1)}{5(0.81)} \\ \geq 0.044(0.4)(1) = 0.018 \geq 0.01 \end{pmatrix}$$

$$C_s = 0.08 = \begin{pmatrix} \leq 0.043 \\ \geq 0.018 \end{pmatrix}$$

Therefore,

$$C_s = 0.043$$

and thus,

$$V = C_s W = 0.043\,(8{,}316) = 358 \text{ kN}$$

6. Vertical force distribution (floor forces), F_x
   Distribution of the base shear is given as a function of C_{vx}

$$C_{vx} = \frac{w_x h_x^{\,k}}{\Sigma w_i h_i^{\,k}}$$

Because $T = 0.81$ s, which is more than 0.5 and less than 2.5, the k-factor should be interpolated unless taken as 2

$$k = 1 + \frac{0.81 - 0.5}{2.5 - 0.5} = 1.16$$

Thus,

$$C_{vx} = \frac{w_x h_x^{\,1.16}}{\Sigma w_i h_i^{\,1.16}}$$

The best way to perform the calculations is by tabulated form, as shown in Example 6-2, Table 1. For demonstration purposes, a sample calculation of the force at floor level 5 is given as

$$w_5 = 1{,}080 \text{ kN}$$

$$h_5 = 15 \text{ m}$$

$$w_5 h_5^{1.16} = 1{,}080(15)^{1.16} = 24{,}986$$

$$\Sigma w_i h_i^{\,1.16} = 168{,}536 \text{ (From Example 6-2, Table 1)}$$

$$C_{v5} = \frac{24{,}986}{168{,}536} = 0.1483$$

Consequently, the force at floor level 5 is given as

$$F_5 = C_{v5} V = 0.1483(358) = 53 \text{ kN}$$

The calculation procedures will be the same for the rest of the forces, with results as tabulated in Example 6-2, Table 1.

Base shear and its distribution are shown in Example 6-2, Fig. 2.

7. Horizontal distribution of forces due to F_x
   The horizontal distribution of each floor force will be similar to the force distribution demonstrated in Example 6-1. Each frame force will be composed of two components, one due to the force applied at the center of rigidity, and another one due to the eccentricity of the force due to an offset of 1.5 m of the center of mass. See Example 6-1 for details.

Floor Level	w_i kN	h_i m	$w_i h_i^{1.16}$ kN-m	F_x kN
8	756	24	30,169	64
7	1,080	21	36,915	78
6	1,080	18	30,870	66
5	1,080	15	24,986	53
4	1,080	12	19,287	41
3	1,080	9	13,815	29
2	1,080	6	8,631	18
1	1,080	3	3,863	8
		$\Sigma =$	168,536	358

Example 6-2, Table 1 Tabulated Calculations of F_x

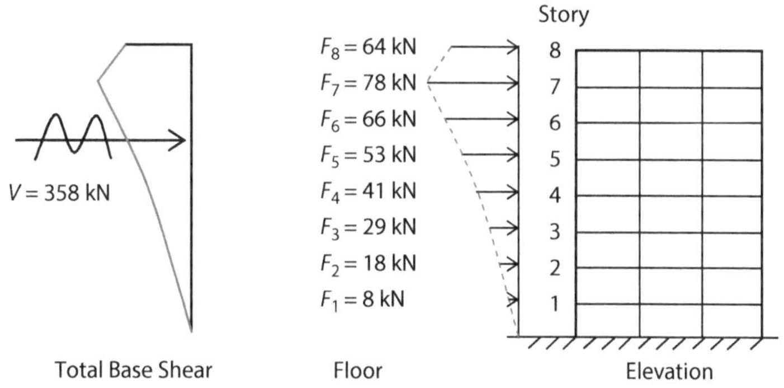

Example 6-2, Figure 2 Vertical distribution of floor forces.

6.7.4 Modal Response Spectrum Analysis

The IBC allows the use of this method as presented by ASCE 7. In ASCE 7, this method is used for structures that do not conform to the requirements of the simplified method or the equivalent lateral force procedure method. In modal response spectrum analysis, structures are modeled as multiple degrees of freedom systems where the resulting frequencies and mode shapes are extracted as illustrated in Chap. 3. This method, of course, is only valid for elastic systems.

For each mode, the spectral forces and displacements are found and then combined as needed (by SABS, SRSS, or CQC, for example). The combined forces are then divided by the response modification factor, R, to obtain the design forces. Similarly, inelastic displacements are amplified by a ductility demand factor, which is explicitly given by the IBC as C_d (see Table 6-A6).

The number of modes to be included in this analysis is the minimum number of modes required to obtain 90 percent of the mass participation of the entire structure mass in each direction. The modal mass participation ratio, MP, for each mode is given as shown in Chap. 3 as

$$MP_n = \Gamma_n^2 \frac{m_n^*}{m_{tot}}$$

where MP_n = modal mass participation ratio in the direction under consideration
Γ_n = participation factor for mode n; $= \phi_n^T MR/\phi_n^T M \phi_n$
m_n^* = modal mass of mode n
m_{to} = total mass of the structure for the direction under consideration

The IBC response spectrum defined in Fig. 6-3 is used according to the given soil profile and proper zone-mapped factors, S_s and S_1. Alternatively, a site-specific response spectrum is permitted if it takes all site characteristics into consideration, including seismic hazard analysis. A damping ratio ζ = 5% is used for building this kind of response spectrum.

Even though not explicitly given in the code, it has been general practice to take the vertical component of the earthquake as two-third of the horizontal component.

In modal analysis procedures, the obtained torsion need not be amplified by the torsion amplification factor A_x. This factor will be defined later.

According to the ASCE 7 standards, the base shear obtained using response spectrum procedure shall not be less than 85 percent of the base shear obtained by the equivalent lateral force procedure (0.85 $C_s W$). In addition, the base shear obtained using equivalent lateral force procedures shall be based on the period obtained from modal analysis, but it shall not exceed the upper limit of the period given in that section, that is, $T_{modal} \le C_u T_a$. Consequently, if the base shear obtained by response spectrum procedure is less than 85 percent of the force obtained by the equivalent lateral force procedure, then all relevant force quantities, such as base shear and internal forces (but not the drifts), obtained from the response spectrum procedure shall be multiplied by the following quantity:

$$\text{Multiplication factor} = \frac{0.85V}{V_t}$$

$$= \frac{0.85C_s W}{V_t}$$

where V = base shear obtained from the equivalent lateral force procedure
V_t = base shear obtained from the response spectrum procedure

The response modification indicates that the code is more concerned about the distribution of the force inside the structure as determined by the response spectrum model. Consequently, the force distribution inside the structure must be maintained with revised values according to the conditions listed earlier.

Example 6-3 A six-story building with a story height equal to 3 m is shown in Example 6-3, Fig. 1. The building will be constructed in a region with mapped acceleration coefficient at short period and at 1-s period equal to S_s = 0.50 and S_1 = 0.15, respectively. The mapped long-period transition period is given as T_L = 8 s. The site is classified as Soil Type C.

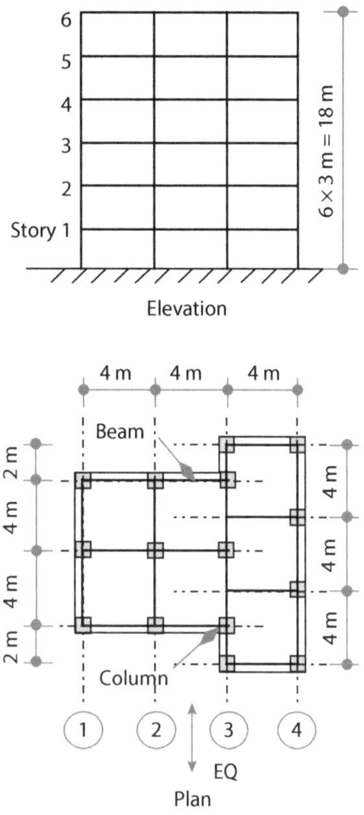

EXAMPLE 6-3, FIGURE 1 Six-story building.

The structural system of the building consists of reinforced concrete moment frames in each direction and spaced as shown in the plan. All columns and all beams are 500 × 500 mm. The material used is concrete grade, $f_c' = 25$ MPa, with reinforcement grade, $f_y = 420$ MPa.

Each floor carries a total effective seismic weight of intensity equal to, $q_D = 10$ kN/m^2.

The modal analysis procedure must be used to find the design seismic force for each of the four frames marked, 1 through 4, on the plan if the earthquake direction is along these frames as also shown in the figure.

Solution Note that the equivalent lateral force procedure may be used.

1. Seismic coefficients F_a, F_v, S_{DS}, and S_{D1}
   The seismic coefficients F_a and F_v are obtained from the IBC tables provided as Tables 6-A3 and 6-A4. $S_s = 0.50$, $F_a = 1.2$, $S_1 = 0.15$, and $F_v = 1.65$.
   Accordingly, the seismic coefficients S_{DS} and S_{D1} are calculated as follows:

$$S_{MS} = F_a S_s = 1.2(0.50) = 0.60$$

$$S_{M1} = F_v S_1 = 1.65(0.15) = 0.248$$

$$S_{DS} = \frac{2}{3} S_{MS} = \frac{2}{3}(0.60) = 0.40$$

$$S_{D1} = \frac{2}{3} S_{M1} = \frac{2}{3}(0.248) = 0.165$$

2. Risk SDC and importance, I_e
   A residential building is in Risk Category I. When referred to Table 6-A2 for Risk Category I and for $S_{DS} = 0.40$ and $S_{D1} = 0.165$, it can be determined that the building is in SDC C. This building is similar to the one in Example 6-2 in that intermediate moment frames may also be used for this building, which has an R-factor equal to 5.
   The importance factor, I_e, is equal to 1.

3. Design response spectrum
   The design response spectrum can be determined from the IBC spectrum, which is a function of the seismic coefficients. The IBC spectrum and its controlling parameters are shown in Example 6-3, Fig. 2(a). These parameters are calculated as follows:

$$S_{DS} = 0.40, \qquad S_{D1} = 0.165 \qquad \text{(as given previously)}$$

$$0.4\, S_{DS} = 0.4\,(0.40) = 0.16$$

$$T_s = \frac{S_{D1}}{S_{DS}} = \frac{0.165}{0.40} = 0.413 \text{ s}$$

$$T_o = 0.2\, T_s = 0.2\,(0.41) = 0.083 \text{ s}$$

$$T_L = \text{given} = 8 \text{ s}$$

   The final response spectrum is shown in Example 6-3, Fig. 2(b).
   To use this spectrum in a computer program, it will be needed to provide pairs of period and acceleration. For example, the following pairs from Example 6-3, Table 1, may be used for computer input.

4. Periods and mode shapes
   To obtain the periods and the mode shapes, the building must be modeled as multiple degrees of freedom as illustrated in Chap. 3. For structures with more than 3DOFs (degree of freedoms), hand calculations become unreasonable and computer analysis will be needed.
   Because modeling is considered an art, it must be handled carefully to capture the dynamic aspects of the problem and to provide accuracy and efficiency. The building may be modeled as a space frame structure if the floors of the building are stiff enough to be

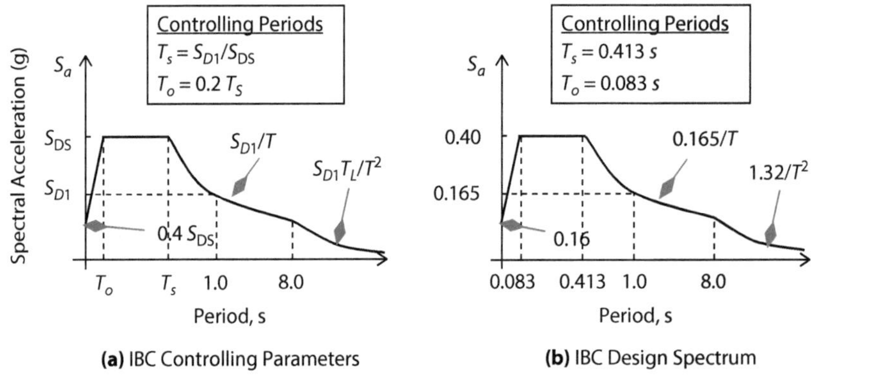

Example 6-3, Figure 2 IBC design response spectrum.

T (s)	0	0.083	0.413	0.5	0.6	0.8	1	2	4	5	8	10
S_a (g)	0.16	0.40	0.40	0.33	0.275	0.206	0.165	0.083	0.041	0.033	0.021	0.013

Example 6-3, Table 1 Design Response Spectrum Pairs

considered rigid. As identified earlier, diaphragms may be idealized as rigid if they consist of concrete slabs or concrete-filled metal decks with span-to-depth ratios of 3 or less that have no horizontal irregularities according to ASCE 7 and be at least 50 mm thick according to ACI. The mass may be lumped and applied as concentrated node masses at the nodes.

If the diaphragm is not stiff enough, it must be modeled as a flexible diaphragm using the finite element model. In this case, the diaphragm is modeled with plate elements or more generally as shell elements. The diaphragm elements are then connected to the beams and columns at their nodes. In this case, the mass is directly applied to the diaphragm shell elements as surface mass density. In practice, most building slabs qualify as rigid diaphragms.

5. Modeling

Example 6-3, Fig. 3, shows the building modeled as a 3D, space frame with lumped masses. The mass will be lumped and allocated to the nodes at the intersection of beams and columns. The supports may be modeled as either hinge or fixed supports as required by design. This example considers fixed supports.

The distribution of the mass over all nodes produces a more accurate and realistic distribution of the mass. However, the lumped masses may be limited to a few nodes as long as the location of the center of mass is preserved. It is also important to displace the actual center of mass a distance equal to 5 percent of the dimension perpendicular to earthquake excitation as required by the code. This is illustrated in the following calculations.

The model shown in Example 6-3, Fig. 3, can easily be entered into any standard structural analysis software with proper numbering of nodes and members. The reference global coordinate system is capital X, Y, and Z as shown in the figure. This is a standard operation these days.

Material: Concrete: $E_c = 4{,}700\sqrt{25} = 23{,}500$ MPa

Geometry of members is calculated for all members with respect to their six degrees of freedom (referenced to their local member axes marked with small letters as shown in

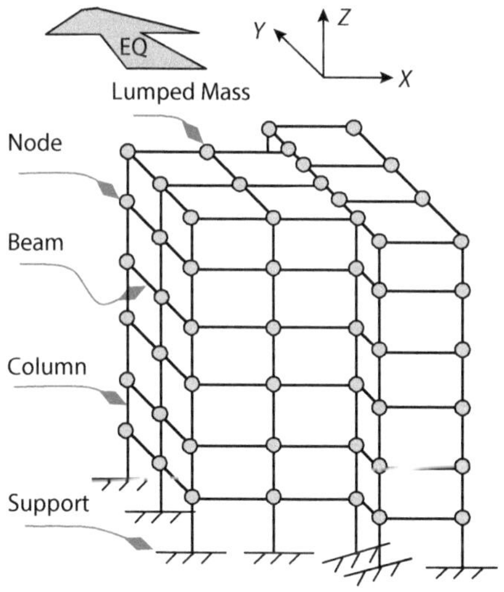

Example 6-3, Figure 3 3D frame model.

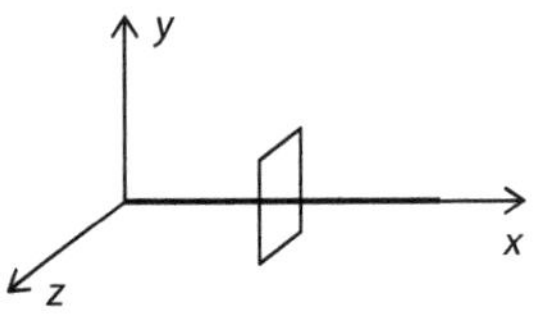

EXAMPLE 6-3, FIGURE 4 Local coordinates.

Example 6-3, Fig. 4). If the dimensions of the section are given as $b \times h$, the required quantities are calculated as follows:

$$A_x = \text{cross-sectional area} = \text{bh}$$

$$A_y = \text{shear area} = \left(\frac{5}{6}\right)\text{bh}$$

$$A_z = \text{shear area} = \left(\frac{5}{6}\right)\text{bh}$$

$$I_x = \text{torsional constant} = \frac{\text{hb}^3}{3}\left(1 - 0.63\frac{b}{h}\right)$$

$$I_y I_z = \text{effective moment of inertia}$$
$$= 0.35\, I_g \text{ for beams and}$$
$$= 0.70\, I_g \text{ for columns}$$

where the gross moment of inertia is taken as $\text{bh}^3/12$ for the major axis and $\text{hb}^3/12$ for the minor axis. The results are shown in Example 6-3, Table 2.

6. Mass
   For consistency of units, the mass is preferably given in kN·s²/m, which is simply the weight divided by the gravitational acceleration. Thus, the mass density of all floors may be calculated in these weight units as

$$m_D = \frac{q_D}{g} = \frac{10}{9.81} = 1.02 \text{ kN·s}^2\text{/m/m}$$

For mass calculation, subdivide the floor into areas $A1$ and $A2$ as shown in Example 6-3, Fig. 5.

For each floor, the subarea and submass, $m1$ and $m2$, are calculated as follows:

$$A1 = 8(8) = 64 \text{ m}^2$$

$$m1 = m_D A1 = 1.02\,(64)$$
$$= 65.28 \text{ kN·s}^2\text{/m}$$

$$A2 = 4(12) = 48 \text{ m}^2$$

Member	Size m	A_x m²	A_y m²	A_z m²	I_x ×10⁻³ m⁴	I_y ×10⁻³ m⁴	I_z ×10⁻³ m⁴
Beams	0.5 × 0.5	0.25	0.208	0.208	7.708	1.823	1.823
Columns	0.5 × 0.5	0.25	0.208	0.208	7.708	3.646	3.646

EXAMPLE 6-3, TABLE 2 Member Properties

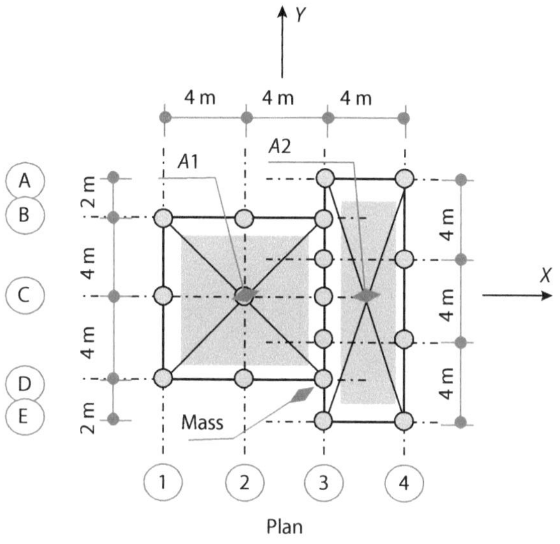

Example 6-3, Figure 5 Mass distribution.

$$m2 = m_D A2 = 1.02(48)$$
$$= 48.96 \text{ kN·s}^2/\text{m}$$
$$m_{\text{tot}} = m1 + m2$$
$$= 65.28 + 48.96$$
$$= 114.24 \text{ kN·s}^2/\text{m}$$

7. Mass allocation

The total mass must be distributed to the nodes in two directions, that is, along directions X and Y, using the methods presented in Example 3-7.

Along direction Y

The mass allocation for this direction is found as follows.

Because the earthquake is acting along direction Y, an accidental eccentricity of 5 percent must be introduced perpendicular to direction Y

$$e_{ac} = 5\% \,(12 \text{ m}) = 0.6 \text{ m}$$

Center of mass, CM, is found by taking the moment of the two floor masses about any axis (e.g., about line 1). Accordingly, CM_1 from line 1 is calculated as

$$\text{CM}_1 = \frac{m1(4)+m2(10)}{m_{\text{tot}}}$$
$$= \frac{65.28(4)+48.96(10)}{114.24} = 6.571 \text{ m}$$

If the mass is considered to be allocated to all 17 nodes, the center of nodes, CN, is found by taking the first statical moment of all nodes about any axis (e.g., about line 1). Accordingly, CN_1 from line 1 is calculated as

$$\text{CN}_1 = \frac{3(4)+7(8)+4(12)}{17} = 6.824 \text{ m}$$

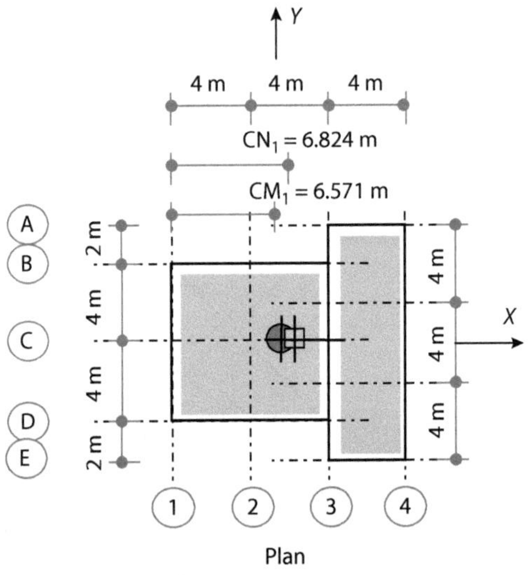

Example 6-3, Figure 6 Mass distribution.

The actual eccentricity of mass, e_{mn}, is as shown in Example 6-3, Fig. 6, that is

$$e_{\text{mn}} = 6.824 - 6.571 = 0.253 \text{ m}$$

Adding accidental eccentricity from above, the final eccentricity becomes

$$e_{\text{tot}} = 0.253 + 0.6 = 0.853 \text{ m}$$

The share of each node from the total mass is calculated in a way similar to the procedures used in finding forces in piles

$$m_i = \frac{m_{\text{tot}}}{N} + \frac{M_{m,\text{cn}}}{\Sigma r_j^2} r_i$$

where m_{tot} = total mass = 114.24 kN·s²/m

$M_{m,\text{cn}}$ = eccentric moment of masses about CN

$M_{m,\text{cn}} = m_{\text{tot}}\, e_{\text{tot}} = 114.24\,(0.853) = 97.447$ kN·s²

N = number of nodes = 17 nodes

Σr_j^2 = polar moment of inertia of masses about CN and is given as

$= 3\,(6.824)^2 + 3\,(6.824 - 4)^2 + 7\,(8 - 6.824)^2 + 4\,(12 - 6.824)^2$

$= 280.471 \text{ m}^2$

Accordingly, the mass allocation is given for each floor as follows:

$$m_i = \frac{114.24}{17} \pm \frac{97.447}{280.471} r_i = 6.72 \pm 0.347\, r_i$$

Masses in Line 1: $m_1 = 6.72 + 0.347\,(6.824) = 9.088$ kN·s²/m

Masses in Line 2: $m_2 = 6.72 + 0.347\,(6.824 - 4) = 7.700$ kN·s²/m

Masses in Line 3: $m_3 = 6.72 - 0.347\,(8 - 6.824) = 6.312$ kN·s²/m

Masses in Line 4: $m_4 = 6.72 - 0.347\,(12 - 6.824) = 4.924$ kN·s²/m

The total masses in one line may be lumped and placed on one node. For example, the total masses in line 3 are $7 \times 6.312 = 44.184$ ($7 \times 0.036 = 0.252$), which may be placed on the node at the intersection of lines C and 3.

The mass can now be entered into the available software program.

It is worth mentioning that the above distribution is not unique. Any distribution that preserves the total mass and the total moment of the mass about the center of nodes, CN, would be valid. For example, an alternative distribution may be obtained by allocating constant mass lumps on each side of CN as explained earlier in Example 3-7.

It is worthwhile pointing out that the center of resistance, CR, and the center of nodes, CN, in this example are the same; however, this is not necessarily the case in general.

Moment of mass about CN is given as before,

$$M_{m,\text{cn}} = m_{\text{tot}} e_{\text{mn}}$$

$$= 114.24\,(0.853) = 97.447 \text{ kN·s}^2$$

$$\Sigma\left|r_j\right| = 3\,(6.824) + 3\,(6.824 - 4) + 7\,(8 - 6.824) + 4\,(12 - 6.824)$$

$$= 57.88 \text{ m}$$

$$\bar{m}_j = \pm\frac{M_{m,\text{cn}}}{\Sigma\left|r_j\right|} = \pm\frac{97.447}{57.88} = \pm\ 1.684$$

$$\Sigma\bar{m}_j = 1.684\,(7 + 4 - 3 - 3) = 8.418$$

hence,

$$\Delta\text{m} = \text{m}_{\text{tot}} \pm \Sigma\bar{m}_j$$

$$= 114.24 - 8.418 = 105.822 \text{ kN·s}^2/\text{m}$$

and hence,

$$m_i = \frac{\Delta m}{N} \pm \bar{m}_i$$

Line 1 & 2:

$$m_{1,2} = \frac{105.822}{17} - 1.684$$

$$= 6.225 - 1.684 = 4.541 \text{ kN·s}^2/\text{m}$$

Line 3 & 4:

$$m_{3,4} = 6.225 + 1.684 = 7.909 \text{ kN·s}^2/\text{m}$$

The reader may want to verify that both total mass and total moment of mass about the center of nodes are preserved.

Along direction X

The mass distribution along direction X will be evenly distributed to the nodes insofar as the CM and the CR lie on the same line C (zero eccentricity). Consequently, the mass allocation for each node will be simply

$$m_i = \frac{m_{\text{tot}}}{N} = \frac{114.24}{17} = 6.72 \text{ kN·s}^2/\text{m}$$

The mass distribution above is for the given direction of the earthquake excitation (along direction Y). For an excitation along direction X, accidental eccentricity must be induced perpendicular to direction X. It should be cautioned that the mass cannot be evenly distributed to the nodes in direction X, even for the given symmetry in this direction. The accidental eccentricity must be introduced to this direction by displacing the center of mass in the perpendicular direction. This is 5% of 12 m = 0.6 m.

8. Base shear, V

   The base shear is found by adding all reactions from the computer output.

9. Force distribution through the structure
   After completing the input and selection of the analysis procedure, including selection of the required modes to be considered for the analysis, one can initiate analysis by software program and receive output that identifies all periods and their mode shapes, all elastic nodal forces, all elastic displacements, all internal moments, shears, and normal forces.

10. Result samples
    The information given previously was entered into the SAP2000 software program for analysis. Key results for the first three modes are given in Example 6-3, Table 3, as a reference for those who would like to try this example in software analysis. Note that the summation of total reactions in any direction is the base shear, in that direction.

11. Check base shear with the equivalent lateral force procedure

    Refer to Example 6-2 for details of calculations.

    Period: $T_a = C_t(h_n)^x = 0.0466\,(18)^{0.9} = 0.628$ s

    Period upper limit: for $S_1 = 0.15$, $\rightarrow$ coefficient $C_u = 1.6$

    $T_u = C_u T_a = 1.6\,(0.628) = 1.00$ s

    Therefore, because $T < T_u$, use calculated period from model,
    $T = 0.743$

    Mass: $M_{tot}\,(114.24)(6) = 685\ \text{kN}\cdot\text{s}^2/\text{m}$ (6 floors)

    Weight: $W = M_{tot}\,g = (685)(9.81) = 6{,}724$ kN (6 floors)

    Using the equivalent lateral force procedure, V, the elastic base shear is calculated at $R = 1$, which is given as

$$C_s = \frac{S_{DS} I_e}{R} = \begin{pmatrix} \le \dfrac{S_{D1} I_e}{RT} \\ \ge 0.01 \end{pmatrix}$$

$$C_s = \frac{0.40\,(1)}{1} = \begin{pmatrix} \le \dfrac{0.165\,(1)}{1\,(0.743)} \\ \ge 0.01 \end{pmatrix}$$

$$C_s = 0.40 = \begin{pmatrix} \le 0.222 \\ \ge 0.01 \end{pmatrix}$$

Mode Number	Period, T (s)	Mode Shape, ϕ	Total Reactions, V_X kN	Total Reactions, V_Y kN
1	0.781	Displacement in direction X	≈ 0	≈ 0
2	0.743	Displacement in direction Y (with slight clockwise torsion)	≈ 0	877
3	0.641	Torsion	≈ 0	401
SRSS (mode 3)			≈ 0	964
SRSS (mode 10)			≈ 0	988

Example 6-3, Table 3 SAP2000 Output of Response Spectrum Analysis

Therefore,

$$C_s = 0.222$$

Thus,

$$V = C_s W = 0.222\,(6{,}724) = 1{,}493 \text{ kN}$$

$$85\% \text{ of base shear} = 0.85\,(1{,}493) = 1{,}269 \text{ kN}$$

Because the base shear obtained from the response spectrum analysis ($V_t = 988$ kN) is less than 85 percent of the base shear obtained from the equivalent lateral force procedure ($0.85\ V = 1{,}269$ kN), the final results obtained from computer analysis for elastic response must be increased to match, $0.85V$, except the drift values

$$V_{\text{elastic}} = \frac{0.85V}{V_t} = \frac{0.85(1{,}493)}{988} = 1.28$$

The final forces for design must then be divided by the R factor. Considering that intermediate concrete frames are used as shown in Step 2 previously ($R = 5$), the final design force for each member will be

$$F_{\text{design}} = \frac{F_{\text{computer}}\ (\text{ratio above})}{R} = F_{\text{computer}}\frac{1.28}{5}$$

$$F_{\text{design}} = 0.26\ F_{\text{computer}}$$

All results from the computer output (except the drift) must be increased according to the increase in the base shear value (multiplied by the factor 0.26). In addition, the drift is only multiplied by $1/R$ (e.g., $1/5 = 0.2$).

6.7.5 Two-Stage Analysis Procedures

The two-stage analysis procedures is a method provided by ASCE 7 to analyze buildings with flexible upper portion and much stiffer lower portion. Such situations arise when buildings comprise very stiff basement walls, or sit on very stiff foundation walls. An example of such arrangement is shown in Fig. 6-7.

In this method, the structure may be separated and analyzed as two independent portions, upper flexible portion that is supported by stiff lower portion as

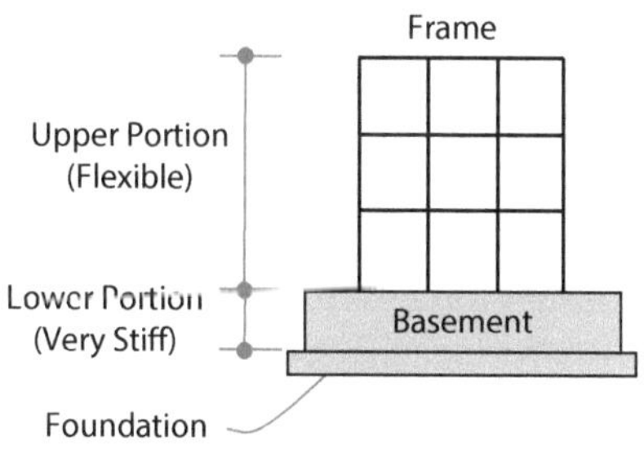

FIGURE 6-7 Flexible tower with stiff basement.

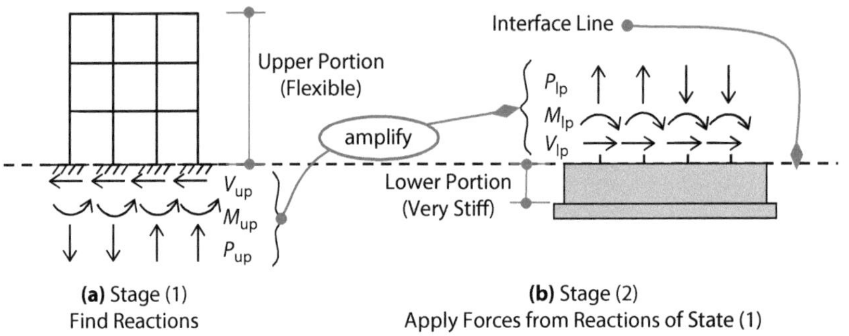

FIGURE 6-8 Two-stage analysis.

shown in Fig. 6-8. In order to use this method, all of the following five conditions shall be satisfied:

a. The stiffness of the lower portions shall be at least 10 times the stiffness of the upper portion, that is,

$$K_{lp} \geq 10\, K_{up}$$

b. The period of the entire structure shall not exceed 1.1 times the period of the upper portion, that is,

$$T_{en} \leq 1.1\, T_{up}$$

c. The upper portion shall be designed as separate structure with its own R and ρ values. Naturally, the analysis and design of the upper portion result in reactions, P_{up}, V_{up}, and M_{up}, as shown in Fig. 6-8(a). These forces and moments shall be amplified and applied to the lower portion as specified in (d).

d. The lower portion shall be designed as separate structure under the application of forces and moments P_{lp}, V_{lp}, and M_{lp}, as shown in Fig. 6-8(b). P_{lp}, V_{lp}, and M_{lp}, are obtained by amplifying the reactions from the upper portions by the ratio of R/ρ of the upper portion to the lower portion, that is,

$$Q_{lp} = \left(\frac{(R/\rho)_{up}}{(R/\rho)_{lp}} \right) Q_{up} \geq Q_{up}$$

where Q_{up} = design reaction elastic forces and moment, P_{up}, V_{up}, and M_{up}, as applicable that are obtained from upper portion analysis and modified by their R and ρ factors

Q_{lp} = design forces and moment, P_{lp}, V_{lp}, and M_{lp}, applied to the lower portion resulting from the amplified upper portion and modified by their R and ρ factors

R = force modification factor as defined earlier

ρ = redundancy factor as will be defined in later sections

e. The upper portion may be analyzed using the equivalent lateral force procedures, or, modal response spectrum procedures as presented in the previous sections, whereas, the lower portion is analyzed by the equivalent lateral force procedures only.

6.7.6 Time-History Analysis

The IBC allows the use of time-history analysis as presented by ASCE 7, Chap. 16. This is the general method used for large, important, and complex structures. Time-history analysis is conducted with numerical methods (see Chap. 5). To include directional effect, the code requires the simultaneous application of earthquake records to both major directions.

Because this type of analysis requires ground motion records, the code also requires use of a suite of pairs of records. These records shall reflect site characteristics and seismic hazard and can either be scaled from actual records or artificially generated (synthetic records).

The code also specifies that if seven records are used in the analysis, the maximum response quantities may be taken as the average of the resulting values. However, if less than seven records are used in the analysis (not less than three), the response values shall be taken as the maximum value of all records for each quantity.

Time-history analysis may be performed using elastic and inelastic structural properties. In the elastic analysis procedure, the design forces and displacements are obtained by modifying the results of the computer output by the factors R and μ. In the inelastic analysis procedure, the design forces and displacements are directly obtained from the analysis. For inelastic analysis, an approved hysteresis model is needed (based on the experimental and analytical results discussed earlier). In addition, the code requires the design to be reviewed by an independent team of experts (e.g., licensed professional engineers).

Example 6-4 A guyed mast with a total height equal to 40 m is shown in Example 6-4, Fig. 1. The mast lies in a seismic zone with mapped spectral acceleration at short period S_S equals to 1.05. The soil at the site is Type B. The structural system of the mast consists of a vertical reinforced concrete

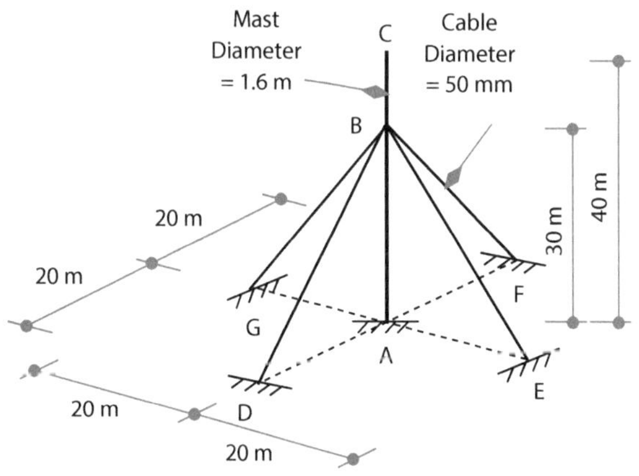

Example 6-4, Figure 1 Guyed tower layout.

shaft with a circular diameter equal to 1.6 m. The mast has a fixed support at end A and is tied to four steel cables (wires) at B. The cables are equally spaced and located at 20 m from the base. The diameter of all cables is 50 mm. The material used is concrete and steel of grades $f_c' = 25$ MPa and $f_y = 420$ MPa, respectively.

Find the design elastic seismic force for this structure if the earthquake direction is along line D-F as shown in the figure.

Solution The cables (wires) are nonlinear structures and do not resist compression forces. As a result, this structure as a whole behaves in a nonlinear manner. This case cannot be solved with the methods of simplified analysis procedure, equivalent lateral force procedure, and modal response spectral analysis because they are only valid for elastic behavior. A time-history analysis is the only option left.

Approximate linear analysis may be performed for cable structures. If the cable is pretensioned, such analysis considers the cable to be an elastic element with average stiffness within the range of operation. This method is widely used in wind engineering for guyed communication towers.

1. Seismic coefficients F_a, S_{DS}, and I_e
   The seismic coefficient, F_a, is obtained from the IBC included as Table 6-A3. For Soil Type *B*, $F_a = 1$. Accordingly, the seismic coefficient S_{DS} is calculated as follows:

   $$S_{MS} = F_a S_s = 1.0(0.42) = 1.05$$

   $$S_{DS} = \frac{2}{3} S_{MS} = \frac{2}{3}(1.05) = 0.70$$

   The importance factor, I_e, is equal to 1.

2. Design ground motion
   Because the system is nonlinear where explicit dynamic analysis is needed, a representative ground motion is needed. According to the IBC, a suite of records must be used for such analysis. The selection of appropriate records for analysis will be addressed in Chap. 11 using synthetic earthquake records. In this example, only one record will be used. For actual problems, the analysis must be repeated under the excitation of the selected suite of different representative records. For demonstration purposes, the El Centro earthquake record portrayed in Fig. 2-5 will be used for this analysis.

   Because the PGA can be arbitrarily assigned to any earthquake record which is usually normalized to one, the PGA must be calibrated to reflect the intensity of the site under consideration. For dynamic analysis, the term effective peak acceleration, or EPA, is usually associated with such analysis because the PGA itself is not a good parameter for analyzing the effect of earthquakes, especially when it comes to inelastic behavior. Recall from Chap. 3 that when the period of the structure tends to zero, the pseudoacceleration of the response spectrum approaches the PGA. The code design response spectrum indicates that the spectrum curves start with a value of 0.40 S_{DS} at zero period as shown in Fig. 6-2. Therefore, in absence of any information on PGA, and for a normalized earthquake record, the 0.40 S_{DS} should be assigned as its PGA.

   Consequently, for this example, the input file for the normalized El Centro earthquake should include an EQ scale factor equal to 0.40 S_{DS} times the gravitational acceleration

   $$EQ_{\text{scale factor}} = 0.40\, S_{DS}\,(g) = 0.40\,(0.70)(9.81) = 2.747 \text{ m/s}^2$$

3. Modeling
   The mast can be modeled as a 3D structure with the mast as a beam element. The cables may be modeled with special nonlinear elements, such as cable elements or tension only link elements. This selection, of course, depends on the software to be used. In either way, the cable elements must not resist any compressive forces. Therefore, the analysis is carried out incrementally, and the stiffness is updated as a function of the stresses in the cables. The stiffness components are turned to zero for any compressive stress or strain that appears in the matrix. Fortunately, these procedures are performed by the software, not by the user.

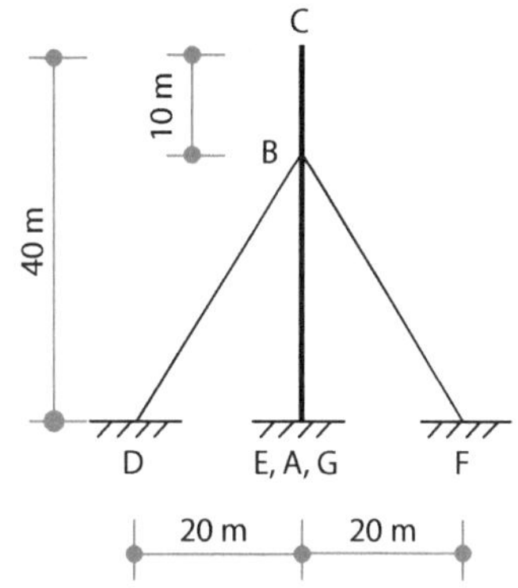

Example 6-4, Figure 2 Guyed tower in 2D representation.

For simplicity and demonstration purposes, the mast will be considered braced in the direction perpendicular to the earthquake direction and, as a result, will be analyzed as a 2D structure in the direction of excitation (along line D-F as shown in Example 6-4, Fig. 2).

The model shown in Example 6-4, Fig. 2, can easily be entered into any standard structural analysis software with the proper numbering of nodes and members. This is a standard operation. For example, if SAP2000 software is used, the, structure in Example 6-4, Fig. 2, may be modeled with five nodes between A and C, connecting four beam elements from joint A to joint C. The cables may be modeled with two nodes each at D and B with cables modeled as a two-hook element for cables D-B and F-B. In SAP2000 software, the hook element is defined to act as a tension-only link. The rest of the input data may be calculated as follows.

Material is given for concrete, $f_c' = 25$ MPa. Thus,

$$E_c = 4{,}700\sqrt{25} = 23{,}500 \text{ MPa}$$

Calculate the geometry of members for all members with respect to their six degrees of freedoms. If the local member axes are as shown in Example 6-4, Fig. 3, with dimensions as diameter d, the required quantities are calculated as follows:

A_x = cross-sectional area = $\pi d^2/4$
A_y = shear area = $\pi d^2/4$
A_z = shear area = $\pi d^2/4$
I_x = torsional constant = $\pi d^4/32$
I_y, I_z = effective moment of inertia is taken as $0.5\,I_g$, judged from the fact that the mast is not a pure beam and yet not heavily loaded enough in the axial direction to qualify as a column ($I_{\text{eff}} = 0.5\,I_g = 0.50\,\pi d^4/64$).

The resulting quantities are shown in Example 6-4, Table 3.

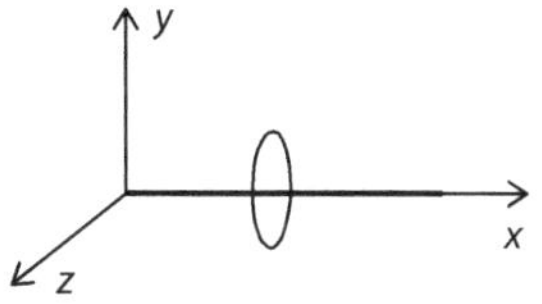

Example 6-4, Figure 3 Local coordinates.

Member	Size m	A_x m²	A_y m²	A_z m²	I_x m⁴	I_y m⁴	I_z m⁴
Beam	1.6 × 1.6	2.01	2.01	2.01	0.643	0.161	0.161
Cable	D = 0.05	0.001963	—	—	—	—	—

EXAMPLE 6-4, TABLE 3 Member Properties

4. Mass
   For consistency of units, the mass is preferably given in kN·s²/m, which is simply the weight divided by the gravitational acceleration. In most commercial software, the mass can be assigned to the member through density of the member, including any additional distributed mass. Consequently,

$$\gamma_{mast} = 25 \text{ kN/m}^3$$
$$\rho_{mast} = 2.5 \text{ kN·s}^2\text{/m/m}^3$$
$$\gamma_{cable} = 77 \text{ kN/m}^3$$
$$\rho_{cable} = 7.8 \text{ kN·s}^2\text{/m/m}^3$$

5. Base shear, V
   The base shear history may be traced from the computer output to find its maximum value.
6. Force distribution through the structure
   After completing the input, selecting the analysis procedure and running the software program, one may obtain the history of all elastic nodal forces, elastic displacements, internal moments shears, and normal forces by consulting the output file.
7 Result samples
   The information from the example above was entered into SAP2000 software. Key results are given for the first three modes in Example 6-4, Table 4, as a reference for readers who want to analyze this example with their own software.
   As noted earlier, approximate linear analysis may be performed for cable structures using average stiffness within the range of operation. Therefore, Example 6-4, Table 4, also includes linear analysis of the structure for comparison purposes. In the linear analysis, the cables are modeled as truss members where they can take compressive forces as well as tension forces. Such compression, of course, will be offset by the pretension force in the cable. It should be noted, however, that the structural period will be different in both cases. Therefore, direct comparison between forces and displacements will not be consistent. The comparison between the two structures will only be feasible if it is viewed on the global level.

	Displacement, mm		Base Shear, kN		Cable Force, kN	
Analysis Case	Min	Max	Min	Max	Min	Max
Linear	–300	266	–1,673	1,243	–888	942
Nonlinear	–262	255	–963	–916	0	913

EXAMPLE 6-4, TABLE 4 SAP2000 Output of Nonlinear Time-History Analysis

Review of Example 6-4, Table 4, indicates that the minimum force in the cables is zero because the cable is slack under contraction. In addition, the base shear is higher in the linear case because the structure will be stiffer. As a result, it has a shorter period, which attracts more forces (as can be seen from the El Centro response spectrum). The relative stiffness of the two structures is also obvious if the displacement and base shear ratios are compared for both cases.

6.7.7 Directional Effect

Depending on the Risk Category, the IBC requires that the effect of earthquake components from the three major directions be considered. For example, for SDC A and B, this requirement is satisfied if the seismic design forces are applied separately and independently in each of the two major orthogonal directions.

SDC C shall conform to SDC B. In addition, the directional effect for structures that have horizontal irregularities (Type 5) may be satisfied by taking 100 percent of the forces from one direction plus 30 percent of the forces from the perpendicular direction. Alternatively, the structures mentioned above may satisfy the directional effect requirement by the simultaneous application of orthogonal pairs of ground motion acceleration histories in a time-history analysis.

For SDC D, E, and F, the directional effect shall satisfy the requirements of SDC C. In addition, the code calls for special requirements of columns and walls that form part of two or more intersecting seismic systems.

In summary, the directional effect in two perpendicular directions (when required) and the vertical direction may be combined as follows.

The effect of both major horizontal components of the earthquake E_x and E_y shall be combined to form two load cases Q_E, such that

$$Q_{E1} = E_x + 0.3\,E_y \qquad \text{case 1}$$

$$Q_{E2} = 0.3\,E_x + E_y \qquad \text{case 2}$$

The effect of the vertical component of an earthquake E_v is calculated from the total vertical seismic force, which is taken as a fraction of the dead load, D. The effect of the vertical component is given as a function of the seismic coefficient S_{DS} as follows:

$$E_v = 0.2\,S_{\text{DS}}D$$

Consequently, the effect of the vertical component of the earthquake will be proportional to the effect of the dead load D by the amount of 0.2 S_{DS}.

Knowing the effect of both horizontal and vertical components including the directional effect, the design earthquake force E or E_m is taken as follows:

$$E = \rho Q_E \pm 0.2\,S_{\text{DS}}D$$

$$E_m = \Omega_o Q_E \pm 0.2\,S_{\text{DS}}D$$

where E = effect of the combined horizontal and vertical components of an earthquake
E_m = estimated maximum force that can develop in the structure due to system overstrength effect
Ω_o = system overstrength factor as defined in Table 6-A6
ρ = reliability/redundancy factor as defined in the following section

Redundancy Factor (ρ)

The redundancy factor, ρ, is intended to encourage redundant force paths in the structure, which is inversely proportional to redundancy. In other words, the code requires larger design seismic forces for less redundant systems. The code assigns two values for the redundancy factor that are outlined in the following two cases.

Case 1: ρ = 1 The redundancy factor is permitted to be 1.0 in the following conditions:

1. Structures in SDC B or C
2. Drift calculations and P-Δ effect
3. Design of nonstructural components
4. Design of nonbuilding structures that are not similar to buildings
5. Design of collector elements, splices, and their connections for which load combinations with overstrength factor are used
6. Design of members or connections where the load combinations of overstrength are required for design
7. Diaphragm loads as determined by the following equation

$$F_{px} = \frac{\sum F_i}{\sum w_i} w_{px} \qquad \text{(outlined in Sec. 6.10.1)}$$

8. Structures with damping systems designed in accordance with ASCE 7 standards.
9. Design of structural walls for out-of-plane forces, including their anchorage

Case 2: ρ = 1.3 The redundancy factor shall be taken as ρ = 1.3 for SDC D, E, or F. However, ρ is permitted to be 1.0 for these seismic design categories if one of the following two conditions is met:

1. Each story resisting more than 35 percent of the base shear in the direction of interest shall comply with Table 6-3.
2. Structures that are regular in plan at all levels, provided that the seismic force-resisting systems consist of at least two bays of seismic force-resisting perimeter framing on each side of the structure in each orthogonal direction at each story resisting more than 35 percent of the base shear. The number of bays for a shear wall shall be calculated as the length of shear wall divided by the story height or two times the length of shear wall divided by the story height for light-framed construction.

Lateral Force-Resisting Element	Requirements
Braced frames	Removal of an individual brace, or connection thereto, would not result in more than a 33 percent reduction in story strength, nor does the resulting system have an extreme torsional irregularity (horizontal structural irregularity Type 1b).
Moment frames	Loss of moment resistance at the beam-to-column connections at both ends of a single beam would not result in more than a 33 percent reduction in story strength, nor does the resulting system have an extreme torsional irregularity (horizontal structural irregularity Type 1b).
Shear walls or wall pier with a height-to-length ratio greater than 1.0	Removal of a shear wall or wall pier with a height-to-length ratio greater than 1.0 within any story, or collector connections thereto, would not result in more than a 33 percent reduction in story strength, nor does the resulting system have an extreme torsional irregularity (horizontal structural irregularity Type 1b).
Cantilever columns	Loss of moment resistance at the base connections of any single cantilever column would not result in more than a 33 percent reduction in story strength, nor does the resulting system have an extreme torsional irregularity (horizontal structural irregularity Type 1b).
Other	No requirements.

Table 6-3 Requirements for Each Story Resisting More Than 35 Percent of the Base Shear

6.8 Load Combinations

The IBC load combinations are given in two groups, namely, general combinations and overstrength combinations. In addition to the other force effect, the load combinations that include seismic effect are given as follows:

1. General Load and Resistance Factor Design (LRFD) combinations. Quantities in { } indicate (or)

$$U = 1.2D + \begin{Bmatrix} 1.0 \\ 0.5 \end{Bmatrix} L + \begin{Bmatrix} 0.7 \\ 0.2 \end{Bmatrix} S + E$$

$$U = 0.9D + E$$

   Live load factor in the combination above is taken as 1 in the case of public assemblies, or when $L > 4.8\ \text{kN/m}^2$.

   Snow load factor in the combination above is taken as 0.7 in the case of roofs of configuration that do not shed snow off of the structure (e.g., saw tooth roofs).

2. Overstrength combinations. This overstrength combination is intended to be used with members and elements as specifically required by the code. For example, this includes but not limited to foundations of cantilever columns, elements supporting discontinuous walls or frames of structures having plan irregularity Type 4 or vertical irregularity Type 4, collector elements and their

connections (in SDC: C to F), batter, and other pile situations. This load combination is given as

$$U = 1.2D + \begin{Bmatrix} 1.0 \\ 0.5 \end{Bmatrix} L + 0.2S + E_m$$

$$U = 0.9D + E_m$$

where D = dead load
L = live load
E = earthquake-induced forces as defined later in this chapter
E_m = maximum possible earthquake-induced forces as defined earlier in this chapter
S = snow load
{ } = quantities in { } indicate (or)

Example 6-5 A one-story open structure with height equal to 5 m is shown in Example 6-5, Fig. 1. The structure will be constructed in a region with a mapped acceleration coefficient at a short period (and at a 1s period) equal to $S_s = 1.44$ and $S_1 = 0.30$, respectively. The long-period transition period $T_L = 4$ s. The site is classified as Soil Type B. The structural system consists of two reinforced concrete moment frames in each direction spaced 8 m in plan. All columns are 0.5 × 0.5 m square columns with pinned connection at the base.

The slab carries dead, live, and snow loads as follows:

$$q_D = 20 \text{ kN/m}^2 \qquad \text{(including weight of columns)}$$

$$q_L = 10 \text{ kN/m}^2$$

$$q_S = 2 \text{ kN/m}^2$$

It is required to find the design forces and moments due to seismic load combinations, including the directional effect of column C4 at the intersection of lines 2 and B shown in Example 6-5, Fig. 1.

Solution Because the structure is regular, the equivalent lateral force procedure may be used if it satisfies the period limitations outlined earlier. This condition will be illustrated in step (2) below.

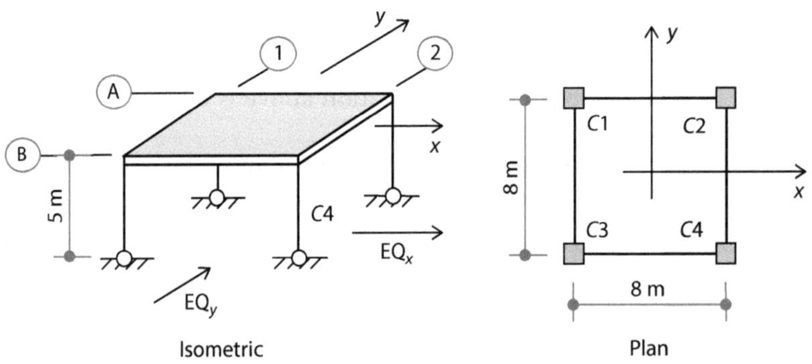

EXAMPLE 6-5, FIGURE 1

Refer to App. 6-1 for IBC tables.

1. Seismic coefficients, F_a, F_v, S_{DS}, and S_{D1}
   The seismic coefficient, F_a and F_v, are obtained from two IBC tables that are included as Tables 6-A3 and 6-A4. For $S_s = 1.44$, $F_a = 1.0$, $S_1 = 0.30$, and $F_v = 1.0$.
   Accordingly, the seismic coefficients, S_{DS} and S_{D1}, are calculated as follows:

$$S_{MS} = F_a S_s = 1.0\,(1.44) = 1.44$$

$$S_{DS} = \frac{2}{3} S_{MS} = \frac{2}{3}(1.44) = 0.96$$

$$S_{M1} = F_v S_1 = 1.0\,(0.30) = 0.30$$

$$S_{D1} = \frac{2}{3} S_{M1} = \frac{2}{3}(0.30) = 0.20$$

2. Risk, SDC, and Importance, I_e
   A structure with low hazard to human lies in Risk Category I. When referred to Table 6-A2 for Risk Category I and an S_{DS} of 0.96 and an S_{D1} of 0.20, it can be determined that the building is in SDC D.
3. Period, T, and response modification factor, R
   The response spectrum controlling period, T_s, and limiting period, T_{limit}, are determined as follows:

$$T_s = \frac{S_{D1}}{S_{DS}} = \frac{0.20}{0.96} = 0.208 \text{ s}$$

$$T_{\text{limit}} = 3.5\,T_s = 3.5\,(0.208) = 0.73 \text{ s}$$

   The approximate period, T_a, is calculated as

$$T_a = C_t (h_n)^x$$

   where building height $h_n = 5$ m, $C_t = 0.044$, and $x = 0.9$ for reinforced moment frames. Consequently, the period is given as

$$T_a = C_t (h_n)^x = 0.044\,(5)^{0.9} = 0.19 \text{ s}$$

   Because, $T_a < T_{\text{limit}}$ and the structure is regular, the equivalent lateral force procedure may be used.
   According to Table 6-A6 ordinary and intermediate moment frames are not allowed for SDC D, but SMFs are permitted. Therefore, if an SMF is selected for the building, the response modification factor, R, is obtained for SMF as $R = 8$.
4. Effective seismic weight, W

$$\text{Area of each floor:} \qquad A = 8(8) = 64 \text{ m}^2$$

   The effective seismic weight of the structure should include dead load and snow load because its intensity is more than 1.44 kN/m^2.
   Therefore, total effective seismic weight of the building is

$$W = (q_D + q_S)\,A$$
$$= (20 + 2)(64) = 1{,}408 \text{ kN}$$

5. Base shear, V
   Since $T < T_L$, then

$$C_s = \frac{S_{DS} I_e}{R} = \begin{pmatrix} \le \dfrac{S_{D1} I_e}{RT} \\ \ge 0.044\, S_{DS} I_e \ge 0.01 \end{pmatrix}$$

$$C_s = \frac{0.96(1)}{8} = 0.12 = \begin{pmatrix} \le \dfrac{0.20(1)}{8(0.19)} = 0.13 \\ \ge 0.044(0.96)(1) = 0.042 \ge 0.01 \end{pmatrix}$$

Therefore,

$$C_s = 0.12$$

As a result,

$$V = C_s W = 0.12\,(1{,}408) = 169 \text{ kN}$$

The total base shear will be applied to the floor because it is only a one-story building.

6. Horizontal distribution of forces (refer to Example 6-1 for details)
   When an accidental eccentricity of 5 percent of perpendicular dimensions is applied, the torsional moment will be

$$M_{\text{tor}} = Fe = 169(0.05)(8) = 68 \text{ kN·m}$$

   Because the frames are identical, their stiffness will be the same. Thus, the force induced in frames A and B will be

$$Q_i = \frac{F}{\Sigma k_j} k_i + \frac{M_{\text{tor}}}{\Sigma k_j r_j^2} k_i r$$

$$Q_i = \frac{169}{2k} k + \frac{68}{k(4)^2(2)} k(4)$$

$$\begin{pmatrix} Q_A \\ Q_B \end{pmatrix} = 85 \pm 9 = \begin{pmatrix} 76 \\ 94 \end{pmatrix} \text{ kN}$$

   Because the structure is doubly symmetric, the forces induced by earthquake excitation in direction Y will be identical to the forces induced by excitation in direction X.
   The resulting forces in the frames are summarized in Example 6-5, Fig. 2.

7. Internal forces
   The internal forces and moments in frame B may be found using any structural analysis method. For this frame, the horizontal reactions are equal, and the solution can easily be obtained by hand calculations. Such analysis yields reactions and moments in the frame as shown in Example 6-5, Fig. 3.
   Because the structure is doubly symmetric, the forces and moments in frame 2 will be the same as in frame B. Consequently, the horizontal forces and moments in Column C4 due to both horizontal earthquake directions, Q_E, will be given as follows:

Direction X:	$P_{\text{Ex}} = 59$ kN,	$V_{\text{Ex}} = 47$ kN	$M_{\text{Ex}} = 235$ kN·m
Direction Y:	$P_{\text{Ey}} = 59$ kN	$V_{\text{Ey}} = 47$ kN	$M_{\text{Ey}} = 235$ kN·m

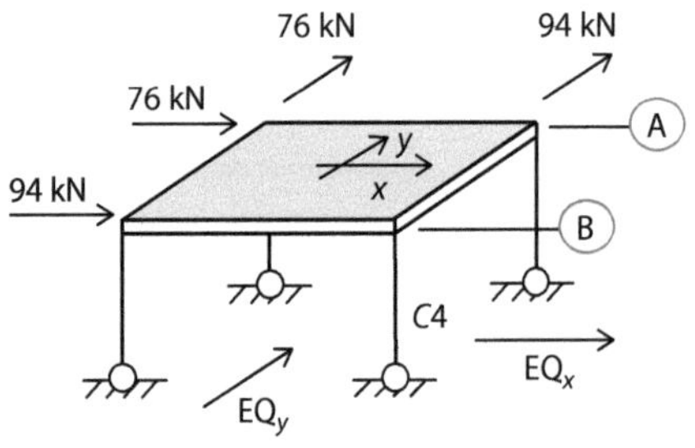

Example 6-5, Figure 2 Frame forces.

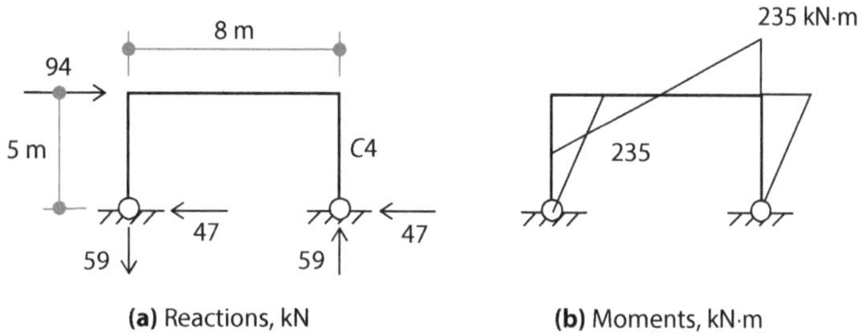

EXAMPLE 6-5, FIGURE 3 Frame B forces and moments.

8. Directional Effect

   Even though such analysis is not required by the code for this type of structures, it will be presented here for demonstration purposes. Combining forces from orthogonal directions is obtained by adding 30 percent of the effect of one direction to 100 percent of the effect from the other direction. This combining effect results in two load cases for the effect of horizontal excitation, Q_{E1} and Q_{E2}, which are expressed as

$$Q_{E1} = E_x + 0.3\,E_y \qquad \text{case 1}$$

$$Q_{E2} = 0.3\,E_x + E_y \qquad \text{case 2}$$

The combining effect is vectorial. The vectorial results of the internal forces from excitation directions X and Y are shown in Example 6-5, Fig. 4(a), and Example 6-5, Fig. 4(b), respectively. The results of the directional combination effect will be as described below.

Case 1:

$$P = P_{\text{Ex}} + 0.3P_{\text{Ey}} = 59 + 0.3(59) = 77 \text{ kN}$$
$$V_x = V_{\text{Ex}} + 0.3V_{\text{Ey}} = 47 + 0.3(0) = 47 \text{ kN}$$
$$V_y = V_{\text{Ex}} + 0.3V_{\text{Ey}} = 0 + 0.3(47) = 14 \text{ kN}$$
$$M_x = M_{\text{Ex}} + 0.3M_{\text{Ey}} = 0 + 0.3(235) = 71 \text{ kN·m}$$
$$M_y = M_{\text{Ex}} + 0.3M_{\text{Ey}} = 235 + 0.3(0) = 235 \text{ kN·m}$$

Case 2:

$$P = 0.3P_{\text{Ex}} + P_{\text{Ey}} = 0.3(59) + 59 = 77 \text{ kN}$$
$$V_x = 0.3V_{\text{Ex}} + V_{\text{Ey}} = 0.3(47) + 0 = 14 \text{ kN}$$
$$V_y = 0.3V_{\text{Ex}} + V_{\text{Ey}} = 0.3(0) + 47 = 47 \text{ kN}$$

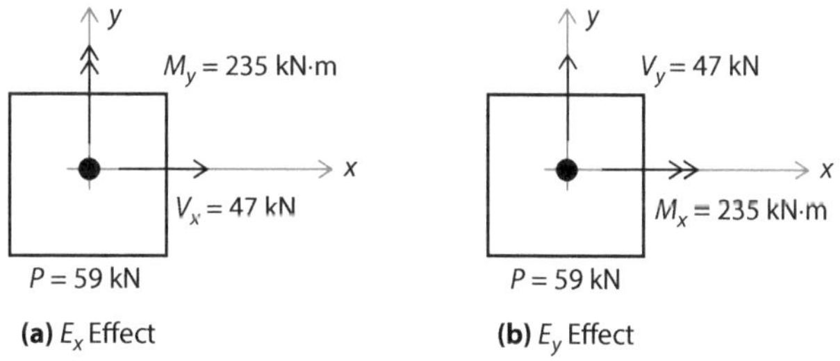

EXAMPLE 6-5, FIGURE 4 Internal actions due to separate excitation.

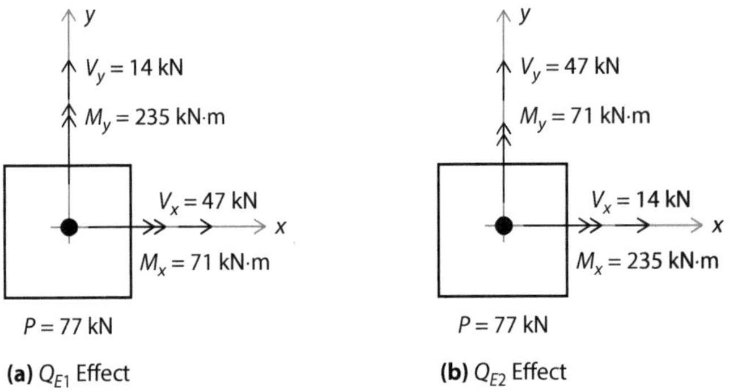

Example 6-5, Figure 5 Directional combination of excitation.

$$M_x = 0.3M_{Ex} + M_{Ey} = 0.3(0) + 235 = 235 \text{ kN·m}$$
$$M_y = 0.3M_{Ex} + M_{Ey} = 0.3(235) + 0 = 71 \text{ kN·m}$$

These results are shown in Example 6-5, Fig. 5.

Because the effect of the vertical component of the earthquake is proportional to the dead load effect, the dead load effect will be evaluated first.

9. Dead load

The dead load effect is determined by finding the distribution of the slab load to the beams and, as a result, to the columns. The distribution of the slab load to frame B is the shaded area marked in Example 6-5, Fig. 6(a). Accordingly, the frame will be subjected to a triangular load with maximum intensity equal to

$$(q_D)_{\text{frame}} = (q_D)_{\text{slab}}\,(4\,m) = 20\,(4) = 80 \text{ kN/m}^2$$

This load is applied to frame B as shown in Example 6-5, Fig. 6(b). The solution to this case of loading is obtained by using any structural analysis method. For example, the solution in Example 6-5, Fig. 6(b), is obtained by using standard software in structural analysis.

In reference to Example 6-5, Fig. 6(b), the internal forces and moments in column C4 that are due to the dead load on frame B are given as follows.

Frame B: $P = 160$ kN $V_x = 20$ kN $M_y = 100$ kN·m

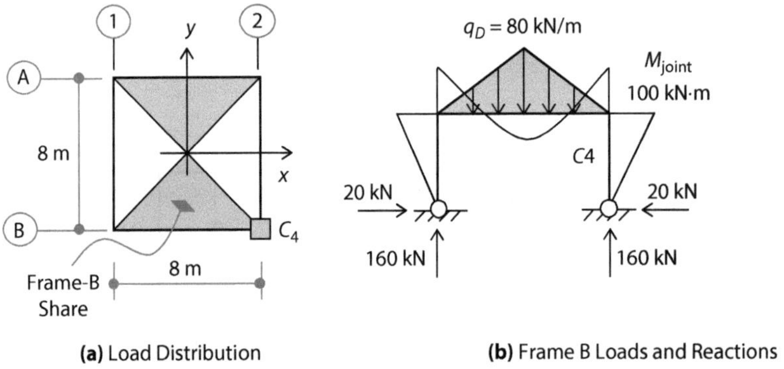

Example 6-5, Figure 6 Dead load distribution to frames.

Column C4 will also incur forces and moments from frame 2's share of the dead load. Because frame B and frame 2 are identical in dimensions, properties, and loading, the internal forces and moments in column C4 that are due to the dead load on frame 2 will be the same. Therefore,

Frame 2: $P = 160$ kN $V_y = 20$ kN $M_x = 100$ kN·m

The total internal forces and moments in column C4 that are due to dead load are shown in Example 6-5, Fig. 7(a). They are calculated as follows:

$$P = 160 + 160 = 320 \text{ kN}$$
$$V_x = 20 \text{ kN} \qquad V_y = 20 \text{ kN}$$
$$M_x = 100 \text{ kN·m} \qquad M_y = 100 \text{ kN·m}$$

The effect of live load and snow load will be proportional to the effect of the dead load because the load distribution will take the same path with a different intensity.

If due to live load, multiply by a factor of $10/20 = 0.5$. Hence,

Frame B: $P = 80$ kN $V_x = 10$ kN $M_y = 50$ kN·m

Frame 2: $P = 80$ kN $V_y = 10$ kN $M_x = 50$ kN·m

If due to snow load, multiply by a factor of $2/20 = 0.1$. Hence,

Frame B: $P = 16$ kN $V_x = 2$ kN $M_y = 10$ kN·m

Frame 2: $P = 16$ kN $V_y = 2$ kN $M_x = 10$ kN·m

The final gravity load effects are shown in Example 6-5, Fig. 7.

10. Vertical earthquake excitation
The vertical earthquake component is given as follows:

$$E_v = 0.2\, S_{DS} D$$
$$= 0.2\,(0.96)\, D = 0.192\, D$$

Therefore, the vertical earthquake component will be equal to the dead load effect multiplied by a factor of 0.16.

Not only does the vertical component of an earthquake induce vertical forces (as is commonly perceived), it also induces shears and moments in the structure. Results are shown in Example 6-5, Fig. 8.

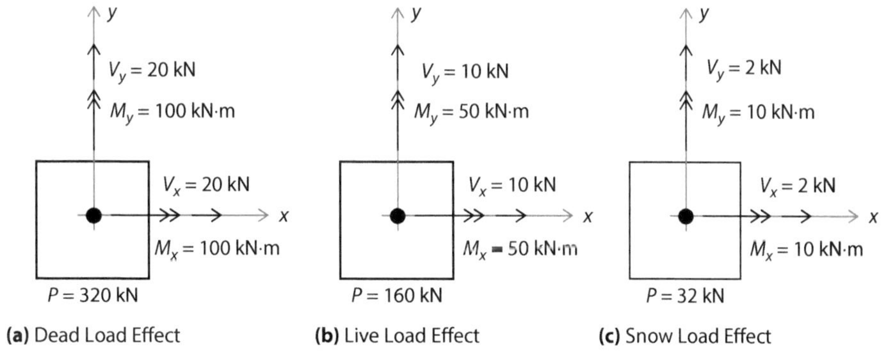

Example 6-5, Figure 7 Gravity load effects in column C4.

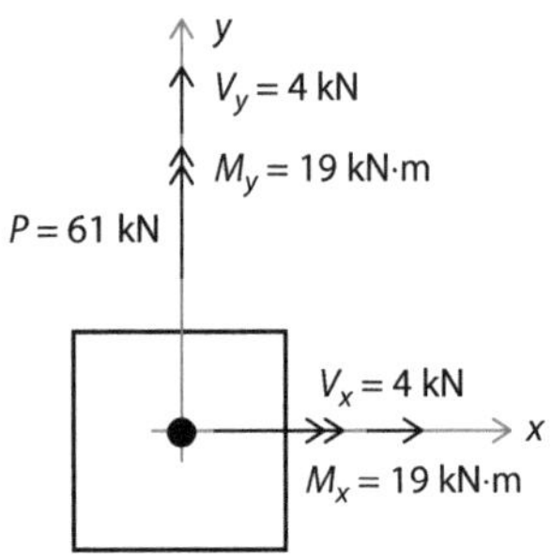

Example 6-5, Figure 8 E_v effect.

11. Horizontal and vertical combination of earthquake effect
The combination of the horizontal and vertical components is given as

$$E = E_h + E_v$$
$$= \rho Q_E + E_v$$

where Q_E represents either case 1, Q_{E1} or case 2, Q_{E2}. Therefore, this combination results in two load cases E_1 and E_2. Because the structure does not satisfy the code requirements for reduction of ρ, as outlined in an earlier section, ρ shall be taken as 1.3 for SDC D. Using ρ = 1.3, the two load cases are given as

$$E_1 = 1.3\,Q_{E1} + E_v$$
$$E_2 = 1.3\,Q_{E2} + E_v$$

The two load cases previously are evaluated by graphical inspection as shown in Example 6-5, Fig. 9. For example, Example 6-5, Fig. 9(a), shows the effect of each of the three earthquake components separately, while Example 6-5, Fig. 9(b), shows their effect combined according to the two equations above.

12. Load Combinations
Because all components from dead, live, snow, and earthquake effect are evaluated, the design load combination can now be evaluated as follows:

$$U = 1.2D + 1.0L + 0.2S + E$$
$$U = 0.9D + E$$

Because the second combination is applied mostly to check uplift problems, the first combination will be more critical for the design of the members. Therefore, this example will only consider the first load case. Furthermore, because the earthquake effect results in two load cases, E_1 and E_2, each combination previously also breaks down into two load cases. The first load combination breaks into the following two load cases:

$$U_1 = 1.2D + 1.0L + 0.2S + E_1$$
$$U_2 = 1.2D + 1.0L + 0.2S + E_2$$

For case U_1,

$$P = 1.2(320) + 1.0(160) + 0.2(32) + 161 = 711 \text{ kN}$$
$$V_x = 1.2(20) + 1.0(10) + 0.2(2) + 65 = 99 \text{ kN}$$
$$V_y = 1.2(20) + 1.0(10) + 0.2(2) + 22 = 56 \text{ kN}$$
$$M_x = 1.2(100) + 1.0(50) + 0.2(10) + 111 = 283 \text{ kN·m}$$
$$M_y = 1.2(100) + 1.0(50) + 0.2(10) + 325 = 497 \text{ kN·m}$$

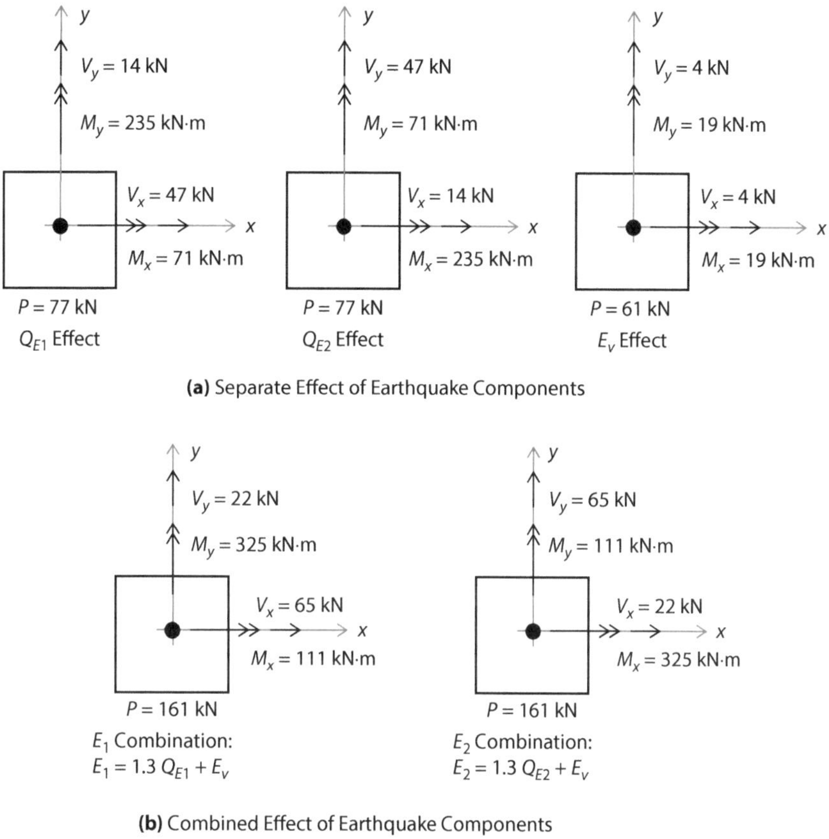

Example 6-5, Figure 9 Combination of earthquake components.

For case U_2,

$$P = 1.2(320) + 1.0(160) + 0.2(32) + 161 = 711 \text{ kN}$$
$$V_x = 1.2(20) + 1.0(10) + 0.2(2) + 22 = 56 \text{ kN}$$
$$V_y = 1.2(20) + 1.0(10) + 0.2(2) + 65 = 99 \text{ kN}$$
$$M_x = 1.2(100) + 1.0(50) + 0.2(10) + 325 = 497 \text{ kN·m}$$
$$M_y = 1.2(100) + 1.0(50) + 0.2(10) + 111 = 283 \text{ kN·m}$$

The column can now be designed using the design method provided by the applicable design code (e.g., the design criterion in the ACI code).

6.9 Definitions and Requirements of Structural Systems

Approved structural systems in most codes are similar. In principle, the structural systems for building structures may be grouped into three major categories, namely, frames, shear walls, and a combination of both (dual systems). For each system, the code assigns values for R and Ω_o and for C_d as given in ASCE 7 Table 12.2-1 for building structures. Each system listed in this table has associated detailing and design requirements. The ACI code provides detailing of reinforced concrete structures for seismic provisions, the AISC code provides detailing of steel elements for seismic provisions. These requirements will be described in the next several chapters.

The systems approved by the IBC code are listed in ASCE 7 Table 12.2-1. Select examples of such popular and frequently used systems are given in App. 6-1 at the end of this chapter. Remember that some of these systems are prohibited in high seismic risk regions. For example, concrete ordinary moment-resisting frames (OMRFs) are prohibited for Seismic Design Categories C, D, E, and F.

For nonbuilding structures, the code assigns values for R, Ω_o, and C_d as given in ASCE 7 Tables 15.4-1 and 15.4-2. Such structures include tanks, silos, chimneys, towers, bins and hoppers, signs, and monuments.

6.10 Special Topics

6.10.1 Diaphragm Design Forces

Floor and roof diaphragms shall be designed to resist an in-plane force, F_{px}, at level x given as

$$F_{px} = \frac{\sum_{x}^{n} F_i}{\sum_{x}^{n} w_i} w_{px} = \begin{pmatrix} \leq 0.4\, S_{DS}\, I_e\, w_{px} \\ \geq 0.2\, S_{DS}\, I_e\, w_{px} \end{pmatrix}$$

Notice that both summations are carried out from the diaphragm under consideration, x, to the top of the building, n. F and w are the forces and weights of the floors as defined in previous sections.

6.10.2 Torsional Effect

Because torsion has been a major cause of failures in previous earthquakes, the design must carefully consider the effect of torsion. The code requires an additional torsional moment to be added to the existing torsion. The additional torsion, which is also known as accidental eccentricity, is introduced to the structure by displacing the actual center of mass by an amount equal to 5 percent of the structural dimension. This 5 percent is given perpendicular to the direction of excitation as illustrated in Fig. 6-9.

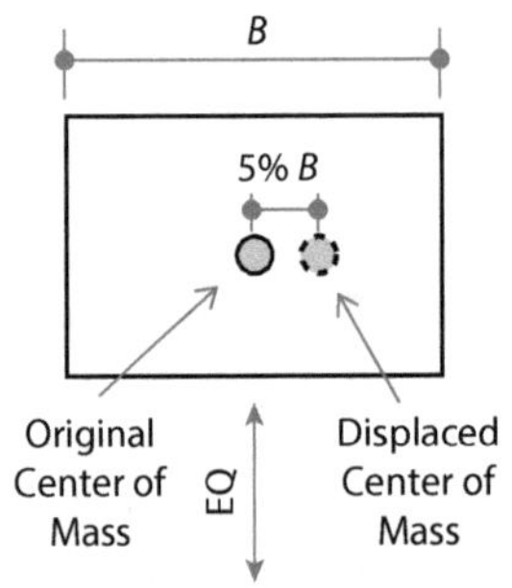

FIGURE 6-9 Displaced center of mass.

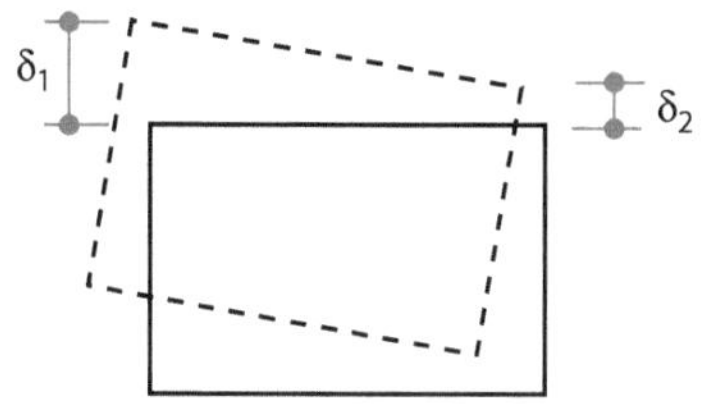

Figure 6-10 Torsional displacements.

For irregular structures, the additional accidental eccentricity mentioned previously shall be amplified by a factor A_x at each level. The amplification factor is given by the following:

$$A_x = \left[\frac{\delta_{\max}}{1.2\,\delta_{\text{avg}}}\right]^2 \leq 3.0$$

where $\delta_{\max}$ and δ_{avg} are the maximum and average displacements given in reference to Fig. 6-10 as

$$\delta_{\max} = \delta_1$$

$$\delta_{\text{avg}} = \frac{\delta_1 + \delta_1}{2}$$

6.10.3 Drift Limitations

Drift limitations under earthquakes are imposed on inelastic displacements. The story drift Δ is defined as the relative displacements between adjacent story displacements δ. The story displacement δ_x, at level x is calculated as follows:

$$\delta_x = \frac{C_d\,\delta_{\text{xe}}}{I_e}$$

where δ_x = Inelastic displacement at level x
δ_{xe} = Elastic displacement due to base shear at level x
C_d = Deflection amplification factor (App. 6-1)
I_e = Importance factor
Δ = Relative inelastic displacement at level x, ($\Delta_x = \delta_{x+1} - \delta_x$)

The value of Δ shall also be modified to include the P-Δ effect, if it exists, such that

$$\Delta_{P\text{-}\Delta} = \frac{\Delta}{1-\theta}$$

where θ is a stability coefficient (to be defined in the next section).

The story drift limitation is given according to the SDC in Table 6-A5.

The drift limitations under earthquakes are almost 10 times its limitations under other service loads.

6.10.4 Structural Separation

The pounding of adjacent structures during an earthquake has caused serious damage to structures in previous earthquakes, especially when floor slabs in adjacent structures meet at different levels. To avoid this pounding effect, the IBC requires that structures be set back a distance δ_M from property lines such that

$$\delta_M = \frac{C_d \delta_{\max}}{I_e}$$

where $\delta_{\max}$ is the maximum elastic displacement, C_d is the displacement amplification factor, and I_e is the importance factor as defined earlier in this chapter.

If structures exist on the same property, the clear distance between them must be at least equal to δ_{MT}, such that

$$\delta_{\mathrm{MT}} = \sqrt{(\delta_{M1})^2 + (\delta_{M2})^2}$$

where δ_{M1} and δ_{M2} are the inelastic displacements of the two adjacent structures as shown in Fig. 6-11. It should be noted that δ_{MT} is simply the total displacement of the two structures combined by SRSS method.

Remember that such separation must be based on the anticipated inelastic displacement of the adjacent structures.

6.10.5 *P*-Δ Effect

The design shall consider the *P*-Δ effect according to a stability coefficient, θ, where this stability coefficient is calculated as follows:

$$\theta = \frac{P_x \Delta I_e}{V_x h_{sx} C_d}$$

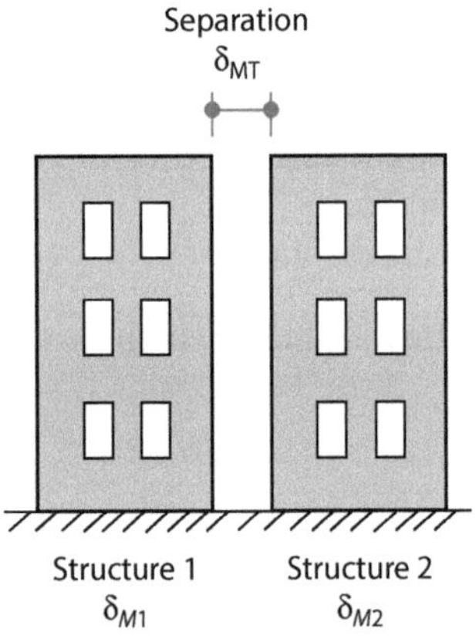

FIGURE 6-11 Structure separation.

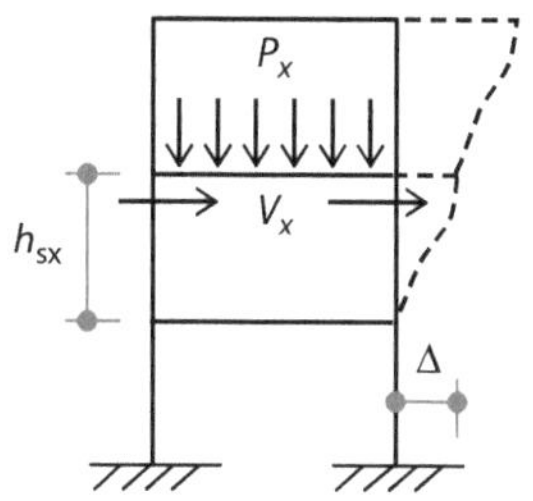

FIGURE 6-12 *P*-Δ effect in story *x*.

where P_x = total vertical unfactored loads above level x
Δ = design story drift (inelastic) as shown in Fig. 6-12
I_e = importance factor for earthquakes as defined earlier
V_x = seismic shear force between level x and level x-1
C_d = deflection amplification factor

The *P*-Δ effect shall not be considered if

$$\theta \leq 0.1$$

The *P*-Δ effect shall be considered if

$$\theta > 0.1$$

However, the structure must be redesigned if

$$\theta \geq \theta_{max}$$

where

$$\theta_{max} = \frac{0.5}{\beta C_d} \leq 0.25$$

where β is the ratio of shear demand to shear capacity. β may be conservatively taken as equal to 1.0.

6.11 Problems

6-1 The structure shown in Prob. 6-1, Fig. 1, is idealized as 2DOF system. The beams are considered rigid. All columns are reinforced concrete having the same cross section with stiffness as shown. The structure lies in a site with seismic parameters according to IBC code as follows:

1. $S_s = 100\%$
   $S_1 = 40\%$
   $T_L = 10$ s
2. Soil Type C

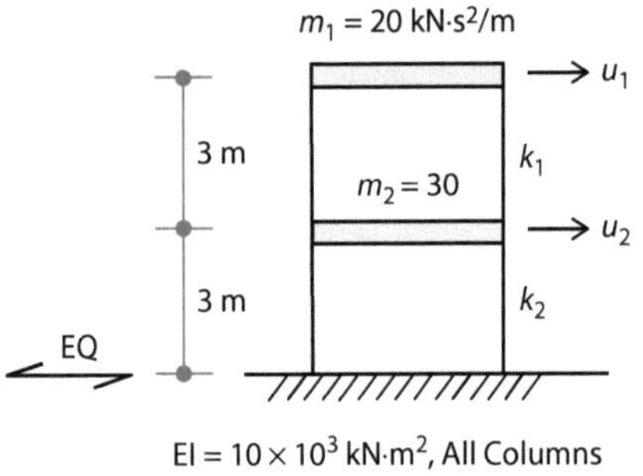

PROBLEM 6-1, FIGURE 1

The structure is required to be designed using modal response spectrum method outlined in IBC code. The frame will be detailed as required by the relevant codes. Accordingly:

a. Find the frequencies and their mode shapes.
b. Find the design story forces.
c. Find the design base shear.
d. Check interstory drift limitations.

NOTE: Use SRSS combination criterion.

APPENDIX 6-1

Appendix 6-1 contains information on seismic parameters presented in tables based on the IBC and ASCE 7.

Risk Category	Description	Seismic Importance Factor, I_e
I	Low hazard to humans, such as agriculture facilities	1
II	Other than I, III, and IV	1
III	Hazardous to humans because of large number of occupants, such as schools and public areas	1.25
IV	Essential facilities that are either important for security or provide major services such as rescue operations, health services, transportation, and communications	1.5

TABLE 6-A1 Importance Factor

S_{DS}	S_{D1}	Risk Category	
		I or II or III	IV
$S_{DS} < 0.167$	$S_{D1} < 0.067$	A	A
$0.167 \leq S_{DS} < 0.33$	$0.067 \leq S_{D1} < 0.133$	B	C
$0.33 \leq S_{DS} < 0.50$	$0.133 \leq S_{D1} < 0.20$	C	D
$0.50 \leq S_{DS}$	$0.20 \leq S_{D1}$	D	D
If: $S_1 \geq 0.75$		E	F
For: $S_s \leq 0.15$ & $S_1 \leq 0.04$, it is permitted to use		A	A

TABLE 6-A2 Seismic Design Category (SDC)

Site Class	Mapped Spectral Acceleration at Short Period				
	$S_s \leq 0.25$	$S_s = 0.50$	$S_s = 0.75$	$S_s = 1.00$	$S_s \geq 1.25$
A	0.8	0.8	0.8	0.8	0.8
B	1.0	1.0	1.0	1.0	1.0
C	1.2	1.2	1.1	1.0	1.0
D	1.6	1.4	1.2	1.1	1.0
E	2.5	1.7	1.2	0.9	0.9
F	Site-specific investigation is required.				

TABLE 6-A3 IBC Seismic Site Coefficient at Short Period, F_a

Site Class	Mapped Spectral Acceleration at a 1-s Period				
	$S_1 \leq 0.1$	$S_1 = 0.2$	$S_1 = 0.3$	$S_1 = 0.4$	$S_1 \geq 0.5$
A	0.8	0.8	0.8	0.8	0.8
B	1.0	1.0	1.0	1.0	1.0
C	1.7	1.6	1.5	1.4	1.3
D	2.4	2.0	1.8	1.6	1.5
E	3.5	3.2	2.8	2.4	2.4
F	Site-specific investigation is required.				

TABLE 6-A4 IBC Seismic Site Coefficient at One Second, F_v

Building	Risk Category		
	I or II	III	IV
Other than masonry shear walls, four stories or less above the base, that have been designed to accommodate drifts	0.025	0.020	0.015
Masonry cantilever shear wall structures	0.010	0.010	0.010
Other masonry shear wall structures	0.007	0.007	0.007
All other buildings	0.020	0.015	0.010

TABLE 6-A5 Drift Limitation Ratio (Δ/h_{sx}) According to the IBC

Structural System	Description of System[1]	R	Ω_o	C_d	System and Height Limitations by Seismic Design Category in Meters[2]				
					A or B	C	D	E	F
Concrete	BSSW[3]	5	2.5	5	NL	NL	48	48	30
	BOSW	4	2.5	4	NL	NL	NP	NP	NP
	NSSW[3]	6	2.5	5	NL	NL	48	48	30
	NOSW	5	2.5	4.5	NL	NL	NP	NP	NP
	SMF	8	3	5.5	NL	NL	NL	NL	NL
	IMF	5	3	4.5	NL	NL	NP	NP	NP
	OMF	3	3	2.5	NL	NP	NP	NP	NP
Steel	EBF (MC)	8	2	4	NL	NL	48	48	30
	EBF (NMC)	7	2	4	NL	NL	48	48	30
	SCBF	6	2	5	NL	NL	48	48	30
	OCBF	3.25	2	3.25	NL	NL	10	10	NP
	SMF	8	3	5.5	NL	NL	NL	NL	NL
	IMF	4.5	3	4	NL	NL	10	NP	NP
	OMF	3.5	3	3	NL	NL	NP	NP	NP
	STMF	7	3	5.5	NL	NL	48	30	NP
Dual-concrete	SSW with SMF	7	2.5	5.5	NL	NL	NL	NL	NL
	SSW with IMF	6.5	2.5	5	NL	NL	48	30	30
	OSW with SMF	6	2.5	5	NL	NL	NP	NP	NP
	OSW with IMF	5.5	2.5	4.5	NL	NL	NP	NP	NP
Dual-steel	EBR with SMF	8	2.5	4	NL	NL	NL	NL	NL
	SCBR with SMF	7	2.5	5.5	NL	NL	NL	NL	NL
	SCBR with IMF	6	2.5	5	NL	NL	10	NP	NP

[1]The following letters are used to identify systems in this table:

CBF	Concentrically braced frames
EBF	Eccentrically braced frames
MC	Moment connection
MF	Moment frames
NMC	Nonmoment connection
SW	Shear walls
TMF	Truss moment frames

System grade and bearing or nonbearing walls are identified with the following prefixes:

B	Bearing walls
I	Intermediate system grade
N	Nonbearing walls
O	Ordinary system grade
S	Special system grade

[2]NL No limitation
NP Not permitted

[3]Special shear walls include precast special shear walls.

Table 6-A6 Selected Popular Building Seismic Systems from ASCE 7

Nonbuilding Structural Type	R	Ω_o	C_d	System and Height Limitations by Seismic Design Category in Meters[1]			
				A and B	C	D	E and F
Cast-in-place silos, stacks, and chimneys having walls continuous to the foundations	3	1.75	3	NL	NL	NL	NL
Trussed towers (freestanding or guyed), stacks, and chimneys	3	2	2.5	NL	NL	NL	NL
Amusement structures and monuments	2	2	2	NL	NL	NL	NL
Inverted pendulum-type structures (except elevated tanks, vessels, bins, and hoppers)	2	2	2	NL	NL	NL	NL
Signs and billboards	3.5	1.75	3	NL	NL	NL	NL
All other self-supporting structures, tanks, or vessels not covered by ASCE 7 or by approved standards that are similar to buildings	1.25	2	2.5	NL	NL	15	15

[1]NL No limitation
NP Not permitted

Table 6-A7 Selected Nonbuilding Seismic Systems from ASCE 7

Structure Type	C_t	x
Steel moment frames free to deflect under seismic forces	0.0724	0.8
Reinforced concrete moment frames free to deflect under seismic forces	0.0466	0.9
Eccentrically braced steel frames	0.0731	0.75
All other structural systems	0.0488	0.75

Table 6-A8 Approximate Period Calculation Coefficients

S_{D1}	≥ 0.4	0.3	0.2	0.15	≤ 0.1
C_u	1.4	1.4	1.5	1.6	1.7

Table 6-A9 Coefficient for Upper Limit on Calculated Period

CHAPTER 7

Seismic Provisions of Reinforced Concrete Structures (ACI 318)

7.1 Introduction

The American Concrete Institute (ACI) code known as ACI 318 is widely recognized for reinforced concrete analysis and design. The International Building Code (IBC) adopted ACI 318 as the IBC official provisions for reinforced concrete design. ACI 318 offers detailed seismic provisions in Chap. 21. ACI 318 has also revised its load combinations in the 2002 version to follow the general trend of load combinations offered by other building codes.

Although ACI 318 does not provide guidelines for force calculations, it does provide the detailing required to achieve the properties of the different reinforced concrete systems that are used and endorsed by the IBC for seismic design. Such systems include ordinary, intermediate, and special moment frames (SMF). They also include ordinary and special shear walls.

The earthquake-relevant load combinations required by the 2002 and later versions of ACI 318 are listed as follows:

$$U = 1.2D + \begin{Bmatrix} 1.0 \\ 0.5 \end{Bmatrix} L + 0.2S + E$$

$$U = 0.9D + E$$

where D = dead load

E = earthquake-induced forces as defined later in this chapter

L = live load

S = snow load

{ } = quantities in { } indicate (or)

Live load factor in the above combination is taken as 1.0 in case of: (1) public assemblies, (2) garages, or (3) when L > 4.8 kN/m^2.

Because the behavior of concrete as a material is greatly affected by confinement, as explained in Chap. 5, the details of stirrups become essential insofar as they provide the confinement requirements for the various systems. Therefore, the following sections provide an in-depth description of the ACI 318 detailing requirements of various structural systems and components.

7.2 Ordinary Moment Frames

Ordinary moment frames (OMF) are assigned an R-factor equal to 3 by the IBC. In order to meet such force reduction, ACI 318 requires these frames to conform to all of its provisions from Chap. 1 through Chap. 20. In addition to these requirements, ACI requires the two following requirements for ordinary frames if used for seismic resistance:

a. Beams: Beams shall have at least two continuous bars at both faces, top and bottom, at the support; and shall be developed from the face of support.

b. Beam-columns: Beam-columns with clear span depth ratio less than or equal to 5 shall be designed for shear as in intermediate moment beam-columns.

The code requires that all design forces and moments of these frames be obtained from second-order elastic analysis. Critical sections are then designed to reach their ultimate state with concrete stress in compression reaching its designated strength, f_c', and steel stress reaching its yield level, f_y. Of course, this design concept is a lower-bound solution of the structure.

Detailing of this type of frame is familiar to most engineers and is given for both beams and beam-columns. Important requirements and detailing of these frames may be highlighted in this section for later use with intermediate and special framing systems and elements.

7.2.1 Ordinary Beams

According to ACI 318, a member in a frame is considered a beam if the intensity of the external-factored axial load, P_u, in the member is limited to

$$P_u \leq 0.1 f_c' A_g$$

where A_g = gross-sectional area of the member

f_c' = compressive strength of concrete

The following reinforcement detailing of ordinary beams is presented.

Main Reinforcement

The main reinforcement in ordinary beams is limited to maximum and minimum values as a function of the shape of the cross section and to the location of its reinforcement. The basic reinforcement ratio is given for a singly reinforced rectangular section as defined in Fig. 7-1. The section has total dimensions of $b \times h$. The reinforcement is placed on the tension side of the beam as shown in Fig. 7-1.

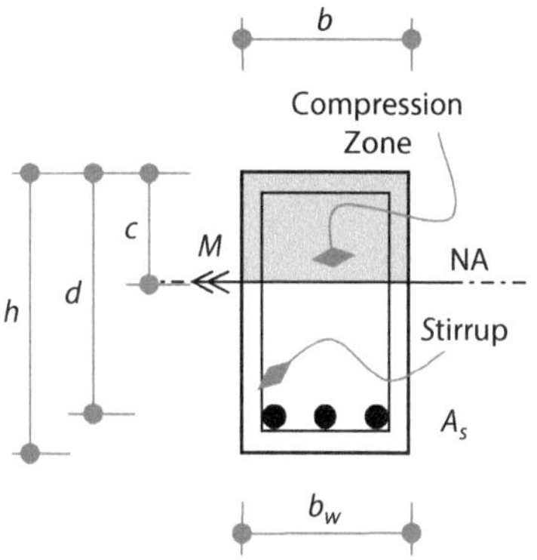

FIGURE 7-1 Singly reinforced rectangular section.

The reinforcement of a singly reinforced rectangular section subjected to pure moment may be given in terms of a basic reinforcement ratio, ρ_o, and a basic reinforcement index, ω_o, which are defined as follows:

$$\rho_o = \frac{A_s}{bd}$$

$$\omega_o = \rho_o \frac{f_y}{f'_c}$$

where A_s = area of reinforcement on the tension side of the cross section
b = width of the compression zone of the rectangular section
d = effective depth of the beam which is defined as the distance between the extreme compression fibers and centroid of steel
f'_c = characteristic strength of concrete
f_y = yield stress of steel reinforcement

The minimum reinforcement ratio for beams is referenced to the web, $\rho_{w,\min}$, which is given as follows:

$$\rho_{w,\min} = \frac{\sqrt{f'_c}}{4 f_y} \geq \frac{1.4}{f_y}$$

Accordingly, minimum area of steel, $A_{s,\min}$, is given in terms of $\rho_{w,\min}$ as

$$A_{s,\min} = \rho_{w,\min} b_w d$$

The ductility of ordinary beams depends mainly on the amount of steel in the section. ACI 318 expresses ductility in terms of the net tensile strain ε_t, which is defined as the net tensile strain in the extreme layer of tension steel, excluding effective prestressing, creep, shrinkage, and temperature. The code considers ductility to be a tension-controlled condition, whereas brittleness is considered to be a compression-controlled condition. In terms of the net tensile strains, these limits are generally given as follows:

$$\varepsilon_t \leq 0.002 \qquad \text{compression-controlled condition}$$

$$\varepsilon_t \geq 0.005 \qquad \text{tension-controlled condition}$$

In order to satisfy ductility requirements in ordinary beams, ACI 318 limits the net strain, ε_t, in beams to a minimum value of 0.005 in order to qualify as a beam with a strength-reduction factor of $\phi = 0.9$. However, the code limits this net strain to a minimum of 0.004 at a reduced strength-reduction factor to be interpolated according to the intensity of the low axial load under $0.1 f'_c A_g$.

Consequently, the maximum basic reinforcement in the section is required to prevent brittle failure and to guarantee ductile failure with ample warning, that is, yielding of the steel before crushing of the concrete. In general, this condition may be expressed in terms of the neutral axis depth or in terms of a maximum basic reinforcement index, ω_o. These limitations may be set at a net tensile strain of 0.005 and 0.004, as defined above.

If the limitations above are referenced to a minimum net tensile strain ($\varepsilon_t = 0.005$), the following limits apply in terms of neutral axis depth c:

$$\frac{c}{d} \leq \frac{3}{8} = 0.375$$

By noting that for a rectangular section, $\dfrac{0.85a}{d} = \omega_o$; the maximum reinforcement limitations may be stated in terms of reinforcement index as follows:

$$\omega_o = \omega_w \leq 0.32\,\beta_1$$

However, if the limitations above are referenced to a minimum net tensile strain ($\varepsilon_t = 0.004$), the following limits apply in terms of neutral axis depth, c

$$\frac{c}{d} \leq \frac{3}{7} = 0.429$$

and in terms of basic reinforcement index, ω_o

$$\omega_o = \omega_w \leq 0.36\,\beta_1$$

In this book, the maximum reinforcement limitations will be set at a net tensile strain of $\varepsilon_t = 0.005$, in order to use the strength-reduction factor at its maximum value of $\phi = 0.9$.

If required, the maximum reinforcement of rectangular sections can be increased if compression steel is placed in the compression zone as shown in Fig. 7-2. If the area of compression steel is designated as A'_s, the reinforcement ratio of compression steel is ρ' and the reinforcement index of compression steel is defined as ω', then the maximum reinforcement of the tension steel may be increased such that

$$(\omega - \omega') \leq 0.32\,\beta_1$$

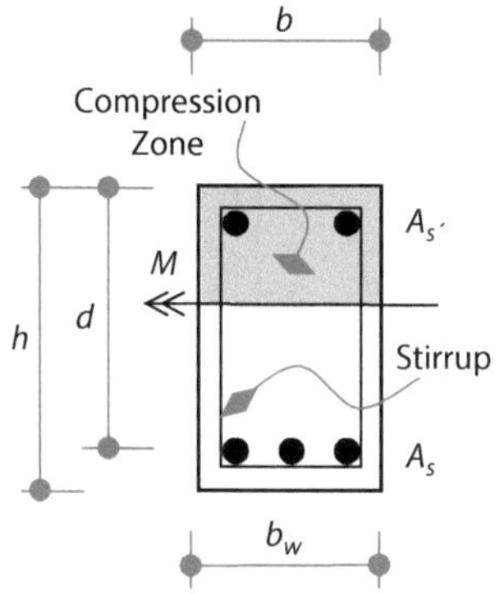

FIGURE 7-2 Doubly reinforced rectangular section.

where

$$\omega = \rho \frac{f_y}{f_c'}; \quad \rho = \frac{A_s}{bd}$$

$$\omega' = \rho' \frac{f_s'}{f_c'}; \quad \rho' = \frac{A_s'}{bd}$$

The condition given previously can still be stated in the general form given for rectangular singly reinforced sections if the reinforcement index is referenced to the web as follows:

in terms of neutral axis depth, c

$$\frac{c}{d} \le \frac{3}{8} = 0.375$$

in terms of web reinforcement index, ω_w

$$\omega_w \le 0.32\,\beta_1$$

where $\omega_w = (\omega - \omega')$

and $\omega_w = \rho_w \dfrac{f_y}{f_c'}; \quad \rho_w = \dfrac{A_{sw}}{b_w d}$

where A_{sw} is the reinforcing steel required to balance the compression of the web, excluding compression steel.

The steel in the compression zone qualifies as double reinforcement if it is only required by design. For example, the steel used as hangers for stirrups in the compression side does not qualify as double reinforcement.

If it exists as part of a flooring system, the beam forms with part of the floor to create a rectangular section with flanges as shown in Fig. 7-3. This combination of the beam and the slab is known as a T-section if the beam is an interior beam with flanges projecting on both sides. If the beam, however, is an exterior beam with a flange on only one side, the section is known as an L-section.

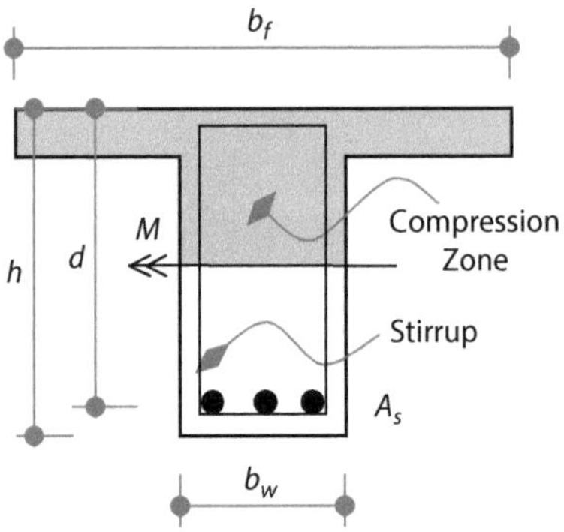

FIGURE 7-3 T- Section.

The participating width of the slab that acts as an effective part of the T-section enabling it to behave as one unit is known as the effective width of the T-section, b_f. For both T and L-sections, the minimum and maximum reinforcement ratios are related to the width of the rectangular beam (i.e., the web width). Consequently, the reinforcement limitations of the T-section may be expressed as follows.

The maximum reinforcement ratio of tension steel for a flanged section is still given by the expression defined for a double reinforcement case

$$\frac{c}{d} \leq \frac{3}{8} = 0.375$$

or

$$\omega_w \leq 0.32\,\beta_1$$

where ω_w is the reinforcement index referenced to the web. This is given in terms of the reinforcement ratio referenced to the web, ρ_w, as follows:

$$\omega_w = \rho_w \frac{f_y}{f_c'}$$

and

$$\rho_w = \frac{A_{sw}}{b_w d}$$

where A_{sw} is the reinforcing steel required to balance the compression of the web, excluding compression in the flanges.

The minimum reinforcement of the T-section is limited to the following:

$$A_{s,\min} = \rho_{w,\min} b_w d$$

where $\rho_{w,\min}$ is the minimum reinforcement ratio as defined before.

As mentioned before, the flange portions of the T-section are part of the flooring system and usually exist as part of the floor's slab. In general, slabs qualify as rigid diaphragms in buildings if the thickness of their solid part is equal to or more than 50 mm.

Slabs are usually designed as a rectangular beam if they are solid slabs or designed as T-sections if they are ribbed slabs. However, they have different minimum reinforcement requirements. The minimum reinforcement of the solid part of the slab is given as a function of the total thickness of the solid part of the slab and is known as shrinkage and temperature steel. Accordingly

$$A_{s,\min} = 0.0018\ \text{bh} \qquad \text{for } f_y = 420\ \text{MPa}$$

$$A_{s,\min} = 0.0020\ \text{bh} \qquad \text{for } f_y = 300\ \text{MPa}$$

where f_y = yield stress of steel reinforcement.

Maximum steel limitations of sections subjected to low intensity of axial loads (i.e., $P_u < 0.1 f_c' A_g$ or $P_u < \phi P_b$) may be addressed by noting from their interaction diagrams that the moment capacity increases with higher axial loads. Thus, a conservative design would design such sections for pure moment without axial load. For sections with low intensity of axial load, tension-controlling strain can be expressed in the following form:

$$\frac{P_n}{f_c' b_w d} = \frac{0.85a}{d} - \omega_w$$

Because the limiting tension-controlled net strain at 0.005 is given as

$$\frac{0.85a}{d} \le 0.32\beta_1$$

The limit of maximum steel in the presence of axial load is set as follows:

$$\omega_w + \frac{P_n}{f_c' b_w d} \le 0.32\beta_1$$

Development of Reinforcement

The development of straight reinforcement bars depends on many factors such as reinforcement size, location, epoxy coating, and the weight of concrete. Development length also depends on whether the bars are in tension or in compression.

Straight Bars in Tension The basic development length of uncoated bottom straight bars embedded in normal weight concrete, ℓ_d, is given by the following expressions, but shall not be less than 300 mm:

$$\frac{\ell_d}{d_b} \ge 0.60 \frac{f_y}{\sqrt{f_c'}} \qquad \text{(for bars with diameter > 19 mm)}$$

$$\frac{\ell_d}{d_b} \ge 0.48 \frac{f_y}{\sqrt{f_c'}} \qquad \text{(for bars with diameter} \le 19\ \text{mm)}$$

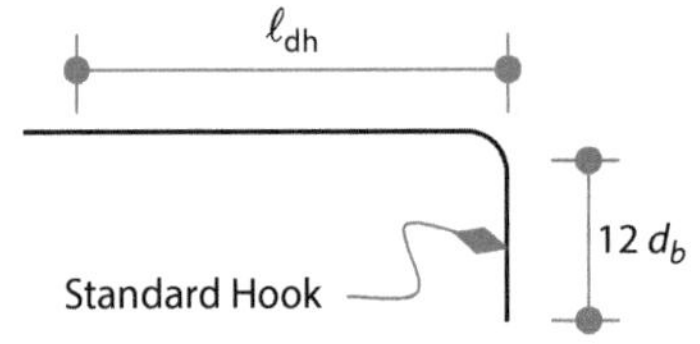

FIGURE 7-4 Basic hook development length.

The development length of top bars must be increased by 30 percent. The top reinforcement is defined as any reinforcement that is placed over 300 mm of fresh concrete. Accordingly, the development length is given as

$$(\ell_d)_{\text{top}} = 1.3(\ell_d)_{\text{bottom}}$$

Standard Hook in Tension The basic hook development length, ℓ_{dh}, is defined in Fig. 7-4. ℓ_{dh} is given as

$$\frac{\ell_{\text{dh}}}{d_b} \geq 0.24 \frac{f_y}{\sqrt{f_c'}}$$

The development length of the hook bar, ℓ_d, is given as the basic length, ℓ_{dh}, multiplied by special modifying factors given in the code.

The development length of uncoated bottom hook bars embedded in normal weight concrete, ℓ_d, is given as the basic hook development length

$$\frac{\ell_{\text{dh}}}{d_b} \geq 0.24 \frac{f_y}{\sqrt{f_c'}}$$

Splices in Tension Minimum lap splices are classified as Class A and Class B splices. In general, and in normal cases, Class B would be required as follows:

$$\ell_{\text{splice}} = 1.3\,\ell_d$$

Bars in Compression The basic development length of bars in compression, ℓ_d, is given by the following expressions:

$$\frac{\ell_d}{d_b} \geq 0.25 \frac{f_y}{\sqrt{f_c'}} \geq 0.04 f_y$$

Splices in Compression Minimum lap splices in compression depend on reinforcement yield stress as follows:

$$\ell_{\text{splice}} = 0.07\, f_y d_b \geq 300 \text{ mm} \qquad (\text{for} \quad f_y \leq 420 \text{ MPa})$$

$$\ell_{\text{splice}} = (0.13 f_y - 24) d_b \geq 300 \text{ mm} \qquad (\text{for} \quad f_y > 420 \text{ MPa})$$

The splice length shall be increased by one-third for concrete of grade, $f_c' \leq 20$ MPa.

Shear Reinforcement

In recent construction practice and due to high labor cost, shear reinforcement is only provided in the form of vertical stirrups. The stirrups are used to resist shear and to play a critical role in seismic design. Stirrups provide the required confinement in seismic design to produce ductile members.

According to ACI 318, shear strength of beams, V_n, is given as

$$V_n = V_c + V_s$$

where V_c = contribution of concrete to shear strength
V_s = contribution of steel to shear strength

For reinforcement limitation in shear, define the following quantity, V_w, as a reference quantity such that

$$V_w = 0.165\sqrt{f'_c}\, b_w d$$

The contribution of concrete and steel to shear strength of beams is given as follows:

$$V_c = V_w$$

$$V_s = A_v f_{yt} \frac{d}{s}$$

where A_v = total area of vertical stirrup legs in the cross section
s = spacing of stirrups along the axis of the beam

The minimum number of stirrups in any section is given in terms of their maximum spacing, s, as follows:

1. For $V_s \le 2V_w$,

$$s_{max} \le \begin{pmatrix} \dfrac{d}{2} \\ 600\text{ mm} \\ \dfrac{A_v f_{yt}}{0.062\sqrt{f'_c}\, b_w} \\ \dfrac{A_v f_{yt}}{0.35\, b_w} \end{pmatrix}$$

2. For $V_s \le 4V_w$,

$$s_{max} \le \begin{pmatrix} \dfrac{d}{4} \\ 300\text{ mm} \\ \dfrac{A_v f_{yt}}{0.062\sqrt{f'_c}\, b_w} \\ \dfrac{A_v f_{yt}}{0.35\, b_w} \end{pmatrix}$$

3. For $V_s > 4V_w$, the cross section is not acceptable for resisting shear and must be changed in dimensions to satisfy this condition.

7.2.2 Ordinary Beam-Columns

According to ACI 318, a member in a frame is considered a column if the member is subjected to pure axial load. The member is considered a beam-column if the member is subjected to combined bending moment and axial load with an intensity of the factored axial load, P_u, exceeding the following:

$$P_u > 0.1 f_c' A_g$$

where f_c' and A_g are as defined before.

The following sections present reinforcement detailing of ordinary columns and beam-columns.

Main Reinforcement

The main reinforcement in ordinary columns and in beam-columns is limited to maximum and minimum values. The reinforcement in beam-columns is expressed as the ratio of the total steel to the gross area of the section and is known as the gross reinforcement ratio, ρ_g. In reference to Fig. 7-5, the gross reinforcement ratio may be expressed as follows:

$$\rho_g = \frac{A_{st}}{A_g}$$

where A_g = gross-sectional area of the member and is equal to $b \times h$ for rectangular columns as shown in Fig. 7-5

A_{st} = total area of reinforcement in the cross section

The limits of gross reinforcement ratio, ρ_g, are given as

$$\rho_g \leq 0.08$$

$$\rho_g \geq 0.01$$

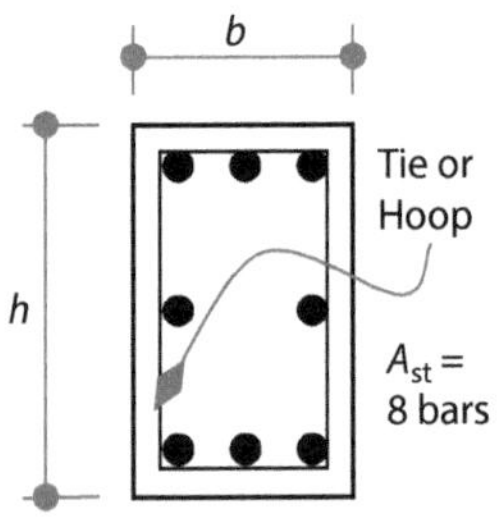

FIGURE 7-5 Rectangular beam column.

Development of Reinforcement

The development of reinforcement is similar to ordinary moment beams.

Shear Reinforcement

In general, internal shear forces in columns are zero. In reality, members are rarely subjected to pure axial loads. Therefore, shear forces usually exist in columns with at least negligible values. In the presence of axial loads, the shear capacity of the concrete is increased to reflect the effect of compressive stresses that decrease the effect of diagonal tension in the member. If the shear is zero or can be safely taken by the concrete alone, minimum ties are required in the rectangular sections and are provided as follows:

1. Minimum diameter of ties is 10 mm
2. Maximum spacing of ties is given as

$$s_{\max} \leq \begin{pmatrix} 16d_b \\ 48d_t \\ b \end{pmatrix}$$

where b = dimension of the shorter side of the member
d_b = diameter of the main reinforcement bars
d_t = diameter of the tie bars

If the shear capacity of the concrete alone in the beam-column is not safe, the ties must also follow the shear provisions of beams as given before.

For spirally reinforced concrete beam-columns as shown in Fig. 7-6, the minimum lateral reinforcement known as spirals is given as the ratio of spirals volume to concrete core volume, ρ_s, such that

$$\rho_s \geq \rho_{s,\min}$$

where ρ_s and $\rho_{s,\min}$ are given in the following expressions:

$$\rho_s = \frac{4A_{\text{sp}}}{\text{dp}}$$

$$\rho_{s,\min} = 0.45\left(\frac{A_g}{A_c} - 1\right)\frac{f_c'}{f_{\text{yt}}}$$

where A_c = area of the concrete core, including diameter of spirals
A_g = gross area of the concrete cross section
A_{sp} = cross-sectional area of spiral
f_c', f_{yt} = designated stresses of concrete and spiral steel
p = spacing of spiral (pitch)
ρ_s = ratio of the volume of spirals to the volume of concrete

The spacing of spirals, also known as the pitch of the spirals, is limited to

$$s_{\max} = 80 \text{ mm}$$

$$s_{\min} = 25 \text{ mm}$$

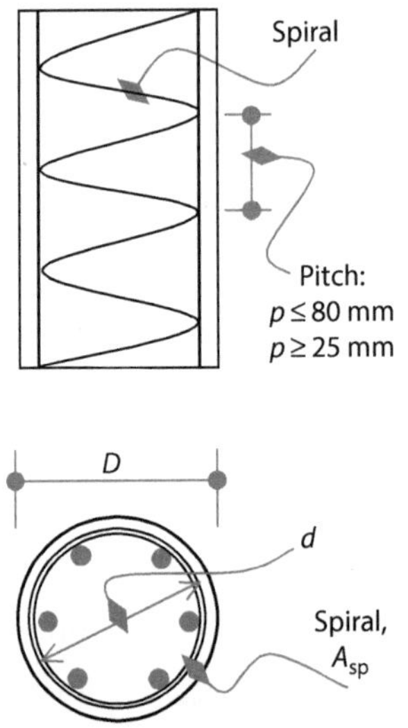

FIGURE 7-6 Spirally reinforced column.

Example 7-1 The frame shown in Example 7-1, Fig. 1, is part of a building that is in a seismic zone that allows the use of ordinary moment frames. The share of the frame from the elastic base shear is calculated as $F_x = 246$ kN.

Design an OMF to resist the given seismic force. Material properties are given as $f_c' = 25$ MPa, $f_y = 420$ MPa.

Solution Even though the basic detailing requirements are listed in this section, the reader is strongly advised to review the basic ACI design philosophy and procedure before solving this example.

The solution will be performed to check three load cases: (1) dead and live load effects; (2) dead, live, and earthquake load effects; and (3) dead and earthquake load effects.

Case 1: Dead and live loads

The load combination is given as

$$U_1 = 1.2D + 1.6L$$

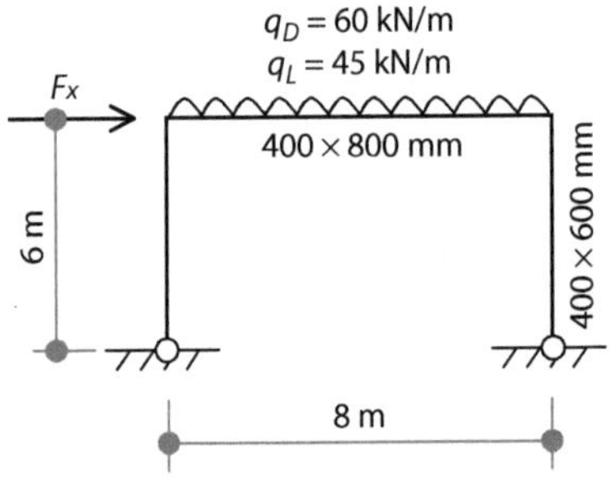

EXAMPLE 7-1, FIGURE 1

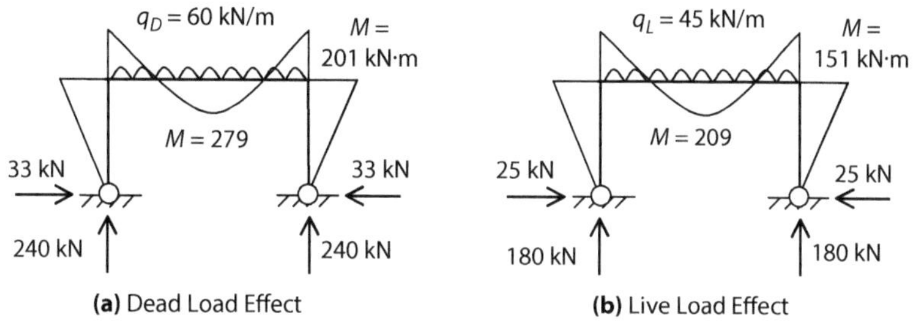

Example 7-1, Figure 2 Frame internal forces.

Obviously, the given frame is a one-time, statically indeterminate structure. Thus, moments, shears, and axial loads may be found by any structural method. For reinforced concrete structures, the moment of inertia for this analysis is given by ACI as follows:

1. Section properties

   Modulus of elasticity: $E_c = 4{,}700\sqrt{25} = 23{,}500$ MPa

   Beam: $A_g = 400\,(800) = 320{,}000$ mm²

   $$I_g = \frac{400(800)^3}{12} = 17{,}067 \times 10^6 \text{ mm}^4$$

   $$I_{\text{eff}} = 0.35\,I_g = 0.35\,(17{,}067 \times 10^6) = 5{,}973 \times 10^6 \text{ mm}^4$$

   Columns: $A_g = 400\,(600) = 240{,}000$ mm²

   $$I_g = \frac{400(600)^3}{12} = 7{,}200 \times 10^6 \text{ mm}^4$$

   $$I_{\text{eff}} = 0.70\,I_g = 0.70\,(7{,}200 \times 10^6) = 5{,}040 \times 10^6 \text{ mm}^4$$

2. Straining actions

   Using the dimensions and properties above, the resulting moments, shears, and normal forces are constructed for both dead and live loads as shown in Example 7-1, Fig. 2.

3. Beam moment

   Check beam action: $0.1 f'_c A_g = 0.1\,(25)(320{,}000)$
   $= 800{,}000$ N (800 kN)

   Ultimate load: $P_u = 1.2P_D + 1.6P_L$
   $= 1.2(33) + 1.6(25) = 80$ kN < limit

   Since

   $$P_u = 80 \text{ kN} < \text{Limit} = 800 \text{ kN}$$

   the member may be designed without the effect of axial load.

   Positive moment region: $M_u = 1.2M_D + 1.6M_L = 1.2(279) + 1.6\,(209) = 669$ kN·m

   $$M_n = \frac{M_u}{\phi} = \frac{669}{0.9} = 744 \text{ kN·m}$$

   $$P_n = \frac{P_u}{\phi} = \frac{80}{0.9} = 89 \text{ kN}$$

Design:
$$\frac{M_n}{f_c'bd^2} = \frac{744\times10^6}{25(400)(740)^2} = 0.136$$

Using design charts in App. 7-1, Sheet 1, the reinforcement index is given as $\omega_o = 0.15$, hence,

$$A_s = \omega_o \frac{f_c'}{f_y} bd = 0.15\left(\frac{25}{420}\right)(400)(740) = 2{,}643 \text{ mm}^2 \ (6\phi25)$$

Check steel limits for tension-controlled action (at $\phi = 0.9$)

Since

$$\frac{P_n}{f_c' b_w d} = \frac{89\times10^3}{25(400)(740)} = 0.012$$

then

$$\omega_w + \frac{P_n}{f_c' b_w d} = 0.15 + 0.012$$
$$= 0.162 < 0.32\beta_1 = 0.272 \text{ ok}$$

Negative moment region:
$$M_u = 1.2M_D + 1.6M_L$$
$$= 1.2(201) + 1.6(151) = 483 \text{ kN·m}$$
$$M_n = \frac{M_u}{\phi} = \frac{483}{0.9} = 536 \text{ kN·m}$$

Design:
$$\frac{M_n}{f_c'bd^2} = \frac{536\times10^6}{25(400)(740)^2} = 0.098$$

Using the design charts in App. 7-1, Sheet 1, the reinforcement index is given as $\omega_o = 0.11$, hence,

$$A_s = \omega_o \frac{f_c'}{f_y} bd = 0.11\frac{25}{420}(400)(740) = 1{,}938 \text{ mm}^2 \ (4\phi25)$$

Check steel limits for tension-controlled action (at $\phi = 0.9$)

$$\omega_w + \frac{P_n}{f_c' b_w d} = 0.11 + 0.012$$
$$= 0.122 < 0.32\beta_1 = 0.272 \text{ ok}$$

4. Beam shear
   Using ϕ 10 bars for stirrups

$$V_u = 1.2V_D + 1.6V_L$$
$$= 1.2(240) + 1.6(180) = 576 \text{ kN}$$
$$V_n = \frac{V_u}{\phi} = \frac{576}{0.75} = 768 \text{ kN}$$

Reference: $V_w = 0.165\sqrt{f'_c}b_w d$

$$= 0.165\sqrt{25}(400)(740) = 246{,}667\text{ N}$$

Concrete strength: $V_c = V_w = 247$ kN

Limits: $2V_w = 494$ kN, $4V_w = 988$ kN

Steel: $V_s = V_n - V_c = 768 - 247 = 521\text{ kN} < 988\text{ kN ok}$

Strength spacing: $V_s = A_v f_{yt}\dfrac{d}{s}$ (use ϕ 10 for stirrups = 78 mm^2)

$$521{,}000 = 156(420)\left(\frac{740}{s}\right)$$

$$s = 93\text{ mm}$$

Maximum spacing: since $V_s > 2V_w$, then

$$s_{max} \le \begin{pmatrix} \dfrac{d}{4} \\ 300\text{ mm} \\ \dfrac{A_v f_{yt}}{0.062\sqrt{f'_c}\,b_w} \\ \dfrac{A_v f_{yt}}{0.35\,b_w} \end{pmatrix} \le \begin{pmatrix} \dfrac{740}{4} \\ 300\text{ mm} \\ \dfrac{156(420)}{0.062\sqrt{25}(400)} \\ \dfrac{156(420)}{0.35(400)} \end{pmatrix} \le \begin{pmatrix} 185 \\ 300 \\ 528 \\ 468 \end{pmatrix}$$

Therefore, use stirrups that are 2 ϕ 10 @ 90 mm.

5. Beam-column moment

Check beam-column action: $0.1 f'_c A_g = 0.1\,(25)(240{,}000)$

$$= 600{,}000\text{ N }(600\text{ kN})$$

Ultimate load: $P_u = 1.2P_D + 1.6P_L$

$$= 1.2\,(240) + 1.6\,(180) = 576\text{ kN} < \text{limit}$$

Because $P_u = 576$ kN < Limit = 600 kN, the member may be designed as a beam. However, for demonstration purposes of beam-column design, the beam will be designed as a beam-column with the effect of axial load.

According to ACI 318, frames under gravity loads may be considered to be braced and, as a result, may be designed as nonsway members. The beam-column may be considered short or long according to the following limit against slenderness ratio:

Check slenderness: $\dfrac{k\ell_u}{r}(<,>)\,34 - 12\left(\dfrac{M_1}{M_2}\right)_s \le 40$

$$\text{Limit} = 34 - 12\left(\frac{M_1}{M_2}\right) = 34 - 12(0) = 34$$

Effective length factor, k:

The alignment chart in App. 7-1, Sheet 7, may be used to find the effective length factor for frames

$$\psi_A = \frac{\Sigma EI/\ell_c}{\Sigma EI/\ell_b} = \frac{5{,}040/6}{5{,}973/8} = 1.13$$

$$\psi_B = \infty \text{ for pinned support}$$

When we use Sheet 7 for a nonsway case, $k = 0.89$.

Slenderness ratio: $$\frac{k\ell_u}{r} = \frac{0.89(6{,}000)}{0.3(600)} = 30 < \text{Limit} = 34$$

Because the slenderness ratio is less than the limit, the beam-column is considered to be short, which means that the P-Δ effect is negligible. In this case, a second-order analysis of moment magnifications is not needed. The design may be carried out using the interaction diagrams in App. 7-1, Sheets 3 through 6. Accordingly,

$$P_u \text{ (already calculated above)} = 576 \text{ kN}$$

$$P_n = \frac{P_u}{\phi} = \frac{576}{0.65} = 886 \text{ kN}$$

$$M_u = 1.2\,M_D + 1.6\,M_L = 1.2(201) + 1.6(151)$$
$$= 483 \text{ kN·m}$$

$$M_n = \frac{M_u}{\phi} = \frac{483}{0.65} = 743 \text{ kN·m}$$

Therefore

$$\frac{P_n}{f'_c bh} = \frac{886 \times 10^3}{25(400)(600)} = 0.148$$

$$\frac{M_n}{f'_c bh^2} = \frac{743 \times 10^6}{25(400)(600)^2} = 0.206$$

Using Sheet 4 for $\gamma = (600 - 120)/600 = 0.8$ results in

$$\rho\mu = 0.4$$

Since

$$\mu = \frac{f_y}{0.85\,f'_c} = \frac{420}{0.85(25)} = 19.76$$

Then

$$\rho = \frac{0.4}{19.76} = 0.02 \rightarrow \begin{pmatrix} <0.08 \\ >0.01 \end{pmatrix}$$

Steel: $A_{st} = \rho bh = 0.02(400)(600) = 4{,}858 \text{ mm}^2$ ($8\,\phi\,28$)

6. Beam-column shear
   Beam-column concrete shear strength is affected by axial load as follows:

$$V_u = 1.2\,V_D + 1.6\,V_L = 1.2(33) + 1.6(25) = 80 \text{ kN}$$

$$V_n = \frac{V_u}{\phi} = \frac{80}{0.75} = 106 \text{ kN}$$

Reference: $$V_w = 0.165\sqrt{f'_c}\, b_w d$$
$$= 0.165\sqrt{25}(400)(540) = 180{,}000 \text{ N}$$

Limits: $2V_w = 360$ kN, $4V_w = 720$ kN

Concrete strength: $V_c = \left(1+0.07\frac{N_u}{A_g}\right)V_w$

$$= \left[1+0.07\left(\frac{576{,}000}{240{,}000}\right)\right]V_w$$

$$= 1.168(180) = 210 \text{ kN}$$

Capacity: $\phi V_c = 0.75(210) = 158 \text{ kN} > V_u$

Because concrete strength is enough, there is no need for stirrups for strength, but minimum stirrups must be provided in columns as follows:

Spacing: $s \le 16\, d_b = 16(28) = 448 \text{ mm}$

$$\le 48\, d_s = 48(10) = 480 \text{ mm}$$

$$\le b = 400 \text{ mm}$$

Therefore, use stirrups that are ϕ 10 @ 400 mm.

Case 2: Dead, live, and seismic forces
Load combination is given as

$$U_2 = 1.2D + 1.0L + 1.0E$$

1. Straining actions
   The seismic force for OMF is given by the elastic force divided by the *R*-factor, which is 3 for this frame as given in App. 6-1

   $$F_x = \frac{F_{\text{el}}}{R} = \frac{246}{3} = 82 \text{ kN}$$

   Using the dimensions and properties above, the resulting moments, shears, and normal forces are constructed for the seismic force as shown in Example 7-1, Fig. 3.

2. Beam moment
   Check beam action: $0.1 f'_c A_g = 0.1(25)(320{,}000)$

   $$= 800{,}000 \text{ N } (800 \text{ kN})$$

   Ultimate load: $P_u = 1.2P_D + 1.0P_L + 1.0P_E$

   $$= 1.2(33) + 1.0(25) + 1.0(41) = 106 \text{ kN} < \text{limit}$$

   Because P_u = 106 kN < Limit = 800 kN, the member may be designed without the effect of axial load.

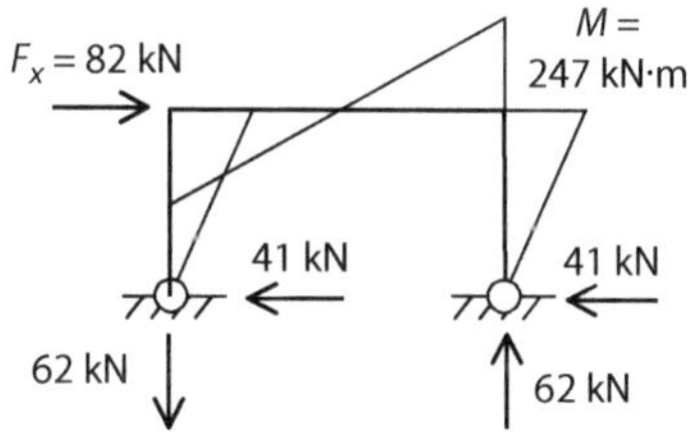

EXAMPLE 7-1, FIGURE 3 Seismic forces.

Positive moment region:

$$M_u = 1.2M_D + 1.0M_L + 1.0M_E$$
$$= 1.2(279) + 1.0(209) + 1.0(0)$$
$$= 53 \text{ kN·m} < \text{case 1}$$

Therefore, there is no need for design. Case 1 controls.

Negative moment region:

$$M_u = 1.2M_D + 1.0M_L + 1.0M_E$$
$$= 1.2(201) + 1.0(151) + 1.0(247)$$
$$= 639 \text{ kN·m} > \text{case 1}$$

$$M_n = \frac{M_u}{\phi} = \frac{639}{0.9} = 710 \text{ kN·m}$$
$$P_n = \frac{P_u}{\phi} = \frac{106}{0.9} = 118 \text{ kN}$$

Design:

$$\frac{M_n}{f'_c bd^2} = \frac{710\times 10^6}{25(400)(740)^2} = 0.13$$

Using the design charts in App. 7-1, Sheet 1, the reinforcement index is given as $\omega_o = 0.14$,

$$A_s = \omega_o \frac{f'_c}{f_y} bd = 0.14\left(\frac{25}{420}\right)(400)(740) = 2{,}647 \text{ mm}^2 \ (6\ \phi\ 25)$$

Check steel limits for tension-controlled action (at $\phi = 0.9$)
Since

$$\frac{P_n}{f'_c b_w d} = \frac{118\times 10^3}{25(400)(740)} = 0.016$$

then

$$\omega_w + \frac{P_n}{f'_c b_w d} = 0.14 + 0.016$$
$$= 0.156 < 0.32\beta_1 = 0.272 \text{ ok}$$

3. Beam shear
   Using ϕ 10 bars for stirrups

$$V_u = 1.2V_D + 1.0V_L + 1.0V_E$$
$$= 1.2(240) + 1.0(180) + 1.0(62)$$
$$= 530 \text{ kN} < \text{case 1}$$

Therefore, there is no need for design. Case 1 controls.

4. Beam-column moment
   Check beam-column action:

$$0.1f'_c A_g = 0.1(25)(240{,}000)$$
$$= 600{,}000 \text{ N } (600 \text{ kN})$$

Ultimate load:

$$P_u = 1.2P_D + 1.0P_L + 1.0P_E$$
$$= 1.2(240) + 1.0(180) + 1.0(62)$$
$$= 530 \text{ kN} > \text{limit}$$

Because $P_u = 530$ kN > Limit = 600 kN, the member may be designed as a beam. However, for demonstration purposes of beam-column design, the beam will be designed as a beam-column with the effect of axial load.

According to ACI 318, frames under lateral loads may not be considered to be braced. As a result, they shall be designed as sway members. In the sway case, the beam-column may be considered to be short or long according to the following limit against slenderness ratio

Check slenderness: $\frac{k\ell_u}{r}(<,>)22$

Effective length factor, k:

To find the effective length factor for frames, the alignment chart in App. 7-1, Sheet 7, may be used again

$$\psi_A = \frac{\Sigma EI/\ell_c}{\Sigma EI/\ell_b} = \frac{5{,}040/6}{5{,}973/8} = 1.13$$

$$\psi_B = \infty \text{ for pinned support}$$

When we use Sheet 7 for the sway case, $k = 2.40$.

Slenderness ratio: $\frac{k\ell_u}{r} = \frac{2.40(6{,}000)}{0.3(600)} = 80 > \text{Limit} = 22$

Because the slenderness ratio is more than the limit, the beam-column is considered long, which means that the P-Δ effect must be considered. In this case, a second-order analysis of moment magnifications is needed. These may be evaluated using the sway magnification factor, δ_s, as defined by ACI 318. According to ACI 318, the design moment is given by the following procedures.

Effective moment of inertia for buckling analysis is different from structural analysis for concrete sections. For buckling analysis, EI may be taken as

$$EI = 0.25\,E_c I_g = 0.25(23{,}500)(7{,}200 \times 10^6)$$
$$= 42{,}300 \times 10^9 \text{ N·mm}^2$$

Critical load: $P_c = \frac{\pi^2 EI}{(k\ell)^2} = \frac{\pi^2(42{,}300)}{(2.40 \times 6)^2} = 2{,}013$ kN

Summation of P_c in the entire floor: $\Sigma P_c = 2{,}013(2) = 4{,}026$ kN

Summation of P_u in the entire floor: $\Sigma P_u = P_{u1} + P_{u2}$

The vertical reactions that are due to seismic forces as given in Example 7-1, Fig. 3, are equal in magnitude and opposite in direction. Therefore, they cancel out in the summation

$$\Sigma P_u = (1.2P_D + 1.0P_L)(2)$$
$$= [1.2(240) + 1.0(180)](2)$$

Moment magnification: $\delta_s = \frac{1}{1 - \frac{\Sigma P_u}{\phi \Sigma P_c}}$ with limits as $\delta_s \le 2.5$, ≥ 1.0

$$\delta_s = \frac{1}{1 - \frac{936}{0.75(4{,}026)}} = 1.45$$

The design moment is given as a combination of the nonsway moments and the sway moments as follows:

$$M_{2ns} = 1.2(201) + 1.0(151) = 392 \text{ kN·m}$$

$$M_{2s} = 1.0(247) = 247 \text{ kN·m}$$

Consequently, the design moment is given as

$$M_u = M_{2ns} + \delta_s M_{2s}$$

$$M_u = 392 + 1.45(247) = 750 \text{ kN·m}$$

Design may be carried out using the interaction diagrams in App. 7-1, Sheets 3 through 6. Accordingly,

$$P_u \text{ (already calculated above)} = 530 \text{ KN}$$

$$P_n = \frac{P_u}{\phi} = \frac{530}{0.65} = 815 \text{ kN}$$

$$M_n = \frac{M_u}{\phi} = \frac{750}{0.65} = 1{,}154 \text{ kN·m}$$

Therefore,

$$\frac{P_n}{f'_c bh} = \frac{815 \times 10^3}{25(400)(600)} = 0.136$$

$$\frac{M_n}{f'_c bh^2} = \frac{1{,}154 \times 10^6}{25(400)(600)^2} = 0.32$$

Using Sheet 4 for $\gamma = (600 - 120)/600 = 0.8$, results in

$$\rho\mu = 0.8$$

since

$$\mu = \frac{f_y}{0.85 f'_c} = \frac{420}{0.85(25)} = 19.76$$

then

$$\rho = \frac{0.8}{19.76} = 0.04 \rightarrow \begin{pmatrix} <0.08 \\ >0.01 \end{pmatrix} \text{ ok}$$

Steel: $A_{st} = \rho bh = 0.04(400)(600) = 7{,}917 \text{ mm}^2$ (16 ϕ 28)

5. Beam-column shear
Beam-column concrete shear strength is affected by axial load as follows:

$$V_u = 1.2V_D + 1.0V_L + 1.0V_E$$

$$= 1.2(33) + 1.0(25) + 1.0(41) = 106 \text{ kN}$$

$$V_n = \frac{V_u}{\phi} = \frac{106}{0.75} = 141 \text{ kN}$$

Reference: $V_w = 0.165\sqrt{f_c'}\, b_w d$

$= 0.165\sqrt{25}(400)(540) = 180{,}000 \text{ N}$

Limits: $2V_w = 360 \text{ kN},\ 4V_w = 720 \text{ kN}$

Concrete strength: $V_c = \left(1 + 0.07\dfrac{N_u}{A_g}\right)V_w$

$$= \left[1 + 0.07\left(\frac{530{,}000}{240{,}000}\right)\right]V_w$$

$$= 1.154(180) = 208 \text{ kN}$$

Capacity: $\phi V_c = 0.75(208) = 156 \text{ kN} > V_u$

Insofar as concrete strength is enough, there is no need for stirrups for strength, but minimum stirrups must be provided in columns as follows.

Spacing:

$$s \le 16d_b = 16(28) = 448 \text{ mm}$$
$$\le 48d_s = 48(10) = 480 \text{ mm}$$
$$\le b = 400 \text{ mm}$$

Therefore, use stirrups that are ϕ 10 @ 400 mm.

Case 3: Dead and seismic forces
The load combination is given as

$$U_3 = 0.9D + 1.0E$$

1. Affected areas
   An inspection reveals that this load case affects only the beam moment at the face. As a result,

$$M_u = 0.9M_D + 1.0M_E$$
$$= 0.9(201) - 1.0(247) = -66 \text{ kN·m}$$
$$M_n = \frac{M_u}{\phi} = \frac{66}{0.9} = 73 \text{ kN·m}$$

Design: $\dfrac{M_n}{f_c' b d^2} = \dfrac{73 \times 10^6}{25(400)(740)^2} = 0.013$

Using the design charts in App. 7-1, Sheets 1 and 2, the reinforcement index is given less than minimum. Thus, use the minimum reinforcement

$$\rho_{o,\min} = 0.0033$$
$$A_s = \rho_{o,\min}\, bd = 0.0033(400)(740) = 977 \text{ mm}^2\ (2\ \phi\ 25)$$

2. Development length
   The bottom bars in the beam must be developed in tension. The development length is given as

$$\frac{\ell_d}{d_b} \ge 0.48\frac{f_y}{\sqrt{f_c'}} = 0.48\frac{420}{\sqrt{25}} = 40.32$$

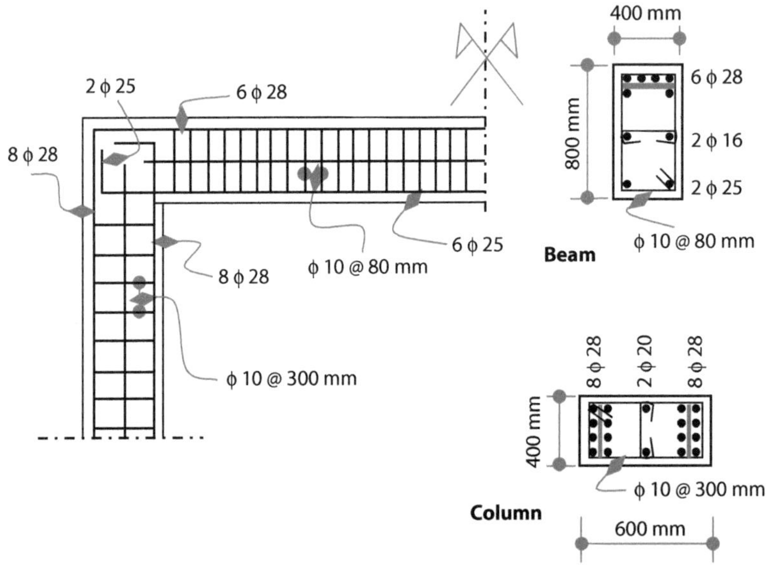

Example 7-1, Figure 4 OMF detailing.

Therefore,

$$\ell_d = 40.32(25) = 1{,}008 \text{ mm}$$

Because column depth is equal to 540 mm the column depth is not enough to accommodate 1,008 mm of straight bar for development. Consequently, hook bar must be used for development. The development length of standard hook bar is given as

$$\frac{\ell_d}{d_b} \geq 0.24\frac{f_y}{\sqrt{f_c'}} = 0.24\frac{420}{\sqrt{25}} = 20.16$$

Therefore,

$$\ell_d = 20.16(25) = 504 \text{ mm}$$

The final design details are shown in Example 7-1, Fig. 4.

7.3 Intermediate Moment Frames

Intermediate moment frames (IMF) are assigned an R-factor equal to 5 by the IBC. To meet such force reduction, ACI 318 requires these frames to conform to its provisions for ordinary moment frames. These frames must also conform to the requirements of this section.

This type of frame has enhanced detailing and design requirements over that of ordinary moment frames. These requirements enhance its ductility and, as a result, increase the allowable force reduction to $R = 5$.

The factored shear forces for both beams and beam-columns of IMF shall be evaluated as the smaller of: (a) the maximum possible shear that can develop in the section due to plastic moments and (b) the maximum of the basic load combination for seismic loading, including factored seismic shear as follows:

$$V_u = \frac{\Sigma M_n}{\ell_n} + \frac{q_u \ell_n}{2}$$

$$\leq 1.2\,V_D + \begin{Bmatrix} 1.0 \\ 0.5 \end{Bmatrix} V_L + 0.2\,V_S + 2.0\,V_E \quad \text{in beams}$$

$$\leq 1.2\,V_D + \begin{Bmatrix} 1.0 \\ 0.5 \end{Bmatrix} V_L + 0.2 V_S + \Omega_o V_E \quad \text{in beam-columns}$$

where D, L, S, E = dead, live, snow, and earthquake loads components
M_n = nominal moment of ACI-318 (plastic moment)
ℓ_n = clear span of the member
q_u = load intensity due to factored gravity loads
Ω_o = overstrength factor

Notice that Ω_o is given as 3 for IMF in the latest ASCE-7.

The components of the second equation above are graphically identified as shown in Fig. 7-7.

Additional requirements are also needed for both beams and beam-columns. For clarity, these requirements will also be presented in graphical forms.

7.3.1 Intermediate Beams

Similar to OMF, intermediate beams are identified with a maximum factored axial load as

$$P_u \leq 0.1 f_c' A_g$$

where f_c' and A_g are as defined before.

For beams in frames, the ends of the beams are potential locations for the development of plastic hinges. Under seismic excitation, there is a high likelihood of developing positive plastic hinge as well as negative plastic hinge. Consequently, the reinforcement limitations are given for main and lateral direction as shown below.

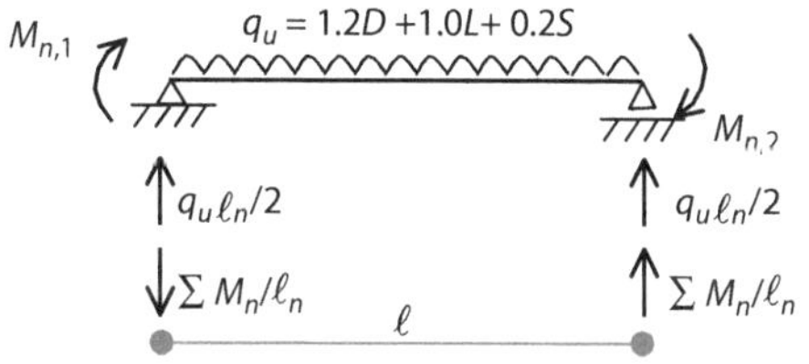

Figure 7-7 Design shear.

Main Reinforcement

The additional main reinforcement design requirements are shown in Fig. 7-8 and explained below.

1. The positive moment at the face of the column must exceed one-third the negative moment strength provided at the face of the supporting column as shown in Detail A.
2. The minimum positive and negative moments at midspan must exceed one-fifth the maximum moment strength provided at the face of the supporting column as shown in Detail B.

Lateral Reinforcement

The additional stirrup requirements shown in Fig. 7-8 may be explained as described below.

The potential plastic hinge region is assumed to extend a distance $2h$ from the face of the support. Therefore, over this distance, $2h$, the maximum spacing of hoops is given in terms of effective depth of the member, d, the diameter of main reinforcement, d_b, the diameter of hoops, d_s, and, 300 mm, as shown in Detail A. The first hoop shall start at least 50 mm from the face of the support as also shown in Detail A.

7.3.2 Intermediate Beam-Columns

Similar to OMF intermediate beam-columns are identified with minimum factored axial load as

$$P_u > 0.1 f'_c A_g$$

where f'_c and A_g are as defined before.

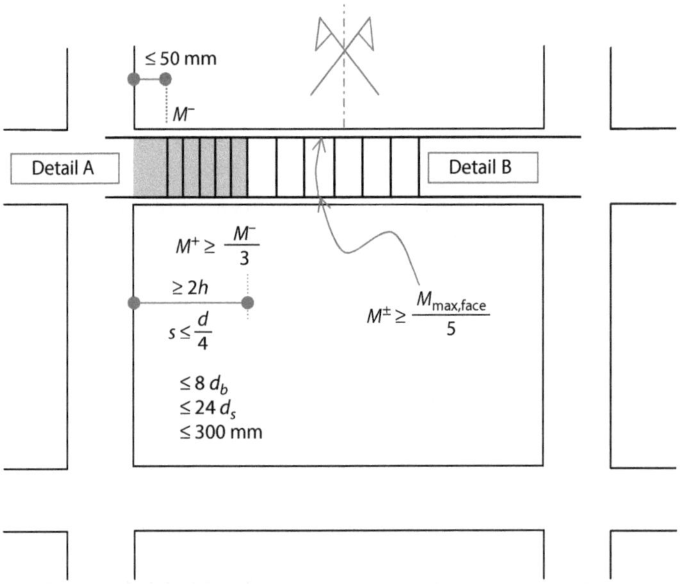

FIGURE 7-8 IMF beam requirements and detailing.

For beam-columns in frames, the ends of the beam-columns are the potential locations for the development of plastic hinges. The following provisions shall also apply on both sides of any potential location of a plastic hinge (e.g., potentially under concentrated loads and in critical moment section inside the beam). Consequently, the reinforcement limitations are given for lateral direction as shown below. There is no additional requirement for main steel other than the OMF requirements.

Lateral Reinforcement

The additional stirrup requirements in Fig. 7-9 may be explained as shown below.

1. The potential plastic hinge region is assumed to extend a distance ℓ_o from the face of the support. This distance is given in terms of the clear height of the column, h_n, dimensions of the column, h or b, and 450 mm, as shown in Detail A.
2. The maximum spacing of the ties in the plastic hinge region is given as s_o and is expressed in terms of the diameter of main reinforcement, d_b, the diameter of ties, d_s, half of the shorter side of the column, b or h, and 300 mm, as shown in Detail B. The first tie shall start at a distance not more than $s_o/2$ as also shown in Detail B.
3. The maximum spacing of ties in the middle of the column (between the plastic hinge regions) shall not exceed twice their spacing, $2s_o$, as shown in Detail C.

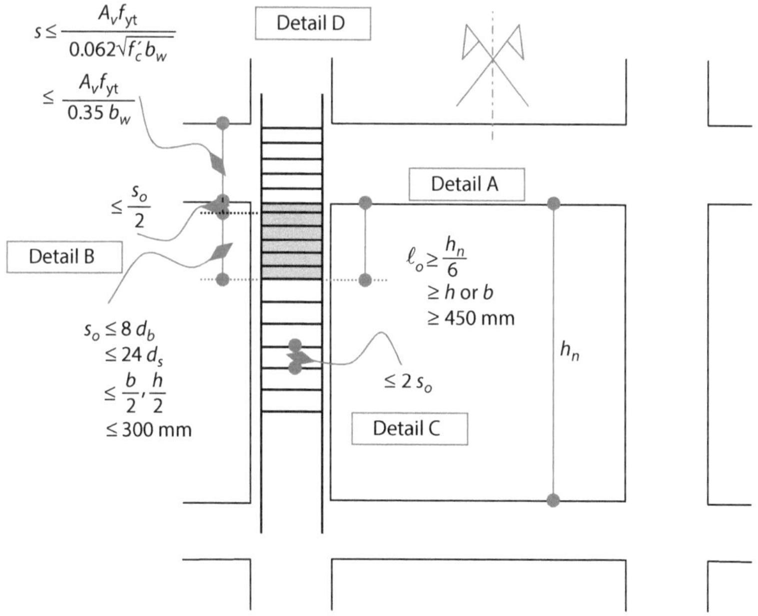

FIGURE 7-9 IMF beam-column requirements and detailing.

4. The ties must continue through the joint. The minimum amount of these ties is expressed in terms of maximum spacing, which is similar to regular beam stirrups in OMF and is also shown in Detail D.

Example 7-2 The frame shown in Example 7-2, Fig. 1, is the same frame given in Example 7-1. The share of the frame from the elastic base shear is given as $F_x = 260$ kN.

Design the same frame as an intermediate moment-resisting frame (IMF) to resist the given seismic force. Given material properties are the same as $f'_c = 25$ MPa and $f_y = 420$ MPa.

Solution Because this example is intended to be solved in conjunction with Example 7-1, the reader is advised to review Example 7-1 before working on this second example. Remember to review the basic ACI design philosophy and procedures before working on these examples.

The solution will involve checking three load cases: (1) dead and live load effects; (2) dead, live, and earthquake load effects; and (3) dead and earthquake load effects.

Case 1: Dead and live loads

Load combination is given as

$$U_1 = 1.2D + 1.6L$$

This load case will be exactly the same as in Example 7-1. See case 1 in Example 7-1 for the solution.

Case 2: Dead, live, and seismic loads

Load combination is given as

$$U_2 = 1.2D + 1.0L + 1.0E$$

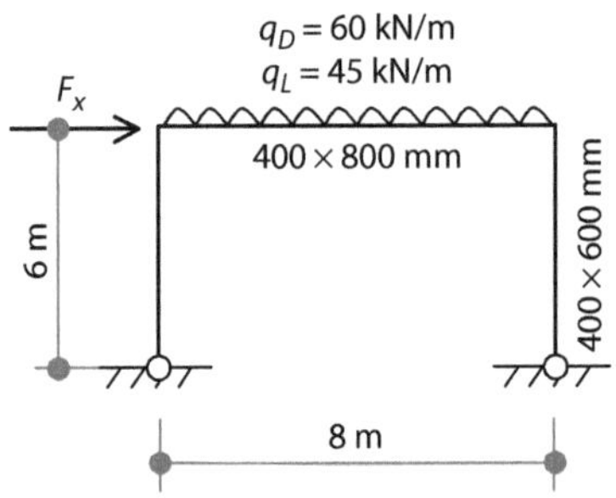

Example 7-2, Figure 1

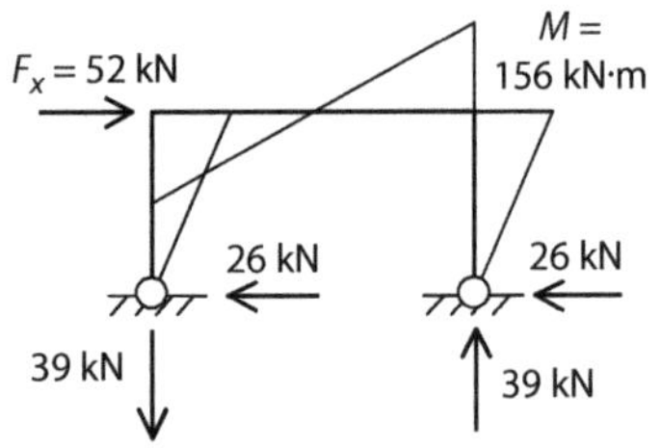

Example 7-2, Figure 2 Seismic forces.

1. Straining actions
   The seismic force for IMF is given by the elastic force divided by the R-factor, which is 5 for this frame as given in App. 6-1. Thus,

$$F_x = \frac{F_{el}}{R} = \frac{260}{5} = 52 \text{ kN}$$

   Using the above dimensions and properties, the resulting moments, shears, and normal forces are constructed for the seismic force as shown in Example 7-2, Fig. 2.

2. Beam moment

   Check beam action: $0.1 f_c' A_g = 0.1(25)(320{,}000)$

$$= 800{,}000 \text{ N (800 kN)}$$

   Ultimate load:

$$P_u = 1.2P_D + 1.0P_L + 1.0P_E$$
$$= 1.2(33) + 1.0(25) + 1.0(26)$$
$$= 91 \text{ kN} < \text{limit}$$

   Since $P_u = 91$ kN < Limit = 600 kN, the member may be designed without the effect of axial load.

   Positive moment region: As in Example 7-1.

   Negative moment region:

$$M_u = 1.2M_D + 1.0M_L + 1.0M_E$$
$$= 1.2(201) + 1.0(151) + 1.0(156)$$
$$= 548 \text{ kN·m} > \text{case 1}$$

$$M_n = \frac{M_u}{\phi} = \frac{548}{0.9} = 609 \text{ kN·m}$$

$$P_n = \frac{P_u}{\phi} = \frac{91}{0.9} = 101 \text{ kN}$$

   Design:

$$\frac{M_n}{f_c' bd^2} = \frac{609 \times 10^6}{25(400)(740)^2} = 0.111$$

   Using design charts in App. 7-1, Sheet 1, the reinforcement index is given as $\omega_o = 0.12$, hence,

$$A_s = \omega_o \frac{f_c'}{f_y} bd = 0.12\left(\frac{25}{420}\right)(400)(740) = 2{,}114 \text{ mm}^2 \ (5\phi 25)$$

   Check steel limits for tension-controlled action (at $\phi = 0.9$)
   Since

$$\frac{P_n}{f_c' b_w d} = \frac{101 \times 10^3}{25(400)(740)} = 0.014$$

   then

$$\omega_w + \frac{P_n}{f_c' b_w d} = 0.12 + 0.014$$
$$= 0.134 < 0.32\beta_1 = 0.272 \text{ ok}$$

Even though 5 ϕ 25 is enough, it is more practical to use an even number. Therefore, use 6ϕ25. For additional requirements for IMF beams, refer to Fig. 7-8 for details.

The positive moment at the face and at midspan must be designed for a minimum value as a function of the negative moment at the joint. Inspection reveals that the maximum design negative moment obtained at the joint is 609 kN·m. Thus,

Moments at face

Positive moment at face: $$M_n^+ = \frac{M_u^-}{3} + \frac{609}{3} = 203 \text{ kN·m}$$

Design: $$\frac{M_n}{f_c'bd^2} = \frac{203\times10^6}{25(400)(740)^2} = 0.037$$

Using design charts in App. 7-1, Sheets 1 and 2, the reinforcement index is given as less than minimum. Thus, use the minimum reinforcement

$$\rho_{w,\min} = 0.0033$$
$$A_s = \rho_{w,\min}\, bd = 0.0033(400)(740) = 966 \text{ mm}^2\ (2\,\phi\,25)$$

Moments at midspan

Both moments at midspan: $$M_n^{\pm} = \frac{M_n^{\text{face}}}{5} = \frac{609}{5} = 122 \text{ kN·m}$$

Since this moment (122) is less than face moment (203), the reinforcement ratio must also be less than the minimum reinforcement ratio. Thus, use the minimum reinforcement

$$\rho_{w,\min} = 0.0033$$
$$A_s = \rho_{w,\min}\, bd = 0.0033(400)(740) = 966 \text{ mm}^2\ (2\,\phi\,25)$$

3. Beam shear
Using ϕ 10 bars for stirrups

$$V_u = \frac{\Sigma M_n}{\ell_n} + V_u(\text{gravity})$$

where $$M_{n1} = 203 \text{ kN}$$
$$M_{n2} = 609 \text{ kN}$$

Gravity load: $$q_u = 1.2\,q_D + 1.0\,q_L$$
$$= 1.2(60) + 1.0(45) = 117 \text{ kN/m}$$

Consequently,

$$V_u \le \frac{203+609}{(8-0.6)} + \frac{117(8)}{2} = 577 \text{ kN}$$

For IMF, V_u shall not be more than the following (see Fig. 7-7)

$$V_u \le 1.2V_D + 1.0V_L + 2.0V_E$$
$$\le 1.2\,(240) + 1.0\,(180) + 2.0\,(39) = 546 \text{ kN}$$

Therefore,

$$V_u = 546 \text{ kN}$$

$$V_n = \frac{V_u}{\phi} = \frac{546}{0.75} = 728 \text{ kN}$$

Reference: $V_w = 0.165\sqrt{f_c'}\, b_w d$

$$= 0.165\sqrt{25}(400)(740) = 246{,}667 \text{ N}$$

Concrete strength: $V_c = V_w = 247$ kN

Limits: $2V_w = 494$ kN, $4V_w = 988$ kN

Steel: $V_s = V_n - V_c$

$$= 728 - 247 = 481 \text{ kN} < 988 \text{ kN ok}$$

Strength spacing: $V_s = A_v f_{yt} \frac{d}{s}$

(use ϕ 10 for stirrups = 78 mm²)

$$481{,}000 = (156)(420)\left(\frac{740}{s}\right)$$

$$s = 101 \text{ mm}$$

Maximum spacing over a distance = $2h = 2\,(800) = 1{,}600$ mm

$$s \le \frac{d}{4} = \frac{740}{4} = 185 \text{ mm}$$

$$\le 8d_b = 8(25) = 200 \text{ mm}$$

$$\le 48d_s = 48(10) = 240 \text{ mm}$$

$$\le 300 \text{ mm}$$

$$\le \frac{A_v f_{yt}}{0.062\sqrt{f_c'}\, b_w} = \frac{156(420)}{0.062\sqrt{25}(400)} = 528 \text{ mm}$$

$$\frac{A_v f_{yt}}{0.35\, b_w} = \frac{156(420)}{0.35(400)} = 468 \text{ mm}$$

Therefore, use stirrups, 2ϕ10 @ 100 mm, over a distance = 1,600 mm from face of column.

4. Beam-column moment
   Check beam-column action: $0.1 f_c' A_g = 0.1(25)(240{,}000)$

$$= 600{,}000 \text{ N (600 kN)}$$

Ultimate load: $P_u = 1.2P_D + 1.0P_L + 1.0P_E$

$$= 1.2\,(240) + 1.0\,(180) + 1.0\,(39)$$

$$= 507 \text{ kN} > \text{limit}$$

Because $P_u = 507$ kN > Limit = 600 kN, the member may be designed as a beam. However, to demonstrate beam-column design, the beam will be designed as a beam-column with the effect of axial load.

According to ACI 318, frames under lateral loads may not be considered braced, and hence, shall be designed as sway members. In the sway case, the beam-column may be considered short or long according to the following limit against slenderness ratio

Check slenderness: $\frac{k\ell_u}{r}(<,>)22$

Effective length factor, k:

To find the effective length factor for frames, use the alignment chart in App. 7-1, Sheet 7

$$\psi_A = \frac{\Sigma EI/\ell_c}{\Sigma EI/\ell_b} = \frac{5{,}040/6}{5{,}973/8} = 1.13$$

$$\psi_B = \infty \text{ for pinned support}$$

Using Sheet 7 for sway case, we find that $k = 2.40$.

Slenderness ratio: $\frac{k\ell_u}{r} = \frac{2.40(6{,}000)}{0.3(600)} = 80 > \text{Limit} = 22$

Because the slenderness ratio is more than the limit, the beam-column is considered to be long, which means that the P-Δ effect must be considered. In this case, second-order analysis of moment magnifications are needed. These may be evaluated using the sway magnification factor, δ_s, as defined by ACI 318. According to ACI 318, the design moment is given by the following procedures.

Effective moment of inertia for buckling analysis is different from structural analysis for concrete sections. For buckling analysis, EI is given as in Example 7-1

$$EI = 0.25\, E_c I_g$$

$$= 0.25(23{,}500)(7{,}200 \times 10^6) = 42{,}300 \times 10^9 \text{ N·mm}^2$$

Critical load: $P_c = \frac{\pi^2 EI}{(k\ell)^2} = \frac{\pi^2(42{,}300)}{(2.40 \times 6)^2} = 2{,}013 \text{ kN}$

Summation of P_c in the entire floor: $\Sigma P_c = 2{,}013(2) = 4{,}026$ kN

Summation of P_u in the entire floor: $\Sigma P_u = P_{u1} + P_{u2}$

The vertical reactions that are due to seismic forces as given in Example 7-2, Fig. 2, are equal in magnitude and opposite in direction. Therefore, they cancel out in the summation

$$\Sigma P_u = (1.2\, P_D + 1.0\, P_L)(2)$$

$$= [1.2\,(240) + 1.0\,(180)](2) = 936 \text{ kN}$$

Moment magnification: $\delta_s = \frac{1}{1 - \frac{\Sigma P_u}{\phi \Sigma P_c}} \begin{pmatrix} \le 2.5 \\ \ge 1.0 \end{pmatrix}$

$$\delta_s = \frac{1}{1 - \frac{936}{0.75(4{,}026)}} = 1.45$$

The design moment is given as a combination of the nonsway moments and the sway moments as follows:

$$M_{2ns} = 1.2(201) + 1.0(151) = 392 \text{ kN·m}$$
$$M_{2s} = 1.0(156) = 156 \text{ kN·m}$$

Consequently, the design moment is given as

$$M_u = M_{2ns} + d_s M_{2s}$$
$$M_u = 392 + 1.45(156) = 618 \text{ kN·m}$$

Design may be carried out using the interaction diagrams in App. 7-1, Sheets 3 through 6. Accordingly,

$$P_u \text{ (already calculated above)} = 507 \text{ kN}$$
$$P_n = \frac{P_u}{\phi} = \frac{530}{0.65} = 780 \text{ kN}$$
$$M_n = \frac{M_u}{\phi} = \frac{618}{0.65} = 951 \text{ kN·m}$$

Therefore,

$$\frac{P_n}{f_c' bh} = \frac{780 \times 10^3}{25(400)(600)} = 0.13$$
$$\frac{M_n}{f_c' bh^2} = \frac{951 \times 10^6}{25(400)(600)^2} = 0.264$$

Using Sheet 4 for: $\gamma = \dfrac{600 - 120}{600} = 0.8$ results in

$$\rho\mu = 0.6$$

Since

$$\mu = \frac{f_y}{0.85 f_c'} = \frac{420}{0.85(25)} = 19.76$$

then

$$\rho = \frac{0.6}{19.76} = 0.03 \rightarrow \begin{pmatrix} <0.08 \\ >0.01 \end{pmatrix} \text{ok}$$

Steel: $A_{st} = \rho bh = 0.03(400)(600) = 7{,}200 \text{ mm}^2 \ (12\ \phi\ 28)$

5. Beam-column shear
Beam-column concrete shear strength is affected by axial load as follows:

$$V_u = \frac{\Sigma M_n}{\ell_n} + V_u \text{ (gravity)}$$

where

$$M_{n1} = 730 \text{ kN}$$
$$M_{n2} = 0 \text{ (pinned support)}$$

Gravity load, $q_u = 0$, in the column.

$$V_u = \frac{730}{6} + 0 = 122 \text{ kN}$$

For IMF, V_u shall not be more than the following (refer to Fig. 7-7 for illustration)

$$V_u \le 1.2\ V_D + 1.0\ V_L + 2.0\ V_E$$

$$\le 1.2\,(33) + 1.0\,(25) + 2.0\,(26) = 117 \text{ kN}$$

Therefore,

$$V_u = 117 \text{ kN}$$

$$V_n = \frac{V_u}{\phi} = \frac{117}{0.75} = 156 \text{ kN}$$

Reference: $$V_w = 0.165\sqrt{f'_c}\,b_w d$$

$$= 0.165\sqrt{25}(400)(540) = 180{,}000 \text{ N}$$

Limits: $2V_w = 360$ kN, $4V_w = 720$ kN

Concrete strength: $$V_c = \left(1 + 0.07\frac{N_u}{A_g}\right)V_w$$

$$= \left[1 + 0.07\left(\frac{507{,}000}{240{,}000}\right)\right]V_w$$

$$= 1.148\,(180) = 207 \text{ kN}$$

Capacity: $$\phi V_c = 0.75(207) = 155 \text{ kN} > V_u$$

Because concrete strength is enough, there is no need for stirrups for strength, but minimum hoops must be provided in columns as follows.

Refer to Fig. 7-8 for details.

Length ℓ_o: $$\ell_o \ge \frac{h_n}{6} = \frac{6{,}000}{6} = 1{,}000 \text{ mm}$$

$$\ge h,\ b = 400 \text{ mm}$$

$$\ge 450 \text{ mm}$$

Therefore,

$$\ell_o = 1{,}000 \text{ mm}$$

Spacing, s_o: $$s_o \le 8d_b = 8(25) = 200 \text{ mm}$$

$$\le 24d_s = 48(10) = 240 \text{ mm}$$

$$\le \frac{b}{2}, \frac{h}{2} = \frac{400}{2} + = 200 \text{ mm}$$

$$\le 300 \text{ mm}$$

$$\le \frac{A_v f_{yt}}{0.35\,b_w} = \frac{156(420)}{0.35(400)} = 468 \text{ mm}$$

Therefore,

$$s_o = 200 \text{ mm}$$
$$2s_o = 400 \text{ mm}$$
$$\frac{s_o}{2} = 100 \text{ mm}$$

Case 3: Dead and seismic loads

Load combination is given as

$$U_3 = 0.9D + 1.0E$$

1. Affected areas
   Inspection reveals that this load case affects only the beam moment at the face. As a result,

$$M_u = 0.9M_D + 1.0M_E$$
$$= 0.9(201) - 1.0(156) = 25 \text{ kN·m}$$

   The result above indicates that the moment will not put tension in the bottom face of the beam. Thus, no steel is needed in the bottom face of the beam for strength.

2. Development length
   For development length, see Example 7-1.
   Final design details are shown in Example 7-2, Fig. 3.

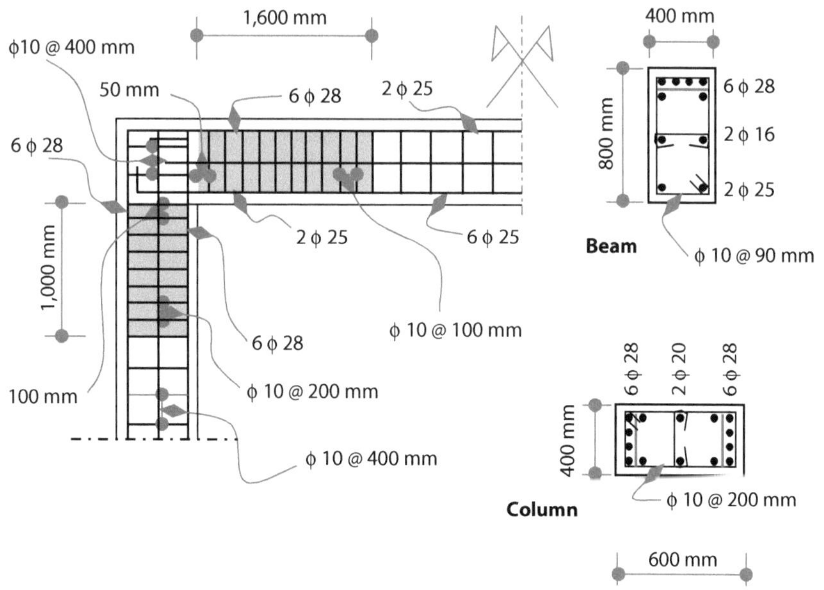

Example 7-2, Figure 3 IMF detailing.

7.4 Special Moment Frames

Special moment frames (SMF) are assigned an R-factor equal to 8 by the IBC. To meet this force reduction, ACI 318 requires these frames to conform to its provisions for ordinary moment frames. These frames must also conform to the requirements of this section.

This type of frame has enhanced detailing and design requirements over that of ordinary and intermediate moment frames. These requirements enhance its ductility and, as a result, increase the allowable force reduction to $R = 8.5$.

This type of frame has stringent requirements in terms of section dimensions, detailing, and design forces. Before addressing these requirements, it is important to understand the following definitions.

Seismic Hook A seismic hook is a bend of at least 135°, with a projecting length inside the core of a rectangular concrete section equal to at least $6d_b$ or 75 mm as shown in Fig. 7-10(a). The bend may be taken 90° in circular hoops.

Crosstie A crosstie is a straight tie that has a seismic hook on one side and a 90° bend on the other. The bend must have at least six times its diameter as shown in Fig. 7-10(b). The crossties are intended to ease the installation of the ties by engaging the seismic hook first and then pushing the bend afterward.

Hoops Hoops are special ties for seismic sections that are classified as continuous hoops and composed of (overlapping) hoops. Continuous hoops consist of one continuous bar, whereas the composed hoops consist of an open stirrup and a crosstie arranged as shown in Fig. 7-11. Continuous hoops provide better support than

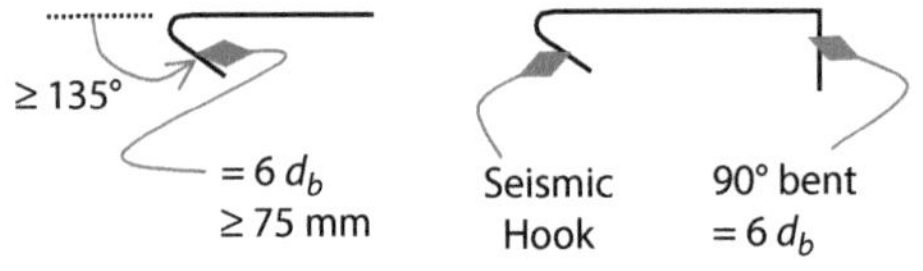

(a) Seismic Hook **(b)** Crosstie

FIGURE 7-10 Seismic ties.

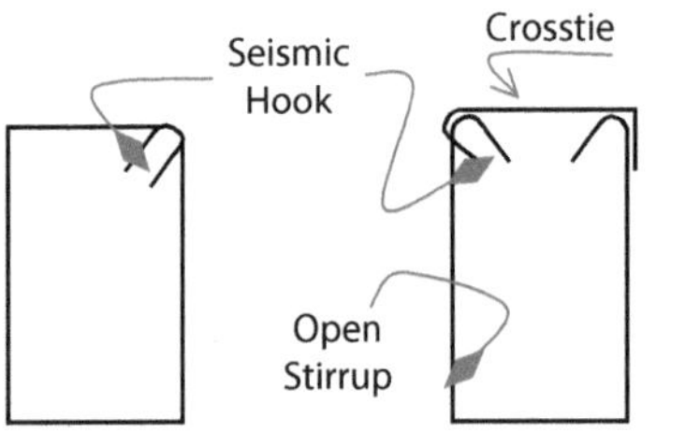

(a) Continuous Hoop **(b)** Composed Hoop

FIGURE 7-11 Hoops.

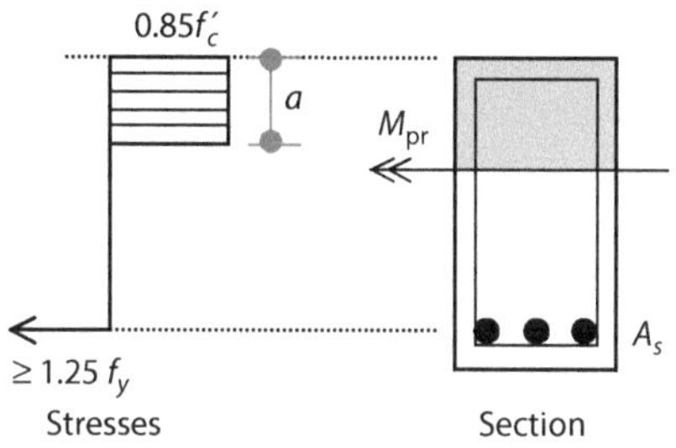

FIGURE 7-12 Probable moment.

composed hoops. However, composed hoops are easier to install. Both types of hoops must be detailed to conform to seismic hook requirements as shown in Fig. 7-11.

Probable Moment, M_{pr} Probable moment, M_{pr}, is the nominal moment as obtained from ACI procedures using stress in steel equal to at least $1.25f_y$ and the strength reduction factor $\phi = 1.0$.

Figure 7-12 shows the stress distribution to be used to obtain this moment. The, $1.25f_y$ stress used in the evaluation of this moment is given as a minimum and not an exact quantity. In general, the ultimate stress of reinforcing steel reaches $1.5f_y$, and should be reflected in this calculation. The most probable moment is an important factor in the design insofar as this moment accounts for the overstrength effect. This action is needed to keep brittle elements elastic as discussed previously in the behavior of structures under seismic excitation.

Similar to previous systems, SMF has requirements for beams and for beam-columns that are presented in the following sections.

7.4.1 Special Beams

Similar to other types of frames, special beams are identified with a maximum-factored axial load as

$$P_u \leq 0.1 f_c' A_g$$

where f_c' and A_g are as defined before.

Special beams have special requirements in terms of design shear, dimensions, main reinforcement and lateral reinforcement as follows.

Design Shear, V_e

The design shear force including seismic effect, V_e, for special beams shall be evaluated according to the maximum probable shear that can develop in the section due to probable moments as defined earlier. This shear force includes the effect of gravity loads and the over-strength effect insofar as the moment is the function of the ultimate strength of reinforcement.

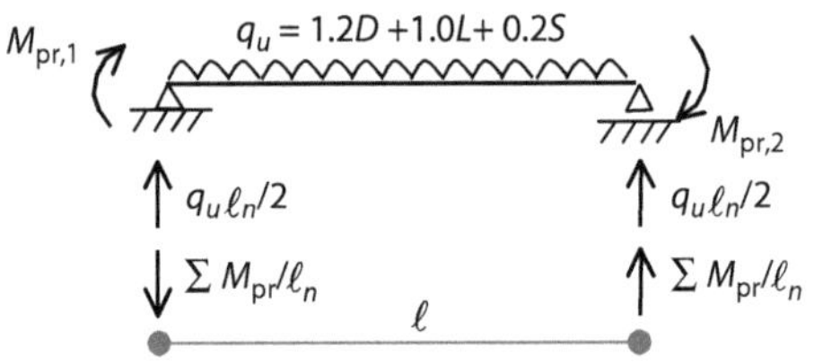

FIGURE 7-13 Design shear.

Consequently, the required design shear strength may be expressed as shown in Fig. 7-13 as

$$V_e = \frac{\Sigma M_{pr}}{\ell_n} + \frac{q_u \ell_n}{2}$$

$$\geq 1.2 V_D + \begin{Bmatrix} 1.0 \\ 0.5 \end{Bmatrix} V_L + 0.2 V_S + V_E$$

where D, L, S, and E = dead, live, snow, and earthquake load components
ℓ_n = clear span of the member
M_{pr} = probable moment of ACI ($f_s \geq 1.25 f_y$)
q_u = load intensity due to factored gravity loads

The design shear above must be resisted by the nominal shear strength of the member, V_n, which consists of concrete contribution (V_c) and steel contribution (V_s) as defined earlier in this chapter. However, the contribution of the concrete shall be ignored under the following two conditions when they occur simultaneously

$$V_c = 0 \quad \text{if} \quad P_u \leq 0.05 f_c' A_g$$

and if $\quad V_E \geq \frac{1}{2} V_e$

where V_E is the shear force due to earthquake effect only.

Dimension Limitations

The additional requirements for dimensions are shown in Fig. 7-14 and explained below.

1. The web width shall not be less than 0.3 times its height but not less than 250 mm as shown in Detail A.
2. The transverse projection of the beam over the supporting columns shall not exceed three-fourths the dimension of the supporting column as shown in Detail B.
3. The clear span of the beam shall not be less than four times its effective depth as shown in Detail C. This requirement is intended to guarantee the flexure behavior of the beam without significant shear deformations (plane section hypothesis).

Main Reinforcement

The additional requirements for main reinforcement design are shown in Fig. 7-14 and explained below.

1. The positive moment at the face of the column must exceed one-half the negative moment strength provided at the face of the supporting column as shown in Detail D.
2. The minimum positive and negative moments at midspan must exceed one-fourth the maximum moment strength provided at the face of the supporting column as shown in Detail E.
3. At least two bars at the top and bottom faces of the beam must be continuous as shown in Detail F.
4. The splice location shall not be less than $2d$ from the face of the support or from the critical section of any plastic hinge as shown in Detail F.
5. Maximum reinforcement ratio, ρ_{max}, shall not exceed 0.025.

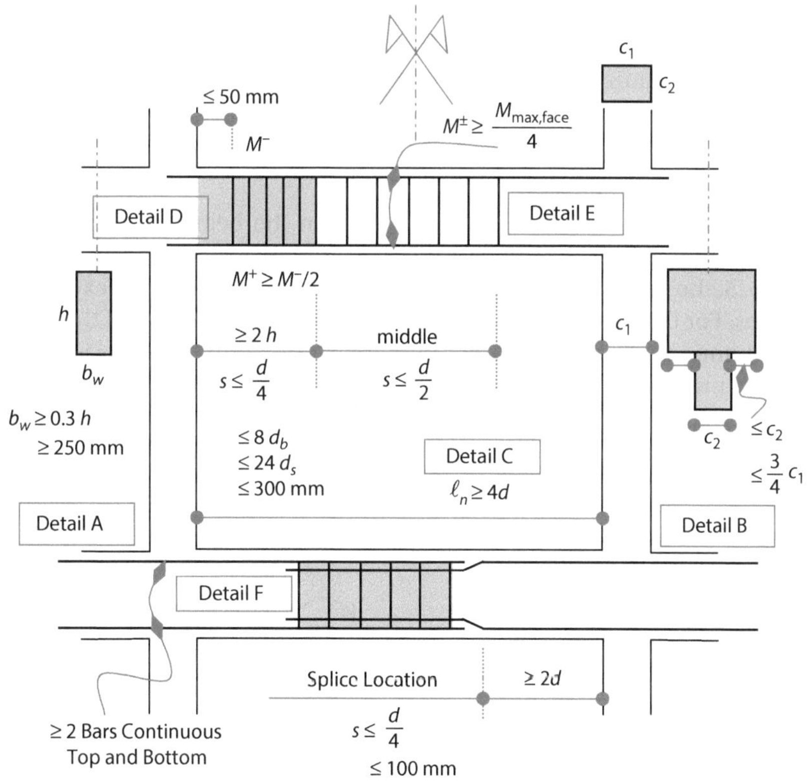

FIGURE 7-14 SMF beam requirement and detailing.

Lateral Reinforcement

The additional requirements for lateral reinforcement are shown in Fig. 7-14 and explained below.

1. The potential plastic hinge region is assumed to extend a distance of $2h$ from the face of the support. Therefore, over this distance, $2h$, the maximum spacing of stirrups is given in terms of the effective depth of the member, d, the diameter of main reinforcement, d_b, the diameter of stirrups, d_s and 300 mm, as shown in Detail D. The first stirrup shall start at least 50 mm from the face of the support as also shown in Detail D.
2. The spacing of stirrups in the middle of the beam shall not exceed $d/2$ as shown in Detail E.
3. The spacing of stirrups over the splice length shall not exceed $d/4$ or 100 mm as shown in Detail F.

7.4.2 Special Beam-Columns

Similar to other types, special beam-columns are identified with a maximum factored axial load as

$$P_u > 0.1 f_c' A_g$$

where f_c' and A_g are as defined before.

Special beam-columns have special requirements in terms of design forces, dimensions, main reinforcement, and lateral reinforcement as follows.

Design Forces

Design Moment To force the plastic hinges in the beams to achieve a weak beam-strong column mechanism as explained under recommended seismic systems in Chap. 5, the flexural strength of the columns must exceed the flexural strength of the beams. For this reason, ACI 318 requires that the summation of the design moments of columns be 20 percent, larger than the summation of the design moments of the girders meeting at the same joint. This requirement may be expressed with reference to Fig. 7-15 as

$$\Sigma M_{nc} \geq (1.2)\, \Sigma M_{nb}$$

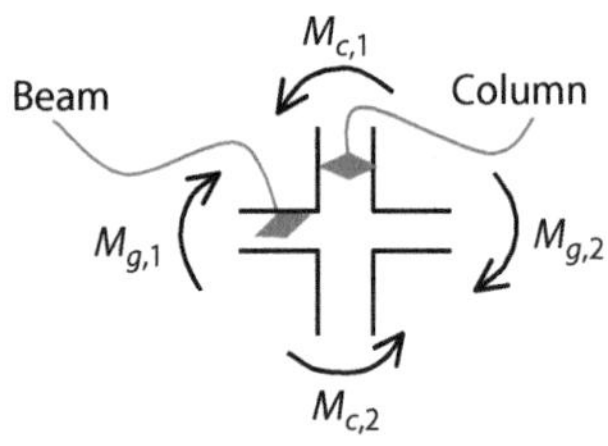

FIGURE 7-15 Joint moments.

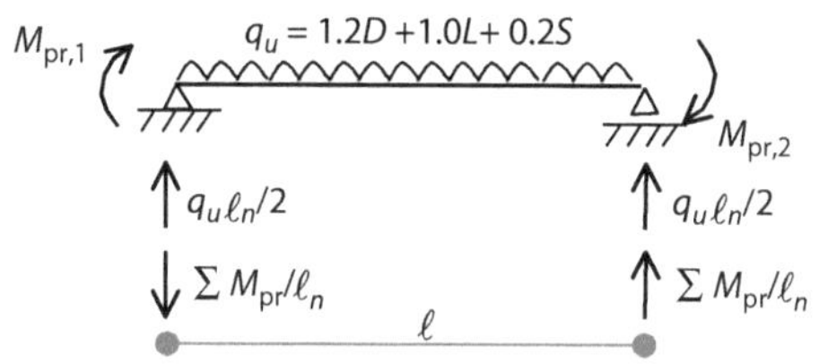

FIGURE 7-16 Design shear.

Design Shear, V_e The design shear force, including seismic effect V_e for special beam-columns, shall be evaluated according to the maximum probable shear that can develop in the section due to probable moments as defined earlier. This shear force includes the effect of gravity loads and the over-strength effect since the moment is a function of the ultimate strength of reinforcement.

Consequently, the required design shear strength may be expressed as shown in Fig. 7-16 as

$$V_e = \frac{\Sigma M_{pr}}{\ell_n} + \frac{q_u \ell_n}{2}$$

However, the design shear shall not be less than the factored maximum seismic load combination, including seismic forces, that is,

$$V_u \geq 1.2V_D + \begin{Bmatrix} 1.0 \\ 0.5 \end{Bmatrix} V_L + 0.2V_S + V_E$$

where D, L, S, and E = dead, live, snow, and earthquake load components
ℓ_n = clear span of the member
M_{pr} = probable moment of ACI ($f_s \geq 1.25 f_y$)
q_u = load intensity due to factored gravity loads

The design shear above must be resisted by the nominal shear strength of the member, V_n, which consists of concrete contribution, V_c, and steel contribution, V_s, as defined earlier in this chapter. However, the contribution of the concrete shall be ignored under the following two conditions when they occur simultaneously

$$V_c = 0$$

$$\text{if} \quad P_u \leq 0.05 f'_c A_g$$

$$\text{and if} \quad V_E \geq \tfrac{1}{2} V_e$$

where V_E is the shear force due to earthquake effect only.

Dimension Limitations

The additional requirements for dimensions are shown in Fig. 7-17, which requires that the width of one side of the section not be less than 0.4 times its other side, but not less than 300 mm as shown in Detail A.

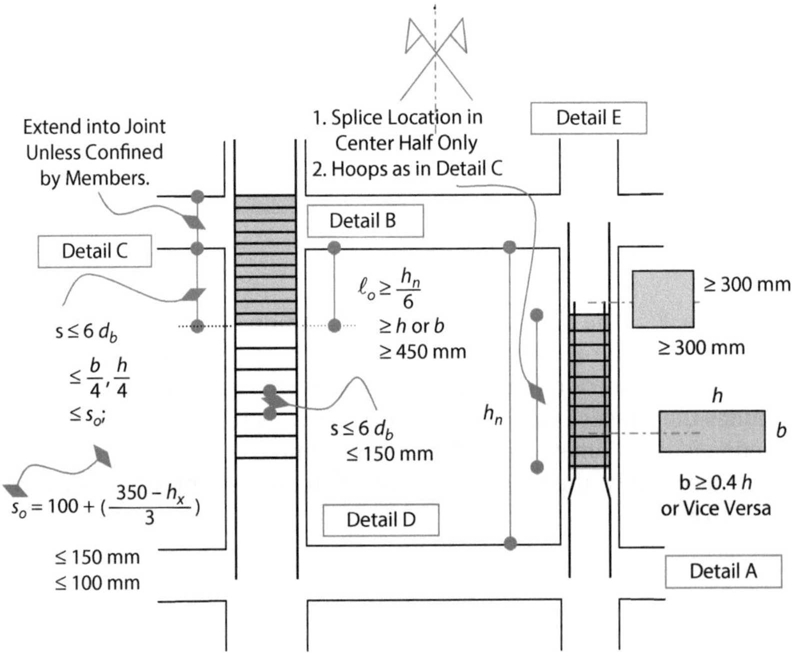

FIGURE 7-17 SMF beam-column requirements and detailing.

Main Reinforcement

The additional requirements for main reinforcement design are shown in Fig. 7-17 and explained below.

1. The gross reinforcement ratio, ρ_g, is limited to a maximum and minimum as follows:

$$\rho_g \leq 0.06 \qquad \text{(preferably} \leq 0.04\text{)}$$

$$\rho_g \geq 0.01$$

2. The splice location shall be limited to the center half of the beam-column to keep the splice outside of the regions of the plastic hinges as shown in Detail E. The spacing of the splice hoop shall be the same as in the plastic hinge region (as shown in Detail C).

Lateral Reinforcement Details

Before addressing these requirements, it is important to understand the following definitions:

1. The distance center to center of the outer hoop is designated as b_c (e.g., b_{c1} in direction 1 and b_{c2} in direction 2 as shown in Fig. 7-18). b_c indicates either b_{c1} or b_{c2} according to the direction under consideration.

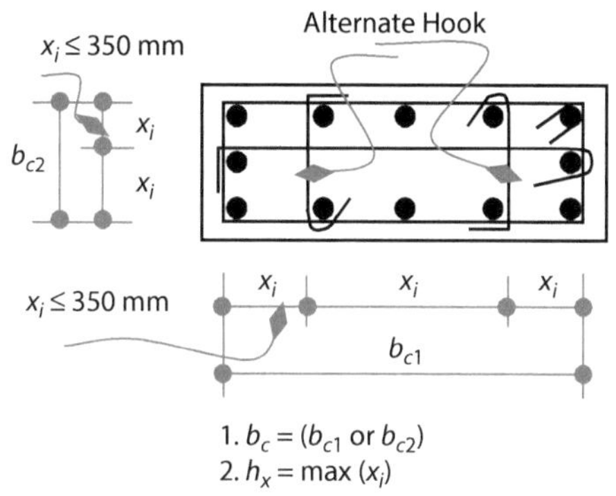

FIGURE 7-18 Hoop dimensions.

2. The distance center to center between crossties is designated as x_i. h_x is given as the maximum value of x_i as shown in Fig. 7-18.
3. h_x shall not exceed 350 mm as shown in Fig. 7-18.
4. All hooks shown in Fig. 7-18 shall be seismic hooks as defined earlier. The 90° bend of consecutive crossties shall be on opposite sides of the column.

The additional requirements for lateral reinforcement are shown in Fig. 7-17 and explained below.

1. The potential plastic hinge region is assumed to be within a distance ℓ_o, from the face of the support. The minimum length of plastic hinge region, ℓ_o, is given as a function of the clear span of the beam-column, the dimensions of the section and, 450 mm, as shown in Detail B. Therefore, over this distance, ℓ_o, the maximum spacing of hoops is given in terms of the diameter of main reinforcement, d_b, the sides of the section and the limit to a quantity, s_o, as defined in Detail C.
2. The amount and spacing of hoops in the plastic hinge region must extend through the joint as shown in Detail C. This extension is waived if the joint is confined by members framing to the joint.
3. The spacing of hoops in the middle of the beam shall neither exceed $6d_b$ nor 150 mm as shown in Detail D.
4. The spacing of hoops over the splice length shall conform to the requirements of hoops in the plastic hinge region that are given in Detail C. The splice reinforcement must also develop as a tension splice.

Minimum Lateral Reinforcement

The minimum hoops required in the potential plastic hinge region for confinement (over the length, ℓ_o) depend on the shape of the section. Minimum hoops for rectangular and circular sections are given as follows.

Rectangular Sections The minimum area of hoops, A_{sh}, in any directions of the rectangular section shall be as

$$A_{sh} \geq 0.3\left(\frac{sb_c f'_c}{f_{yt}}\right)\left(\frac{A_g}{A_{ch}} - 1\right)$$

$$\geq 0.09\left(\frac{sb_c f'_c}{f_{yt}}\right)$$

where A_{ch} = area of concrete core, including hoops, which is given as
$= (b_{c1} + d_h)\cdot(b_{c2} + d_h)$
A_g = gross area of the concrete section
A_{sh} = area of hoops for shear
b_c = center-to-center dimension of the hoop in the direction under consideration as shown in Fig. 7-19
d_h = diameter of hoops
f'_c = characteristic strength of concrete
f_{yt} = yield stress of hoop reinforcement
s = spacing of hoops along the member

Circular and Spiral Sections The minimum volume of hoops, ρ_s, in circular sections with hoops or with spirals shall be as

$$\rho_s \geq 0.12\frac{f'_c}{f_{yt}}$$

$$\geq 0.45\left(\frac{A_g}{A_{ch}} - 1\right)\frac{f'_c}{f_{yt}}$$

where A_{ch} = area of concrete core, including hoops
A_g = gross area of the concrete section
f'_c = characteristic strength of concrete
f_{yt} = yield stress of hoop reinforcement

Concrete Cover Protection

The maximum allowable concrete cover must be 100 mm. In other words, when concrete cover exceeds 100 mm, additional hoops are needed within the outer 100 mm cover spaced at 300 mm.

7.4.3 Special Joints

As noted in the previous section, the beam-column hoops in the plastic hinge region must be extended into the joint area as shown in Fig. 7-17. However, there are additional requirements for joints in SMF in terms of design forces and detailing requirements.

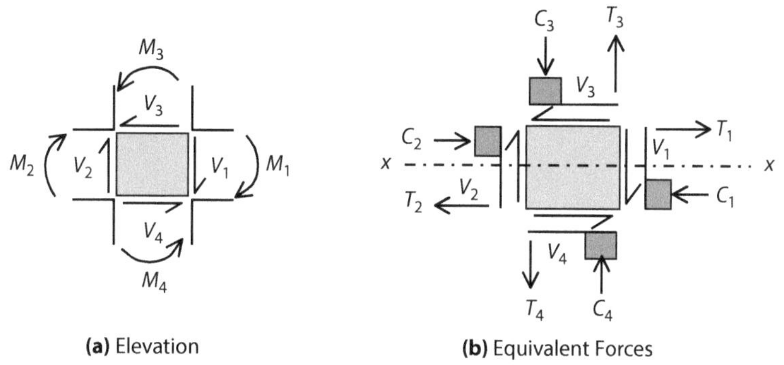

Figure 7-19 Joint forces.

Force Requirements Design forces of the joints must be based on the probable moment, M_{pr}, which is based on factored yield stress, f_y', such that

$$f_y' \geq 1.25 f_y$$

Consequently, the design forces will consist of direct shear and forces from moments. As an example, for the state of internal shear and moment shown in Fig. 7-19(a), the design shear force at face *x-x* in Fig. 7-19(b) will be as follows:

$$V_u = V_3 - T_1 - C_2$$

This shearing force must be resisted by the nominal shear capacity of the joint, which is given according to the number of sides of the joints that are framing into members. Accordingly, the nominal shear capacity is given as

$$V_n = 1.7\sqrt{f_c'}\,A_j \quad \text{(four sides framing)}$$
$$V_n = 1.2\sqrt{f_c'}\,A_j \quad \text{(three sides framing)}$$
$$V_n = 1.0\sqrt{f_c'}\,A_j \quad \text{(others)}$$

where A_j is the effective area of the joint cross section.

Additional Requirements The reinforcement bar sizes, d_b, that pass through the joint must not exceed 5 percent of the adjacent column dimension as shown in Fig. 7-20

$$d_b \leq 0.05\, h_{\text{col}}$$

Development of Reinforcement

The development of reinforcement in special beam-columns is given as a function of a basic hook development length, ℓ_{dh}, as defined in Fig. 7-21. ℓ_{dh} is given as

$$\ell_{\text{dh}} = 0.19\frac{f_y}{\sqrt{f_c'}}d_b \geq \begin{pmatrix} 8d_b \\ 150\,\text{mm} \end{pmatrix}$$

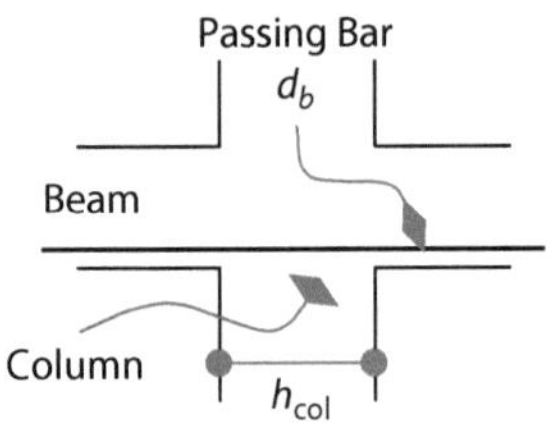

FIGURE 7-20 Joint detailing.

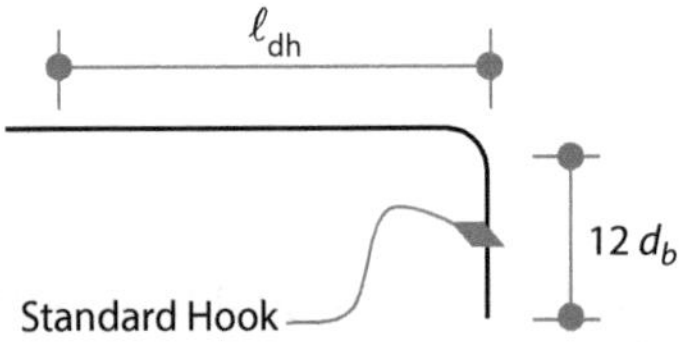

FIGURE 7-21 Basic hook development length.

The development length of straight bars is given as a function of the basic hook development length for top and bottom reinforcement. This development length is also a function of the confinement of concrete. The development length, ℓ_d, of straight bars for different conditions is given by the following expressions.

Confined Concrete

Development length in confined concrete is defined for top and bottom reinforcement. The top reinforcement is defined as any reinforcement that is placed over 300 mm of fresh concrete. Accordingly, the development length is given as
Top bars

$$\ell_d = 3.25\ell_{dh} \geq 0.602\frac{f_y}{\sqrt{f_c'}}d_b \geq \begin{pmatrix} 26\,d_b \\ 488\text{ mm} \end{pmatrix}$$

Bottom bars

$$\ell_d = 2.5\ell_{dh} \geq 0.463\frac{f_y}{\sqrt{f_c'}}d_b \geq \begin{pmatrix} 20\,d_b \\ 375\text{ mm} \end{pmatrix}$$

Example 7-3 The frame shown in Example 7-3, Fig. 1, is the same frame given in Examples 7-1 and 7-2. The share of the frame from the elastic base shear is given as

$$F_x = 272\text{ kN}$$

Design the same frame as an SMF to resist the given seismic forces. Material properties are given as the same

$$f_c' = 25\text{ MPa and } f_y = 420\text{ MPa}$$

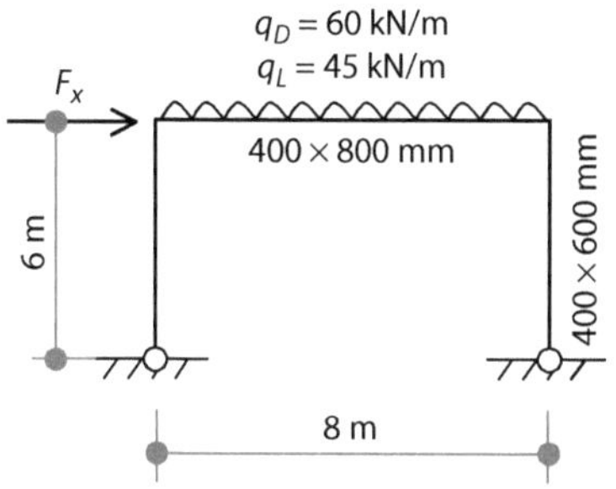

Example 7-3, Figure 1

Solution Because this example is intended to be solved in conjunction with Example 7-1, the reader is advised to solve Example 7-1 before working on this example. Be sure to review the basic ACI 318 design philosophy and procedures before solving this example.

The solution will be performed to check three load cases: (1) dead and live load effects; (2) dead, live, and earthquake load effects; and (3) dead and earthquake load effects.

Case 1: Dead and live loads
Load combination is given as

$$U_1 = 1.2D + 1.6L$$

This load case will be exactly the same as in Example 7-1. See case 1 in Example 7-1 for the solution of this case.

Case 2: Dead, live, and seismic forces
Load combination is given as

$$U_2 = 1.2D + 1.0L + 1.0E$$

1. Straining actions
   The seismic force for SMF is given as the elastic force divided by the *R*-factor, which is 8 for this frame as given in App. 6-1

$$F_x = \frac{F_{el}}{R} = \frac{272}{8} = 34 \text{ kN}$$

   Using the dimensions and properties above, the resulting moments, shears, and normal forces are constructed for the seismic force as shown in Example 7-3, Fig. 2.

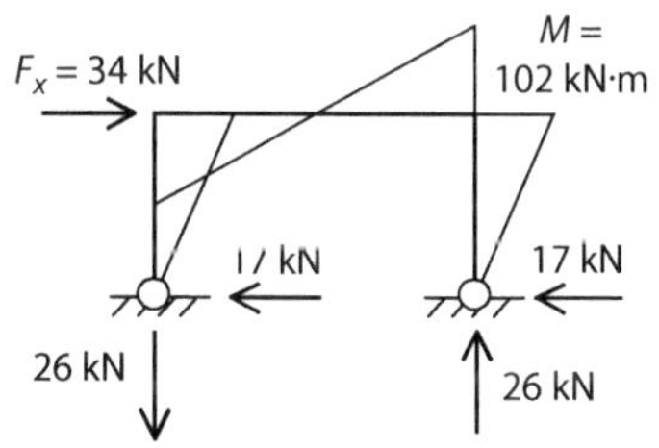

Example 7-3, Figure 2 Seismic forces.

2. Beam moment

Check beam action: $0.1 f_c' A_g = 0.1(25)(320{,}000)$

$= 800{,}000$ N (800 kN)

Ultimate load: $P_u = 1.2P_D + 1.0P_L + 1.0P_E$

$= 1.2(33) + 1.0(25) + 1.0(17) = 81$ kN < limit

Because $P_u = 81$ kN < Limit = 800 kN, the member may be designed without the effect of axial load.

The dimensions of special beam must satisfy the following.

Minimum width: $b, h \geq 250$ mm

Section limitations: $\geq 0.3\,h = 0.3(800) = 240$ mm

The section width equals 400 mm and satisfies the limitations above.

Positive moment region: As in Example 7-1.

Negative moment region: $M_u = 1.2M_D + 1.0M_L + 1.0M_E$

$= 1.2(201) + 1.0(151)) + 1.0(102)$

$= 494$ kN·m > case 1

$$M_n = \frac{M_u}{\phi} = \frac{494}{0.9} = 549 \text{ kN}\cdot\text{m}$$

$$P_n = \frac{P_u}{\phi} = \frac{81}{0.9} = 90 \text{ kN}$$

Design:

$$\frac{M_n}{f_c' b d^2} = \frac{549 \times 10^6}{25(400)(740)^2} = 0.10$$

Using design charts in App. 7-1, Sheet 1, the reinforcement index is given as $\omega_o = 0.11$,

$$A_s = \omega_o \frac{f_c'}{f_y} bd = 0.11\left(\frac{25}{420}\right)(400)(740) = 1{,}938 \text{ mm}^2 \ (4\,\phi 25)$$

Check steel limits for tension-controlled action (at $\phi = 0.9$)
Since

$$\frac{P_n}{f_c' b_w d} = \frac{90 \times 10^3}{25(400)(740)} = 0.012$$

then

$$\omega_w + \frac{P_n}{f_c' b_w d} = 0.11 + 0.012$$

$$= 0.122 < 0.32\beta_1 = 0.272 \text{ ok}$$

For additional requirements for SMF beams, refer to Fig. 7-14.

The positive moment at the face and at midspan must be designed for a minimum as a function of the negative moment at the joint. By inspection, the maximum design negative moment obtained at the joint is 549 kN·m.

Moments at face

Positive moment at face: $$M_n^+ = \frac{M_n^+}{2} = \frac{549}{2} = 275 \text{ kN·m}$$

Design: $$\frac{M_n}{f_c' bd^2} = \frac{275 \times 10^6}{25(400)(740)^2} = 0.050$$

Using design charts in App. 7-1, Sheets 1 and 2, the reinforcement index is given as less than minimum. Thus, use minimum reinforcement

$$\rho_{w,\min} = 0.0033$$

$$A_s = \rho_{w,\min} bd = 0.0033(400)(740) = 966 \text{ mm}^2 \ (2\ \phi\ 25)$$

Moments at midspan

Both moments at midspan: $$M_n^- = \frac{M_n^{\text{face}}}{4} = \frac{549}{5} = 138 \text{ kN·m}$$

Design: $$\frac{M_n}{f_c' bd^2} = \frac{138 \times 10^6}{25(400)(740)^2} = 0.025$$

Using design charts in App. 7-1, Sheets 1 and 2, the reinforcement index is given as less than minimum. Thus, use minimum reinforcement

$$\rho_{w,\min} = 0.0033$$

$$A_s = \rho_{w,\min} bd = 0.0033(400)(740) = 966 \text{ mm}^2 \ (2\ \phi\ 25)$$

3. Beam shear
   Using ϕ 10 (#3) bars for stirrups

$$V_u = 1.2V_D + 1.0V_L + 1.0V_E$$
$$= 1.2(240) + 1.0(180) + 1.0(26)$$
$$= 494 \text{ kN}$$

For SMF, V_u shall not be less than the following (refer to Fig. 7-14 for illustration)

$$V_e \geq \frac{\Sigma M_{\text{pr}}}{\ell_n} + V_u \text{ (gravity)}$$

where

$$M_{n1} \approx 1.25 M_{n1} = 1.25(549) = 686 \text{ kN·m}$$
$$M_{n2} \approx 1.25 M_{n2} = 1.25(275) = 344 \text{ kN·m}$$

Gravity load: $$q_u = 1.2q_D + 1.0q_L$$

$$= 1.2(60) + 1.0(45) = 117 \text{ kN/m}$$

Consequently,

$$V_e \geq \frac{686 + 344}{(8 - 0.6)} + \frac{117(8)}{2} = 607 \text{ kN}$$

Therefore, $V_e = 607$ kN

$$V_n = \frac{V_e}{\phi} = \frac{607}{0.75} = 809 \text{ kN}$$

Reference: $$V_w = 0.165\sqrt{f_c'}b_w d$$

$$= 0.165\sqrt{25}(400)(740) = 246{,}667 \text{ N}$$

Concrete strength must be ignored if both of the following conditions exist.

First limit: If $P_u \leq 0.05 f_c' A_g = 0.05(25)(320{,}000) = 400{,}000$ N

Second limit: If $V_E \geq ½ V_e = ½\,(607) = 303$ kN

First limit: $P_u = 81$ kN $\leq$ limit = 400 kN exists

Second limit: $V_E = 26$ kN < limit = 303 kN does not exist

Because only one of the conditions above exists, the concrete shear strength may be used for resistance, hence,

Concrete strength: $V_c = V_w = 247$ kN

Limits: $2V_w = 494$ kN, $4V_w = 988$ kN

Steel: $V_s = V_n - V_c = 809 - 247 = 562$ kN < 988 kN ok

Strength spacing: $$V_s = A_v f_{yt}\frac{d}{s} \text{(use } \phi \text{ 10 for stirrups} = 78 \text{ mm}^2\text{)}$$

$$562{,}000 = 156\,(420)\left(\frac{740}{s}\right)$$

$$s = 86 \text{ mm}$$

Maximum spacing over a distance = $2h = 2(800) = 1{,}600$ mm

$$s \leq \frac{d}{4} = \frac{740}{4} = 185 \text{ mm}$$

$$\leq 8\,d_b = 8(25) = 200 \text{ mm}$$

$$\leq 48\,d_s = 48(10) = 240 \text{ mm}$$

$$\leq 300 \text{ mm}$$

$$\leq \frac{A_v f_{yt}}{0.062\sqrt{f_c'}\,b_w} = \frac{156(420)}{0.062\sqrt{25}(400)} = 528 \text{ m}$$

$$\leq \frac{A_v f_{yt}}{0.35\,b_w} = \frac{156(420)}{0.35(400)} = 468 \text{ mm}$$

Therefore, use stirrups that are 2ϕ10 @ 80 mm over a distance = 1,600 mm from the face of the column.

4. Beam-column moment
   Check beam-column action: $0.1 f_c' A_g = 0.1(25)(240{,}000)$

$$= 600{,}000 \text{ N (600 kN)}$$

Ultimate load:
$$P_u = 1.2P_D + 1.0P_L + 1.0P_E$$
$$= 1.2(240) + 1.0(180) + 1.0(26)$$
$$= 494 \text{ kN} > \text{limit}$$

Because $P_u = 494$ kN > Limit = 600 kN the member must be designed as a beam-column with the effect of axial load.

The dimensions of special beam-column must satisfy the following:

Minimum width: $b, h \geq 400$ mm

Section limitations: $b \geq 0.4\,h = 0.4(600) = 240$ mm

The section width equals 400 mm and satisfies the limitations above.

According to ACI 318, frames under lateral loads may not be considered braced and shall be designed as sway members. In the sway case, the beam-column may be considered short or long according to the following limit against slenderness ratio:

Check slenderness: $\dfrac{k\ell_u}{r}(<,>)22$

Effective length factor, k:

To find the effective length factor of frames, use the alignment chart in App. 7-1, Sheet 7

$$\psi_A = \frac{\Sigma EI/\ell_c}{\Sigma EI/\ell_b} = \frac{5{,}040/6}{5{,}973/8} = 1.13$$
$$\psi_B = \infty \text{ for pinned support}$$

Using Sheet 7, sway case, $k = 2.40$

Slenderness ratio:
$$\frac{k\ell_u}{r} = \frac{2.40(6{,}000)}{0.3(600)} = 80 > \text{Limit} = 22$$

Because the slenderness ratio is more than the limit, the beam-column is considered long, which means that the P-Δ effect must be considered. In this case, second-order analyses of moment magnifications are needed. These may be evaluated using the sway magnification factor, δ_s, as defined by ACI 318. According to ACI 318, the design moment is given by the procedures described below.

Effective moment of inertia for buckling analysis is different from structural analysis for concrete sections. For buckling analysis, EI is given as in Example 7-1

$$EI = 0.25E_cI_g = 0.25(23{,}500)(7{,}200 \times 10^6)$$
$$= 42{,}300 \times 10^9 \text{ N·mm}^2$$

Critical load:
$$P_c = \frac{\pi^2 EI}{(k\ell)^2} = \frac{\pi^2(42{,}300)}{(2.40\times 6)^2} = 2{,}013 \text{ kN}$$

Summation of P_c in the entire floor: $\Sigma P_c = 2{,}013(2) = 4{,}026$ kN

Summation of P_u in the entire floor: $\Sigma P_u = P_{u1} + P_{u2}$

Note that the vertical reactions due to seismic forces as given in Example 7-2, Fig. 2, are opposite and cancel out in the summation

$$\Sigma P_u = (1.2P_D + 1.0P_L)(2)$$
$$= [1.2(240) + 1.0(180)](2) = 936 \text{ kN}$$

Moment magnification: $$\delta_s = \frac{1}{1 - \dfrac{\Sigma P_u}{\phi \Sigma P_c}} \begin{pmatrix} \le 2.5 \\ \ge 1.0 \end{pmatrix}$$

$$\delta_s = \frac{1}{1 - \dfrac{936}{0.75(4{,}026)}} = 1.45$$

The design moment is given as a combination of the nonsway moments and the sway moments as follows:

$$M_{2ns} = 1.2(201) + 1.0(151) = 392 \text{ kN·m}$$

$$M_{2s} = 1.0(102) = 102 \text{ kN·m}$$

Therefore, the design moment is given as follows:

$$M_u = M_{2ns} + \delta_s M_{2s}$$

$$M_u = 392 + 1.45(102) = 534 \text{ kN·m}$$

The beam-column moment must be 20 percent larger than girder moments

$$\Sigma M_{nc} \ge 6/5\, \Sigma M_{nb} = 6/5\,(549) = 659 \text{ kN·m}$$

Therefore,

$$M_u = 659 \text{ kN·m}$$

Design may be accomplished using interaction diagrams given in App. 7-1, Sheets 3 through 6. Accordingly,

$$P_u \text{ (already calculated above)} = 494 \text{ kN}$$

$$P_n = \frac{P_u}{\phi} = \frac{494}{0.65} = 760 \text{ kN}$$

$$M_n = \frac{M_u}{\phi} = \frac{659}{0.65} = 1014 \text{ kN·m}$$

Therefore

$$\frac{P_n}{f_c' bh} = \frac{760 \times 10^3}{25(400)(600)} = 0.127$$

$$\frac{M_n}{f_c' bh^2} = \frac{1{,}014 \times 10^6}{25(400)(600)^2} = 0.281$$

Using Sheet 4 for $\gamma = \dfrac{600-120}{600} = 0.8$, results in

$$\rho\mu = 0.66$$

Since

$$\mu = \frac{f_y}{0.85 f_c'} = \frac{420}{0.85(25)} = 19.76$$

then

$$\rho = \frac{0.66}{19.76} = 0.033 \rightarrow \begin{pmatrix} <0.08 \\ >0.01 \end{pmatrix} \text{ ok}$$

Steel: $A_{st} = \rho bh = 0.033(400)(600)$

$$= 7{,}620 \text{ mm}^2 \ (12\ \phi\ 28)$$

5. Beam-column shear
   Beam-column concrete shear strength is affected by axial load as follows:

$$V_u = 1.2V_D + 1.0V_L + 1.0\ V_E$$
$$= 1.2(33) + 1.0(25) + 1.0(17) = 82 \text{ kN}$$

For SMF, V_u shall not be less than the following (refer to Fig. 7-16 for illustration)

$$V_e \geq \frac{\Sigma M_{pr}}{\ell_n} + V_u(\text{gravity})$$

where

$$M_{n1} \approx 1.25M_{n1} = 1.25(659) = 824 \text{ kN·m}$$

$$M_{n2} = 0 \text{ kN·m (pinned support)}$$

Gravity load: $q_u = 0$ in the column

Consequently,

$$V_e \geq \frac{824}{(6-0.4)} + 0 = 147 \text{ kN}$$

Therefore,

$$V_e = 147 \text{ kN}$$
$$V_n = \frac{V_e}{\phi} = \frac{147}{0.75} = 196 \text{ kN}$$

Reference: $V_w = 0.165\sqrt{f_c'}\, b_w d$

$$= 0.165\sqrt{25}(400)(540) = 180{,}000 \text{ N}$$

Concrete strength must be ignored if both of the following conditions exist.

First limit: If $P_u \leq 0.05 f_c' A_g = 0.05(25)(240{,}000) = 300{,}000$ N

Second limit: If $V_E \geq \frac{1}{2} V_e = \frac{1}{2}(147) = 74$ kN

First limit: $P_u = 494$ kN $\geq$ limit $= 300$ kN does not exist

Second limit: $V_E = 17$ kN $\leq$ limit $= 74$ kN does not exist

Because none of the conditions above exists, the concrete shear strength may be used for resistance.

Concrete strength $V_c = \left(1+0.07\frac{N_u}{A_g}\right)V_w$

$$= \left[1+0.07\left(\frac{494{,}000}{240{,}000}\right)\right](180)$$

$$= 1.144(180) = 206 \text{ kN}$$

Capacity: $\phi V_c = 0.75(206) = 154 \text{ kN}$

Limits: $2V_w = 309 \text{ kN},\ 4V_w = 618 \text{ kN}$

Steel: $V_s = V_n - V_c = 196 - 206 = -v_e \text{ kN} < 618 \text{ kN}$

Strength spacing: Because V_s is negative, there is no need for stirrups for strength.

Minimum stirrups and detailing in columns are given as follows (refer to Fig. 7-17 for details)

Length, ℓ_o: $\ell_o \geq \frac{h_n}{6} = \frac{6{,}000}{6} = 1{,}000 \text{ mm}$

$$\geq h, b = 400 \text{ mm}$$

$$\geq 450 \text{ mm}$$

Therefore,

$$\ell_o = 1{,}000 \text{ mm}$$

Spacing in the potential plastic hinge region (i.e., over length, ℓ_o)

$$s \leq 6d_b = 6(25) = 150 \text{ mm}$$

$$\leq \frac{b}{4}, \frac{h}{4} = \frac{400}{4} = 100 \text{ mm}$$

$$\leq s_o = 100 + \frac{350-300}{3}$$

$$= 117 \text{ mm } (> 100 \text{ mm}, < 150 \text{ mm})$$

Therefore, $s = 100$ mm, and must extend into the joint.

Spacing in middle of the member

$$s \leq 6\,d_b = 6\,(25) = 150 \text{ mm}$$

$$\leq 150 \text{ mm}$$

Minimum hoops in potential plastic hinge region (i.e., over length, ℓ_o) in the transverse direction

$$A_{sh} = 0.3\left(\frac{sb_c f'_c}{f_{yt}}\right)\left(\frac{A_g}{A_{ch}} - 1\right)$$

$$\geq 0.09\left(\frac{sb_c f'_c}{f_{yt}}\right)$$

$$A_{sh} = 0.3\left(\frac{100(510)(25)}{420}\right)\left(\frac{400(600)}{320(520)} - 1\right) = 403 \text{ mm}^2$$

$$\geq 0.09\left(\frac{100(510)(25)}{420}\right) = 273 \text{ mm}^2$$

Therefore, use six legs of ϕ 10 hoops.
In the longitudinal direction

$$A_{sh} = 403\ \text{mm}^2 \frac{310}{510} = 245\ \text{mm}^2$$

Therefore, use four legs of ϕ 10 hoops.

Case 3: Dead and seismic forces

Load combination is given as $U_3 = 0.9D + 1.0E$

1. Affected areas
Inspection reveals that this load case affects only the beam moment at the face

$$M_u = 0.9M_D + 1.0M_E$$

$$= 0.9(201) - 1.0(102) = 79\ \text{kN·m}$$

This result indicates that the moment will not put tension in the bottom face of the beam. Thus, no steel for strength is needed in the bottom face of the beam.

2. Development length
For development length, see Example 7-1.
The final design details are shown in Example 7-3, Fig. 3.

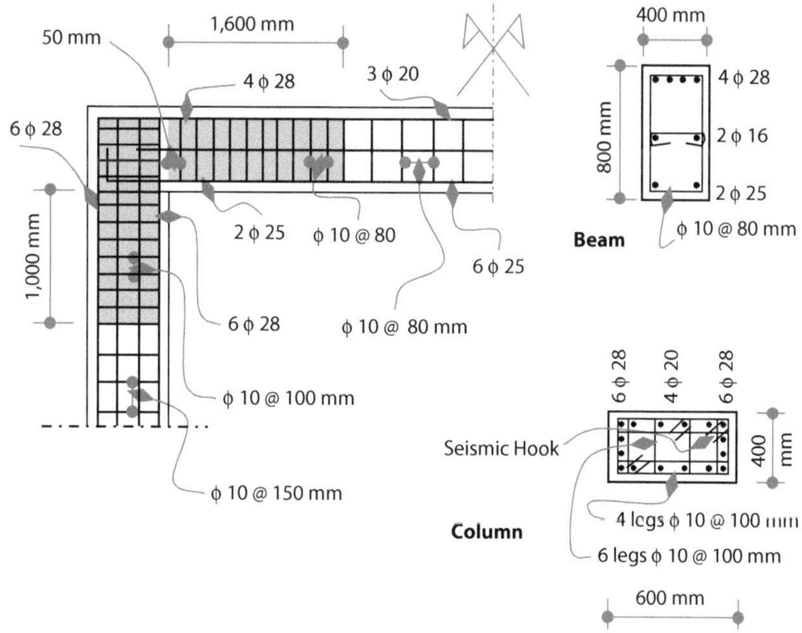

Example 7-3, Figure 3 SMF detailing.

7.5 Ordinary Shear Walls

Similar to the case of ordinary moment frames, ordinary shear walls (OSW) are designed and detailed according to ACI 318 for nonseismic provisions that are given in Chap. 1 through Chap. 20.

Shear walls are classified into tall walls and squat walls. Walls are considered tall if their span/depth ratio satisfies the beam theory hypothesis of plane section, which implies that shear deformations are negligible; thus, the strain distribution is linear across the section. In principle, this behavior is usually guaranteed if the span/depth ratio is 4 or more.

Walls are considered squat if they do not satisfy the plane section hypothesis, which is characterized by significance of shear deformations. In this case, shear deformations must be considered where the walls are analyzed as deep beams. Recent developments in truss models, also known as strut and tie models, offer an attractive solution for such walls. In principle, the behavior of squat walls is not ductile, because it is dominated by shear behavior.

Tall walls are considered excellent systems in resisting lateral loads because they have large strength and rigidity. In addition, they exhibit reasonable ductility that makes them also attractive for seismic resistance. Tall walls may be analyzed and designed as beam-columns. However, they have detailing that differs from beam-column detailing requirements.

Ordinary walls inherit enough ductility to allow them to be designed for response modification factors as specified by the IBC seismic code. Because axial loads affect ductility, as noted in Chap. 5, shear walls are also classified as bearing and nonbearing shear walls. For example, the IBC assigns a value of R-factor equal to 4 for ordinary bearing walls and a value of R-factor equal to 5 for ordinary nonbearing walls.

Before proceeding, consider the necessary wall components shown in Fig. 7-22. ACI allows the effective depth of the wall, when needed for calculations, to be taken as 80 percent of the wall length

$$d = 0.8\ell_w$$

Force Requirements

The force requirements for ordinary walls are analogous to those for beams. If the following quantity is defined as a reference quantity, V_h, such that

$$V_h = 0.165\sqrt{f_c'}\,hd$$

where d is conservatively taken as 80 percent of the wall length ($d = 0.8\,\ell_w$) as shown in Fig. 7-22. The maximum allowable shear strength, V_n of the wall is taken as

$$V_n \leq 5V_h$$

The shear strength is then calculated as in beams as

$$V_c = V_h$$

$$V_s = A_v f_y \frac{d}{s}$$

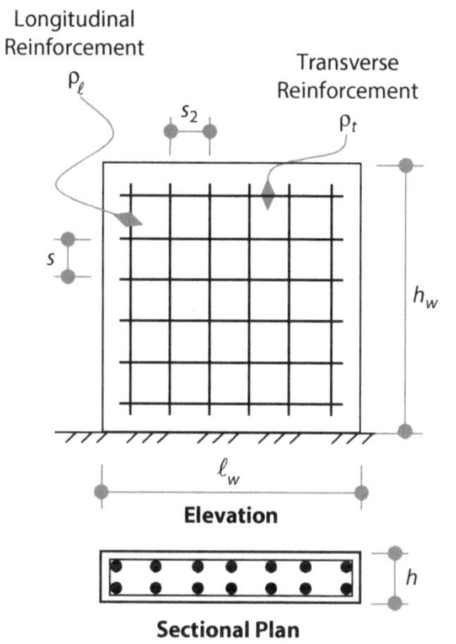

FIGURE 7-22 Ordinary shear wall components.

Reinforcement Requirements

The minimum reinforcement requirements of ordinary walls are given as a function of the amount of factored external shear in the section.

1. For $V_u < 0.5\,\phi\,V_h$

   Transverse reinforcement

$$\rho_\ell = 0.0020 \quad \text{for bars with} \begin{pmatrix} d_b \leq 16 \text{ mm} \\ f_y \geq 420 \text{ MPa} \end{pmatrix}$$
$$= 0.0025 \quad \text{otherwise}$$

   Longitudinal reinforcement

$$\rho_\ell = 0.0012 \quad \text{for bars with} \begin{pmatrix} d_b \leq 16 \text{ mm} \\ f_y \geq 420 \text{ MPa} \end{pmatrix}$$
$$= 0.0015 \quad \text{otherwise}$$

   with max spacing of both directions

$$s_{\max} = 3h$$
$$\leq 450 \text{ mm}$$

2. For $V_u \geq 0.5\,\phi V_h$

Transverse reinforcement

$$\rho_t = 0.0025$$

with max spacing
$$s_{\max} = \frac{\ell_w}{5}$$
$$\leq 3h$$
$$\leq 450 \text{ mm}$$

The longitudinal reinforcement is given as a function of the transverse reinforcement as follows:

$$\rho_\ell = 0.0025 + 0.5\left(2.5 - \frac{h_w}{\ell_w}\right)(\rho_t - 0.0025) \leq \rho_t$$
$$\geq 0.0025$$

with max spacing:
$$s_{2,\max} = \frac{\ell_w}{3}$$
$$\leq 3h$$
$$\leq 450 \text{ mm}$$

Example 7-4 The residential building shown in Example 7-4, Fig. 1, is a bearing-wall type building subjected to a horizontal earthquake excitation in direction y. The building is six stories with 200 mm solid, one-way slabs spanning in direction x. The slabs are supported by 10 shear walls as shown in the same figure.

Each wall has 300 mm width with the following material properties

$$f_c' = 30 \text{ MPa}, \qquad f_y = 420 \text{ MPa}, \qquad I_{\text{eff}} = 0.5\, I_g$$

The following loading has to be considered in the design of the building:

1. Own weight (slabs and shear walls)
2. Superimposed dead load, $q_{\text{SDL}} = 2 \text{ kN/m}^2$
3. Live load, $q_L = 3 \text{ kN/m}^2$
4. Wind load, $q_w = 1 \text{ kN/m}^2$
5. Seismic loading according to the IBC with the following parameters:
   a. Mapped acceleration coefficient at short period and at 1-s period equal to $S_s = 0.50$ and $S_1 = 0.12$, respectively
   b. The site is classified as Soil Type C
   c. The long-period transition period, $T_L = 8$ s

A. If the gravity loads are assumed to be totally carried by the walls in direction y and their reactions on the walls are uniformly distributed over the wall cross section, design one interior shear wall (#3 on the plan) using ACI 318 provisions.

B. Consider the IBC provisions to include torsional effect, drift limitations, and the P-Δ effect.

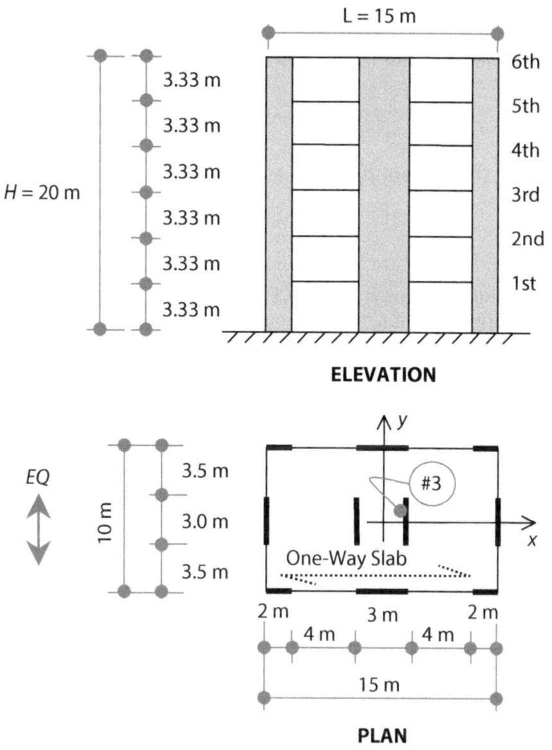

Example 7-4, Figure 1 Building layout.

Solution The example will be solved by carrying out design for the seismic load case ($1.2D + 1.0L + 1.0E$), and then checking for the basic load ($1.2D + 1.6L$). Since the building is regular with overall height less than 48 m, the equivalent lateral force procedure may be used.

1. Seismic coefficients F_a, F_v, S_{DS}, and S_{D1}

   The seismic coefficients, F_a and F_v, are obtained from IBC Tables 6-A3 and 6-A4, App. 6-1: $S_s = 0.50$, $F_a = 1.2$, $S_1 = 0.12$, and $F_v = 1.68$.

   Accordingly, the seismic coefficients S_{DS} and S_{D1} are calculated as follows:

$$S_{MS} = F_a S_s = 1.2(0.50) = 0.60$$

$$S_{M1} = F_v S_1 = 1.68(0.12) = 0.20$$

$$S_{DS} = \frac{2}{3} S_{MS} = \frac{2}{3}(0.60) = 0.40$$

$$S_{D1} = \frac{2}{3} S_{M1} = \frac{2}{3}(0.20) = 0.13$$

2. Group, category, and importance, I

   A residential building in Risk Category I has an S_{DS} of 0.40 and an S_{D1} of 0.13, according to Table 6-A2, App. 6-1. Based on this information, the building is in Seismic Design Category C.

   The importance factor is $I = 1$.

3. Period, T; response modification factor, R; and coefficient of displacement, C_d

   Response spectrum controlling period, T_s,

$$T_s = \frac{S_{D1}}{S_{DS}} = \frac{0.13}{0.4} = 0.33$$

   The approximate period, T_a, is calculated as

$$T_a = C_t(h_n)^x$$

   where building height, h_n, is 20 m. Coefficients C_t and x are obtained from App. 6-1, Table 6-A8, as $C_t = 0.0488$ and $x = 0.75$. Consequently, the period is given as

$$T_a = C_t(h_n)^x = 0.0488\,(20)^{0.75} = 0.46 \text{ s}$$

   In reference to Table 6-A6, App. 6-1, ordinary bearing shear walls are allowed for SDC C without height limitation. Therefore, if an ordinary bearing shear wall is selected for the building, the response parameters are obtained for BOSW as $R = 4$ and $C_d = 4$.

4. Gravity loads

   Slabs: $q_D = \gamma h = 25(0.2) = 5 \text{ kN/m}^2$

   $q_{SDL} = 2 \text{ kN/m}^2$

   $q_L = 3 \text{ kN/m}^2$

   Walls: $q_{wall} = \gamma b = 25(0.3) = 7.5 \text{ kN/m}^2$

5. Mass weight (see Chap. 6)

$$W = D + 0.25 \text{ warehouse} + \text{partitions} + \text{equipment}$$

   where $D = \text{slabs} + \text{walls}$

$$D = 7(10)(15)(6) + 7.5\,(26)(20)$$
$$= 6{,}300 + 3{,}900 = 10{,}200 \text{ kN}$$

   Because there are no warehouse loads, partitions, or equipment given, then

$$W = 10{,}200 \text{ kN}$$

6. Horizontal seismic force (base shear) (see Chap. 6)

   According to the parameters obtained above,

$$C_s = \frac{S_{DS} I}{R} = \begin{pmatrix} \le \dfrac{S_{D1} I}{RT} & \text{for } T \le T_L \\ \ge 0.044\, S_{DS} I_e \ge 0.01 & \text{for } S_1 < 0.6 \end{pmatrix}$$

$$C_s = \frac{0.4(1)}{4} = \begin{pmatrix} \le \dfrac{0.13\,(1)}{4(0.46)} \\ \ge 0.044(0.4)(1) \ge 0.01 \end{pmatrix}$$

$$C_s = 0.10 = \begin{pmatrix} \le 071 \\ \ge 0.0176 \end{pmatrix}$$

Therefore,

$$C_s = 0.071$$

As a result,

$$V = C_s W = 0.071(10{,}200) = 724 \text{ kN}$$

7. Wind effect

Total wind force: $F_w = 1(15)(20) = 300 \text{ kN} < V = 724 \text{ kN}$
Therefore, seismic force controls design.

8. Distribution of base shear to floors (see Chap. 6)

$$F_x = C_{vx} V$$

where
$$C_{vx} = \frac{w_x h_x^k}{\Sigma w_i h_i^k}$$

Since

$$T = 0.46 < 0.5$$

then $k = 1$.

Because floor weights are the same, the equation above reduces to

$$C_{vx} = \frac{h_x^1}{\Sigma h_i^1}$$

and

$$F_x = \frac{h_x}{\Sigma h_i} V$$

where
$$\Sigma h_i = 3.333 + 6.667 + 13.333 + 16.667 + 20$$
$$= 70 \text{ m}$$

As a result, the floor force is given as

$$f_x = h_x \frac{724}{70} = 10.343 h_x \text{ kN/m}$$

The distribution given previously is a triangle with zero value at the base. Therefore, the force may be treated as a distributed triangular load as shown in Example 7-4, Fig. 2, with intensity at the top equal to

$$f_{x,\text{top}} = \frac{724(2)}{20} = 72.4 \text{ kN/m}$$

9. Torsional effect (see Chap. 6)

Because the building is symmetric, the only torsion that exists is the accidental torsion

$$M_{\text{tor}} = V e_{\text{min}}$$
$$= 724(0.05)(15)$$
$$= 543 \text{ kN·m}$$

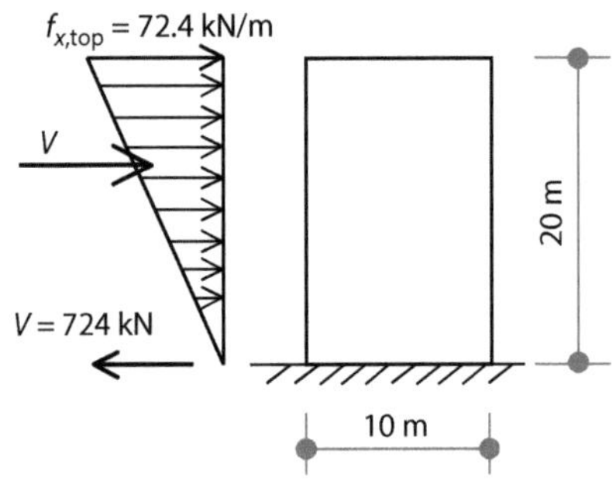

Example 7-4, Figure 2 Building force.

Distribution of torsional moment to shear walls is given as

$$Q_i = \frac{M_{\text{tor}}}{\Sigma k_j r_j^{\,2}} k_i r_i$$

where $\Sigma k_j r_j^{\,2} = k(1.5^2 + 7.5^2)(2) = 117\text{ k}$

Therefore,

$$Q_3 = \frac{543}{117\text{ k}} k(1.5) = 7\text{ kN}$$

10. Vertical seismic force (see Chap. 6)

$$E_v = 0.2S_{\text{DS}}\,D$$
$$= 0.2(0.4)(1)\,D = 0.08\,D$$

Total vertical load: $Q_v = 0.08(10{,}200) = 816\text{ kN}$

11. Redundancy factor (see Chap. 6)

    For SDC C, the redundancy factor is $\rho = 1$.

12. Wall share from seismic force (see Chap. 6)

    Horizontal force acting at two-third from the base

$$V_E = \frac{V}{4} + Q_3$$
$$= \frac{724}{4} + 7$$
$$= 181 + 7 = 188\text{ kN}$$

Vertical force acting at center

This force is proportional to dead load which is calculated in the next step as 2,340 kN. As a result,

$$P_E = 0.08\,P_D$$
$$= 0.08(2{,}340) = 187\text{ kN}$$

The effect of these forces is shown in Example 7-4, Fig. 3.

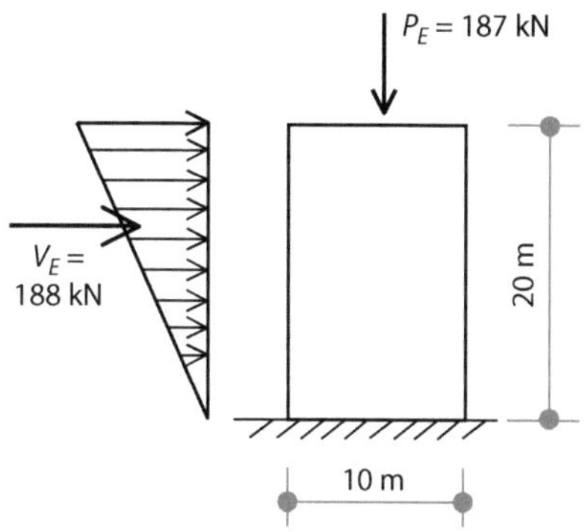

Example 7-4, Figure 3 Wall 3, seismic forces.

13. Design forces

Axial loads

$$P_D = (q_D + q_{SDL})\ \text{(tributary area) (number of floors)} + \text{own weight}$$
$$= (5+2)[(10)(1.5{+}3)](6) + 7.5\,(3)(20) = 2{,}340\ \text{kN}$$
$$P_L = (q_L)\ \text{(tributary area) (number of floors)}$$
$$= (3)[(10)(1.5{+}3)](6) = 810\ \text{kN}$$
$$P_E = 187\ \text{kN} \qquad \text{(from step above)}$$

Maximum shear force at the base

$$V_D = 0, \qquad V_L = 0, \qquad V_E = 188\ \text{kN}$$

Maximum bending moment at the base

$$M_D = 0,\ M_L = 0$$
$$M_E = 188\left(\frac{2}{3}\right)(20) = 2{,}507\ \text{kN·m}$$

14. Flexure design

Check beam-column action

$$\text{Limit} = 0.1\,f_c'A_g = 0.1(30)(300)(3{,}000) = 2{,}700 \times 10^3\ \text{N}$$
$$P_u = 1.2P_D + P_L + P_E = 1.2(2{,}340) + 810 + 187 = 3{,}805\ \text{kN}$$

Because P_u is more than the limit above, the wall shall be designed as beam-column, including the effect of axial load (see section on moment frames)

$$P_n = \frac{P_u}{\phi} = \frac{3{,}805}{0.65} = 5{,}854\ \text{kN}$$
$$M_u = 1.2\,M_D + M_L + M_E = 1.2(0) + 0 + 2{,}507 = 2{,}507\ \text{kN·m}$$
$$M_n = \frac{M_u}{\phi} = \frac{2{,}507}{0.65} = 3{,}857\ \text{kN·m}$$

Therefore,

$$\frac{P_n}{f'_c bh} = \frac{5{,}854 \times 10^3}{30(300)(3{,}000)} = 0.217$$

$$\frac{M_n}{f'_c bh^2} = \frac{3{,}857 \times 10^6}{30(300)(3{,}000)^2} = 0.048$$

Using App. 7-1, Sheet 4, for $\gamma = 0.8$ results in

$$\rho\mu < 0 \qquad \text{(use minimum steel)}$$

since

$$\mu = \frac{f_y}{0.85 f'_c} = \frac{420}{0.85(30)} = 16.47$$

then

$$\rho = \frac{0.0}{16.47} = 0.0 (< 0.01, \text{therefore, use } 0.01)$$

Steel: $A_{st} = \rho bh = 0.01(300)(3{,}000) = 9{,}000 \text{ mm}^2$

$= 4{,}500 \text{ mm}^2$ each side; use 5 ϕ 25 each face

15. Shear design

The shear strength of walls is usually controlled by minimum shear reinforcement. Therefore, it is good practice to provide the wall with minimum reinforcement first and then to check its capacity accordingly.

Reference: $V_h = 0.165\sqrt{f'_c}\, hd$, note that $d = 0.8\, \ell_w$

$$V_h = 0.165\sqrt{30}\,(300)\,\{(0.8)(3{,}000)\} = 651{,}000 \text{ N}$$

$$\phi V_h = 0.75(651) = 488 \text{ kN}$$

$$0.5\,\phi V_h = 244 \text{ kN}$$

$$V_u = 1.2\, V_D + V_L + V_E = 1.2\,(0) + 0 + 188 = 188 \text{ kN}$$

$$V_n = \frac{V_u}{\phi} = \frac{188}{0.75} = 251 \text{ kN}$$

Because $V_u = 188$ kN is $< \phi V_h = 488$ kN, and $< 0.5\,\phi V_h = 244$ kN, the minimum reinforcement is given as follows.

Transverse reinforcement

$$\rho_t = 0.0020 \qquad \text{spaced at } s \text{, such that}$$

$$s_{max} \leq \frac{\ell_w}{5} = \frac{3{,}000}{5} = 600 \text{ mm}$$

$$\leq 3\,h = 3\,(300) = 900 \text{ mm}$$

$$\leq 450 \text{ mm}$$

$$A_s = 0.0020(300)(1{,}000) = 600 \text{ mm/m}$$

Use 2 layers 4 ϕ 10/m each

Longitudinal reinforcement

$$\rho_\ell = 0.0012 \text{ spaced at } s_2\text{, such that}$$

$$s_{2,\max} = \frac{\ell_w}{3} = \frac{3{,}000}{3} = 1{,}000 \text{ mm}$$

$$\leq 3\,h = 3(300) = 900 \text{ mm}$$

$$\leq 450 \text{ mm}$$

$$A_s = 0.0012(300)(1{,}000) = 360 \text{ mm}^2/\text{m}$$

Use 2 layers 3 ϕ 10/m each

Final details are shown in Example 7-4, Fig. 4.

16. Check load combination of dead and live loads alone

This load case requires the check of wall safety about the weak axis for strength and stability

$$P_u = 1.2\,P_D + 1.6\,P_L$$

$$P_u = 1.2(2{,}340) + 1.6(810) = 4{,}104 \text{ kN}$$

Strength: As a column, shear reinforcement will be ignored. As a result,

$$P_n = 0.8(0.85\,f_c'A_g + A_{st}f_y)$$

$$P_n = 0.8[0.85\,(30)(300)(3{,}000) + (9{,}000)(420)]$$

$$= 21{,}384 \times 10^3 \text{ N}$$

$$\phi P_n = 0.65(21{,}384) = 13{,}900 \text{ kN} > P_u = 4{,}104 \text{ kN ok}$$

Stability:

$$E_c = 4{,}700\sqrt{30} = 25{,}750 \text{ MPa}$$

$$I_g = \frac{3(0.3)^3}{12} = 0.00675 \text{ m}^4$$

$$EI = 0.25\,E_cI_g$$

$$= 0.25(25{,}750{,}000)(0.00675) = 43{,}450 \text{ kN·m}^2$$

Critical Load:

$$P_c = \frac{\pi^2 EI}{(k\ell)^2} = \frac{\pi^2(43{,}450)}{(1\times 3.333)^2} = 38{,}606 \text{ kN}$$

$$\phi P_c = 0.75(38{,}606) = 28{,}954 \text{ kN} > P_u = 4{,}104 \text{ kN ok}$$

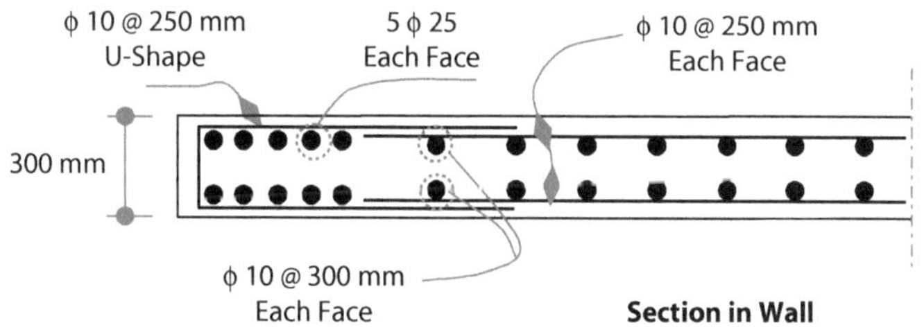

Example 7-4, Figure 4 Ordinary shear wall detailing.

17. Drift limitations

For this example, the deflection may be calculated by conjugate beam method as shown in Example 7-4, Fig. 5.

Because the load is linear, the moment will be a third-degree parabola which has its centroid at one-fifth from base

$$\Delta_S = \frac{(\frac{1}{4} M_b H)(\frac{4}{5} H)}{EI}$$

where

$$E = 4{,}700\sqrt{f_c'}$$
$$= 4{,}700\sqrt{30} = 25{,}740 \text{ MPa}$$
$$I = 0.5 I_g$$
$$= 0.5 \frac{0.3(3)^3}{12} = 0.3375 \text{ m}^4$$

Therefore,

$$\delta_S = \frac{\left(\frac{1}{4}(2{,}507)(20)\right)\left(\frac{4}{5}(20)\right)}{25.75 \times 10^6\,(0.3375)} = 0.023 \text{ m (23 mm)}$$
$$\delta_M = C_d \delta_S = 4(23) = 92 \text{ mm}$$

IBC drift limitations are given in App. 6-1, Table 6-A5, for Risk Category 1 as, 0.020 H. Therefore,

$$\delta_{all} = 0.020\,H = 0.020(20{,}000) = 400 \text{ mm}$$
$$\delta_M = 92 \text{ mm ok}$$

18. P-Δ effect

Calculate stability coefficient, θ:

$$\theta = \frac{P_x \Delta I_e}{V_x h_{sx} C_d} = \frac{(2{,}340 + 810)(0.092)(1)}{188(20)(4)} = 0.02$$

Because $\theta = 0.02 < 0.1$, the P-Δ effect needs not be considered.

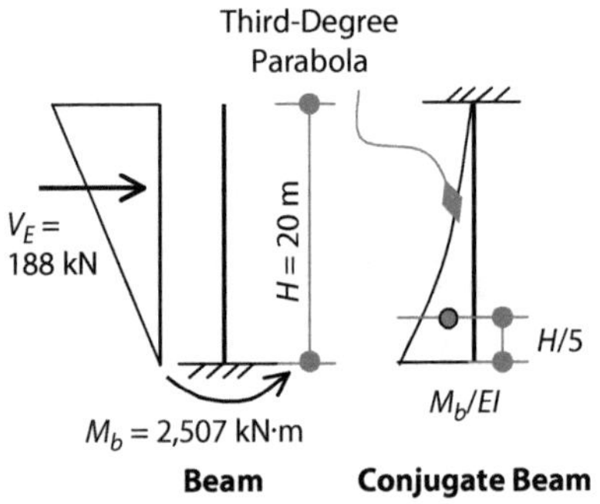

EXAMPLE 7-4, FIGURE 5 Wall deflection.

7.6 Special Shear Walls

Special shear walls (SSW) have more stringent requirements that increase their ductility, thereby allowing more force reduction. SSW are listed in the IBC with an assigned R-factor equal to 5 for bearing walls and 6 for nonbearing walls.

SSW may be designed with or without openings. Such walls will be covered in the following sections.

Similar to its provisions for ordinary shear walls, the ACI code allows the effective depth of the wall, when needed for calculations, to be taken as 80 percent of the wall length

$$d = 0.8\ell_w$$

7.6.1 Special Shear Walls without Openings

SSW components are shown in Fig. 7-23. They consist of web ($b_w\ell_w$) with or without flanges. If present, the effective width, b_f, of the flange is given as

$$\begin{aligned} b_f &= b_w + 0.25h_w \qquad \text{(on each side of the web)} \\ &\le \text{spacing to adjacent flanges, } s_w \\ &\le \text{actual dimensions} \end{aligned}$$

The end regions of SSW are known as boundary elements, regardless of whether or not they have flanges. These elements are classified as either boundary elements or special boundary elements (SBE). The need for SBE is discussed later in this section.

The requirements of SSW are given in terms of force requirements, reinforcement requirements, and boundary element requirements.

Force Requirements

The force requirements for SSW are different from beams and ordinarily walls. For convenience, define the following quantity as a reference quantity, V_{cv}, such that

$$V_{cv} = 0.165\sqrt{f_c'}\,A_{cv}$$

where A_{cv} is the horizontal gross-sectional area of the wall web as shown in Fig. 7-23, that is, $A_{cv} = b_w\ell_w$

The force requirements are defined for the shear strength of the special walls, V_n, which is given as a function of the height-to-length ratio, (h_w/ℓ_w) as shown in Fig. 7-24. Accordingly, the shear strength is given as

$$V_n = A_{cv}(\alpha_c\sqrt{f_c'} + \rho_t f_y)$$

where A_{cv} (as defined above) is $b_w\ell_w$

$$\alpha_c = \begin{pmatrix} 0.165 & \text{for } \dfrac{h_w}{\ell_w} \ge 2 \\ 0.25 & \text{for } \dfrac{h_w}{\ell_w} \le 1.5 \end{pmatrix}$$

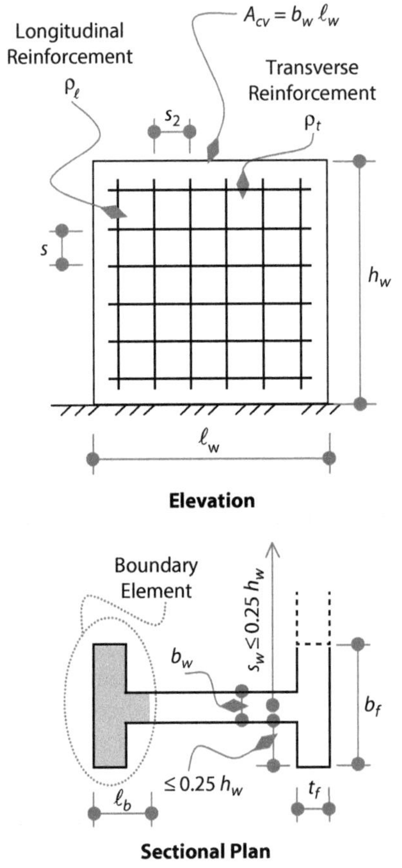

FIGURE 7-23 Special shear wall components.

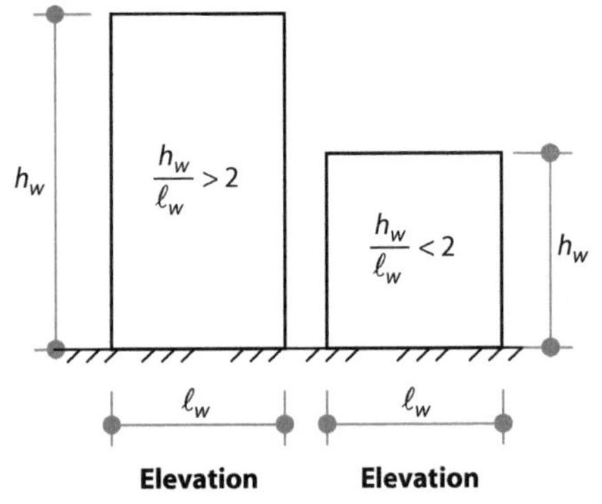

FIGURE 7-24 Tall versus squat wall.

For values of h_w/ℓ_w between 2 and 1.5, α_c may be linearly interpolated. Note that α_c = 0.165 gives concrete shear capacity similar to beams and ordinary shear walls.

For values of h_w/ℓ_w less than 2, reinforcement shall be

$$\rho_\ell \geq \rho_t$$

In order to maintain reasonable distribution of SSW in the building, the code puts limit on the maximum shear capacity of a group of walls that resist common force. For this reason, the total capacity of group of walls resisting common shear force, ΣV_n, shall be limited to four times their capacities, ΣV_{cv}. However, for each individual wall, the capacity, V_n, shall be limited to five times V_{cv}, that is,

$$\Sigma V_n \leq 4\,\Sigma V_{cv} \qquad \text{(for all walls together)}$$

$$V_n \leq 5\,V_{cv} \qquad \text{(for each individual wall)}$$

Reinforcement Requirements

Similar to ordinary shear walls, the minimum reinforcement requirements of special shear walls are given as a function of the amount of factored external shear in the section.

1. For $V_u \leq 0.5\,\phi V_{cv}$

   Transverse reinforcement ratios shall not be less than the following:

$$\rho_t = 0.0020 \quad \text{for bars with} \begin{pmatrix} d_b \leq 16\text{ mm} \\ f_y \geq 420\text{ MPa} \end{pmatrix}$$

$$= 0.0025 \quad \text{otherwise}$$

   Longitudinal reinforcement

$$\rho_\ell = 0.0012 \quad \text{for bars with} \begin{pmatrix} d_b \leq 16\text{ mm} \\ f_y \geq 420\text{ MPa} \end{pmatrix}$$

$$= 0.0015 \quad \text{otherwise}$$

   Spacing of all reinforcement is limited to 450 mm

$$s \leq 450\text{ mm}$$

$$s_2 \leq 450\text{ mm}$$

2. For $V_u > \frac{1}{2}\,\phi V_{cv}$

   Horizontal and vertical reinforcement ratios shall not be less than 0.0025

$$\rho_{t,\min} = 0.0025$$

$$\rho_{\ell,\min} = 0.0025$$

Spacing of all reinforcement is still limited to 450 mm

$$s \leq 450 \text{ mm}$$

$$s_2 \leq 450 \text{ mm}$$

If $V_u > V_{cv}$, then two curtains of reinforcement must be used.

Development length at the plastic hinge region shall be increased by 25 percent

$$(\ell_d)_{\text{plastic hinge}} = 1.25(\ell_d)_{\text{ordinary cases}}$$

Boundary Element Requirements

SBEs are needed to provide additional ductility at the ends of the walls. These are the critical sections because they are the most strained sections. ACI 318 suggests two methods to evaluate the need for such SBE: the displacement method and the stress method.

The dimensions of the boundary elements are defined in Fig. 7-25 as height, h_b, and depth (or length), ℓ_b. The boundary elements may have the same thickness of the web, may be enlarged, or may be flanged as shown in Fig. 7-25.

Regardless of the type of the boundary element, or the method of evaluation, the length (depth) of the boundary element, ℓ_b, is given as

$$\ell_b = c - 0.1\ell_w$$

$$\geq \frac{c}{2} t_f + 300 \text{ mm}$$

where c is the distance from the neutral axis to the extreme compression fibers. c is evaluated at ultimate state under the action of P_u and M_n.

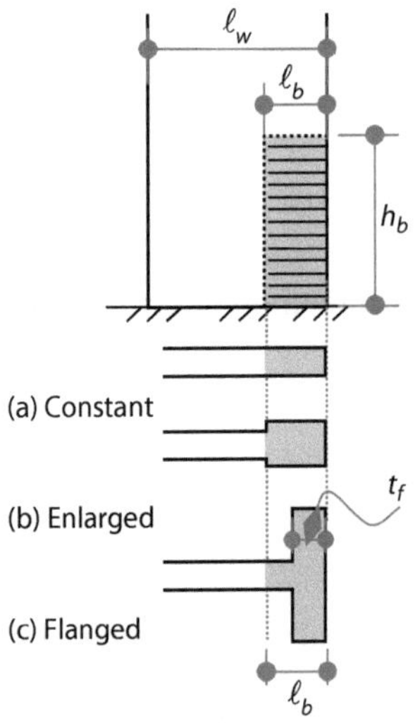

Figure 7-25 Types of boundary element.

As mentioned earlier, the need for SBE may be determined by using a displacement method or a stress method as follows.

1. Displacement method

   The displacement method only applies to shear walls that are continuous in the vertical direction and that have only one critical section for moment and axial load. For shear walls that satisfy these conditions, the SBE are needed if

$$c \geq \frac{\ell_w}{600(\delta_u/h_w)}$$

   For calculation purposes only

$$\frac{\delta_u}{h_w} \geq 0.007$$

   where δ_u is design displacement.

   The height of the SBE, h_b, as shown in Fig. 7-26 is given as

$$h_b \geq \ell_w$$
$$\geq \frac{M_u}{4V_u}$$

2. Stress method

   This method is the more general method and can be used without the restrictions imposed on the displacement method as given previously.

   According to this method, the SBE is needed if

$$f_{\text{cm}} > 0.2 f_c'$$

   where f_{cm} is the stress in the extreme compression fibers calculated using gross concrete section and the linear elastic model under factored loads

$$f_{\text{cm}} = \frac{p_u}{A_g} + \frac{M_u}{I_g} X_M$$

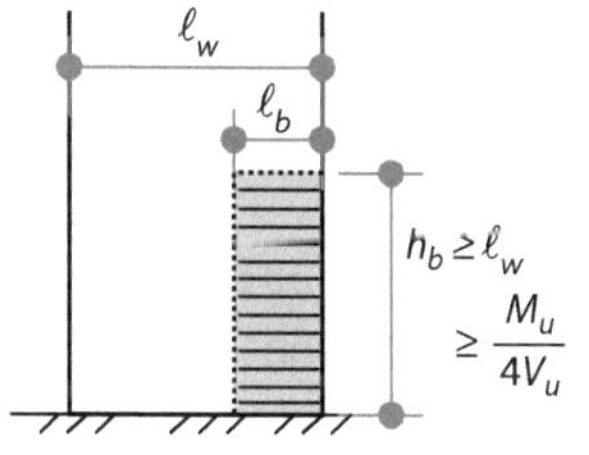

Figure 7-26 SBE displacement method.

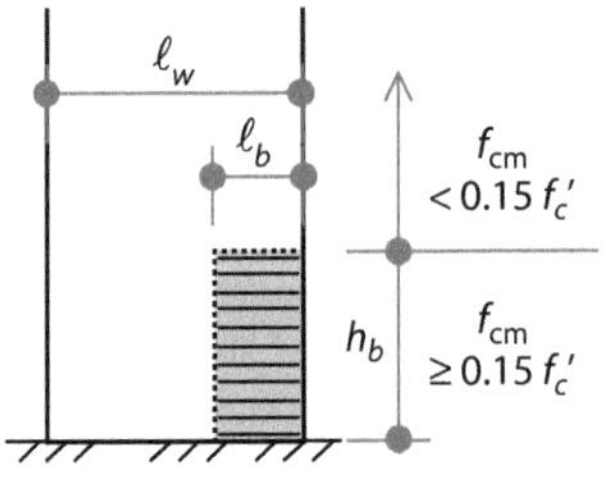

FIGURE 7-27 SBE stress method.

The stress, f_{cm}, as calculated above is a nominal stress used as an indicator, not an actual stress, because it is based on ultimate quantities where the behavior is nonlinear at this stage.

The height of the SBE h_b, must extend in the region where the stress, f_{cm}, exceeds $0.15 f_c'$ as shown in Fig. 7-27, where

$$f_c \geq 0.15 f_c'$$

Detailing of Boundary Elements

1. If SBE is needed

The transverse reinforcement and detailing of the needed SBE shall be as follows.

The minimum area of hoops, A_{sh}, in both directions of the boundary element section shall be as

$$A_{sh} \geq 0.09 s b_c \frac{f_c'}{f_{yt}}$$

The maximum spacing of the transverse reinforcement is given as

$$\begin{aligned} s_{max} &= 6 d_b \\ &\leq \frac{b}{4}, \frac{h}{4} \\ &\leq s_o \end{aligned}$$

where

$$\begin{aligned} s_o &= 100 \text{ mm} + \left(\frac{350 - h_x}{3} \right) \\ &\leq 150 \text{ mm} \\ &\leq 100 \text{ mm} \end{aligned}$$

Notations and definitions are the same as given for the special beam-column and as shown in Fig. 7-28.

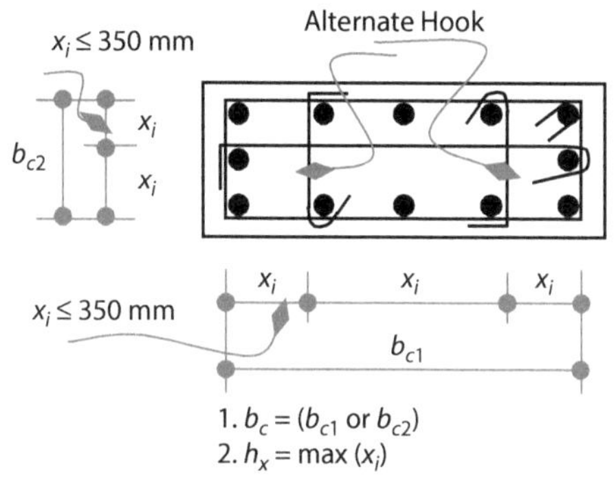

FIGURE 7-28 Detailing of boundary element.

Similar to the special beam-column, the maximum allowable concrete cover must be 100 mm. In other words, when concrete cover exceeds 100-mm additional hoops are needed within the outer 100 mm cover and spaced at 300 mm.

The SBE lateral reinforcement details must extend into the supporting system at the foundation level as shown in Fig. 7-29. Figure 7-29(a) shows the wall supported by a stiff foundation wall, whereas Fig. 7-29(b) shows the wall supported directly by the foundation. The figure indicates that the SBE shall extend in the wall foundation at least the development length of the main bars passing through the SBE. If the wall is supported directly by the foundation, the SBE shall extend at least 300 mm in the foundation.

The web reinforcement must be anchored (developed) inside the SBE. The development length for seismic members is given in the section on special beam-columns. The anchorage length is portrayed in Fig. 7-29.

2. If SBE is not needed

   If the SBE is not needed according to the conditions given in the previous section, the following requirements shall apply:

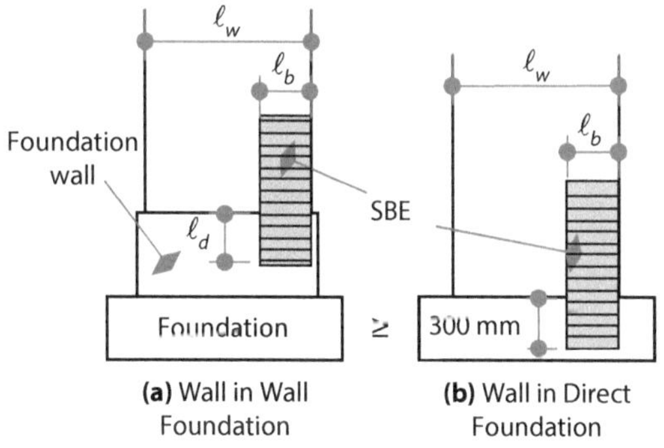

FIGURE 7-29 Extension of special boundary element into foundations.

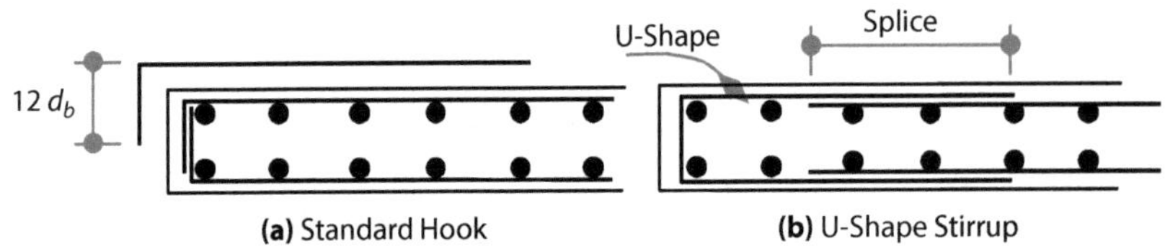

FIGURE 7-30 Detailing of reinforcement in SBE.

A. If, $V_u \leq 0.5\ V_{cv}$:

No requirements are needed.

B. If, $V_u > 0.5\ V_{cv}$

The transverse reinforcement must have a standard hook as in Fig. 7-30(a) or U-shape stirrups as shown in Fig. 7-30(b).

C. If the longitudinal reinforcement at the edge of the wall, ρ_ℓ, exceeds the following:

$$\rho_\ell \frac{2.8}{f_y}$$

then the boundary element must meet the following requirements:

- The boundary element must be provided with length, ℓ_b, such that

$$\ell_b = c - 0.1\ell_w$$
$$\geq \frac{c}{2}$$

- The maximum spacing of crossties inside the boundary element as given in Fig. 7-31 is limited to

$$h_x \leq 350 \text{ mm}$$

- The crossties shown in Fig. 7-31 must be installed with alternate hooks.
- The maximum spacing of hoops, s, inside the boundary element is given as

$$s \leq 200 \text{ mm}$$

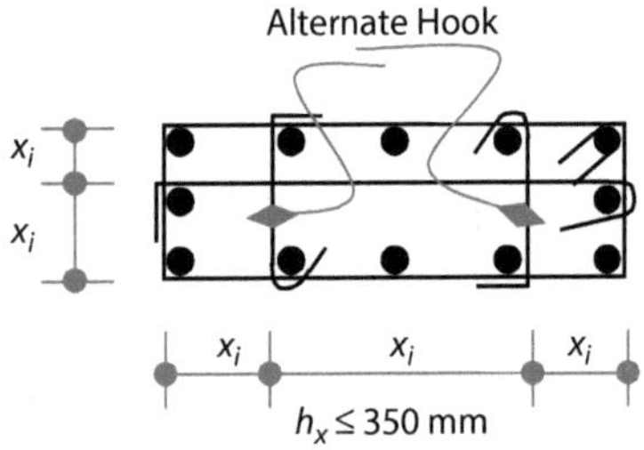

FIGURE 7-31 Detailing of boundary element.

Example 7-5 Repeat Example 7-4 for this example, using special shear walls instead of ordinary shear walls. For convenience, the problem is stated again in this section.

The residential building shown in Example 7-5, Fig. 1, is a bearing wall–type building subjected to a horizontal earthquake excitation in direction y. The building is six stories with 200 mm solid one-way slabs spanning in direction x. The slabs are supported by ten shear walls as shown in the same figure. Each wall has a 300 mm width with the following material properties:

$$f_c' = 30 \text{ MPa}, \quad f_y = 420 \text{ MPa}, \quad I_{\text{eff}} = 0.5I_g$$

The following loading has to be considered in the design of the building:

1. Own weight (slabs and shear walls)
2. Superimposed dead load: $q_{\text{SDL}} = 2 \text{ kN/m}^2$
3. Live load: $q_L = 3 \text{ kN/m}^2$
4. Wind load: $q_w = 1 \text{ kN/m}^2$
5. Seismic loading according to the IBC with the following parameters:
   a. Mapped acceleration coefficient at short period and at 1-s period equal to $S_s = 0.50$ and $S_1 = 0.12$, respectively
   b. The site is classified as Soil Type C
   c. The long-period transition period, T_L, equals 8 s

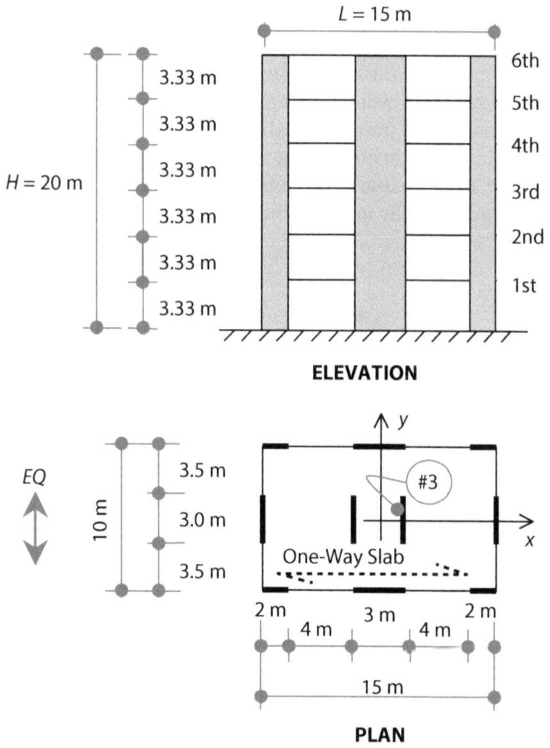

EXAMPLE 7-5, FIGURE 1 Building layout.

If the gravity loads are assumed to be totally carried by the walls in direction *y* and their reactions on the walls are uniformly distributed over the wall cross section, then design one interior shear wall (#3 on the plan) using ACI 318 provisions for SSW.

Solution Special bearing shear walls are assigned an *R*-value of 5 in the IBC as given in Chap. 6. However, for comparison purposes, and in order to avoid repetition of the same calculations in Example 7-4, an *R*-value of 4 will be used in this example. The selection of an *R*-value of 4 for special bearing shear walls is conservative. After all, we design structures to meet minimum requirements for seismic effect.

The basic design procedures will repeat exactly steps 1 through 18 of Example 7-4. The solution of this example begins with step 19.

19. Length of boundary elements

The length of the boundary element is a function of the neutral axis depth, *c*, as shown in Example 7-5, Fig. 2. Therefore, the depth, *c*, needed is calculated by using P_u as required by ACI.

To calculate *c*, the equilibrium of forces shown in Example 7-5, Fig. 2, requires that

$$P_u = C_c + C_s - T$$

where C_c and C_s are compression in concrete and steel, respectively, and *T* is the tension force in the tension steel.

$$P_u = 0.85 f_c' ab + A_s' f_s' - A_s f_y$$

In the equation above, the quantity *a* and the stress in the compression steel are found either by iteration or by considering the compatibility and equilibrium equations as shown in Chap. 5. However, it will be easier and more conservative to neglect the compression steel because this neglect produces higher values of *c*. (This is obvious from the equation above.) Because shear reinforcement is ignored in this calculation, ignoring the compression steel is considered a reasonable compromise. The reader is encouraged to do refined analysis and make comparisons. By ignoring the compression steel, the above equation reduces to

$$P_u = 0.85 f_c' ab - A_s f_y$$

$$3{,}805 \times 10^3 = 0.85(30)(a)(300) - (4{,}500)(420)$$

hence,

$$a = 744 \text{ mm}$$

$$c = \frac{a}{\beta_1} = \frac{744}{0.85} = 875 \text{ mm}$$

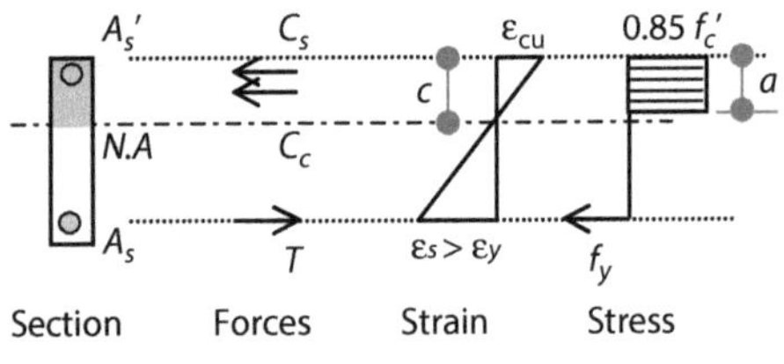

Example 7-5, Figure 2

Therefore, the length of the boundary element,

$$\begin{aligned}\ell_b &= c - 0.1\ell_W \\ &= 875 - 0.1(3{,}000) = 575 \text{ mm} \\ &\geq \frac{c}{2} = \frac{875}{2} = 438 \text{ mm}\end{aligned}$$

Therefore, use $\ell_b = 580\,\text{mm}$.

20. Check the need for SBE

    Using the stress method

$$A_g = 300\,(3{,}000) = 900{,}000 \text{ mm}^2$$

$$I_g = \frac{300\,(3{,}000)^3}{12} = 675{,}000 \times 10^6 \text{ mm}^4$$

$$f_{\text{cm}} = \frac{p_u}{A_g} + \frac{M_u}{I_g} X_M$$

$$f_{\text{cm}} = \frac{3{,}805 \times 10^3}{900 \times 10^3} + \frac{2{,}507 \times 10^6}{675{,}000 \times 10^6}(1{,}500)$$

$$f_{\text{cm}} = 4.23 + 5.57 = 9.8 \text{ MPa}$$

Since

$$f_{\text{cm}} = 9.8 > 0.20 f_c' = 0.20\,(30) = 6 \text{ MPa}$$

then SBE is needed.

The SBE must extend over the area where the stress is in excess of $0.15 f_c' = 0.15\ (30) = 4.5$ MPa. Analysis of the, f_{cm}, equation above reveals that the axial load alone produces, 4.25 MPa. Therefore, the SBE must extend over the entire wall (i.e., $h_b = 20$ m).

21. Detailing of SBE

    Minimum hoops:

    The maximum spacing of the transverse reinforcement is given as

$$\begin{aligned}s_{\max} &= 6\,d_b = 6(25) = 150 \text{ mm} \\ &\leq \frac{b}{4}, \frac{h}{4} = \frac{300}{4} = 75 \text{ mm} \\ &\leq s_o\end{aligned}$$

where

$$\begin{aligned}s_o &= 100 \text{ mm} + \left(\frac{350 - h_x}{3}\right) \\ &\leq 150 \text{ mm} \\ &\geq 100 \text{ mm}\end{aligned}$$

Because s_o must be larger than 100 mm, the 75 mm calculated for spacing, s, is the controlling factor, and there is no need to calculate, s_o.

Maximum spacing of hoops inside the SBE is 75 mm.

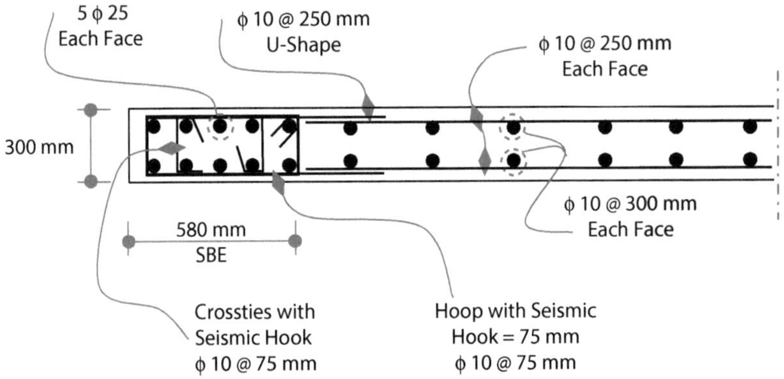

Example 7-5, Figure 3 Special shear wall detailing.

Minimum amount of hoops:

- In the transverse direction

$$A_{sh} \geq 0.09 s b_c \frac{f'_c}{f_{yt}} = 0.09(75)(600)\left(\frac{30}{420}\right)$$
$$= 289\,\text{mm}^2\ (4\phi 10)$$

- In the longitudinal direction: Use the same equation above or by proportionality to dimensions

$$A_{sh} = 289\left(\frac{260}{600}\right) = 125\ \text{mm}^2\ (2\phi 10)$$

Details are shown in Example 7-5, Fig. 3.

7.6.2 Special Shear Walls with Openings

Because the presence of openings in shear walls creates regions of disturbed stress flow, the wall cannot be treated as a single beam element as in the case of shear walls without openings. The exact solution will be difficult and may not be available. However, engineering solutions do exist in practice and are given in forms of modeling.

Two popular schemes of modeling shear walls are the finite element model and the equivalent frame model. Figure 7-32 shows the concept of such modeling.

In finite element modeling, the wall is divided into finite elements using either the plate element or the more popular shell element. Most structural software programs offer such elements in their libraries. Even though finite element models are considered similar to an exact solution if the material properties can be correctly included, these types of models generate tremendous amounts of data to process. This makes the finite element model less attractive in engineering practice.

The equivalent frame model involves less modeling effort and provides reasonable results with engineering a sense of wall behavior. Internal moments, shears, normal forces, and other straining actions may be produced by such models.

Nevertheless, ACI 318 requires additional precautions for walls with openings in terms of shear strength and protection of wall segments between openings. ACI 318 defines the region between two horizontal openings or between an opening and the

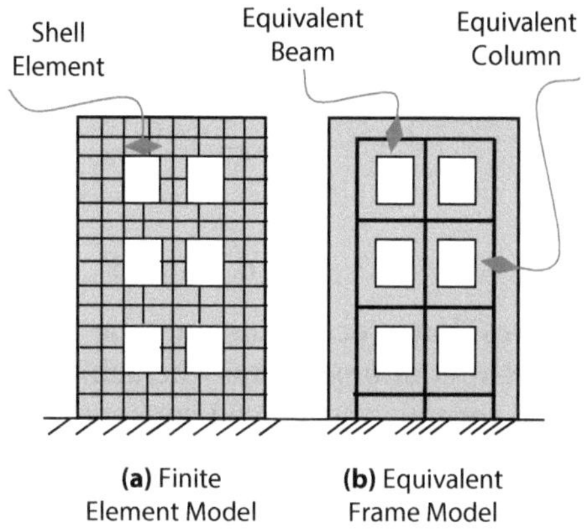

FIGURE 7-32 Modeling schemes of shear walls.

edge of the wall as a vertical segment, or a pier, as shown in Fig. 7-33. In addition, the code defines the region between two vertical openings or between an opening and the edge of the wall as horizontal segments. This is also shown in the same figure.

If the horizontal cross-sectional area of the pier is designated as A_{cp} such that

$$A_{cp} = b_p \ell_p$$

then the following quantity is defined as a reference for further calculations

$$V_{cp} = 0.165\sqrt{f_c'}A_{cp}$$

The equations for walls without openings given in the previous section still apply to walls with openings. For these, we replace the ratio (h_w/ℓ_w) in the given equations with the maximum ratio of the wall, and each of its segments (h_w/ℓ_w) becomes the following

$$\frac{h_w}{\ell_w} = \max\left(\frac{h_w}{\ell_w}, \frac{h_p}{\ell_p}\right)$$

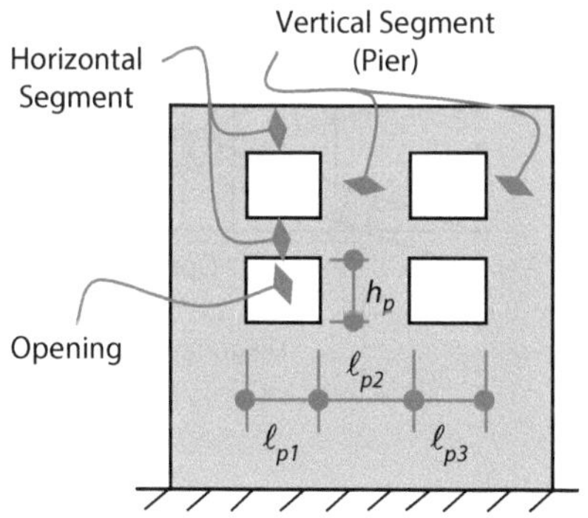

FIGURE 7-33 Openings in shear walls.

Additional requirements for walls with openings are given in terms of maximum shear strength such that

$$V_{n,\text{pier}} \leq 5V_{\text{cp}}$$

and

$$\Sigma(V_n)_{\text{piers}} \leq 4\,\Sigma V_{\text{cp}}$$

The code also requires that precautions must be taken to assure sound force path around openings. Such requirements may be achieved, for example, using capacity design concepts. The recently introduced concept of the strut-and-tie model in the ACI 318 may also help in this area. For this purpose, the next section covers special detailing for horizontal piers, which are defined as coupling beams.

7.7 Coupling Beams

Experience from previous earthquakes indicates that the horizontal segments defined in the previous section are susceptible to shear failure. Many examples in many earthquakes reflect such vulnerability of these elements. For this reason, the code requires that these elements, known as coupling beams, be reinforced with special detailing with or without crossing diagonals as shown in Fig. 7-34.

With or without diagonals, the detailing of the coupling beam shall satisfy the requirements of SMF as given earlier. Because the diagonals are not effective in shallow coupling beams, the diagonals may only be used as follows.

Diagonal trusses shall not be used if

$$\frac{\ell_n}{d} \geq 4$$

Diagonal trusses may be used as required by analysis if

$$\frac{\ell_n}{d} < 4$$

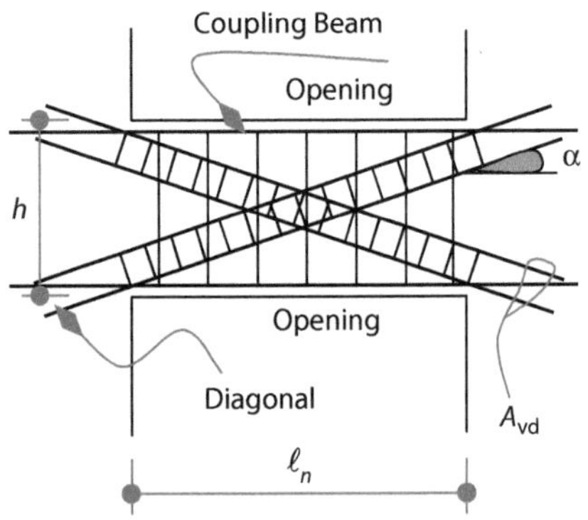

FIGURE 7-34 Coupling beams.

Diagonal trusses shall be used if

$$\frac{\ell_n}{d} < 2$$

and if

$$V_u > 2V_{cw}$$

where d = effective depth of coupling beam
ℓ_n = clear span of coupling beam as in Fig. 7-34
V_{cw} = a reference value for shear strength defined as
$= 0.165\sqrt{f_c'}\,A_{cw}$
A_{cw} = concrete cross-sectional area of the coupling beam
$= b_w h$, (b_w = width of coupling beam)

7.7.1 Detailing of Coupling Beams with Diagonals

The detailing of coupling beams with diagonals is given for both the coupling beams and the diagonals as follows.

Diagonals

Historically, diagonals are required to be enclosed in hoops with minimum dimensions given in terms of the coupling beam width, b_w, as shown in Figs. 7-35 and 7-36. Due to difficulty of fabrication and construction, ACI, provide alternative diagonals without hoops that can be used. In this book, diagonals without hoops will be presented, leaving diagonals with hoops for the reader to see in ACI if interested.

The nominal shear strength of the two diagonals may be calculated as follows:

$$V_n = 2A_{vd} f_y \sin\alpha$$
$$\leq 5\,V_{cw}$$

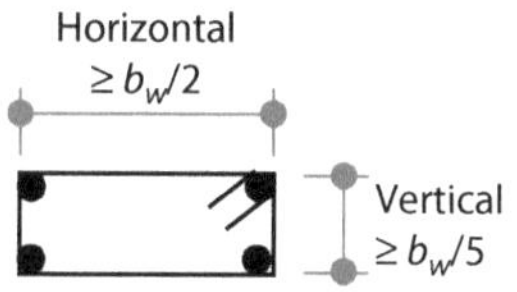

FIGURE 7-35 Dimensions of diagonals.

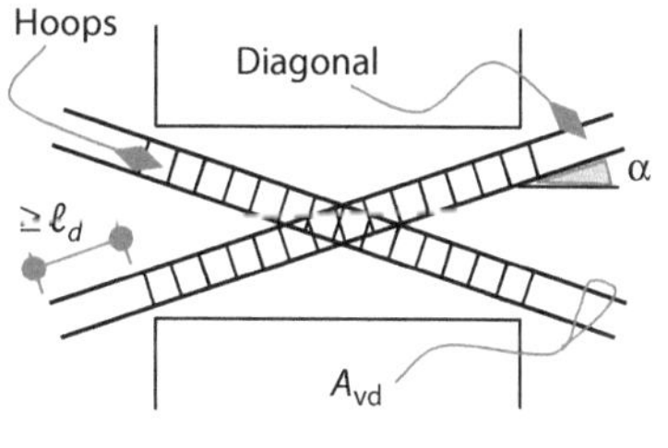

FIGURE 7-36 Coupling beams.

where α = the horizontal angle shown in Fig. 7-36
A_{vd} = area of longitudinal bars as shown in Fig. 7-36
f_y = yield strength of reinforcement
V_{cw} = shear strength reference quantity as defined earlier

The diagonals must be placed in two directions symmetric about the center of the coupling beam. Each direction must have at least four bars placed in two layers or more. The diagonals must be embedded in the wall with development length calculated using yield stress equals to $1.25f_y$, that is,

$$(\ell_d)_{\text{diagonals}} = 1.25\,(\ell_d)_{\text{ordinary cases}}$$

Coupling Beams

The minimum transverse reinforcement of coupling beams with diagonals are given similar to SMF, that is,

$$A_{sh} \geq 0.3\left(\frac{sb_c f_c'}{f_{yt}}\right)\left(\frac{A_g}{A_{ch}} - 1\right)$$
$$\geq 0.09\left(\frac{sb_c f_c'}{f_{yt}}\right)$$

The maximum spacing of the horizontal reinforcement, s_2, is govern by the presence of hoops as shown in Fig. 7-37(a).

$$s_2 \leq 200 \text{ mm}$$

The maximum spacing of the vertical reinforcement (hoops), s, is governed by the size of the diagonals bar size, $d_{b,\text{diag}}$, as shown in Fig. 7-37(b).

$$s \leq 6\,d_{b,\text{diag}}$$
$$\leq 150 \text{ mm}$$

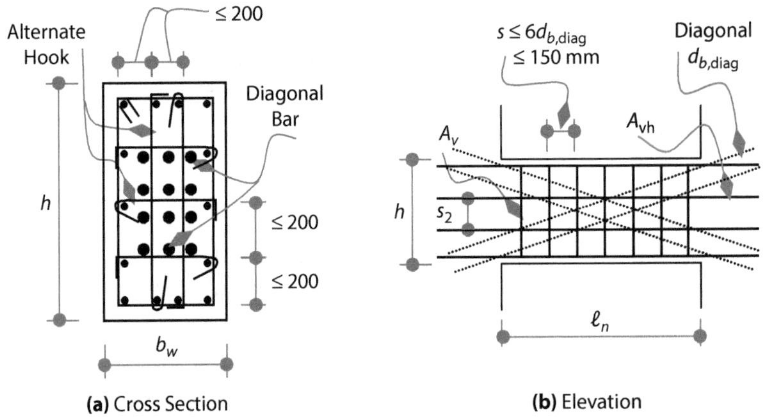

FIGURE 7-37 Coupling beams details.

Recall also that deep beam requires minimum horizontal and vertical reinforcement as follows:

$$A_v = 0.0025\, b_w s$$

$$A_{vh} = 0.0025\, b_w s_2$$

7.8 Diaphragms and Trusses

Diaphragms are horizontal slabs that function to distribute the earthquake-induced forces to the framing system.

7.8.1 Structural System

Diaphragms may be considered rigid and behave as a beam in their plane if they meet minimum thickness requirements. The components of rigid diaphragms are shown in Fig. 7-38, which shows a plan of a diaphragm with an earthquake force acting as shown in the same figure. The edges of the diaphragm perpendicular to the earthquake excitation are defined as struts, chords, ties, or boundary elements. These boundary elements act as flanges of a steel I-section under flexure action. They are designed to resist force, P_u, that is given as follows:

$$P_u = \frac{Q_u}{2} + \frac{M_u}{\ell}$$

where M_u = maximum bending moment resulting from simple beam action in resisting the seismic force as shown in Fig. 7-38

Q_u = any acting force perpendicular to earthquake excitation as shown in Fig. 7-38

The rigidity of the diaphragm depends on the flooring system. For example, if the diaphragm consists of a solid slab as shown in Fig. 7-39 (a), the minimum thickness of the diaphragm to be considered rigid is 50 mm.

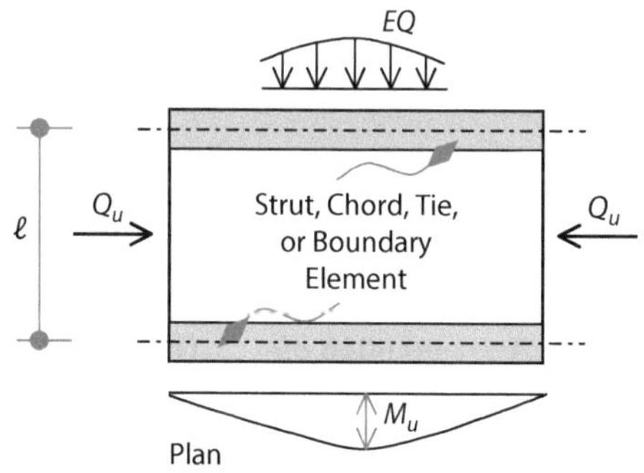

Figure 7-38 Rigid diaphragm action.

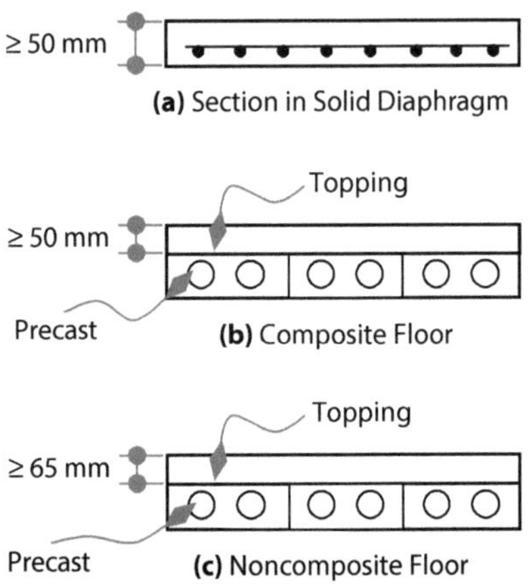

FIGURE 7-39 Rigid diaphragm action.

If the flooring system consists of topping slabs on precast floors, the minimum thickness of the topping slabs in the case of composite action is 50 mm, whereas the minimum thickness of the noncomposite action is 65 mm as shown in Fig. 7-39(b) and (c).

If the diaphragm does not meet the rigidity requirements, the diaphragm is considered flexible and must be modeled with the structure to include its behavior. As an example, the flexible diaphragm may be modeled as a plate or shell element using finite element models.

7.8.2 Shear Strength

The shear strength capacity of the diaphragm is limited to maximum values as follows:

$$V_n \le 4V_{cv}$$

where V_{cv} = the reference shear strength defined for shear walls
V_n = nominal shear strength of the diaphragm, which is given as follows.

$$V_n = A_{cv}(0.165\sqrt{f_c'} + \rho_t f_y) \quad \text{(for solid diaphragms)}$$

$$V_n = A_{cv}\,\rho_t f_y \quad \text{(for topping diaphragms)}$$

The equations above indicate that the concrete shear capacity, V_c, is ignored for topping. The minimum reinforcement of the diaphragm in its web is given as the same for slabs

$$A_{s,\min} = 0.0018\,bh \quad (\text{for } f_y = 420 \text{ MPa})$$

$$A_{s,\min} = 0.0020\,bh \quad (\text{for } f_y = 300 \text{ MPa})$$

With maximum spacing: $s \le 450$ mm

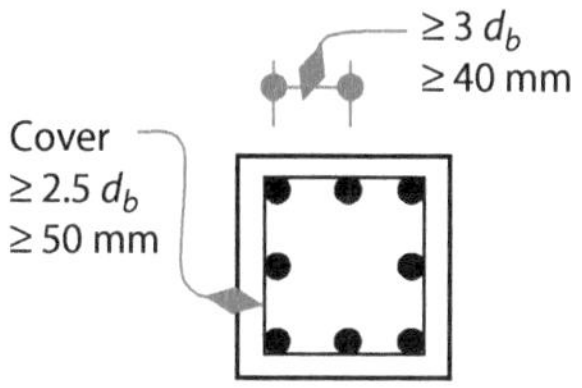

FIGURE 7-40 Chord details.

7.8.3 Diaphragm Chords and Truss Members

As explained earlier, the diaphragm structural system is considered a beam with its chords acting as a flanged section. Therefore, the chords will be subjected to axial force, P_u, as given in the previous section.

The chord is designed as a column (truss member) to resist the force, P_u, with minimum hoops as follows:

$$s \leq \frac{A_v f_{yt}}{0.35\, b_w}$$

Additional detailing requirements are needed if the stress in the chord under the force P_u, and using E and A_g, exceeds $0.20 f_c'$. This condition is similar to the SBE requirement in special shear walls.

If the stress exceeds the $0.20 f_c'$, then minimum lateral reinforcement and spacing requirements shall follow the details of special beam-columns of SMF as given earlier.

Additional spacing limitations are given as shown in Fig. 7-40.

The requirements above also apply to truss members.

7.9 Foundations

The design of foundations shall follow the general procedures provisions for regular design. As failure of foundations is catastrophic, they are considered to be part of the critical path in the force transfer system and must be designed to be stronger than the superstructure. In other words, the plastic hinges in the structure must be forced to form in the columns and walls while keeping the foundations in the elastic range. This philosophy is similar to shear design in the main members. Consequently, the requirements and detailing of foundations may be given as follows.

7.9.1 Strength Requirements

In order to force the plastic hinge in the column, the foundations must be designed for the probable moment, M_{pr}, as shown in Fig. 7-41(a). The probable moment is given as defined in SMF as follows:

$$M_{pr} = f(\geq 1.25 f_y)$$

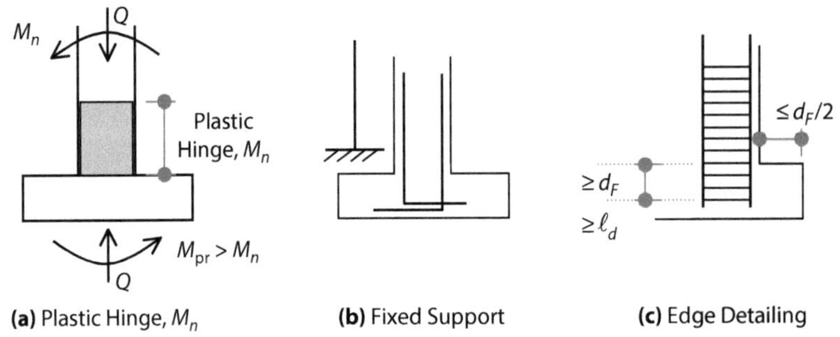

Figure 7-41 Foundation detailing.

7.9.2 Detailing Requirements

Special detailing is required for the connections between the columns and the foundations, which are given as follows:

1. In fixed supports of columns, the reinforcement must be bent toward the center of the foundation. Such detail is shown in Fig. 7-41(b).
2. If a member with special detailing lies within a distance ($d_F/2$) as shown in Fig. 7-41(c), the special detailing must continue into the foundations at least a distance of d_F, where d_F is the effective depth of the foundation.
3. If uplift forces exist as in pile foundations, the top reinforcement of the foundation as shown in Fig. 7-42 must satisfy minimum reinforcements of beams.

$$\rho_{o,\min} = \frac{\sqrt{f'_c}}{4 f_y} \geq \frac{1.4}{f_y}$$

The minimum reinforcement of foundations follows the provisions of slabs. Therefore, the requirement above is more stringent than regular foundation requirements.

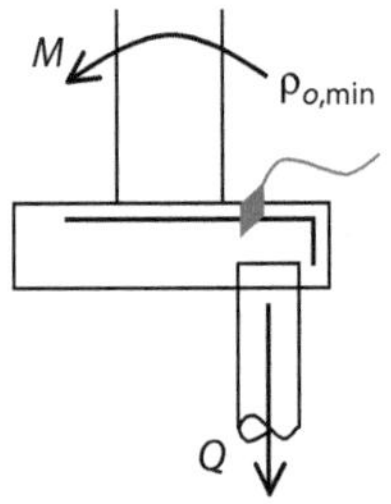

Figure 7-42 Uplift force detailing.

7.10 Precast Concrete

For many years, precast and prestressed concrete was excluded from seismic codes. During the past, there was a general perception that precast concrete was not suited for seismic resistance because of its weak connectivity. As a result, it would not hold together during ground shaking. As discussed in Chap. 5, research and experimental work prove that precast concrete can be used for seismic resistance if it is designed to behave according to the philosophy of seismic behavior and design.

ACI 318 first introduced precast concrete as seismic-resistant elements in its 2002 edition. ACI provisions present three precast systems that can be used as earthquake-resistant structures: precast special moment frames, precast intermediate shear walls, and precast special shear walls.

7.10.1 Precast Special Moment Frames

To qualify as SMF, special moment precast frames must satisfy all the requirements of the cast-in-situ SMF. In addition, precast frames are classified as having either ductile connections or strong connections. The ductile connections are intended to behave as monolithic concrete, whereas strong connections are intended to remain elastic during plastification of the member during the development of mechanisms. The requirements of each type of these connections are given in the following sections.

Precast Special Frames with Ductile Connections

In precast special frames with ductile connections, the shear friction capacity of the connections, V_n, shall exceed twice the maximum probable shear force, V_e, that can develop at the face of the connection

$$V_n \geq 2V_e$$

where V_e = the maximum probable shear that can develop in the section as defined for the special beams and special beam-columns in earlier sections

V_n = nominal shear friction capacity, which is calculated according to ACI provisions as follows:

$$V_n = A_{vf} f_y \mu$$

and μ is given for normal weight concrete according to Table 7-1

Concrete placed monolithically	1.4
Concrete placed against hardened concrete with surface intentionally roughened	1.0
Concrete placed against hardened concrete not intentionally roughened	0.6
Concrete anchored by studs or stud-like devices	0.7

TABLE 7-1 Coefficient of Friction for Normal Weight Concrete (μ)

Precast Special Frames with Strong Connections

In precast special frames with strong connections, the following additional requirements must be satisfied:

1. The span depth ratio must exceed four times for regions of potential plastic hinge developments.
2. Primary reinforcement shall be continuous across the connections and must be developed outside both strong connections and potential plastic hinge regions.
3. Connection strength shall satisfy the following:
   a. Beam-to-beam connections

$$\phi S_n \geq S_e$$

where S_e = maximum probable force in flexure, shear, or axial that can develop in the connections due to design mechanism of inelastic behavior
S_n = nominal strength of the connections in flexure, shear, or axial
ϕ = strength reduction factor as applicable

   b. Column-to-column connections:

$$\phi S_n \geq 1.4\, S_e$$

where S_e, S_n, and ϕ are as defined above.

In addition, the design moment ϕM_n and the design shear strength ϕV_n of the connections shall not be less than the following:

$$\phi M_n \geq 0.4\, M_{\text{pr}} \quad \text{(of the column within its height)}$$

$$\phi V_n \geq V_e \quad \text{(as determined for cast-in-place special frames)}$$

7.10.2 Precast Intermediate Shear Walls

The IBC does not define precast intermediate shear walls in its seismic systems. However, ACI 318 defines precast intermediate shear walls in its seismic section without any reference on their use. The only requirements given by ACI 318 for precast intermediate shear walls are for the connections between the panels and also between the panels and the foundations.

For design philosophy where yielding is intended in the connections, the code restricts yielding to reinforcement or steel elements. For design philosophy where connections are not intended to yield, the connections shall develop strength, S_n, such that

$$S_n \geq 1.5\, S_y$$

where S_y is the yield strength of the connection based on f_y for moment, shear, or axial load.

7.10.3 Precast Special Shear Walls

Precast special shear walls shall satisfy all the requirements of both cast-in-place special shear walls and precast intermediate shear walls as laid out in previous sections.

7.11 Nonseismic-Resisting Systems

Because the entire structure undergoes maximum inelastic displacements, Δ_M, all structural members including those not intended for earthquake resistance must be capable of accommodating such large displacements. Therefore, the nonseismic elements must also be detailed to sustain these large seismic demands. In addition, the nonseismic elements will also be subjected to a large P-Δ effect, which must be addressed.

7.11.1 General Requirements (A)

The general requirements in this section are required if detailed analysis of the effect of inelastic displacements are not considered. These requirements must be met if the induced moments and shears due to δ_u exceed the design strengths

$$\begin{Bmatrix} M \\ V \end{Bmatrix}_{\delta u} > \begin{Bmatrix} M \\ V \end{Bmatrix}_n$$

Beam Requirements

If the gravity-factored axial loads do not exceed $0.1 f_c' A_g$, then

$$P_u \le 0.1 f_c' A_g$$

where f_c' and A_g are as defined before.

The maximum main reinforcement ratio shall be limited to

$$\rho_{o,\max} = 0.025$$

Beam-Column Requirements

If gravity-factored axial loads exceed $0.1 f_c' A_g$, then

$$P_u > 0.1 f_c' A_g$$

where f_c' and A_g are as defined before.

The member must be detailed as in special beam-columns in terms of minimum hoops areas, maximum spacing, and connection requirements. Such detailing is given under the section of special moment resisting frames.

7.11.2 General Requirements (B)

The general requirements given in (A) may be relaxed if the induced moments and shears that are due to δ_u do not exceed the design strengths that is

$$\begin{Bmatrix} M \\ V \end{Bmatrix}_{\delta u} < \begin{Bmatrix} M \\ V \end{Bmatrix}_n$$

The following requirements are relaxed.

The maximum spacing of lateral reinforcement over the entire length of the member shall be as follows:

If $P_u > 0.1 f_c' A_g$,
then $s \le s_o$
$s \le 6\, d_b$
$s \le 150$ mm

If $P_u > 0.35\, P_o$,
then $s \le s_o$;

and the minimum hoops are relaxed to the following:

Rectangular Sections

The minimum area of hoops, A_{sh}, in any direction of the rectangular section shall be as shown below:

$$A_{sh} \ge \frac{1}{2}\left\{0.3\left(\frac{sb_c f_c'}{f_{yt}}\right)\left(\frac{A_g}{A_{ch}} - 1\right)\right\}$$

$$\ge \frac{1}{2}\left\{0.09\left(\frac{sb_c f_c'}{f_{yt}}\right)\right\}$$

where A_{ch} = area of concrete core including hoops, which is given as $(b_{c1} + d_h) \cdot (b_{c2} + d_h)$
A_g = gross area of the concrete section
A_{sh} = area of hoops for shear
b_c = center-to-center dimension of the hoop in the direction under consideration as shown in Fig. 7-19
d_h = diameter of hoops
f_c' = characteristic strength of concrete
f_{yt} = yield stress of hoop reinforcement
s = spacing of hoops along the member

Circular and Spiral Sections

The minimum volume of hoops, ρ_s, in circular sections with hoops or with spirals shall be as shown below.

$$\rho_s \ge \frac{1}{2}\left(0.12\frac{f_c'}{f_{yt}}\right)$$

$$\ge 0.45\left(\frac{A_g}{A_{ch}} - 1\right)\frac{f_c'}{f_{yt}}$$

where A_{ch} = area of concrete core, including hoops
A_g = gross area of the concrete section
f_c' = characteristic strength of concrete
f_{yt} = yield stress of hoop reinforcement

APPENDIX 7-1

Sheet 1

Sheet 1 helps the design of singly reinforced concrete rectangular beams and slabs subjected to pure bending moment.

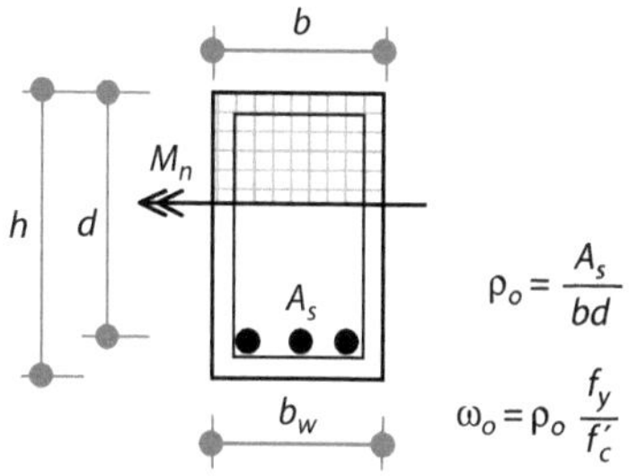

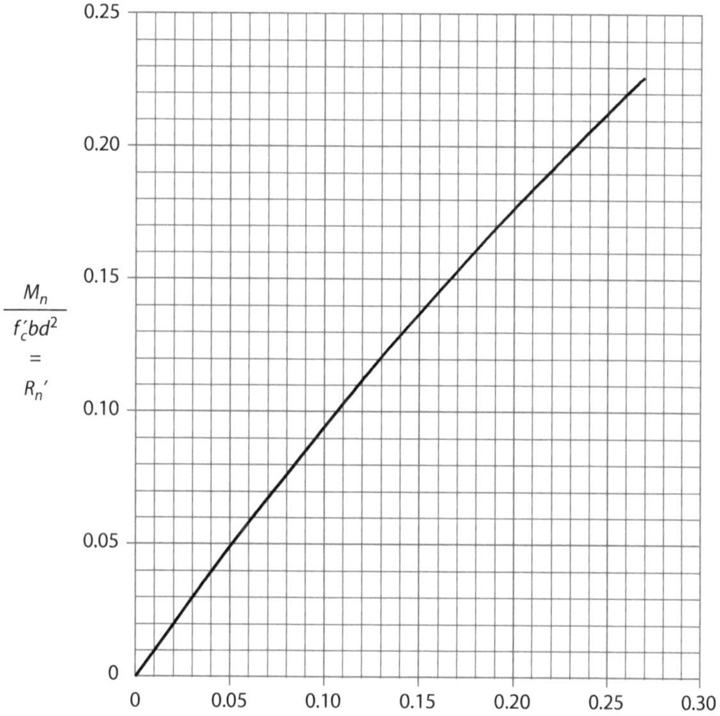

$$\frac{M_n}{f'_c bd^2} = \omega_o(1 - 0.59\omega_o)$$

$$\omega_o = 0.85 - \sqrt{(0.85)^2 - 0.85(2)R'_n}$$

$$A_s = \omega_o \frac{f'_c}{f_y} bd$$

$$d_{conv} = 2.5\sqrt{\frac{M_n}{f'_c b}}$$

$$A_{s,conv} = 0.18\frac{f'_c}{f_y} bd_{conv}$$

Sheet 2

Reinforcement properties and details.

f_y	f'_c	E_c	β_1	$\rho_{w,min}$	$\omega_{w,min}$	$\rho_{w,max}$	$\omega_{w,max}$	$\rho_{w,max}$	$\omega_{w,max}$
MPa	MPa	MPa	—	Ratio		at $\varepsilon_t = 0.005$ at 0.32 b_1		at $\varepsilon_t = 0.004$ at 0.36 b_1	
280	20	21,000	0.85	0.0050	0.070	0.0194	0.272	0.0218	0.306
	25	23,500	0.85	0. 0050	0.056	0.0243	0.272	0.0273	0. 306
	30	25,750	0.85	0. 0050	0.047	0.0291	0.272	0.0327	0. 306
	35	27,800	0.81	0.0053	0.043	0.0324	0.259	0.0364	0.291
420	20	21,000	0.85	0.0033	0.070	0.0129	0.272	0.0145	0.306
	25	23,500	0.85	0.0033	0.056	0.0161	0.272	0.0182	0.306
	30	25,750	0.85	0.0033	0.047	0.0194	0.272	0.0218	0.306
	35	27,800	0.81	0.0035	0.043	0.0216	0.259	0.0243	0.291

Reinforcement Limitations in Slabs			
f_y MPa	ρ_{conv} Ratio	$\rho_{w,min}$ (See Below)	$\rho_{w,max}$ Ratio
300	Same as beams	$A_{s,min} = 0.0020$ bh	Same as beams
420	Same as beams	$A_{s,min} = 0.0018$ bh	Same as beams

Reinforcement Schedule												
Bar	**Dia**	**Mass**	**Nominal Area, mm²**									
ϕ	**mm**	**kg/m**	**1**	**2**	**3**	**4**	**5**	**6**	**7**	**8**	**9**	**10**
ϕ 6	6	0.222	28	56	84	112	140	168	196	224	252	280
ϕ 8	8	0.395	50	100	150	200	250	300	350	400	450	500
ϕ 10	10	0.617	78	156	234	312	390	468	546	624	702	780
ϕ 12	12	0.888	113	226	339	452	565	678	791	904	1017	1130
ϕ 14	14	1.210	154	308	462	616	770	924	1078	1232	1386	1540
ϕ 16	16	1.580	201	402	603	804	1005	1206	1407	1608	1809	2010
ϕ 18	18	2.000	254	508	762	1016	1270	1524	1778	2032	2286	2540
ϕ 20	20	2.470	314	628	942	1256	1570	1884	2198	2512	2826	3140
ϕ 22	22	2.980	380	760	1140	1520	1900	2280	2660	3040	3420	3800
ϕ 25	25	3.850	491	982	1473	1964	2455	2946	3437	3928	4419	4910
ϕ 28	28	4.830	616	1232	1848	2464	3080	3696	4312	4928	5544	6160
ϕ 32	32	6.310	804	1608	2412	3216	4020	4824	5628	6432	7236	8040

Sheet 3

Interaction Diagram, $\gamma = 0.9$

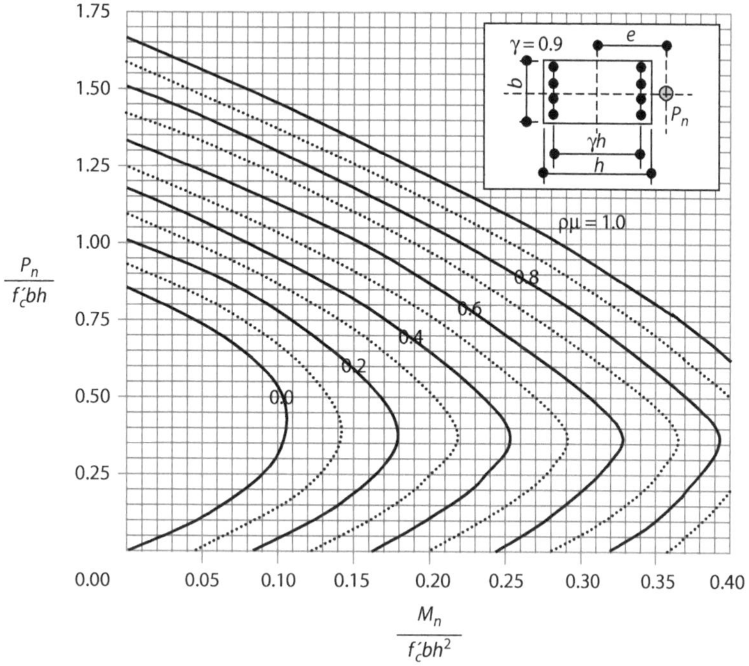

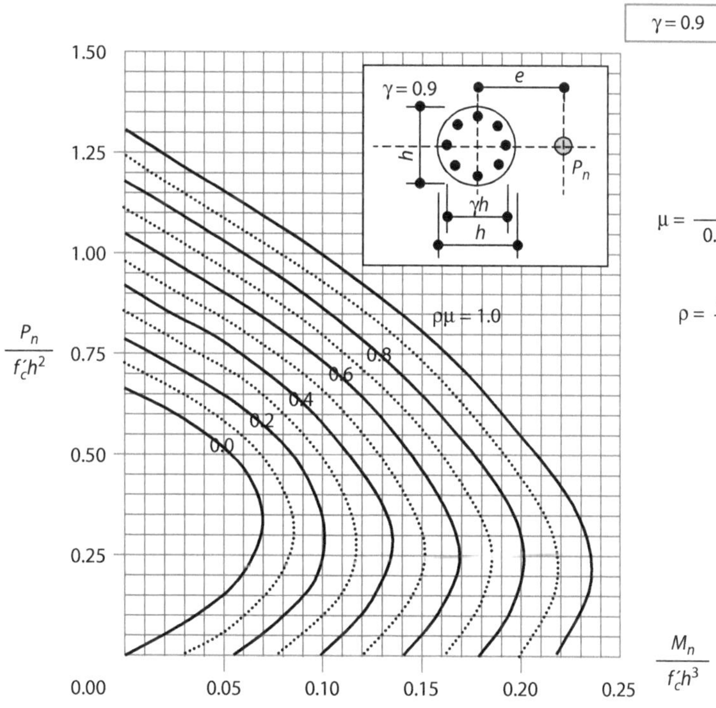

$$\mu = \frac{f_y}{0.85\, f'_c}$$

$$\rho = \frac{A_{st}}{A_g}$$

Sheet 4

Interaction Diagram, $\gamma = 0.8$

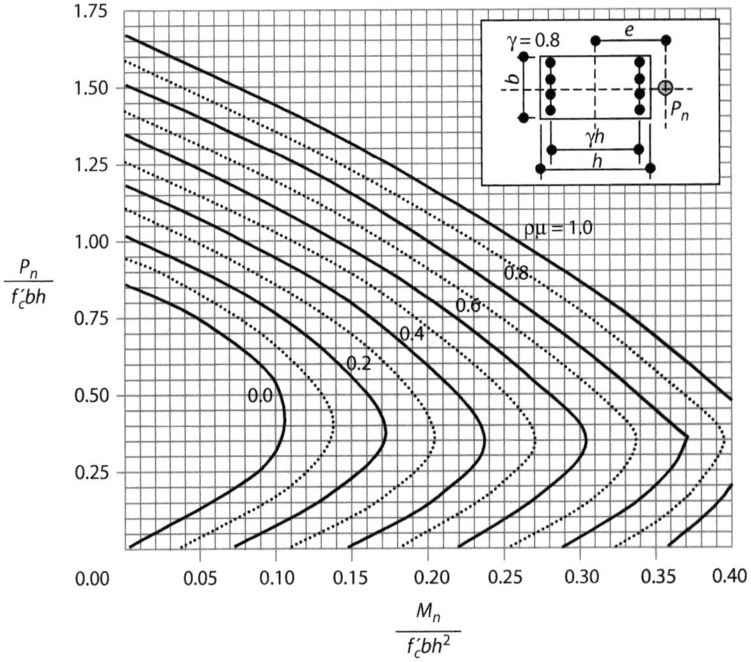

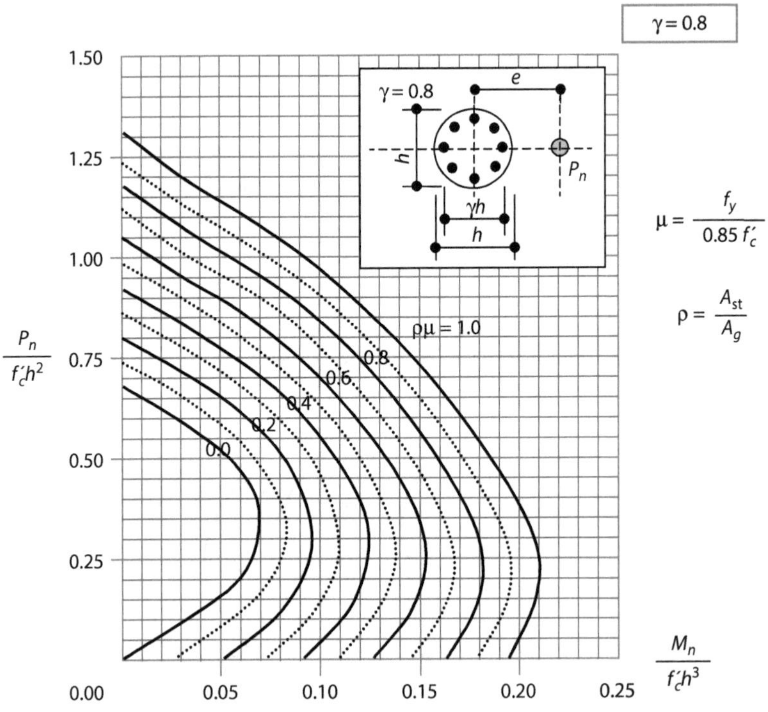

$$\mu = \frac{f_y}{0.85\, f_c'}$$

$$\rho = \frac{A_{st}}{A_g}$$

Sheet 5

Interaction Diagram, $\gamma = 0.7$

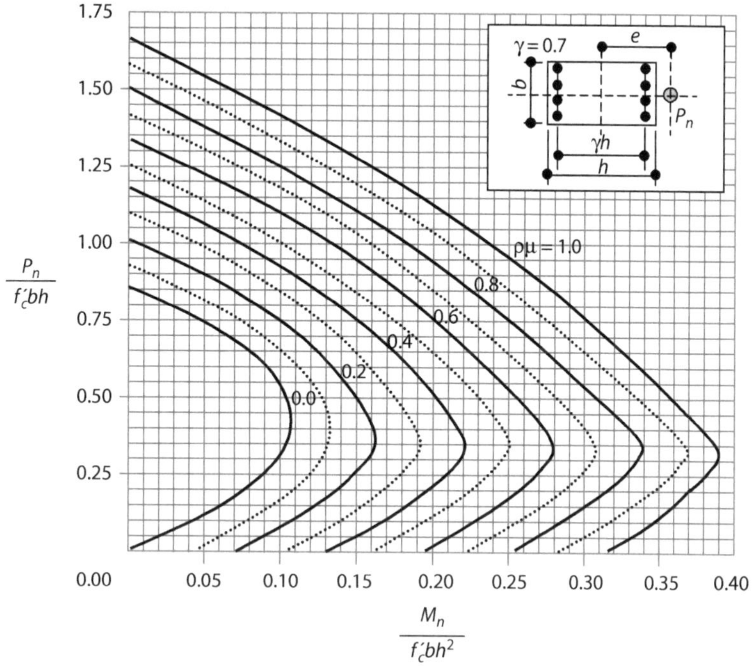

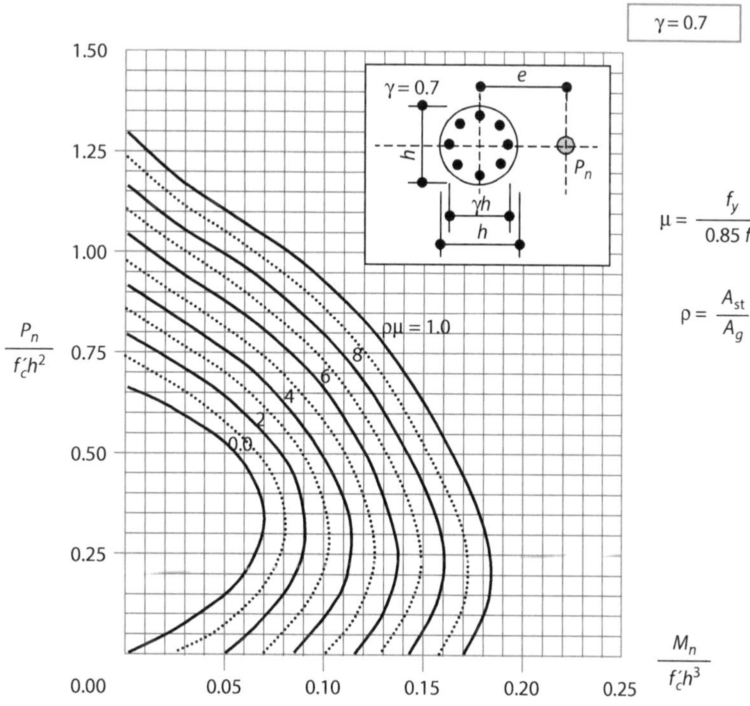

$$\mu = \frac{f_y}{0.85 f'_c}$$

$$\rho = \frac{A_{st}}{A_g}$$

Sheet 6

Interaction Diagram, $\gamma = 0.6$

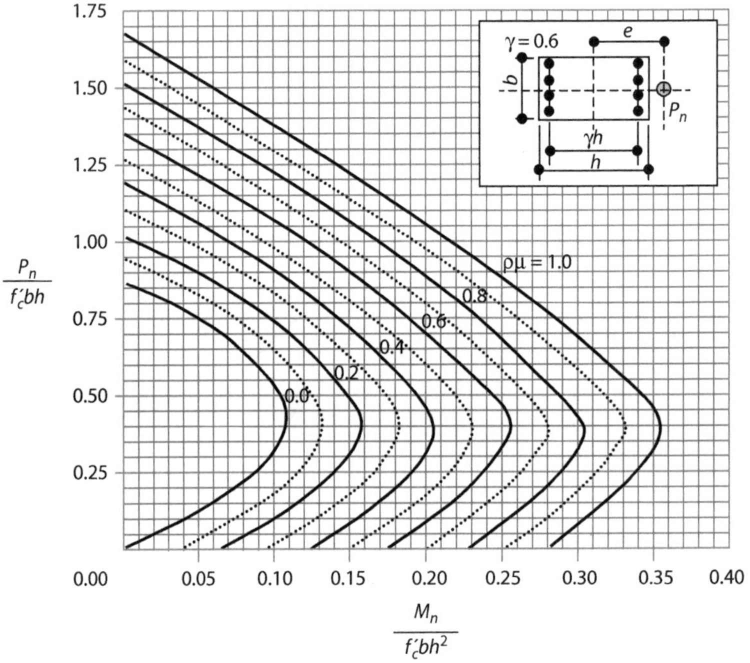

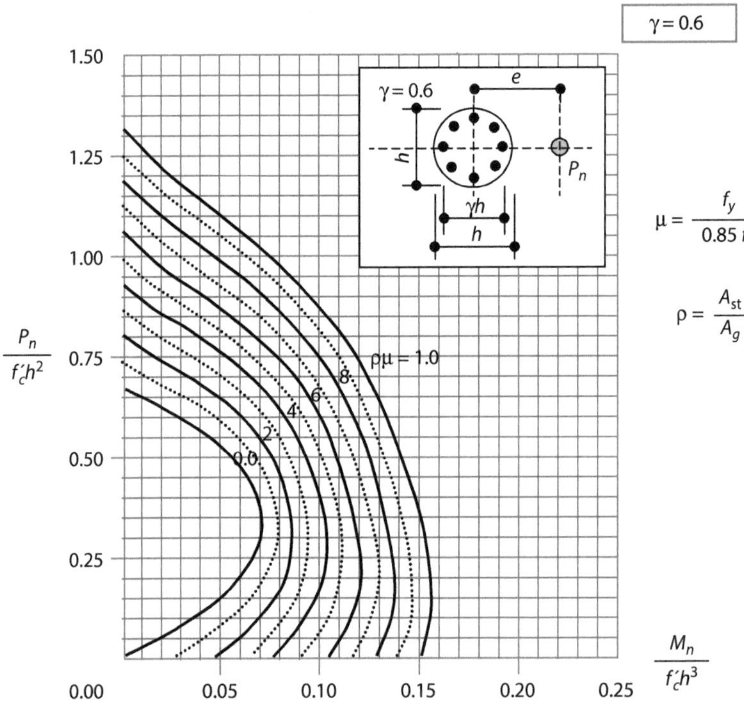

$$\mu = \frac{f_y}{0.85 f'_c}$$

$$\rho = \frac{A_{st}}{A_g}$$

Sheet 7

Effective Length Factor by Alignment Chart

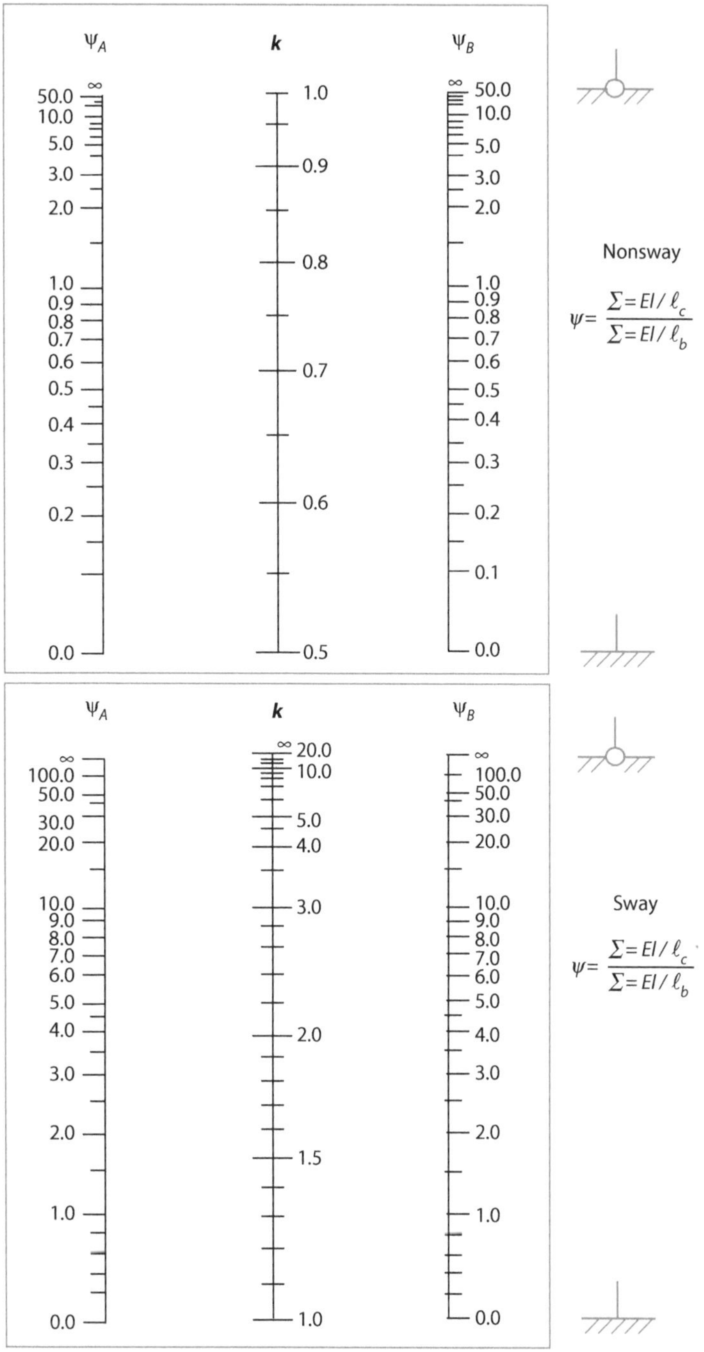

CHAPTER 8

Introduction to AISC Seismic Provisions for Structural Steel Buildings

8.1 Introduction

The excellent ductility and energy dissipation capacity of steel as a material makes it ideal for earthquake-resistant structures. However, as explained in Chap. 5, steel components are often slender owing to the high cost of material. Slender members and components of steel structures exhibit less ductile behavior and premature failures when their members or components lose their stability. To achieve the required ductility and energy dissipation capacity, such premature failures must be prevented, which can be accomplished by special detailing of steel elements, such as bracing and stiffening of elements at local and global levels.

Details of the steel provisions are given by the American Institute of Steel Construction (AISC) in the Seismic Provisions for Structural Steel Buildings (referred to in this chapter as the provisions). These details are so overwhelming that would need a separate book to cover adequately. This chapter is merely a description of the various systems and highlights the major requirements of the provisions. Full description and details of the seismic requirements of the steel systems are found in load and resistance factor design (LRFD) specifications and the supplemental seismic provisions referenced at the end of the book. Seismic requirements are given in separate provisions entitled Seismic Provisions for Structural Steel Buildings, which is an extension to their main LRFD specifications of steel design.

It should be emphasized again that this book does not cover every required detail of the various systems defined by the provisions as mentioned above, and hence, the reader is advised to follow the detailed description of every system as directly given in the provision if needed in real-life practice.

8.2 General Requirements

The AISC standard has general requirements that apply to load combinations, material, and connections for all structural systems. These requirements are summarized as follows:

Load Combinations

The load combinations in the AISC standard are required to follow the load combinations given by the applicable seismic code under consideration. However, when the applicable seismic code does not provide values for the force modification factor, R, and for overstrength factor, Ω_o, AISC requires the use of these values from the American Society of Civil Engineers, ASCE 7.

Material

Steel intended for seismic-resisting elements where inelastic deformations are expected shall be limited to a maximum yield stress, F_y, of 345 MPa. In ordinary systems, the maximum yield stress, F_y, is increased to be limited to 380 MPa. In addition, maximum yield stress, F_y, shall be limited to 450 MPa in columns of some systems as will be indicated later in specific systems.

When required by the standard, the strength of members and connections shall be designed for an expected yield stress, F_{ye}, and expected tensile strength, F_{te}, that is multiplied by amplification factors, R_y and R_t, respectively, such that

$$F_{ye} = R_y F_y$$

$$F_{te} = R_t F_u$$

R_y and R_t are given by AISC for different material types in AISC Table A2.1. Sample of popular values are given in Table 8-A1 at the end of this chapter.

Demand Critical Welds

When designated as demand critical weld, the weld shall be made with filler metals meeting the requirements specified in AWS D1.8/D1.8M Clause 6.3.

These requirements include minimum specified yield strength, minimum specified tensile strength, minimum elongation, and minimum Charpy V-Notch (CVN) toughness of the weld metal as given in Table 8-A2.

Slenderness Requirements

In order to achieve high levels of ductility requirements during an earthquake, steel sections must sustain high levels of plastic deformations without loss of stability. To maintain stability in beams and columns while plastic hinges develop in the structure, slenderness limitations are needed to prevent local buckling and global buckling. Local buckling is prevented by limiting width-to-thickness ratio of section components. Global buckling in beams (lateral torsional buckling) is prevented by adequate bracing. These requirements are given as follows.

Local Buckling It has been a tradition to consider that compact steel sections are capable of developing plastic hinges. Research has shown that the traditional compactness while permits development of plastic hinges in plastic design, does not fulfill the requirement of large ductility demand in seismic regions. Therefore, AISC seismic provisions introduce two levels of compactness for seismic behavior, namely, moderate ductility, and high ductility.

The limiting width-to-thickness ratios to achieve either moderate or high ductility requirements are given in AISC Table D1.1. Examples of these limitations are given in Tables 8-A3 and 8-A4 at the end of this chapter.

Lateral Torsional Buckling In order to achieve ductility requirements, lateral bracing of beams shall be provided with maximum values of L_b for high ductility and moderate ductility as follows.

For moderate ductility,

$$L_b \leq 0.17\, r_y \frac{E}{F_y}$$

For high ductility,

$$L_b \leq 0.086\, r_y \frac{E}{F_y}$$

where E = modulus of elasticity of steel
F_y = yield stress of steel
r_y = radius of gyration about the y-axis (weak axis)

Special Bracing at Plastic Hinge Locations

The top and bottom flanges of the expected plastic hinge location shall be provided with lateral supports at their ends. The lateral supporting elements shall be designed for a force P_u, equals to 6 percent of the flange expected nominal strength as follows:

$$P_u = 0.06 \frac{R_y F_y Z}{h_o}$$

where F_y = specified yield stress of the beam
R_y = yield stress amplification factor given in Table 8-A1
Z = plastic section modulus of the beam
h_o = distance between flange centroids

Protected Zones

Protected zones are defined as areas of members or connections in which limitations to fabrication and attachments apply. These zones are important to the seismic performance of the structure during earthquake excitation and, hence, need to be protected from harmful effect. For example, in protected zones, studs or attachments that penetrate the beam flange are prohibited; piping ducting is also prohibited. These zones will be identified in the systems where applicable.

Column Splices

Column splices location depends on their type. Complete-joint-penetration groove weld splices shall be located at least a distance equals to or greater than

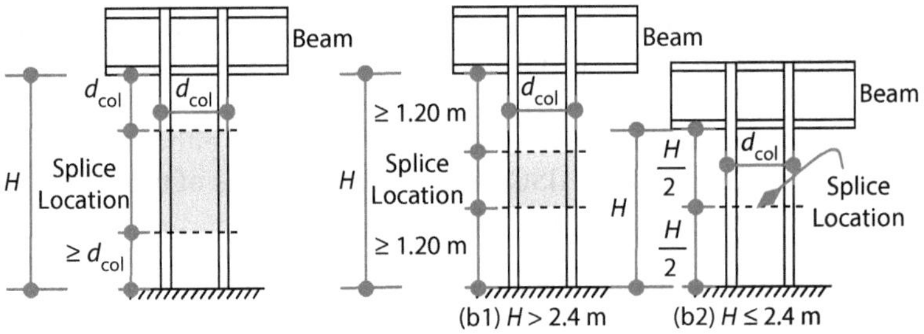

(a) Complete-Joint-Penetration Groove Weld Splices

(b) Other Than Complete-Joint-Penetration Groove Weld Splices

Figure 8-1 Location of column splices.

the column depth as shown in Fig. 8-1(a). Other types of splices shall be located at least at 1.20 m from the beam-to-column flange connections. The splice shall be located in the mid height if the height of the clear column is less or equal to 2.40 m as shown in Fig. 8-1(b).

The column splices axial and flexure strength shall not be less than the strength required by the amplified seismic load combinations.

The column splices shear strength shall not be less than the strength required by the shear developed by the plastic hinging of the least column section, which may be expressed as follows:

$$V_{n,\text{required}} \geq \frac{M_{pc}}{H_c}$$

where H_c is the clear height of the column.

8.3 Structural Systems

Four groups of structural systems, as listed below from G1 to G4, are considered in the AISC standards to resist seismic effect. Each group of these four categories includes subsystems as follows.

G1: Moment-frame systems:

1. Ordinary moment frames (OMF)
2. Intermediate moment frames (IMF)
3. Special moment frames (SMF)
4. Special truss moment frames (STMF)
5. Ordinary cantilever column systems (OCCS)
6. Special cantilever column systems (SCCS)

G2: Braced-frame and shear-wall systems:

1. Ordinary concentrically braced frames (OCBF)
2. Special concentrically braced frames (SCBF)
3. Eccentrically braced frames (EBF)
4. Buckling-restrained braced frames (BRBF)
5. Special plate shear walls (SPSW)

G3: Composite moment-frame systems:

1. Composite ordinary moment frames (C-OMF)
2. Composite intermediate moment frames (C-IMF)
3. Composite special moment frames (C-SMF)
4. Composite partially restrained moment frames (C-PRMF)

G4: Composite braced-frame and shear-wall systems:

1. Composite ordinary braced frames (C-OBF)
2. Composite special concentrically braced frames (C-SCBF)
3. Composite eccentrically braced frames (C-EBF)
4. Composite ordinary shear walls (C-OSW)
5. Composite special shear walls (C-SSW)
6. Composite plate shear walls (C-PSW)

Brief description with basic requirements of some of the popular systems above will be presented hereinafter. However, composite systems will not be addressed in this book. For complete details and requirements, the reader may refer to AISC seismic provisions.

In addition, the reader is advised to refer to AISC provisions for systems requirements when they are used in conjunction with seismic isolation systems. These requirements are not addressed in this book.

8.3.1 Ordinary Moment Frames

Ordinary moment frames (OMF) are expected to withstand minimal inelastic deformation in their members and connections. The code permits fully restrained (FR) as well as partially restrained (PR) moment connections to be used in OMF. To achieve this expectation, the code has additional requirements for moment connections and continuity plates.

FR Moment Connections

The required design strength of the connections for FR moment connections shall be based on the expected yield stress, that is,

$$M_n \geq 1.1\, R_y M_p$$

where M_p = plastic moment of the beam
M_n = required flexure strength of the connection
R_y = expected yield stress factor as specified in Table 8-A1

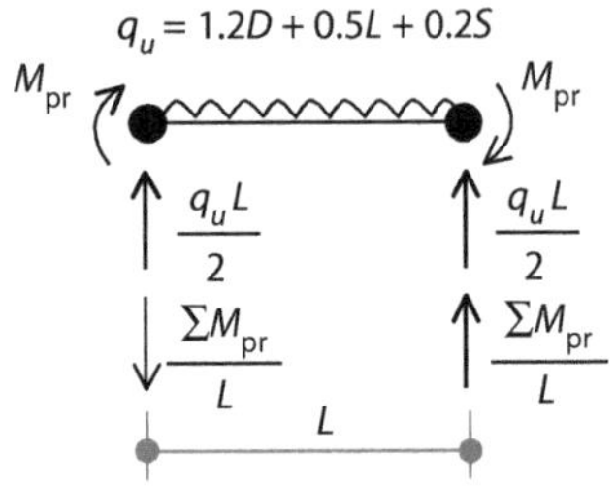

FIGURE 8-2 Shear design requirements.

This requirement is intended to force the plastic deformation in the beam while keeping the connection in its elastic range.

Shear Strength Requirements The connection must be designed to resist the factored gravity loads and the shear resulting from the maximum probable moment, M_{pr}. The design shearing force, V_u, is expressed with reference to Fig. 8-2 as follows:

$$V_u = V_u(q_u) + V_u(M_{pr})$$
$$= \frac{q_u L}{2} + \frac{\Sigma M_{pr}}{L}$$

where

$$M_{pr} = 1.1\, R_y M_p Z$$

where M_p = plastic moment = $F_y Z$
R_y = yield stress amplification factor given in Table 8-A1
Z = plastic section modulus of the beam

For PR moment connections, see AISC code.

Demand Critical Welds Regions

Complete-joint-penetration groove welds of beam flanges to columns are demand critical welds.

8.3.2 Intermediate Moment Frames

Intermediate moment frames (IMF) are expected to provide limited inelastic deformation capacity in their members and connections. The requirements of intermediate moment frames to achieve such capacity are given in terms of beam-to-column connections, lateral support of beams, and strength requirements. These requirements are, of course, in addition to the requirements of ordinary moment frames.

Slenderness of Beams

Beams shall satisfy the width-thickness ratio and bracing limitations of moderate ductility requirement as pointed out earlier in this chapter.

Protected Zones

Regions of plastic hinges at the end of beams are designated as protected zones.

Beam-to-Column Connections

Ductility Requirements The connection must be capable of sustaining an interstory drift angle of at least 0.02 rad.

Flexural Strength Requirements The connections shall be designed for flexure moment determined at the column face, equal to at least 80 percent of the nominal plastic moment of the connected beam at its specified interstory drift angle.

Shear Strength Requirements The shear strength requirements shall be similar to the requirements of OMF given in the previous section.

Demand Critical Welds Regions

The following welds are demand critical welds regions:

1. Groove welds at column splices.
2. Welds at column-to-base plate connections.
3. Complete-joint-penetration groove welds of beam flanges and beam webs to columns.

8.3.3 Special Moment Frames

Special moment frames (SMF) are expected to achieve significant inelastic rotation capacity. The requirements of SMF are given in terms of beam-to-column connections, panel zones of beam-to-column connections, lateral support of beams, and strength requirements.

Column-Beam Moment Ratios

The summation of the plastic moments of the columns at any joint must exceed the summation of the plastic moments of the associated beams. With reference to Fig. 8-3, this condition may be expressed as follows:

$$\frac{\Sigma M^*_{pc}}{\Sigma M^*_{pb}} > 1.0$$

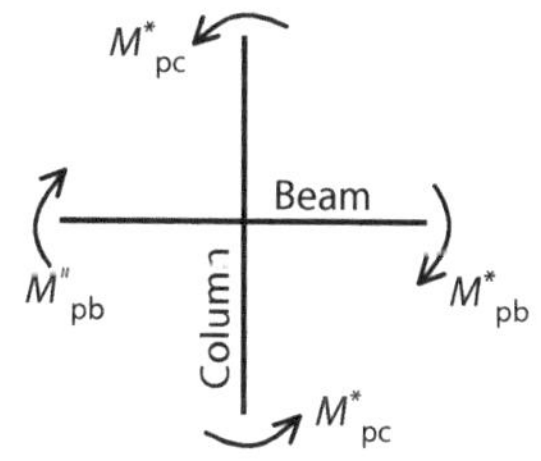

FIGURE 8-3 Joint moments.

The column moment, M^*_{pc} shall be calculated as follows:

$$M^*_{pc} = Z_c\left(F_{yc} - \frac{P_{uc}}{A_g}\right)$$

where A_g = gross-sectional area of the column
F_{yc} = specified yield stress of the column
P_{uc} = factored axial load in the column
Z_c = plastic section modulus of the column

The beam moment, M^*_{pb}, shall be calculated as follows:

$$M^*_{pb} = 1.1\,R_y F_{yb} Z_b + M_{uv}$$

where F_{yb} = specified yield stress of the beam
M_{uv} = the additional moment due to shear amplification from the location of the plastic hinge to the column centerline
R_y = yield stress amplification factor given in Table 8-A1
Z_b = plastic section modulus of the beam

Note that this requirement is in line with the strong-column weak-beam design philosophy.

Slenderness of Beams

Beams shall satisfy the width-thickness ratio and bracing limitations of high ductility requirement as pointed out earlier in this chapter.

Protected Zones

Regions of plastic hinges at the end of beams are designated as protected zones.

Beam-to-Column Connections

Ductility Requirements The connection must be capable of sustaining an interstory drift angle of at least 0.04 rad.

Flexural Strength Requirements The connections shall be designed for flexure moment, determined at the column face, equal to at least 80 percent of the nominal plastic moment of the connected beam at its specified interstory drift angle.

Shear Strength Requirements The shear strength requirements shall be similar to the requirements of ordinary and intermediate moment frame given in previous sections.

Panel Zone The thickness of the panel zone and each of its doubler plates shall be at least equal to the following:

$$t \geq \frac{d_z + w_z}{90}$$

where d_z and w_z are defined in Fig. 8-4. If doublers are attached to the web by plug welds, only the total thickness of all plates together need to satisfy the condition given previously.

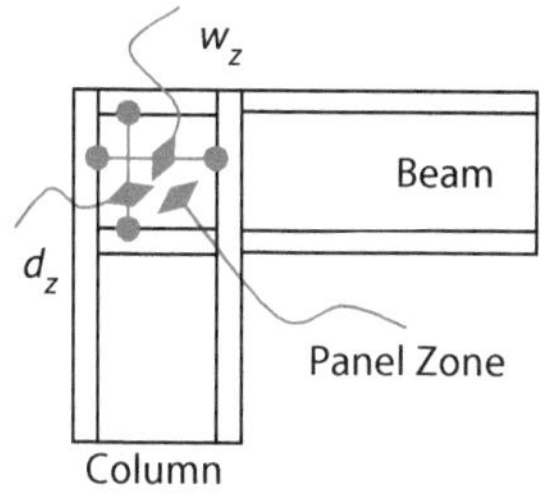

FIGURE 8-4 Dimensions of panel zone.

Lateral Support of Column Flanges

Column flanges shall be laterally braced at the level of both top and bottom flanges of the beam. The required strength of this lateral bracing shall be designed for a force equal to 2 percent of the beam flange nominal strength ($F_y b_f t_f$).

Demand Critical Welds Regions

The following welds are demand critical welds regions:

1. Groove welds at column splices.
2. Welds at column-to-base plate connections.
3. Complete-joint-penetration groove welds of beam flanges and beam webs to columns.

8.3.4 Special Truss Moment Frames

Special truss moment frames (STMF) are expected to experience significant inelastic deformation within a specially designed segment of the truss. STMF dissipate energy by flexure yielding of the top and bottom chords and by axial yielding of the diagonals of a part of the frame called special segment. The special segment is defined as follows.

The STMF and its components are shown in Fig. 8-5. STMF consists of vertical columns with horizontal truss in between that acts as a beam. The middle portion of the

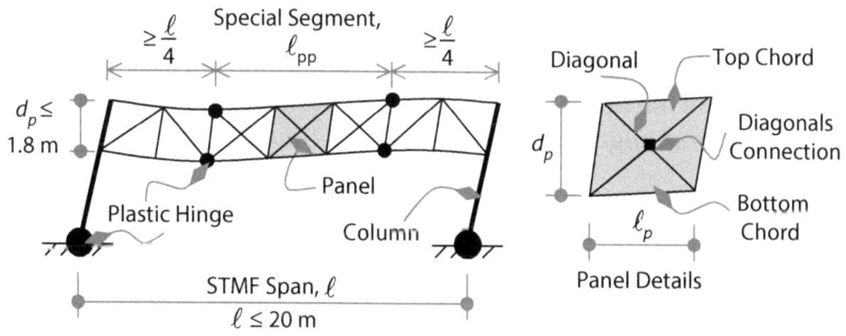

FIGURE 8-5 Special truss moment frame (STMF).

truss is known as a special segment that consists of either Vierendeel panels or cross diagonals (X-shaped) connected at their intersection. Both types may not be mixed in the same truss.

In principle, the special segment is expected to yield under seismic excitation while the rest of the truss remains elastic. This philosophy is in line with the capacity design philosophy in Sec. 5.13.

The STMF shall satisfy the following limitations.

Dimension Limitations

Refer to Fig. 8-5 for the following:

Overall span,

$$\ell \leq 20.0 \text{ m}$$

Panel depth,

$$d_p \leq 1.8 \text{ m}$$

Special Segments

Layout The special segment must be located within the quarter points of the truss span with the following limitations.

Special segment length,

$$\ell_{ss} \geq 0.1\ell$$
$$\ell_{ss} \leq 0.5\ell$$

Panel dimensions,

$$\ell_p \geq 0.67\, d_p$$
$$\ell_p \leq 1.50\, d_p$$

Top and bottom chords of the panel shall be identical and shall provide at least 25 percent of the required vertical shear strength.

Diagonals of the panel shall be made of rolled flat bars and must be interconnected at their intersections with a force Q_n equal to

$$Q_n \geq 0.25\, P_{nt}$$

where P_{nt} is the nominal tensile strength of the diagonal members. However, the required axial strength of the diagonals due dead and live loads shall not exceed $0.03\, F_y A_g$.

The tensile strength, ϕP_n, shall exceed 2.2 times the required strength, that is

$$\phi\, P_n \geq 2.2\, P_u$$

where $P_n = F_y A_g$
$\phi = 0.9$

Refer to AISC provisions for the expected vertical shear strength of special segments.

Bolted connections are not permitted within the panels of the special segments.

Slenderness of Special Segments

Chord members and diagonals shall satisfy the width-thickness ratio and bracing limitations of high ductility requirement as pointed out earlier in this chapter.

Protected Zones

Regions of plastic hinges at the end of chord members are designated as protected zones.

Bracing of Trusses

The top and bottom chords of the special segments must be laterally braced at both ends of the segment. These braces shall be designed for a force P_u as follows:

$$P_u = 0.06\, R_y F_y A_f$$

where F_y = specified yield stress of the beam
R_y = yield stress amplification factor given in Table 8-A1
A_f = gross area of the flange of the special segment chord member

The columns at the levels of top and bottom chords shall be braced. These braces shall be designed for a force P_u as follows:

$$P_u = 0.02\, R_y P_{nc}$$

where P_{nc} is the nominal strength of the chord at the end.

Demand Critical Welds Regions

The following welds are demand critical welds regions:

1. Groove welds at column splices.
2. Welds at column-to-base plate connections.

8.3.5 Ordinary Cantilever Column Systems

Ordinary cantilever column systems (OCCS) are expected to provide minimal inelastic drift capacity through flexural yielding of the columns.

The columns shall be designed using the load combinations including the amplified seismic forces. The required axial load strength, P_{rc}, shall not exceed 15 percent of the available strength, P_c, for these load combinations only, that is,

$$P_{rc} \le P_c$$

Demand Critical Welds Regions

There are no regions of demand critical welds.

8.3.6 Special Cantilever Column Systems

Special cantilever column systems (SCCS) are expected to provide limited inelastic drift capacity through flexural yielding of the columns.

The columns shall be designed using the load combinations including the amplified seismic forces. The required axial load strength, P_{rc}, shall not exceed 15 percent of the available strength, P_c, for these load combinations only, that is,

$$P_{rc} \leq P_c$$

Slenderness of Columns

The columns shall satisfy the width-thickness ratio of high ductility requirement. However, the bracing limitations shall only satisfy the moderate ductility requirement for beams as pointed out earlier in this chapter.

Protected Zones

Regions of plastic hinges at the base of the columns are designated as protected zones.

Base Plate

Axial Strength The axial strength of the base shall not be less than the amplified load combination; neither less than the axial strength of the column splices.

Shear Strength The shear strength of the base shall not be less than the shear required for the column splice.

Flexural Strength The flexural strength of the base shall be the lesser of (1) the amplified load combinations, or (2) 1.1 $R_y F_y Z$.

Demand Critical Welds Regions

The following welds are demand critical welds regions:

1. Groove welds at column splices.
2. Welds at column-to-base plate connections.

8.3.7 Ordinary Concentrically Braced Frames

Ordinary concentrically braced frames, OCBF, are expected to experience limited inelastic deformations. OCBF include frames with eccentricities less than the beam depth. Several basic requirements are summarized in the following paragraphs.

X-bracing, V-bracing, and inverted V-bracing types are allowed in this system. K-bracing type is not permitted in this OCBF system. Examples of accepted configurations of concentric bracing are shown in Fig. 8-6.

Slenderness of Columns

Braces shall satisfy the width-thickness ratio of moderate ductility requirement.

V-braced and Inverted V-braced Frames

One set of lateral braces are required at the point of intersection of the braces.

The beams shall be continuous at the brace connection. The strength of beams shall be based on the code load combinations assuming that braces provide no support

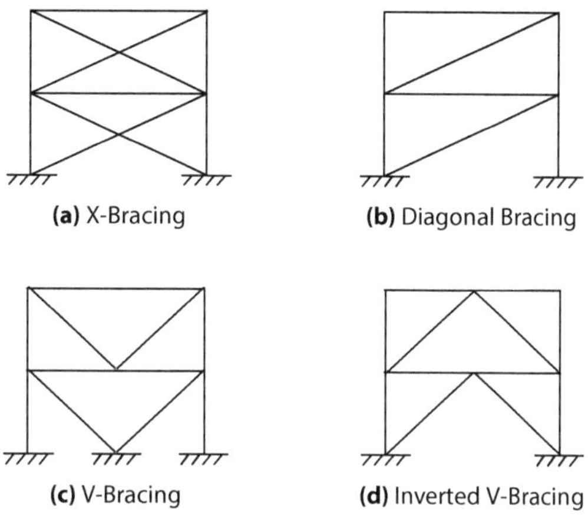

FIGURE 8-6 Concentrically braced frame configurations.

under dead and live loads. For seismic load combinations, the members shall be designed assuming the following forces in the bracings.

1. The force in braces in tension shall be assumed to have the least of the following forces:
   a. Expected yield strength, $R_y F_y A_g$
   b. Amplified seismic load combinations, or
   c. The maximum force that can develop by the system.
2. The force in braces in compression shall be assumed to have a force equal to $0.3\,P_n$.

The slenderness ratio of the braces shall be limited to

$$\frac{K\ell}{r} \leq 4\sqrt{\frac{E}{F_y}}$$

where K = effective length factor
ℓ = unsupported length of the brace
r = critical radius of gyration

Diagonal Brace Connections

Diagonal brace connections shall be designed for the tension forces of the brace using the amplified seismic load. However, the required strength need not exceed the following strengths.

In tension: The expected yield strength, $R_y F_y A_g$.

In compression: The lesser of expected yield strength, $R_y F_y A_g$, or the expected critical buckling strength, $1.14\,F_{cre} A_g$, where $R_y F_y$ is used to calculate F_{cre} instead of F_y in F_{cr} equations of AISC specifications.

Demand Critical Welds Regions

There are no regions of demand critical welds.

8.3.8 Special Concentrically Braced Frames

Special concentrically braced frames (SCBF) are expected to provide significant inelastic deformations, primarily due to tension in tension brace and due to plastic hinge formation in buckled compression brace. SCBF include frames with eccentricities less than the beam depth. Several basic requirements are summarized in the following paragraphs.

X-bracing, V-bracing, and inverted V-bracing types are allowed in this system. K-bracing type is not permitted in this SCBF system. Refer to Fig. 8-6 for accepted configurations of concentric bracing.

Slenderness of Columns

The columns and braces shall satisfy the width-thickness ratio of high ductility requirement. However, the beams shall only satisfy the moderate ductility requirement.

Protected Zones

The protected zones for braces in SCBF are defined as follows:

1. The center one-quarter of the brace.
2. Regions adjacent to the connections in braces at distance equals to the depth of the brace on each side.
3. Elements that connect braces to beams and columns.

Strength Requirements

The strength of columns, beams, and connections shall be based on the amplified seismic forces. The effect of the horizontal forces shall be taken as the larger of the following two cases of analysis:

1. Braces, in tension and in compression, resist forces corresponding to their expected strength.
2. Braces in tension resist forces corresponding to their expected strength, while braces in compression resist forces according to their expected postbuckling strength.

The analysis above shall be based on the following conditions:

1. Tension and compression are determined without gravity loads
2. Expected strength in tension $= R_y F_y A_g$
3. Expected strength in compression = the lesser of $R_y F_y A_g$, or the expected critical buckling strength, $1.14\, F_{cre}\, A_g$, where $R_y F_y$ is used to calculate F_{cre} instead of F_y in F_{cr} equations of AISC specifications
4. Expected postbuckling strength ≤ 0.3 expected strength in compression

Column strength need not exceed the least of the following:

1. Amplified seismic forces.
2. Resistance of foundation to overturning uplift.
3. Nonlinear analysis force results.

Beam-to-Column Connections

The beam-to-column connections at the brace location shall conform to the following:

1. If the connection is simple, it shall accommodate a rotation of 0.025 rad.
2. If the connection is a moment connection, it shall be designed for the lesser of the following:
   a. The expected moment capacity of the beam, 1.1 R_yF_yZ.
   b. The sum of the expected moment capacity of columns, $\Sigma\,(R_yF_yZ)$.

Brace Connections

The brace connections shall be designed for tension, compression, and flexure without interaction as follows:

1. Brace in tension: The connection strength shall be the lesser of
   a. The brace force obtained from analysis.
   b. The expected yield strength of the brace, $R_yF_yA_g$
      where A_g = Gross sectional area of the brace
      F_y = Specified yield stress of the brace
      R_y = Yield stress amplification factor as specified earlier
2. Brace under compression: The connection strength shall be:

   The expected critical buckling strength, 1.14 $F_{cre}A_g$, where R_yF_y is used to calculate F_{cre} instead of F_y in F_{cr} equations of AISC specifications.
3. Buckled braces: The brace connection shall be designed to withstand the flexure forces or rotations imposed by brace buckling as follows:
   a. The flexure capacity of the connections shall be design for the expected flexure capacity of the brace, 1.1 R_yF_yZ.
   b. The rotation capacity of the connections shall accommodate the design story drift.

Diagonal Braces

Diagonal braces shall satisfy the following:

1. The slenderness of the braces shall be limited to $\frac{KL}{r} \leq 200$.
2. The brace effective net area shall not be less than the brace gross area.
3. The minimum yield strength of brace reinforcement shall not be less than the brace yield strength.

4. The connections of brace reinforcement shall have strength to develop the expected reinforcement strength on each side of the reduced section.
5. See AISC provisions for built-up braces.

Brace Requirements The braces shall be deployed alternately along any line of braces. A line of braces is defined as a single line, or, parallel lines within 10 percent offset of the building dimension perpendicular to the line of braces.

The tension brace as shown in Fig. 8-7 shall carry 30 to 70 percent of the total horizontal force. This requirement is waived if the strength of each brace in compression is more than that required due to the amplified load combinations stipulated by the applicable seismic code.

V- and Inverted V-Type Braces

The beams intercepted by braces away from beam-to-column connections as shown in Fig. 8-8, shall meet the following requirements:

1. The intersecting beam with the braces must be continuous.
2. Beams shall be braced to satisfy moderate ductility requirements.

Demand Critical Welds Regions

The following welds are demand critical welds regions:

1. Groove welds at column splices.
2. Welds at column-to-base plate connections.
3. Welds at beam-to-column connections.

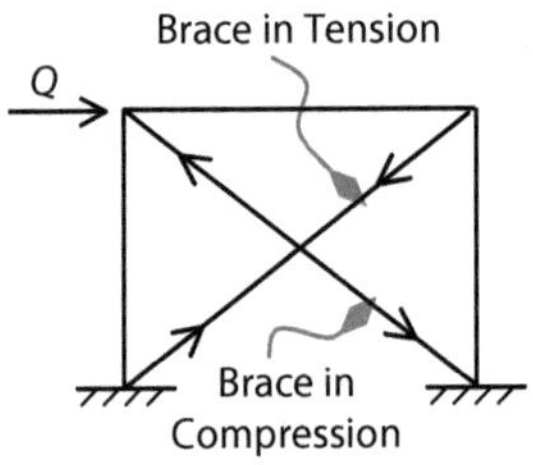

Figure 8-7 Special concentrically braced frame.

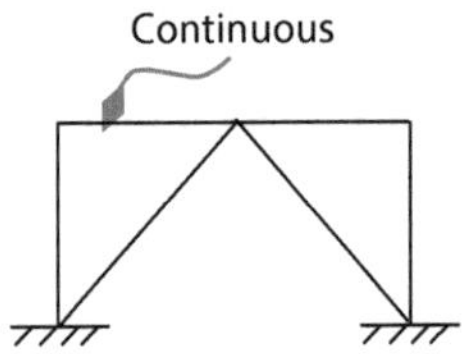

Figure 8-8 SCBF inverted V-type.

8.3.9 Eccentrically Braced Frames

Eccentrically braced frames (EBF) are expected to experience significant inelastic deformations through shear or flexure yielding in their links. EBF dissipate energy by shear and flexure yielding of the link members.

As bracing elements cause obstruction of the bays they occupy, EBF are generally used to create room for entrances such as doors inside the building. To make such space, the braces in this system do not meet at a common point as they would in V-type, for example. Instead, the braces intersect the beams at different points. Because of this arrangement, EBF are generally less stiff than concentrically braced frames.

In EBF, the braces are configured to form stiff triangulation with part of the beams, leaving the other part as a link. For example, Fig. 8-9 shows some arrangements that create qualified links. In principle, the links are expected to yield under seismic excitation while the rest of the members, including beams and braces, remain elastic. This philosophy is in line with the capacity design philosophy given in Sec. 5.13.

Strength Requirements

The strength of columns, beams outside links, braces, and connections shall be based on the amplified seismic forces. The effect of the horizontal forces shall be taken assuming the forces at the end of the links correspond to adjusted link shear strength, $\bar{V}_n$, which is defined as follows:

$$\bar{V}_n = 1.25\, R_y V_n \qquad \text{(for I-shaped links)}$$

$$\bar{V}_n = 1.40\, R_y V_n \qquad \text{(for box links)}$$

where V_n is the shear strength of the link as defined later in this section.

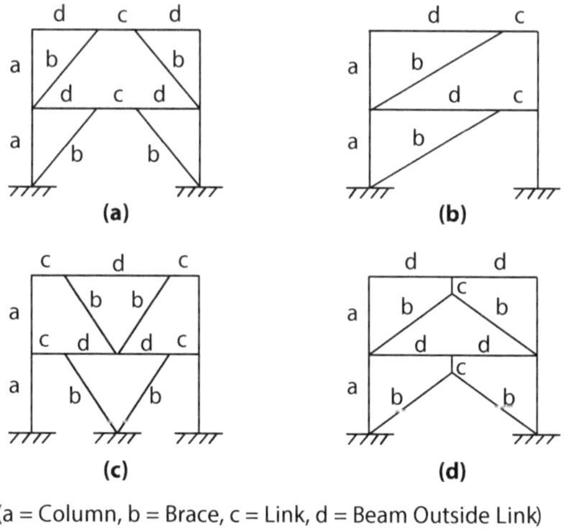

Figure 8-9 Eccentrically braced frames.

Slenderness Requirements

The columns shall satisfy the width-thickness ratio of high ductility requirement. However, the bracing limitations shall only satisfy the moderate ductility requirement for beams as pointed out earlier in this chapter. Where beams outside link regions are composed of different sections from the link, they shall satisfy the moderate ductility requirement.

Protected Zones

Links in EBFs are designated as protected zones.

Beam-to-Column Connections

The beam-to-column connections at the brace location shall conform to the following:

1. If the connection is simple, it shall accommodate a rotation of 0.025 rad.
2. If the connection is a moment connection, it shall be designed for the lesser of the following:
   a. The expected moment capacity of the beam, 1.1 R_yF_yZ.
   b. The sum of the expected moment capacity of columns, $\Sigma\,(R_yF_yZ)$.

General Link Requirements

The links shall satisfy the following general requirements:

1. Links shall be W-shape sections, built-up, I, or built-up box sections.
2. Links shall satisfy the width-thickness requirements of high ductility.
3. Webs of the links shall be one piece without doublers or penetrations.

Link Rotation Angle The link rotation angle, γ_p, is defined as the deviation of the ends of the link from the line connecting the end of the link with the rest of the beam. As an illustration γ_p is shown in Fig. 8-10 with the elastic triangle attached to the link is considered to displace as a rigid piece. All deformations given in Fig. 8-10 are plastic deformations which are assumed to dominate the deformations. Thus, the

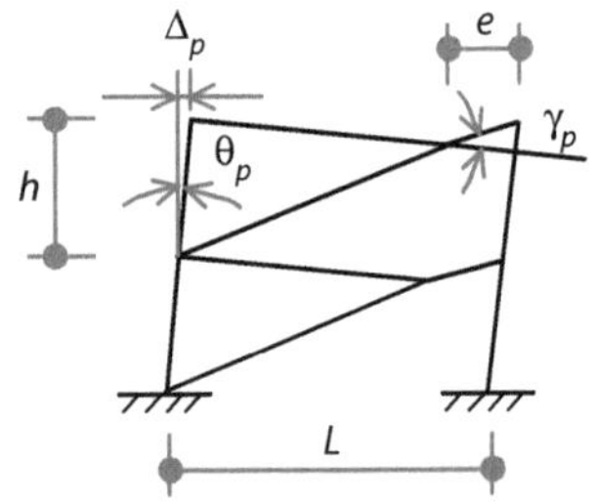

FIGURE 8-10 Link rotation angle.

elastic deformations are excluded. Figure 8-10 demonstrates that the following relations hold:

$$\theta_p = \frac{\Delta_p}{h}$$

$$\gamma_p = \left(\frac{L}{e}\right)\theta_p$$

The link rotation angle shall not exceed the following values:

$$\gamma_p \le 0.08 \text{ rad} \quad \text{for} \quad e \le \frac{1.6 M_p}{V_p}$$

$$\gamma_p \le 0.02 \text{ rad} \quad \text{for} \quad e \ge \frac{2.6 M_p}{V_p}$$

where M_p = nominal plastic moment of the link
V_p = nominal shear strength of the link

Linear interpolations shall be used for values in between.

Link Shear Strength The shear strength of the link is given as a function of the level of axial load P_u in the member. The strength capacity shall satisfy the usual condition:

$$\phi_v V_n = V_u$$

where V_u = factored external loads
ϕ_v = resistance factor for shear = 0.90
V_n = shear strength which is taken as the lesser of shear yielding, or flexure yielding as follows:

a. Shear yielding capacity, V_p

$$V_n = V_p$$

where

$$V_p = 0.6\,F_y A_{\text{lw}} \quad \text{for} \quad \frac{P_u}{P_c} \le 0.15$$

and

$$V_p = 0.6\,F_y A_{\text{lw}} \sqrt{1 - \left(\frac{P_u}{P_c}\right)^2} \quad \text{for} \quad \frac{P_u}{P_c} > 0.15$$

where A_{lw} = summation of area of the link webs
F_y = specified yield stress of the link
P_u = LRFD factored external axial load
$P_c = P_y$
P_y = LRFD nominal axial yield strength $= F_y A_g$

b. Flexure yielding capacity

$$V_n = \frac{2M_p}{e}$$

where M_p = plastic moment of the link, which equals to

$$M_p = F_y Z \quad \text{for} \quad \frac{P_u}{P_c} \leq 0.15$$

$$M_p = F_y Z\left(\frac{1 - P_u/P_c}{0.85}\right) \quad \text{for} \quad \frac{P_u}{P_c} > 0.15$$

e = length of the link as shown in Fig. 8-11

Link Length The length of the link depends on the level of axial load and shear force in the link. If ρ' is defined as follows:

$$\rho' = \left(\frac{P_u/P_c}{V_u/V_c}\right)$$

where P_u = LRFD factored external axial load
$P_c = P_y$ = LRFD nominal axial yield strength $= F_y A_g$
V_u = LRFD factored external shear force
$V_c = V_y$ = LRFD nominal shear yield strength $= 0.6\, F_y A_{lw}$

Consequently, the link length shall be limited to the following:

$$e \leq \frac{1.6 M_p}{V_p} \quad \text{for} \quad \rho' \leq 0.5$$

$$e \leq \frac{1.6 M_p}{V_p}(1.15 - 0.3\,\rho') \quad \text{for} \quad \rho' > 0.5$$

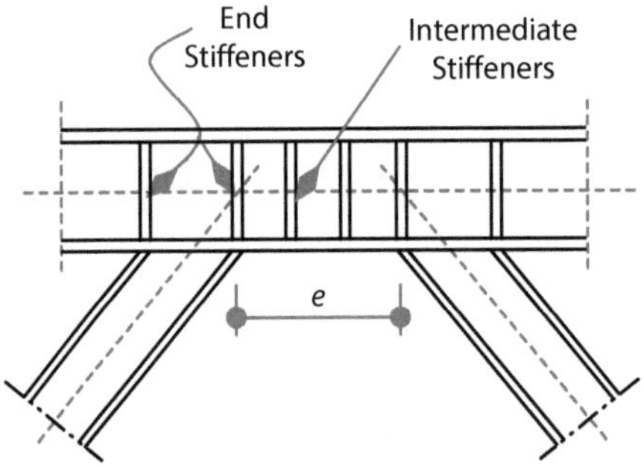

FIGURE 8-11 Link stiffeners.

Link Stiffeners for I-Shapes Both end and intermediate stiffeners, as shown in Fig. 8-11, must be provided in the link region. The stiffeners shall satisfy the following requirements.

End Stiffeners End stiffeners shall be provided on both sides of the web. The width and thickness of the end stiffeners shall meet the following limitations:

$$b_{st} \geq b_f - 2t_w$$

$$t_{st} \geq (0.75t_w, 10 \text{ mm})$$

where b_f = width of the link flange
b_{st} = width of both stiffener
t_{st} = thickness of the stiffener
t_w = thickness of the link web

Intermediate Stiffeners Intermediate stiffeners shall be provided for short links as follows:

a. For $e \leq \dfrac{1.6M_p}{V_p}$

The inelastic response in this case is dominated by shear yielding. The spacing of the stiffeners is given in terms of the link rotation angle, γ_p, as follows:

$$\text{For } \gamma_p \geq 0.08, \qquad s \leq 30t_w - \frac{d}{5}$$

$$\text{For } \gamma_p \leq 0.02, \qquad s \leq 52t_w - \frac{d}{5}$$

Linear interpolations shall be used for values in between.

b. For $e \left(\geq \dfrac{2.6M_p}{V_p} \text{ and } \leq \dfrac{5M_p}{V_p} \right)$

The inelastic response in this case is dominated by flexure yielding. Stiffeners shall be placed from each end of the link at a distance s, where s is given as

$$s = 1.5\, b_f$$

c. For $e \left(\geq \dfrac{1.6M_p}{V_p} \text{ and } \leq \dfrac{2.6M_p}{V_p} \right)$

The inelastic response in this case results from a combination of shear and flexure yielding. The stiffeners shall be provided as in a. and b. given previously.

d. For $e \geq \dfrac{5M_p}{V_p}$

No stiffeners are required.

The intermediate stiffeners shall be provided on both sides of the link if the link is 635 mm or deeper. Only one-side stiffeners are required for depths less than 635 mm. One-side stiffeners shall satisfy the following thickness requirements:

$$b_{st} \geq \frac{b_f}{2} - t_w$$

$$t_{st} \geq (t_w,\ 10\text{ mm})$$

where b_f = width of the link flange
b_{st} = width of the stiffener
d = overall depth of the link
e, M_p, V_p = as defined earlier
t_{st} = thickness of the stiffener
t_w = thickness of the link web

Link Stiffeners for Box Shapes There are requirements for stiffeners of box-shaped links analogous to the stiffeners of I-shaped links. The reader may refer to AISC provisions for details.

Bracing of Links The top and bottom flanges of the link shall be provided with lateral supports at their ends. The lateral supporting elements shall be designed for a force P_u equal to 6 percent of the flange nominal strength such as

$$P_u = 0.06\frac{R_y F_y Z}{h_o}$$

where F_y = specified yield stress of the beam
R_y = yield stress amplification factor given in Table 8-A1
Z = plastic section modulus of the beam

Link-to-Column Connection Link-to-column connections shall be FR moment connections.

Demand Critical Welds Regions

The following welds are demand critical welds regions:

1. Groove welds at column splices.
2. Welds at column-to-base plate connections.
3. Welds at beam-to-column connections.
4. Welds attaching link to columns directly.
5. Web to flange connections in built-up links.

8.3.10 Buckling-Restrained Braced Frames

Buckling-restrained braced frames (BRBF) are expected to provide significant inelastic deformations. BRBF shall be designed to direct inelastic deformations primarily to brace yielding in tension and in compression. Consequently, these frames are fabricated with special braces that can sustain large inelastic deformations under compression without buckling.

Because this chapter is intended as an introduction to steel systems, and not intended to given entire details about steel systems, the reader may refer to the AISC provisions for details of the requirements of these frames. It may be noted that this system is analogous to special concentrically braced frames (SCBR) presented earlier.

8.3.11 Special Plate Shear Walls (SPSW)

Special plate shear walls (SPSW) consist of horizontal beams designated as horizontal boundary elements (HBE) and vertical columns designated as vertical boundary elements (VBE) as shown in Fig. 8-12. The panels bounded by these boundary elements are filled with either solid plates, Fig. 8-12(a), or perforated plates, Fig. 8-12(b), connected to the boundary elements as shown. If the panel contains any openings, the openings shall be bounded by intermediate boundary elements extending to the full width and height of the panel.

SPSW are expected to provide significant inelastic deformation capacity primarily through yielding of webs in tension field action, and also through formation of plastic hinges at the end of their HBEs.

The boundary elements shall have minimum stiffness (in terms of moment of inertia) about an axis perpendicular to the plane of the web as follows:

$$\text{VBE:} \quad I_c \geq 0.0031 \frac{t_w h^4}{L}$$

$$\text{HBE:} \quad I_b \geq 0.0031 \frac{t_w L^4}{h}$$

where L, h, and t_w are as shown in Fig. 8-12.

The VBE panel zone next to the top and base HBE shall comply with the requirements of the special moment frame panel zones.

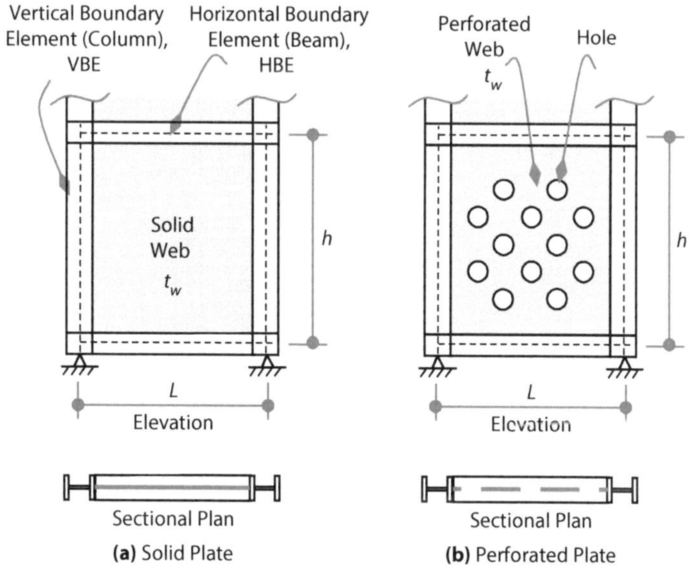

FIGURE 8-12 Special plate shear wall components.

Strength Requirements

The webs of SPSWs shall not be considered in resistance of gravity forces. The required strength of HBEs and VBEs shall be based on the load combinations of amplified seismic forces.

The HBE-to-VBE connection moment ratio shall satisfy the special moment frames beam-to-column connection requirements without consideration of the effect of the webs.

As mentioned earlier, plate webs are analyzed and designed according to tension field action concept. In this concept, the plate is expected to buckle under compression directions forming diagonals in tension in the perpendicular direction. The strength depends on the type of the plate, namely, solid or perforated.

Solid Webs The shear strength capacity of solid webs is given with reference to Fig. 8-13 as follows:

$$V_n = 0.42\,F_y t_w L_{cf} \sin 2\alpha$$

The angle α is permitted to be taken as 40°, or calculated as follows:

$$\tan^4 \alpha = \frac{1+\dfrac{t_w L}{2A_c}}{1+t_w h\left(\dfrac{1}{A_b}+\dfrac{h^3}{360\,I_c L}\right)}$$

where A_b and A_c are the cross-sectional areas of the HBE and VBE respectively.

The strength reduction factor, ϕ, shall be taken as 0.9.

Perforated Webs Perforated webs, when used, shall have regular pattern of circular holes of uniform diameters spaced evenly over the entire web plate as shown in Fig. 8-14. The edge of holes shall have maximum surface roughness of 13 microns.

The spacing of the holes S_{diag}, shall be at least equal to $1.67D$, where D is the hole diameter as shown in Fig. 8-14. However, the distance between the first hole and web connections to HBEs and VBEs shall be between D and $D + 0.75\,S_{diag}$.

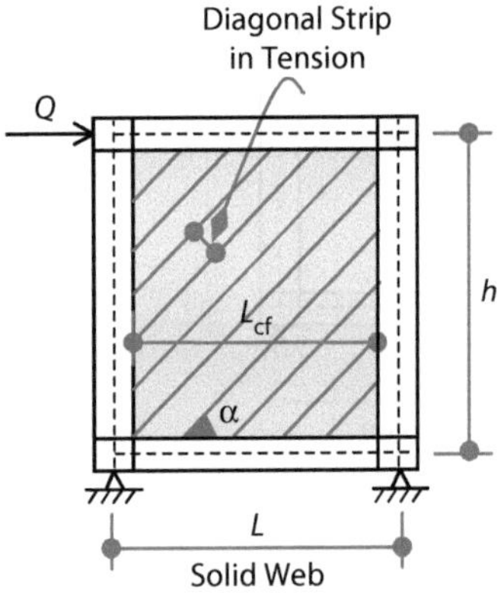

FIGURE 8-13 Strip model.

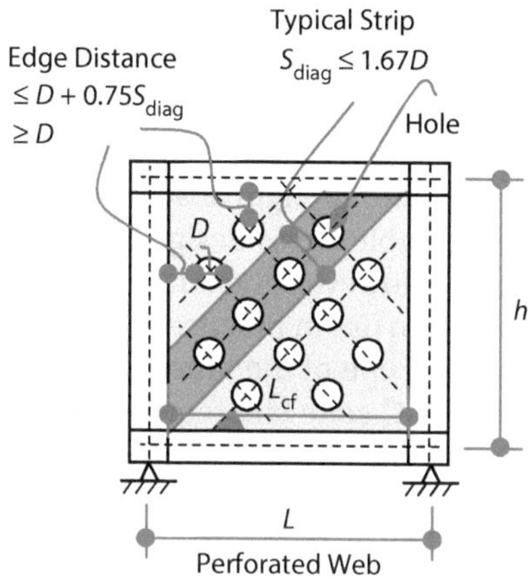

FIGURE 8-14 Strip model.

The shear strength capacity of perforated webs is given with reference to Fig. 8-14 as follows:

$$V_n = 0.42\, F_y t_w L_{cf}\left(1 - \frac{0.7D}{S_{diag}}\right)$$

Slenderness Requirements

HBE, VBE, and intermediate boundary elements shall satisfy the width-thickness ratio of high ductility requirement; however, the bracing limitations shall only satisfy the moderate ductility requirement for beams as pointed out earlier in this chapter.

Protected Zones

The protected zones for SPSWs are defined as follows:

1. The webs of SPSW.
2. Elements that connect webs to HBEs and VBEs.
3. Plastic hinge regions at the end of HBEs with a minimum length equals to the depth of beam.

Demand Critical Welds Regions

The following welds are demand critical welds regions:

1. Groove welds at column splices.
2. Welds at column-to-base plate connections.
3. Welds at HBE-to-VBE connections.

More specific details of the SPSW systems are found in AISC provisions.

8.4 Allowable Stress Design Approach

The modern philosophy of earthquake design that is presented in this book depends on the inelastic behavior of the structure in its ultimate state. Naturally, this philosophy is consistent with the ultimate strength design approach which is known as the LRFD. Therefore, using the allowable stress design (ASD) approach defeats the purpose of applying such concepts. However, the AISC provisions permit the use of the ASD approach as it may still be in use in practice.

To remain within the scope of the ultimate strength concept in conjunction with the ASD approach, the AISC code still requires the structures to be designed for the combinations of factored loads when seismic effect is included. To overcome this difficulty on the stress side, the code has raised the allowable stresses by a factor of 1.7 to resist the combinations of the factored loads.

APPENDIX 8-1

This appendix to Chap. 8 contains some selected of the relevant AISC seismic provisions presented in tables for illustration.

	Application	R_y	R_t
1	Hot-rolled structural shapes bars:		
	• ASTM A36/A36M Gr. 250	1.5	1.2
	• ASTM A572/A572M Gr. 345, or, 380	1.1	1.1
2	Hollow structural sections (HSS):		
	• ASTM A500/A500M Gr. B or C, ASTM A501	1.4	1.3
3	Pipes:		
	• ASTM A53/A53M	1.6	1.2
4	Plates, strips, and sheets:		
	• ASTM A36/A36M	1.3	1.2
	• ASTM A572/A572M Gr. 345, or, 380, ASTM A588/A588M	1.1	1.2
5	Steel reinforcement:		
	• ASTM A615/A706	1.25	1.25

TABLE 8-A1 Expected Strength Factors, R_y and R_t

Property	Classification	
	480 MPa	550 MPa
Min yield strength (MPa)	400	470
Min tensile strength (MPa)	480	550
Min elongation	22%	19%
Min CVN toughness	27 J @ −18°C	

TABLE 8-A2 Filler Metal Classification for Demand Critical Welds

	Description of Element	Width-to-Thickness Ratio	Limiting Width-to-Thickness Ratios	
			High Ductility, λ_{hd}	Moderate Ductility, λ_{md}
Unstiffened Elements	Flanges of rolled or built-up I-shaped Channels and tees Single or double angles in continuous contact	$\frac{b}{t}$	$0.30\sqrt{E/F_y}$	$0.38\sqrt{E/F_y}$
	Flange of H-pile sections	$\frac{b}{t}$	$0.45\sqrt{E/F_y}$	not applicable
	Stems of tees	$\frac{d}{t}$	$0.30\sqrt{E/F_y}$	$0.38\sqrt{E/F_y}$

Table 8-A3 Limiting Width-to-Thickness Ratios for High and Moderate Ductility, λ_{hd} and λ_{md}, for Unstiffened Elements

	Description of Element	Width-to-Thickness Ratio	Limiting Width-to-Thickness Ratios	
			High Ductility, λ_{hd}	Moderate Ductility, λ_{md}
Stiffened Elements	Webs of rolled or built-up I-shapes used as beams or columns, except for beams in SMF and IMF as below	$\frac{h}{t_w}$	For $\frac{P_u}{\phi_c P_y} \leq 0.125$:	For $\frac{P_u}{\phi_c P_y} \leq 0.125$:
			$2.45\sqrt{\frac{E}{F_y}} * \left[1 - 0.93\frac{P_u}{\phi_c P_y}\right]$	$3.76\sqrt{\frac{E}{F_y}} * \left[1 - 2.75\frac{P_u}{\phi_c P_y}\right]$
			For $\frac{P_u}{\phi_c P_y} > 0.125$:	For $\frac{P_u}{\phi_c P_y} > 0.125$:
			$077\sqrt{\frac{E}{F_y}} * \left[2.93 - \frac{P_u}{\phi_c P_y}\right]$ $\geq 1.49\sqrt{E/F_y}$	$1.12\sqrt{\frac{E}{F_y}} * \left[2.33 - \frac{P_u}{\phi_c P_y}\right]$ $\geq 1.49\sqrt{E/F_y}$
	I-shaped beams in SMF, where $\frac{P_u}{\phi_c P_y} \leq 0.125$	$\frac{h}{t_w}$	$2.45\sqrt{E/F_y}$	$2.45\sqrt{E/F_y}$
	I-shaped beams in IMF, where $\frac{P_u}{\phi_c P_y} \leq 0.125$	$\frac{h}{t_w}$	$3.76\sqrt{E/F_y}$	$3.76\sqrt{E/F_y}$
	Webs of H-pile sections	$\frac{h}{t_w}$	$0.94\sqrt{E/F_y}$	not applicable

Table 8-A4 Limiting Width-to-Thickness Ratios for High and Moderate Ductility, λ_{hd} and λ_{md}, for Stiffened Elements

CHAPTER 9

Design of Earthquake-Resistant Bridges (AASHTO Code)

9.1 Introduction

Bridges are generally addressed in separate codes than those for building structures. Provisions and design requirements for bridges in the United States are published by the American Association of State Highway and Transportation Officials (AASHTO). This chapter provides a brief review of bridge components before discussing the seismic provisions and design procedures of the AASHTO code.

Figure 9-1 shows a typical cross section in a conventional modern highway bridge. The bridge usually consists of three main groups: superstructure, substructure, and foundations. The superstructure usually contains the deck system that carries the traffic. The deck system is the main slab with or without supporting girders. The girders usually sit on elastomeric pads, which are considered part of the superstructure. If the slab is designed without girders, it is directly supported by the substructure.

The substructure lies below the superstructure and contains the supporting system of the superstructure. Substructure includes cap beams (if needed), columns, and/or piers. The substructure sits on top of the foundation system required by the site, a typical shallow foundation or a deep pile foundation.

Seismic design of conventional bridges usually assumes that the bridge mass is distributed in the superstructure, which is assumed to be capable of transferring the load to the substructure. The substructure is assumed to carry and resist the seismic forces and transfer their effect to the foundations.

Modern conventional bridges may be seismically isolated at the border between the superstructure and the substructure at the elastomeric pad level. In these days, it is common for a seismic isolation system to use special elastomeric pads specifically designed for this purpose.

Seismic design of conventional bridges described above is covered by AASHTO; however, unconventional bridges are not. Instead, unconventional bridges need special dynamic treatment that depends on the characteristics of the bridge under consideration.

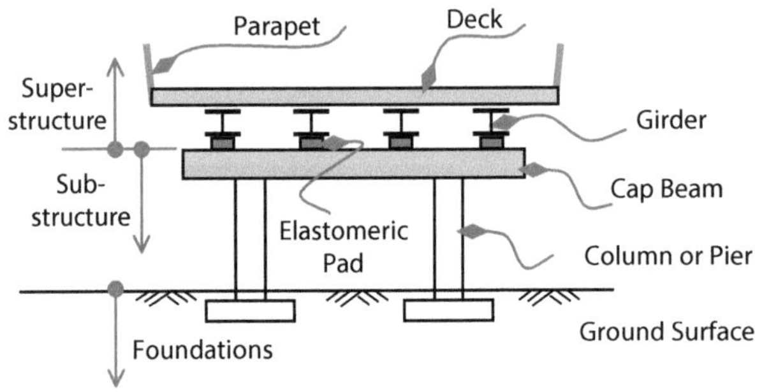

FIGURE 9-1 Typical cross section of conventional modern bridge.

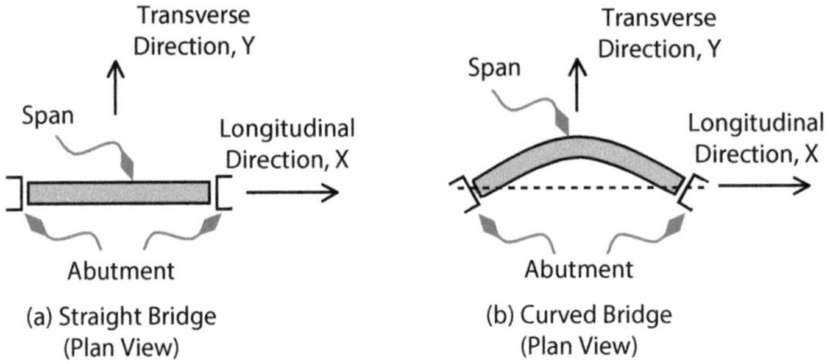

FIGURE 9-2 Major directions of seismic excitations.

Seismic evaluation of bridges usually considers two horizontal directions of seismic excitations. One direction is along the axis between abutments of the bridge. This is known as the longitudinal direction (shown in Fig. 9-2). The other direction is perpendicular to the first one and is known as the transverse direction. The effect of the two directions is then combined. This will be discussed in greater detail in this chapter.

9.2 AASHTO Procedures for Bridge Design

AASHTO provisions are based on the National Earthquake Hazards Reduction Program (NEHRP) administered by the Federal Emergency Management Agency (FEMA) in the United States. In fact, this program is the origin of all modern seismic codes. AASHTO standard procedures apply to conventional steel and concrete girder and box girders with multiple spans of less than or equal to 150 m each. Suspension bridges, cable-stayed bridges, arches, and movable bridges are not covered by AASHTO standards and require individual treatment by specialized engineers.

AASHTO divides the United States into seismic hazard contour lines that plot spectral acceleration coefficients based on a 7 percent probability of exceedance in 75 years. This probability implies a return period of 1,000 years. Accordingly, three coefficients are listed in separate contour maps specified as peak ground

Seismic Zone	1	2	3	4
Acceleration coefficient, S_{D1}	≤ 0.15	≤ 0.30 > 0.15	≤ 0.50 > 0.30	> 0.50

Table 9-1 AASHTO Seismic Zones and Coefficients

acceleration coefficient, PGA, short-period spectral acceleration coefficient, S_S, and long-period spectral acceleration coefficient, S_1. Linear interpolation may be used to obtain values in between.

The coefficients mentioned above PGA, S_s, and S_1 listed in seismic hazard maps are measured at bedrock. The effects of the site characteristics are considered by using modifying factors, F_{pga}, F_a, and F_v, for zero period, short-period range, and long-period range, respectively. Values of F_{pga}, F_a, and F_v are listed in Tables 9-A3, 9-A4, and 9-A5 in App. 9-1 at the end of this chapter. As shown in those tables, sites are classified into six classes of A, B, C, D, E, and F implying that there are six distinct response spectra in this classification. Accordingly, the corresponding response spectral acceleration coefficients at the site A_s, S_{DS}, and S_{D1} are given as follows:

$$A_s = F_{pga}\text{PGA}$$
$$S_{DS} = F_a S_S$$
$$S_{D1} = F_v S_1$$

In addition, AASHTO classifies seismic performance of bridges into four categories according to four seismic zones. These zones are identified according to the spectral acceleration coefficient for long-period range at the site, S_{D1}, as listed in Table 9-1.

AASHTO requires jurisdiction authorities to consider social/survival and security/defense requirements and then classify their bridges into the important categories of critical, essential, or other. For example, essential bridges should be functional during and after an earthquake.

In addition to this classification, AASHTO defines a regular bridge as a bridge with six spans or less and without any abrupt or unusual change in mass, stiffness, or geometry along its spans. The geometry and stiffness requirements of regularity for bridges are given in Table 9-2. Curved bridges are considered regular as long as the subtended

Parameter	Value				
Number of spans	2	3	4	5	6
Maximum subtended angle for a curved bridge	90°	90°	90°	90°	90°
Maximum span length ratio from span to span	3	2	2	1.5	1.5
Maximum bent/pier stiffness ratio from span to span	—	4	4	3	2

Table 9-2 Regular Bridge Requirements

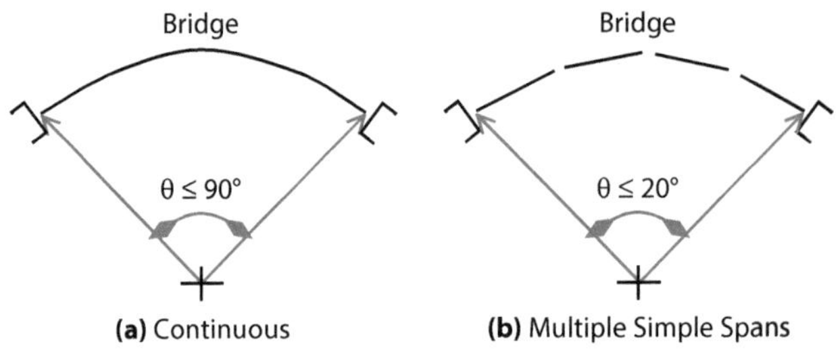

FIGURE 9-3 Regular curved bridges.

angle is less than or equal to 90° as shown in Fig. 9-3. Curved bridges composed of multiple simple spans are considered regular as long as the subtended angle is less than or equal to 20°.

Dynamic analysis of bridges is not required for the type of single-span bridges shown in Fig. 9-4. However, AASHTO requires the connections between the bridge and the abutments to be designed for minimum forces to guarantee that the bridge is tied well during an earthquake.

AASHTO requires dynamic analysis for multispan bridges with two spans or more, as shown in Fig. 9-5. The method of analysis (procedures) depends on the bridge regularity, seismic importance category, and number of spans. AASHTO specifies four methods of analysis for seismic analysis and design of bridges: Uniform load (UL) method, single-mode (SM) spectral method, multimode (MM) method, and time-history (TH) analysis method. Table 9-3 summarizes the method of analysis required for seismic evaluation of the general bridge layout shown in Fig. 9-5.

The UL method is a simplified method that models the bridge a single degree of freedom (SDOF) system with lumped mass.

The SM method assumes that the bridge only responds in the first mode. In the MM method, the bridge is analyzed using modal superposition (response spectrum) as given in Chap. 3.

Table 9-3 indicates that single bridges and bridges in Seismic Zone 1 do not need seismic analysis. However, they do need to satisfy detailing requirements in terms of connections and seismic seats. These will be explained in the following sections.

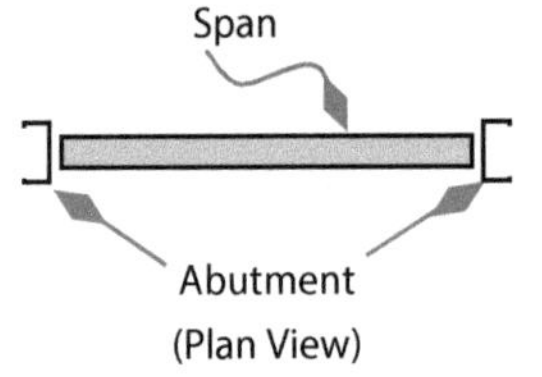

FIGURE 9-4 Single-span bridge.

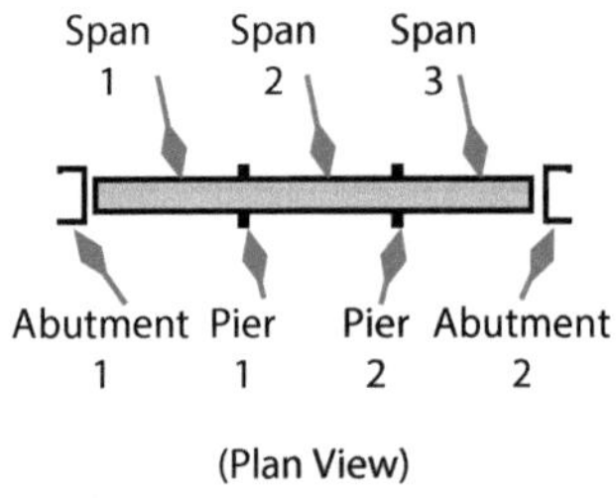

FIGURE 9-5 Multispan bridge.

Seismic Zone	Single-Span Bridges	Multispan Bridges					
		Other		Essential		Critical	
		Regular	Irregular	Regular	Irregular	Regular	Irregular
1	No seismic analysis required	No seismic analysis required					
2		SM/UL	SM	SM /UL	MM	MM	MM
3		SM/UL	MM	MM	MM	MM	TH
4		SM/UL	MM	MM	MM	TH	TH

Table 9-3 Method of Analysis for Seismic Effect

9.3 Response Spectra

The response spectra in AASHTO are defined for six site classes: A, B, C, D, E, and F. The design spectral acceleration for each site class is given as A_s, S_{DS}, and S_{D1} for zero period, for short period, and for long period, respectively. The response spectra, A_s, S_{DS}, and S_{D1}, are taken as given before

$$A_s = F_{pga}\text{PGA}$$
$$S_{DS} = F_a S_S$$
$$S_{D1} = F_v S_1$$

Once A_s, S_{DS}, and S_{D1} are defined for the site, the response spectrum can be constructed as shown in Fig. 9-6, which is expressed mathematically as follows:

Controlling periods

$$T_s = \frac{S_{D1}}{S_{DS}}, \qquad T_o = 0.2T_s$$

For $T_m \le T_o$, the elastic seismic coefficient, C_{sm}, is given as

$$C_{sm} = A_s + (S_{DS} - A_S)\left(\frac{T_m}{T_o}\right)$$

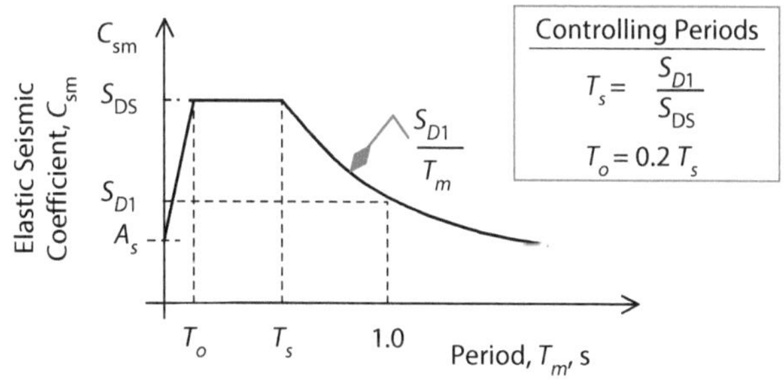

Figure 9-6 AASHTO response spectra.

For T_m, between T_o and T_s, the elastic seismic coefficient, C_{sm}, is given as

$$C_{sm} = S_{DS}$$

For $T_m \geq T_s$, the elastic seismic coefficient, C_{sm}, is given as

$$C_{sm} = \frac{S_{D1}}{T_m}$$

The six classifications of soil classes follow, in principle, the response spectra curves defined by Seed et al., as explained in Chap. 3 (Fig. 3-13). The spectra curves reflect the fact that the response spectra are simply a function of the site characteristics. At this point, AASHTO soil classification may be summarized in broad terms as follows:

1. Soil Type A: Hard rock
2. Soil Type B: Rock
3. Soil Type C: Very dense soil and soft rock
4. Soil Type D: Stiff soil profile
5. Soil Type E: Soft soil profile
6. Soil Type F: Needs special evaluation

9.4 Single-Span Bridges

Single-span bridges as shown in Fig. 9-7 are waived from dynamic analysis. However, AASHTO does require the connections between the abutments and the superstructure to be designed to resist seismic-induced forces and displacements to ensure that the bridge remains tied together during an earthquake.

The connection force, F, shall be taken as a minimum as

$$F = A_s W_T$$

where A_s = spectral acceleration coefficient as defined earlier
W_T = tributary weight of the bridge acting at the connection due to permanent loads including any assumed live load to exist during an earthquake

In addition, the bridge must also meet the displacement requirements. These are given in terms of minimum seismic seat length as described in later sections.

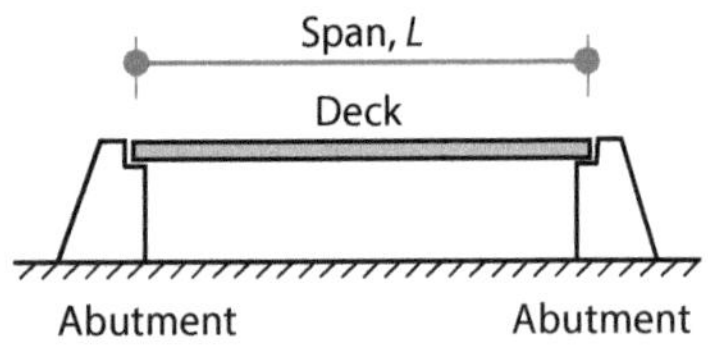

FIGURE 9-7 Longitudinal section in single-span bridge.

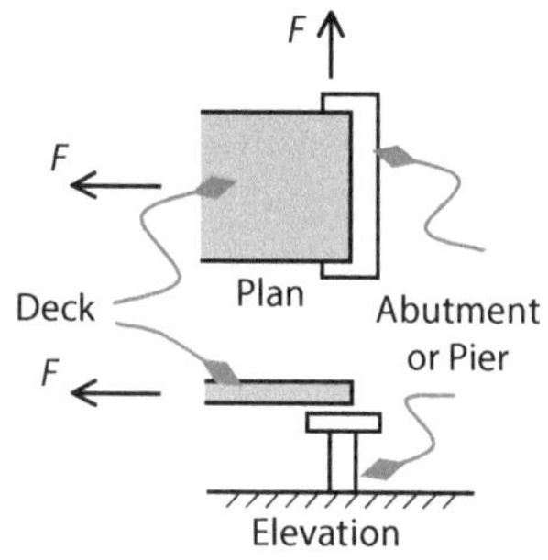

FIGURE 9-8 Force requirements.

9.5 Bridges in Seismic Zone 1

Bridges in Seismic Zone 1 are waived from dynamic analysis. Of the bridge main components, the abutments and the piers do not need to be analyzed for any seismic forces. However, AASHTO does require the connections between the superstructure and between either the abutments or the piers to be designed to resist seismic-induced forces and displacements to ensure that the bridge remains tied together during an earthquake.

The force requirements of the connections between the deck and between either abutments or piers must be designed to transfer a seismic force, F, in both longitudinal direction and transverse direction as shown in Fig. 9-8. The force, F, depends on the acceleration coefficient, A_s, as follows:

$$F_{\text{tot}} = 0.15(D + \text{SDL} + L) \qquad (\text{for } A_s < 0.05)$$
$$F_{\text{tot}} = 0.25(D + \text{SDL} + L) \qquad (\text{for } A_s \geq 0.05)$$

where F_{tot} = design force for all connections holding the segment (L_T) as shown in Fig. 9-9
D = dead load over a tributary area of segment (L_T)
L = live load assumed to exist during earthquake over a tributary area of segment (L_T)
SDL = superimposed dead load over a tributary area of segment (L_T)
A_s = spectral acceleration coefficient

The bridge must also meet the displacement requirements. These are given in terms of minimum seismic seat length as described in later sections.

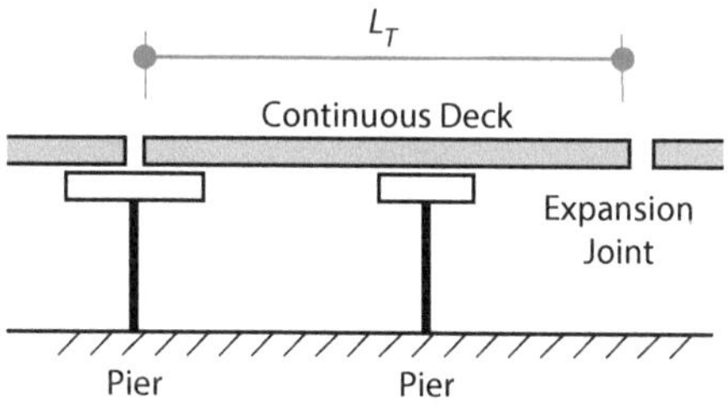

FIGURE 9-9 Tributary area of loads.

9.6 Bridges in Seismic Zone 2

Bridges in Seismic Zone 2 must be evaluated with dynamic analysis using the methods defined in Table 9-2: UL, SM, MM, and TH.

The design forces of the superstructure and substructures may be reduced by the force-reduction factor, R, for a selected system. Values of the R-factor are listed in AASHTO code and are also listed in App. 9-1 at the end of this chapter. Unlike building codes, AASHTO provisions assign force reductions for the main structure that are different than those assigned for the foundation.

For the main structure components, including the superstructure, substructure, and all components except foundations, the design force, F_d, is given as the elastic-induced force, F_{el}, divided by the R factor given in AASHTO table

$$F_d = \frac{F_{el}}{R}$$

For the foundations, other than pile bents or retaining walls, the design force may be reduced by only half the reduction of the main structure

$$F_d = \frac{F_{el}}{R/2} \leq F_{el}$$

This difference in reduction given above is intended to force the formation of plastic hinges in the substructure and to keep the foundation elastic during the excitation (as explained in previous chapters).

9.7 Bridges in Seismic Zones 3 and 4

Bridges in Seismic Zones 3 and 4 must also be evaluated with dynamic analysis using methods defined in Table 9-2: UL, SM, MM, and TH.

Similar to bridges in Seismic Zone 2, the design forces of the superstructure and substructures may be reduced by the force-reduction factor, R, for a selected system. Therefore, for the main components of a structure, including superstructure, substructure, and all components except foundations, the design force, F_d, is given as the elastic-induced force, F_{el}, divided by the R factor listed in AASHTO table

$$F_d = \frac{F_{el}}{R}$$

Unlike bridges in Seismic Zone 2, the design forces of the foundations may not be reduced

$$F_d = F_{el}$$

This difference in reduction is intended to force the formation of plastic hinges in the substructure and to keep the foundation elastic during the excitation (as explained in previous chapters). Because Seismic Zones 3 and 4 expose the bridges to greater hazard, the code does not allow any reduction in the force in the foundation.

In these zones, the code also requires design shear forces to be based on the maximum probable moments, including the overstrength effect as explained in previous chapters. For this purpose, the shear forces must be calculated using an increased strength of flexure capacity to include the overstrength effect by increasing the nominal capacity as follows:

$$M_{os} = 1.30\,M_n \qquad \text{(for concrete structures)}$$

$$M_{os} = 1.25\,M_p \qquad \text{(for steel structures)}$$

where M_n = nominal strength of reinforced concrete section
M_{os} = overstrength moment (most probable moment)
M_p = nominal plastic moment of steel section

9.8 Methods of Analysis

As noted earlier, AASHTO specifies four methods of analysis to find earthquake-induced forces and displacements: the UL method, the SM spectral method, the MM method, and the TH analysis method. These methods are presented in the following sections.

For both directions, the weight of the bridge per unit length is defined as $w(x)$. This includes the weight of the superstructure, the substructure, the foundations, and any added dead loads. In general, the live load is not included in this weight. However, the live load or portion of that may be considered as part of the weight of the bridge in special cases as required by the authority of jurisdiction.

The method of analysis depends on the presence of intermediate expansion joints in the bridge. Because they induce discontinuity in the bridge deck, they must be taken into consideration. Accordingly, bridges may be classified into continuous bridges (without intermediate expansion joints) and discontinuous bridges (with intermediate expansion joints). The two types will be treated separately in the following sections.

9.8.1 Uniform Load Method

The uniform load method models the bridge with its mass lumped at the location of the maximum displacement that results from application of the uniform load to the bridge in each direction. Therefore, for the purpose of finding the period of the structure, the bridge is modeled as an SDOF system. However, the resulting earthquake-induced forces are applied to the structure as uniform forces.

The earthquake-induced forces in the bridge are obtained using the response spectra given in Fig. 9-6. In general, the bridge is analyzed in two perpendicular directions as mentioned earlier: in the longitudinal direction along its major axis between abutments and in a perpendicular transverse direction.

Continuous Bridges

Because continuous bridges are characterized by the absence of intermediate expansion joints, the deck of the bridge will consist of one piece from abutment to abutment. The bridge must be analyzed in both longitudinal and transverse directions.

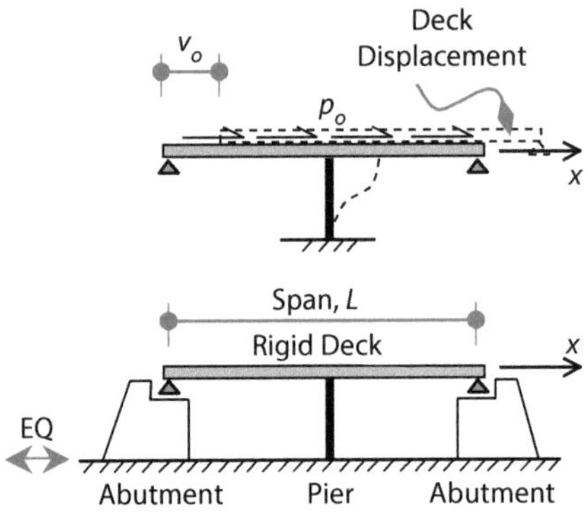

FIGURE 9-10 Longitudinal excitation.

Longitudinal Direction In the longitudinal direction, the bridge deck is considered rigid. As a result, the deck and all of its supporting elements will experience the same displacement. The analysis proceeds by applying an arbitrary uniform force, p_o, along the bridge deck in the direction of excitation and then finding the resulting displacement, v_o. For convenience, this force may be taken as 1. Because this displacement (deflected shape) is considered the first mode of vibration, the earthquake-induced forces will be proportional to this deflected shape. For example, if a bridge is supported by two abutments and one pier, and the bridge is free to move at the abutments as shown in Fig. 9-10, then the deck displacement due to a unit uniform force will be as shown in the same figure. Because the bridge deck is considered to be rigid, the deck displacement along the deck will be constant and equal to v_o.

The period of the bridge is given by the following expression:

$$T_m = 2\pi\sqrt{\frac{M}{K}} = 2\pi\sqrt{\frac{W}{gK}}$$

where T_m = period of the bridge in seconds
M = total mass of the bridge
W = total weight of the bridge
K = stiffness of the bridge
g = ground gravitational acceleration

The total weight and total stiffness of the bridge are calculated as follows:

$$W = \int w(x)\,dx$$

$$K = \frac{p_o L}{v_o}$$

where p_o and v_o are as defined before and shown in Fig. 9-10.

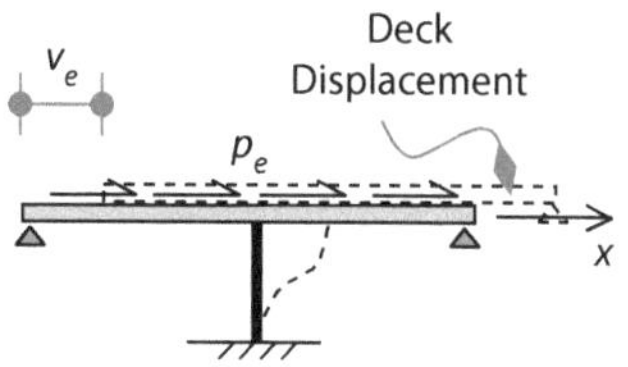

Figure 9-11 Earthquake-induced forces.

The seismic-induced force per unit length of the bridge, p_e, is given by the following expression:

$$p_e = \frac{C_{sm}W}{L}$$

where C_{sm} = elastic response spectrum coefficient as defined in Fig. 9-6
W = total weight of the bridge as defined previously
L = total length of the bridge

By applying the force p_e to the bridge, the internal forces and displacements are calculated in the substructure as shown in Fig. 9-11. The earthquake-induced elastic displacement of the bridge, v_e, can be scaled from v_o by the factor p_e/p_o

$$v_e = \frac{p_e}{p_o} v_o$$

The resulting earthquake-induced forces are resisted by the substructure and consequently transferred to the foundations. For example, the forces in the bridge shown in Fig. 9-10 are resisted by the intermediate pier alone without the abutments. If, however, the bridge is tied at the abutments, the abutments are considered part of the supporting system, and the forces must be resisted by the abutments and the pier according to their stiffness. If the abutment has large stiffness, then the bridge period would tend to zero. Thus, the force is taken as the limits of the response spectrum as applicable.

Transverse Direction In the transverse direction, the bridge deck is not considered rigid. As a result, the deck and all of its supporting elements will experience a displacement that is a function of the bridge stiffness. The analysis proceeds by applying an arbitrary uniform force, p_o, to the bridge deck in the direction of excitation. For convenience, this force may be taken as 1. Because of this unit force, the deck will deflect in the transverse direction with a maximum displacement, $v_{s,max}$. Because this displacement represents the displacement of an SDOF system, the mass may be lumped at this location. For example, if a bridge is supported by two abutments and one pier where the bridge is not free to slide at the abutments as shown in Fig. 9-12, the deck displacement due to unit uniform force will be as shown in the same figure. The mass is then lumped at the location of the maximum displacement as shown.

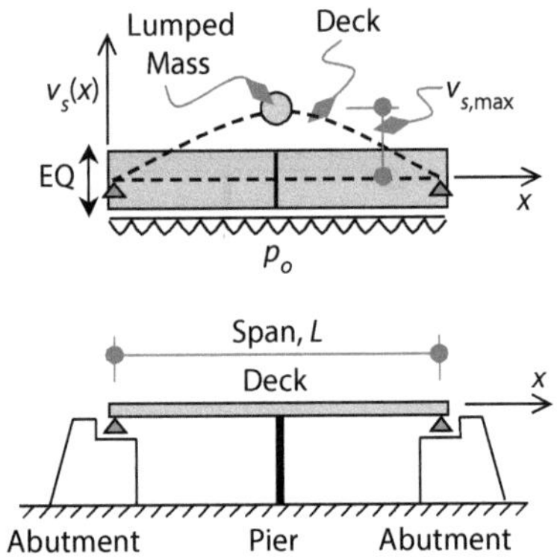

FIGURE 9-12 Transverse excitation.

Accordingly, the period of the bridge is given by the same expression as before

$$T_m = 2\pi\sqrt{\frac{M}{K}} = 2\pi\sqrt{\frac{W}{gK}}$$

where T_m = period of the bridge in seconds
M = total mass of the bridge
W = total weight of the bridge
K = total stiffness of the bridge
g = ground gravitational acceleration

The total weight and total stiffness of the bridge is calculated before

$$W = \int w(x)\,dx$$

$$K = \frac{p_o L}{v_{s,\max}}$$

The seismic-induced force per unit length of the bridge will be a uniform force, p_e, as shown in Fig. 9-13 and will be given by the same expression as in the case of the longitudinal direction

$$p_e = \frac{C_{sm} W}{L}$$

where C_{sm} = elastic response spectrum coefficient as defined in Fig. 9-6
W = total weight of the bridge as defined previously
L = total length of the bridge

By applying this force, p_e, to the bridge, the internal forces and displacements are calculated in the substructure accordingly. The actual earthquake-induced elastic displacement, v_e, is also obtained, which will not be constant in this case.

As in the case of longitudinal excitation, these induced forces are only resisted by the substructure. For example, the forces in the bridge shown in Fig. 9-13 are resisted

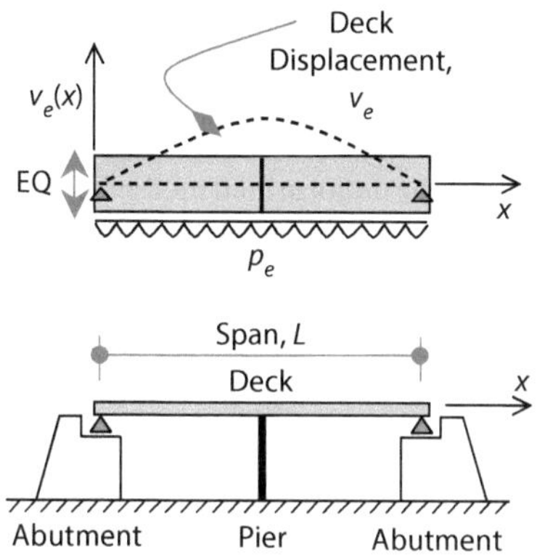

FIGURE 9-13 Earthquake-induced forces.

by both abutments and the intermediate pier. In the transverse direction, the abutments are considered rigid. As a result, they act as a restrictive object to the deck. Because this behavior may be modeled as a hinge support of the deck, the abutments take part of the seismic forces as reactions.

Note that this method can overestimate the lateral forces at the abutments by up to 100 percent in the transverse direction. This occurs with the assumption that the seismic force, p_e, is uniform along the bridge. In reality, this force is equal to p_e at the location of maximum displacement and goes to zero at the abutments following the mode of vibration. Therefore, if this conservatism is undesired, the bridge should be analyzed using the single-mode spectral method (to be covered in the next section).

Example 9-1 A three-span bridge is supported by two abutments, A1 and A2, and two intermediate piers, P1 and P2, with dimensions as shown in Example 9-1, Fig. 1. In the longitudinal direction, the deck is free to slide at both abutments. Each pier consists of three circular concrete columns

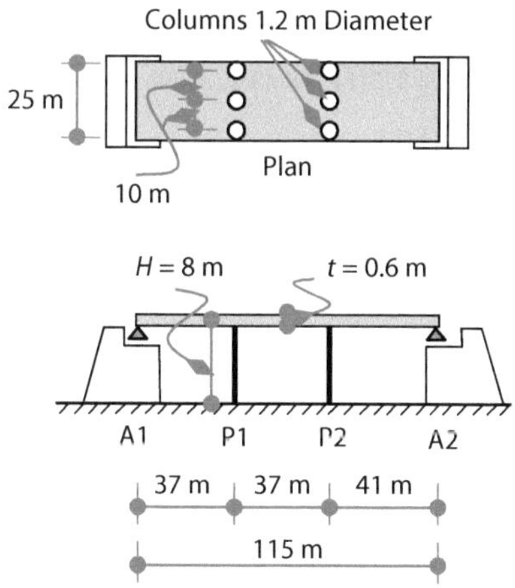

EXAMPLE 9-1, FIGURE 1 Bridge layout.

of grade, $f'_c = 25$ MPa. The columns are 1.2 m in diameter and directly support the deck with total fixity at both deck end and foundation end. The bridge weight per unit length, $w(x)$, is constant and equal to, $w_o = 375$ kN/m.

If the bridge lies in a region with seismic coefficient of PGA = 0.20, $S_S = 0.50$, and $S_1 = 0.20$; and is founded on Soil Type C, use the uniform load method to find the earthquake-induced elastic forces, moments, and displacements in the bridge due to longitudinal and transverse directions separately.

Solution

Part 1: Longitudinal direction

1. Pier properties

   Because the deck is free to slide at both abutments, the piers (three columns each) become the only seismic supporting elements. Because the columns are fixed at top and bottom as shown in Example 9-1, Fig. 2, their stiffness is obtained as follows:

$$E_c = 4{,}700\sqrt{f'_c} = 4{,}700\sqrt{25} = 23{,}500 \text{ MPa}$$

$$I_c = \frac{\pi D^4}{64} = \frac{\pi(1.2)^4}{64} = 0.101788 \text{ m}^4$$

   Column stiffness is evaluated for fixed-fixed ends as

$$k_{\text{col}} = \frac{12E_cI_c}{H^3} = \frac{12(23.5\times10^6)(0.101788)}{(8)^3} = 56{,}063 \text{ kN/m}$$

   The total stiffness of the piers is equal to the summation of the stiffness of each one

$$k_p = k_{\text{col}}\,(\text{No.}) = 56{,}063\,(3) = 168{,}189 \text{ kN/m}$$

$$k_{\text{tot}} = k_p\,(\text{No.}) = 168{,}189\,(2) = 336{,}378 \text{ kN/m}$$

2. Displacement, v_o

$$v_o = \frac{p_oL}{k_{\text{tot}}} = \frac{1(115)}{336{,}378} = 342\times10^{-6} \text{ m}$$

3. Weight and stiffness

$$W = \int w(x)dx = 375(115) = 43{,}125 \text{ kN}$$

$$K = \frac{p_oL}{v_o} = \frac{1(115)}{342\times10^{-6}} = 336{,}257 \text{ kN/m (Note that it is calculated before.)}$$

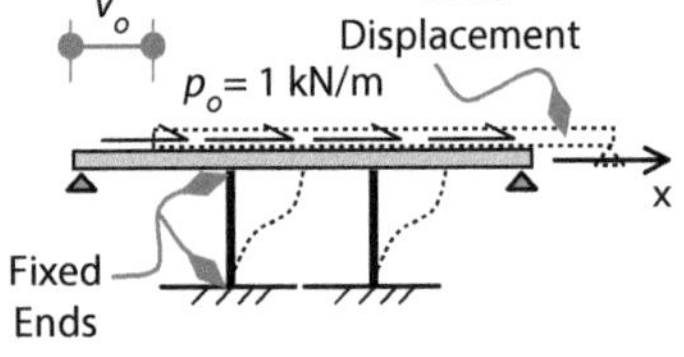

EXAMPLE 9-1, FIGURE 2 Unit force application.

4. Period

$$T_m = 2\pi\sqrt{\frac{W}{gK}} = 2\pi\sqrt{\frac{43{,}125}{9.81(336{,}257)}} = 0.718 \text{ s}$$

5. Elastic response spectrum coefficient, C_{sm}
   Site factors are obtained from tables in App. 9-1 for Soil Class C as follows:

$$F_{pga} = 1.2, \qquad F_a = 1.2, \qquad F_v = 1.6$$

hence,

$$A_s = F_{pga}\text{PGA} = 1.2(0.2) = 0.24$$
$$S_{DS} = F_a S_S = 1.2(0.5) = 0.60$$
$$S_{D1} = F_v S_1 = 1.6(0.2) = 0.32$$
$$T_s = \frac{S_{D1}}{S_{DS}} = \frac{0.32}{0.60} = 0.533$$
$$T_o = 0.2T_s = 0.2(0.533) = 0.107$$

Since $T_m > T_s$, then

$$C_{sm} = \frac{S_{D1}}{T_m} = \frac{0.32}{0.718} = 0.446$$

6. Force per unit length of the bridge

$$p_e = \frac{C_{sm}W}{L} = \frac{0.446(375)(115)}{115} = 167.25 \text{ kN/m}$$

7. Pier forces and displacements

Total force in the deck:	$F_{tot} = p_e L = 167.25(115) = 19{,}234$ kN
Force on top of each pier:	$F_p = F_{tot}/2 = 19{,}234/2 = 9{,}617$ kN
Force on top of each column:	$F_{col} = F_p/3 = 9{,}617/3 = 3{,}206$ kN
Moment in each column:	$M_{col} = F_{col}(4) = 3{,}206(4) = 12{,}824$ kN·m
Elastic displacement:	$v_e = \frac{p_e}{p_o}v_o = \frac{167.25}{1}(342\times10^{-6}) = 0.057$ m

The results are shown in Example 9-2, Fig. 3.

Part 2: Transverse direction

1. Pier properties
   The properties will be the same as noted in Part 1

$$E_c = 23{,}500 \text{ MPa}$$
$$I_c = 0.101{,}788 \text{ m}^4$$

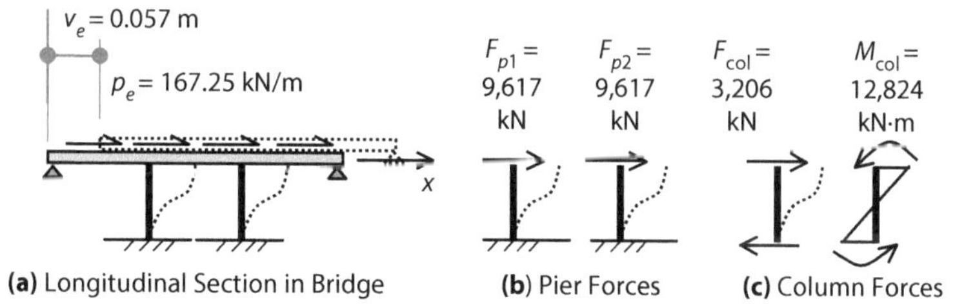

Example 9-1, Figure 3 Earthquake-induced forces and displacements.

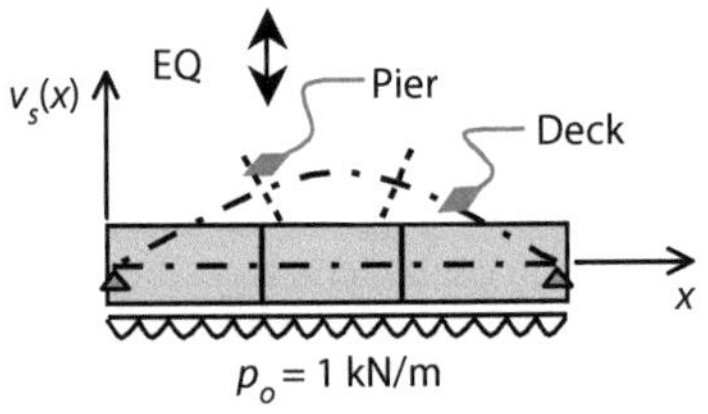

EXAMPLE 9-1, FIGURE 4 Unit force application.

2. Displacement, $v_s(x)$
The displacement in the transverse direction will not be easy to evaluate as in the case of longitudinal excitation. In this case, the deck is considered to be flexible and supported by two abutments and two piers as shown in Example 9-1, Fig. 4. Therefore, the system is highly statically indeterminate. Therefore, in order to evaluate $v_s(x)$, the bridge needs to be modeled as a 3D frame structure, and using structural analysis software becomes unavoidable.

Modeling, of course, is considered an art in addition to an element of engineering. For this analysis, the bridge may be modeled as a space frame as shown in Example 9-1, Fig. 5. The deck is modeled as a longitudinal beam with abutments as pinned supports at each end. The columns are modeled as vertical beams that are supported as required. To preserve the plane section hypothesis of beam theory in the deck, rigid links are required to connect the deck with the columns as shown in Example 9-1, Fig. 5. Because the maximum displacement of the deflected shape is needed, it is important to insert as many nodes as practical along the bridge axis. In this example, one node is inserted between the piers and abutments for this purpose.

A unit force, $p_o = 1$ kN/m, may be applied to the deck as shown in Example 9-1, Fig. 6. This will result in a deflected shape, $v_s(x)$, as also shown in the same figure.

The space frame described above is added as input to SAP software along with the node arrangements shown in Example 9-1, Fig. 5, and using the material and section properties as defined earlier.

The resulting forces and displacements due to p_o are taken directly from the computer output (SAP) as shown in Example 9-1, Tables 1 and 2, respectively.

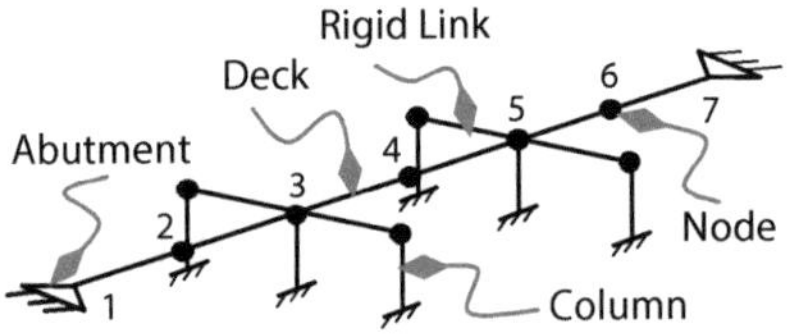

EXAMPLE 9-1, FIGURE 5 Space frame model.

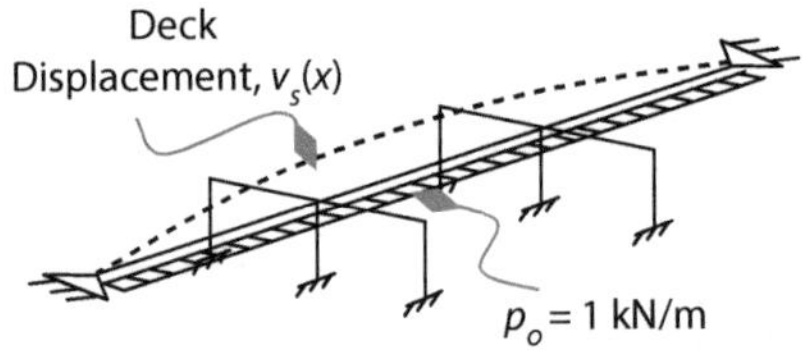

EXAMPLE 9-1, FIGURE 6 Unit force application.

Support	Pier 1			Pier 2			Abutment 1	Abutment 2
Force, kN	12.58			13.35			44.23	44.84
Column	**1**	**2**	**3**	**1**	**2**	**3**	–	–
Force, kN	4.19	4.19	4.19	4.45	4.45	4.45	–	–

Example 9-1, Table 1 Forces and Reactions due to Unit Force, $p_o = 1$ kN/m

Node Number	1	2	3	4	5	6	7
v_s ($\times 10^{-6}$ m)	0	45	77	91	82	50	0

Example 9-1, Table 2 Displacements, v_s, due to Unit Force, $p_o = 1$ kN/m

As can be seen in Example 9-1, Table 2, the maximum displacement along the deck takes place at node 4. Therefore, the maximum displacement is taken as $v_{s,\max} = 91 \times 10^{-6}$ m.

3. Weight and stiffness

$$W = \int w(x)dx = 375(115) = 43{,}125 \text{ kN}$$

$$K = \frac{p_o L}{v_{s,\max}} = \frac{1(115)}{91 \times 10^{-6}} = 1.264 \times 10^6 \text{ kN/m}$$

4. Period

$$T_m = 2\pi\sqrt{\frac{W}{gK}} = 2\pi\sqrt{\frac{43{,}125}{9.81(1.264 \times 10^6)}} = 0.371 \text{ s}$$

5. Elastic response spectrum coefficient, C_{sm}

   Similar to the longitudinal direction, site factors are obtained from tables in App. 9-1 for Soil Class C. These factors and acceleration coefficients will be the same which are repeated here for clarity of presentation as follows:

$$F_{\text{pga}} = 1.2, \qquad F_a = 1.2, \qquad F_v = 1.6$$

   hence,

$$A_s = F_{\text{pga}}\text{PGA} = 1.2(0.2) = 0.24$$

$$S_{\text{DS}} = F_a S_S = 1.2(0.5) = 0.60$$

$$S_{D1} = F_v S_1 = 1.6(0.2) = 0.32$$

$$T_s = \frac{S_{D1}}{S_{\text{DS}}} = \frac{0.32}{0.60} = 0.533$$

$$T_o = 0.2T_s = 0.2(0.533) = 0.107$$

   Since T_m is between T_o and T_s, then

$$C_{\text{sm}} = S_{\text{DS}} = 0.60$$

6. Force per unit length of the bridge

$$p_e = \frac{C_{\text{sm}} W}{L} = \frac{0.60(375)(115)}{115} = 225 \text{ kN/m}$$

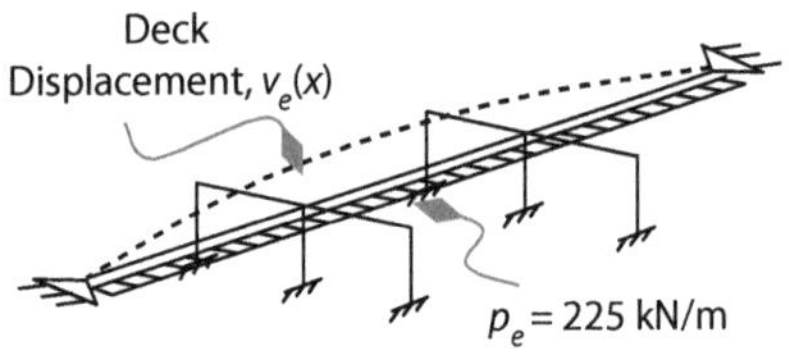

Example 9-1, Figure 7 Seismic force application.

Support	Pier 1			Pier 2			Abutment 1	Abutment 2
Force, kN	2,831			3,004			9,952	10,089
Column	**1**	**2**	**3**	**1**	**2**	**3**	–	–
Force, kN	943	943	943	1,001	1,001	1,001	–	–

Example 9-1, Table 3 Forces due to $p_e = 225$ kN/m

Node Number	1	2	3	4	5	6	7
v_e (mm)	0	10	17	20	18	11	0

Example 9-1, Table 4 Displacements, V_e, due to $p_e = 225$ kN/m

7. Pier forces and displacements
   The forces and displacements due to seismic force are found by applying the force, p_e, to the bridge using the same space frame model used to find p_e as shown in Example 9-1, Fig. 7. The resulting forces and displacements may be scaled from the first run with p_o by the ratio p_e/p_o. Since p_o is taken equal to 1, the results in Example 9-1, Tables 1 and 2, are multiplied by p_e = 225. The results are given in Example 9-1, Tables 3 and 4.

   The moments in the columns are also available from the computer output. For this example, the moments may also be calculated for each column directly from statics as follows:

   Pier P1: $M_{col} = F_{col}(4) = 943\ (4) = 3{,}772$ kN·m

   Pier P2: $M_{col} = F_{col}(4) = 1{,}001\ (4) = 4{,}004$ kN·m

   The results for pier P1 are shown in Example 9-1, Fig. 8.

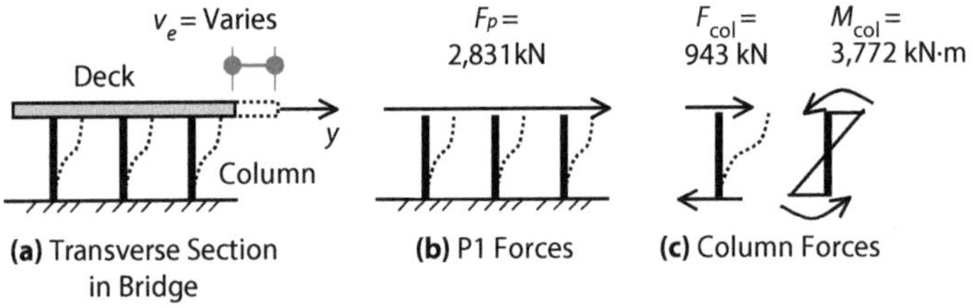

Example 9-1, Figure 8 Earthquake-induced forces.

Discontinuous Bridges

Because discontinuous bridges are characterized by the presence of intermediate expansion joints, the deck of the bridge is divided into separate segments between the expansion joints. The bridge still needs to be analyzed in both longitudinal and transverse directions.

Longitudinal Direction In the longitudinal direction, the deck segments between expansion joints are considered rigid: each segment and all of its supporting elements will experience the same displacement. Therefore, each segment is analyzed independently of the other segments. Accordingly, each segment will have its own weight, period, and induced forces. Therefore, the analysis for each segment proceeds by applying an arbitrary uniform force, p_o, along the bridge segment in the longitudinal direction and by finding the resulting displacement, v_o.

For example, if a bridge is supported by two abutments and three piers where the bridge is free to slide at the abutments as shown in Fig. 9-14, then the deck is analyzed as three independent segments: segment 1, segment 2, and segment 3. The bridge will have three analysis schemes that yield three periods (T_1, T_2, and T_3), three values of independent seismic-induced forces (p_{e1}, p_{e2}, and p_{e3}), and three independent seismic-induced displacements (v_{e1}, v_{e2}, and v_{e3}). The procedures will be identical to the procedures outlined for the continuous bridges and illustrated by Example 9-1.

Transverse Direction In the transverse direction, the bridge segments will not be independent at the location of the expansion joints as in the case of longitudinal directions. The expansion joints are usually detailed to accommodate the expansions in the longitudinal direction only. The effect of expansion joints in the transverse direction will be the reduction of the bridge stiffness because they act as intermediate hinges. Because this effect is addressed in the computer structural model, the procedures can be followed exactly as outlined for continuous bridges and as illustrated in Example 9-1.

9.8.2 Single-Mode Spectral Method

The single-mode spectral method assumes that the bridge only responds in the first mode. The earthquake-induced forces in the bridge are obtained using the response spectra given in Fig. 9-6. In general, the bridge is analyzed in the two perpendicular directions mentioned earlier, the longitudinal direction along its major axis between abutments and the perpendicular transverse direction.

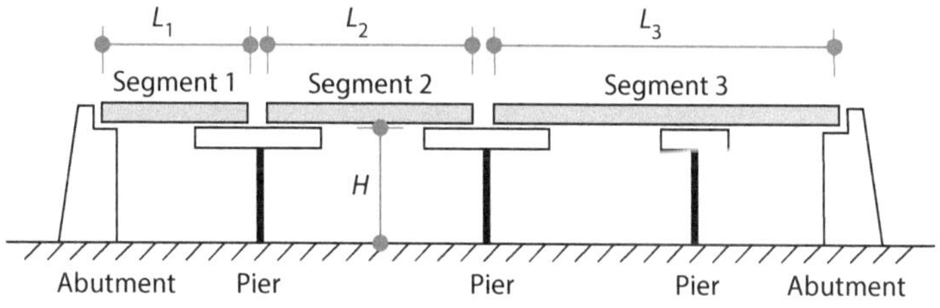

Figure 9-14 Longitudinal excitation.

As in the uniform load method, the weight of the bridge per unit length is defined as $w(x)$. This includes the weight of the superstructure, the substructure, the foundations, and any added dead loads. In general, the live load is not included in this weight. However, the live load or a portion of it may be considered as part of the weight of the bridge in special cases as required by the authority of jurisdiction.

As in the uniform load method, and for the purpose of using this method, bridges are classified into continuous and discontinuous bridges depending on the presence of expansion joints.

Continuous Bridges

Continuous bridges are characterized by the absence of intermediate expansion joints: the deck of a continuous bridge consists of one piece from abutment to abutment. The bridge is analyzed in both longitudinal and transverse directions as noted earlier.

Longitudinal Direction In the longitudinal direction, the bridge deck is considered rigid: the deck and its supporting elements will experience the same displacement. The analysis is similar to the uniform load method by applying an arbitrary uniform force, p_o, along the bridge deck in the direction of excitation. For convenience, this force may be taken as 1. The period and the elastic earthquake-induced forces in the bridge are given as a function of the resulting displacement, $v_s(x)$. For example, if a bridge is supported by two abutments and one pier where the bridge is free to slide at the abutments as shown in Fig. 9-15, the deck displacement due to unit uniform force will be as shown in the same figure.

AASHTO defines three parameters of α, β, and γ to be used in the calculations of the period and the induced forces in the bridge. These parameters are defined as follows:

$$\alpha = \int v_s(x)dx \ (\text{m}^2)$$
$$\beta = \int w(x)\, v_s(x)dx \ (\text{kN·m})$$
$$\gamma = \int w(x)\, v_s^2(x)dx \ (\text{kN·m}^2)$$

where $v_s(x)$ = displacement of the deck due to unit uniform force, p_o
$w(x)$ = weight of the bridge per unit length as defined previously

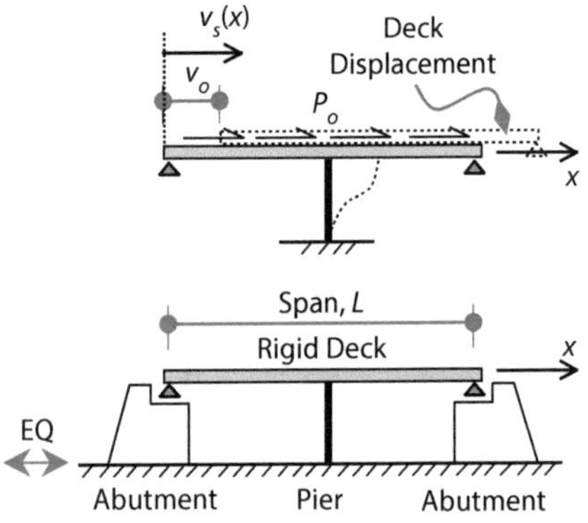

FIGURE 9-15 Longitudinal excitation.

The period of the bridge is given by the following expression:

$$T_m = 2\pi\sqrt{\frac{\gamma}{p_o g\alpha}}$$

where T_m = period of the bridge in seconds
α, γ = parameters as given above
g = ground gravitational acceleration
p_o = unit uniform force applied to the bridge as shown in Fig. 9-15

The seismic-induced force per unit length of the bridge is given by the following expression:

$$p_e = \frac{\beta C_{sm}}{g} w(x)v_s(x)$$

where β, γ = parameters as given above
C_{sm} = elastic response spectrum coefficient as defined in Fig. 9-6
$v_s(x)$ = displacement of the deck due to unit uniform force, p_o
$w(x)$ = weight of the bridge per unit length as defined previously

Because the bridge deck is considered to be rigid, the deck displacement along the deck will be constant and equal to v_o. If v_o is substituted in the parameters above, the following results are obtained:

$$\alpha = \int v_s(x)dx = \int v_o dx = v_o L$$
$$\beta = \int w(x)v_s(x)dx = \int w(x)\, v_o dx = v_o W$$
$$\gamma = \int w(x)v_s^2(x)dx = \int w(x)\, v_o^2 dx = v_o^2 W$$

Inserting the quantities above in the period equation, the period of the bridge is given by the following expression:

$$T_m = 2\pi\sqrt{\frac{\gamma}{p_o g\alpha}} = 2\pi\sqrt{\frac{v_o^2 W}{p_o g v_o L}} = 2\pi\sqrt{\frac{W v_o}{g p_o L}}$$
$$= 2\pi\sqrt{\frac{W}{gK}}$$

Note that the expression above reduces to the expression of the period given by the uniform load method.

The seismic-induced force per unit length of the bridge reduces to the following expression:

$$p_e = \frac{\beta C_{sm}}{\gamma} w(x)v_s(x) = \frac{(v_o W)C_{sm}}{(v_o^2 W)} w(x)v_o$$

or,

$$p_e = \frac{C_{sm} W}{L}$$

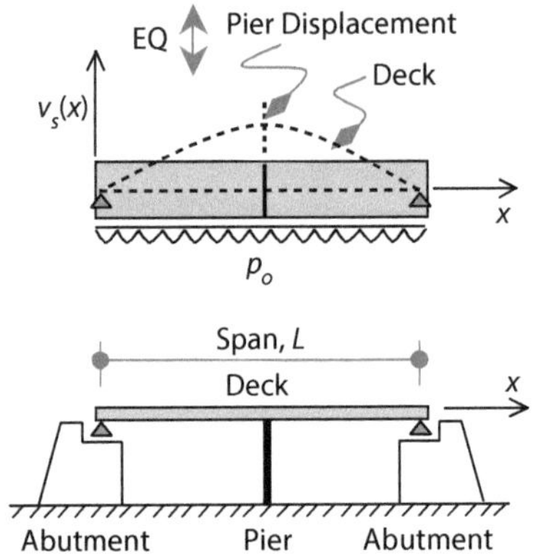

FIGURE 9-16 Transverse excitation.

Note also that the expression for the induced forces above reduces to the expression of the uniform load method. Therefore, for excitation in the longitudinal direction, both the uniform load method and single-mode spectral method are exactly the same. As a result, it seems that there is no need for the single-mode spectral method. In fact, the reason for that is historical because the single-mode spectral method appeared in AASHTO provisions first. Because the single-mode spectral method tends to be cumbersome, the uniform load method was introduced as a simplification, and the single-mode spectral method for longitudinal excitation became redundant.

Transverse Direction In the transverse direction, the bridge deck is not considered rigid, the deck and its supporting elements will experience a displacement that is a function of the bridge stiffness. Similar to the uniform load method, the analysis applies an arbitrary uniform force, p_o, to the bridge deck in the direction of excitation. For convenience, this force may be taken as 1. Similar to longitudinal direction, the period and the elastic earthquake-induced forces in the bridge are given as a function of the resulting displacement, $v_s(x)$. This displacement (deflected shape) is considered the first mode of vibration, and the earthquake-induced forces will be proportional to this shape. As an example, if a bridge is supported by two abutments and one pier where the bridge is fixed at the abutments as shown in Fig. 9-16, the deck displacement due to unit uniform force will be as shown in the same figure.

The procedures for the transverse direction will be the same for the longitudinal direction, taking into consideration that the resulting deflected shape is not constant. Instead, the shape is curved according to the stiffness of both deck and substructure.

As shown in the equation above p_e is proportional to v_s because v_s is assumed to be the first mode of vibration. By applying this force, p_e, to the bridge, the internal forces and displacements are calculated in the substructure accordingly. The actual earthquake-induced elastic displacement, v_e, is also obtained. In this case, the displacement will not be constant.

The resulting earthquake-induced forces are applied to the bridge proportional to the displacement, $v_s(x)$, as this is the assumed mode shape of vibration. The application of p_e proportional to $v_s(x)$ is more realistic than applying it uniformly as in the uniform load method. Consequently, the single-mode spectral method is considered superior to the uniform load method in the transverse direction. In the transverse direction, the abutments are considered to be rigid and act as restrictive objects to the deck. This behavior may be modeled as pinned support of the deck. As a result, the abutments take a portion of the seismic forces as reactions.

Example 9-2 Use the single-mode spectral method to analyze the bridge given in Example 9-1. For convenience, the example is repeated in this section.

A three-span bridge is supported by two abutments, A1 and A2, and two intermediate piers, P1 and P2, with dimensions as shown in Example 9-2, Fig. 1. The deck is free to slide at both abutments. Each pier consists of three circular concrete columns of grade, $f_c' = 25$ MPa. The columns are 1.2 m in diameter and directly support the deck with total fixity at both the deck end and the foundation end. The bridge weight per unit length, $w(x)$, is constant and equal to, $w_o = 375$ kN/m.

If the bridge lies in a region with a seismic coefficient of PGA = 0.20, $S_S = 0.50$, and $S_1 = 0.20$; and is founded on Soil Type C, use the single-mode spectral method to find the earthquake-induced elastic forces, elastic moments and elastic displacements in the bridge due to longitudinal and transverse directions separately.

Solution

Part 1: Longitudinal direction

Because both uniform load method and single-mode spectral method are shown to be the same, there is no need to use the single-mode spectral method for this direction. Refer to Example 9-1 for analysis of this part.

Part 2: Transverse direction

1. Pier properties

   The properties of the piers will be the same as given in Example 9-1

$$E_c = 23{,}500 \text{ MPa}$$

$$I_c = 0.101788 \text{ m}^4$$

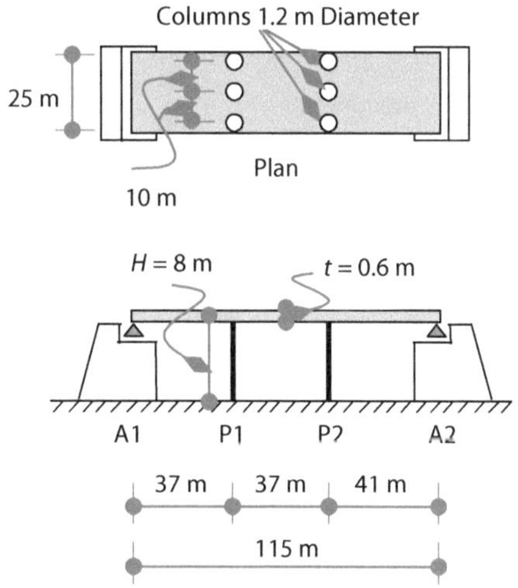

Example 9-2, Figure 1 Bridge layout.

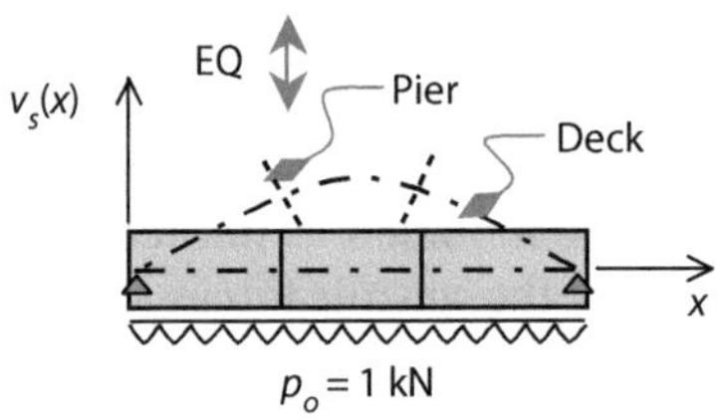

EXAMPLE 9-2, FIGURE 2 Unit force application.

2. Displacement, $v_s(x)$
Displacement in the transverse direction will not be as easy to evaluate as longitudinal excitation. In this case, the deck is considered to be flexible and is supported by two abutments and two piers as shown in Example 9-2, Fig. 2. Therefore, the system is highly statically indeterminate. In order to evaluate $v_s(x)$, the bridge needs to be modeled as a 3D frame structure. As a result, the use of structural analysis software becomes unavoidable.

The structure may be analyzed with the same space frame model used in Example 9-1 (and repeated here for convenience as shown in Example 9-2, Fig. 3). The deck is modeled as a longitudinal beam with abutments as pinned supports at each end. The columns are modeled as vertical beams supported as required. In order to preserve the plane section hypothesis of beam theory in the deck, rigid links are required to connect the deck with the columns as shown in Example 9-2, Fig. 3. Because the deflected shape needs to be integrated to obtain the parameters of α, β, and γ, it is important to insert as many nodes as practical along the bridge axis. In this example, one node is inserted between the piers and abutments for this purpose.

A unit force of, $p_o = 1$ kN/m, may be applied to the deck as shown in Example 9-2, Fig. 4. This will result in a deflected shape, $v_s(x)$, as shown in the same figure.

The space frame described above is entered into SAP software with the node arrangements shown in Example 9-2, Fig. 3, using the material and section properties that are defined earlier.

The resulting displacements from the software program output (SAP) are shown in Example 9-2, Table 1. These are the same displacements that occurred in the Example 9-1 run.

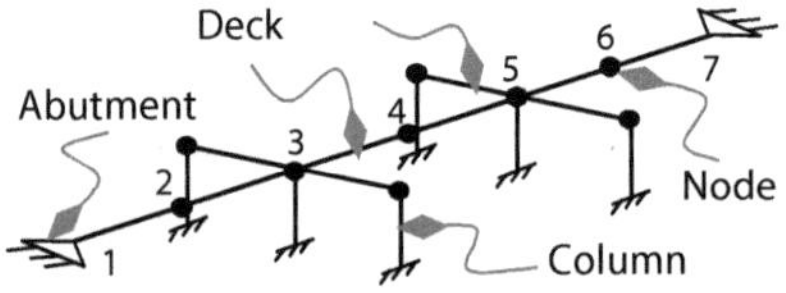

EXAMPLE 9-2, FIGURE 3 Space frame model.

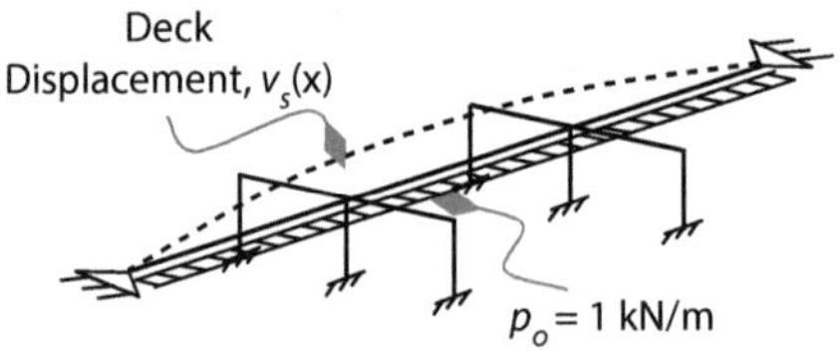

EXAMPLE 9-2, FIGURE 4 Unit force application.

Node Number	1	2	3	4	5	6	7
v_s (×10^{-6} m)	0	45	77	91	82	50	0

EXAMPLE 9-2, TABLE 1 Displacements, v_s, due to Unit Force, p_o = 1 kN/m

3. Parameters α, β, and γ
   The parameters α, β, and γ may be obtained by numerical integration using the nodal displacements shown in Example 9-2, Table 1. The following numerical integration results are obtained using an Excel spreadsheet:

$$\alpha = \int v_s(x)dx = \text{numerical integration} = 0.0066 \text{ m}^2$$
$$\beta = \int w(x)v_s(x)dx = \text{numerical integration} = 2.475 \text{ kN.m}$$
$$\gamma = \int w(x)v_s^2(x)dx = \text{numerical integration} = 0.000174 \text{ kN.m}^2$$

   Period:

$$T_m = 2\pi\sqrt{\frac{\gamma}{p_o g\alpha}} = 2\pi\sqrt{\frac{0.000174}{1\,(9.81)\,(0.0066)}} = 0.326 \text{ s}$$

4. Elastic response spectrum coefficient, C_{sm}
   Similar to the uniform load method, site factors are obtained from tables in App. 9-1 for Soil Class C. These factors and acceleration coefficients will be the same which are repeated here for clarity of presentation as follows:

$$F_{pga} = 1.2, \quad F_a = 1.2, \quad F_v = 1.6$$

   hence,

$$A_s = F_{pga}\text{PGA} = 1.2(0.2) = 0.24$$
$$S_{DS} = F_a S_S = 1.2(0.5) = 0.60$$
$$S_{D1} = F_v S_1 = 1.6(0.2) = 0.32$$
$$T_s = \frac{S_{D1}}{S_{DS}} = \frac{0.32}{0.60} = 0.533$$
$$T_o = 0.2T_s = 0.2(0.533) = 0.107$$

   Since T_m is between T_o and T_s, then

$$C_{sm} = S_{DS} = 0.60$$

5. Force per unit length of the bridge

$$p_e = \frac{\beta C_{sm}}{\gamma} w(x)v_s(x) = \frac{2.475\,(0.60)}{0.000174}(375)v_s$$
$$= 3{,}200{,}431\ v_s(\text{kN/m})$$

6. Pier forces and displacements
   Earthquake-induced forces and displacements are found by applying the force, p_e, to the bridge using the same space frame model used to find p_e as shown in Example 9-2, Fig. 5. The resulting displacements from the software program output at the nodes as defined in Example 9-2, Fig. 3, are given in Example 9-2, Table 2.
   The resulting forces and displacements are also obtained from the computer output as shown in Example 9-2, Tables 2 and 3, as follows:

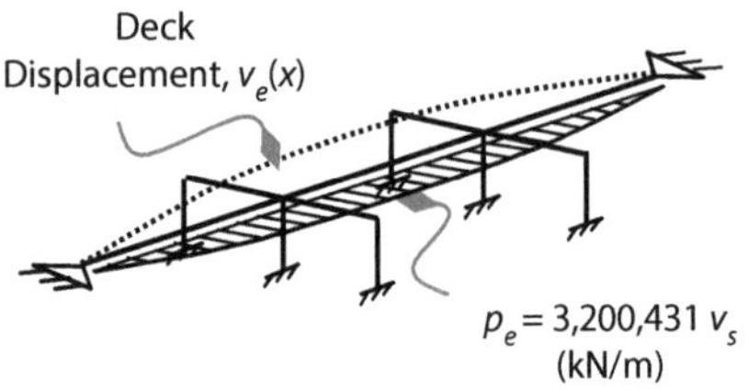

Example 9-2, Figure 5 Seismic force application.

Support	Pier 1			Pier 2			Abutment 1	Abutment 2
Force (kN)	2,842			3,024			7,561	7,693
Column	**1**	**2**	**3**	**1**	**2**	**3**	–	–
Force (kN)	947	947	947	1,008	1,008	1,008	–	–

Example 9-2, Table 2 Forces and Reactions due to Force, $p_e = 3{,}200{,}431\ v_s$ (kN/m)

Node No.	1	2	3	4	5	6	7
v_e (mm)	0	10	18	20	19	11	0

Example 9-2, Table 3 Displacements, v_e, due to $p_e = 3{,}200{,}431\ v_s$ (kN/m)

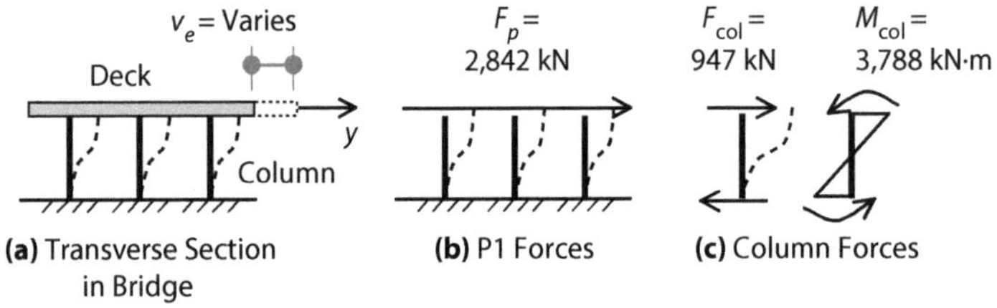

Example 9-2, Figure 6 Earthquake-induced forces.

The moments in the columns are also available from the computer output. For this example, the moments may also be calculated for each column directly from statics as follows:

Pier P1: $M_{col} = F_{col}(4) = 947\ (4) = 3{,}788$ kN·m

Pier P2: $M_{col} = F_{col}(4) = 1{,}008\ (4) = 4{,}032$ kN·m

The results for pier P1 are shown graphically in Example 9-2, Fig. 6.

Comparing results from the uniform load method and the single-mode spectral method reveals that the results are fairly close, except for the reactions at the abutments. As mentioned previously, the uniform load method tends to overestimate abutment reactions up to 100 percent. In this example, the ratio is given as follows:

Abutment 1: $F_{UL}/F_{SM} = 9{,}952/7{,}561 = 1.32$ (32% higher)

Abutment 2: $F_{UL}/F_{SM} = 10{,}089/7{,}693 = 1.31$ (31% higher)

Sinusoidal Method for Continuous Bridges

As noted in the previous section, the evaluation of earthquake-induced forces and displacements in the transverse direction of continuous bridges can be cumbersome. The procedures require the analyst to conduct three stages of analysis procedures that include two runs of a space frame program with an intermediate stage of numerical integration. These procedures may be simplified using the sinusoidal method developed by the author for continuous bridges. For more details on the development, refer to the published paper on this subject as listed in the bibliography (Armouti, 2002).

This method uses the beam on elastic foundation model in conjunction with the minimum total potential concept to evaluate both v_s and v_e. Therefore, the earthquake-induced forces are obtained by simplified procedures. Because the beam first mode is sinusoidal as presented in Chap. 3, the sinusoidal method uses this fact by considering the vibration mode to be sinusoidal.

For the development of this method, consider the bridge given in Example 9-1, which is supported by two abutments, A1 and A2, and two piers, P1 and P2. In this method, the deflected shape of the bridge in the transverse direction is assumed to be sinusoidal as shown in Fig. 9-17. Accordingly, the displacement of the deck, $v_s(x)$, may be expressed in the following form:

$$v_s(x) = v_{so} \sin\frac{\pi x}{L}$$

where v_{so} is the maximum amplitude of the half sine wave at the middle of the bridge.

The supporting piers may now be considered elastic supports to the bridge with an equivalent stiffness that depends on their boundary conditions. For piers that consist of columns only, the stiffness is given as

$$k_p = \Sigma\frac{12EI_c}{H^3} \quad \text{(for fixed-fixed end columns)}$$

$$k_p = \Sigma\frac{3EI_c}{H^3} \quad \text{(for free-fixed end columns)}$$

where H = height of columns

I_c = moment of inertia of the columns in the pier

The abutments are considered to be pinned supports in the transverse direction as presented earlier.

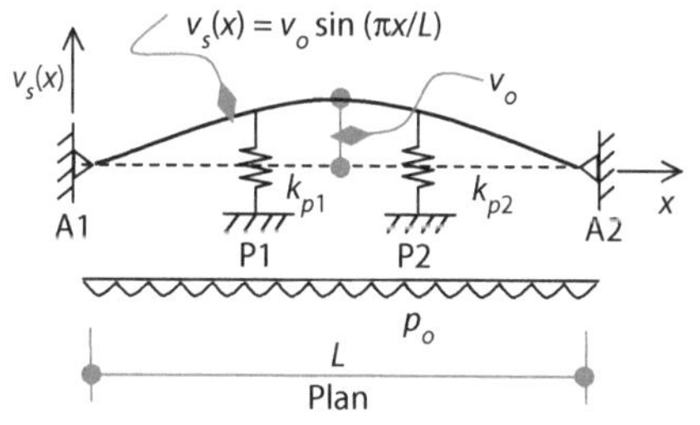

Figure 9-17 Application of load, p_o.

The principle of minimum total potential is used to calculate v_{so}. The total potential, π, is given as the summation of the strain energy, U, and the negative of the work done by external forces, W_{po}. The strain energy is the result of the deformation of the deck, U_d, and the deformation of the piers, U_p. If the integrals in the following expressions are defined over the entire length of the bridge, L, then the total potential energy is expressed as

$$\pi = U - W$$

where $U = U_d + U_p$

The strain energy of a beam in flexure is evaluated as

$$U_d = \int \frac{1}{2} EI_d (v'')^2\, dx$$
$$= \frac{EI_d}{2} \int (v_{so})^2 \left(\frac{\pi}{L}\right)^4 \sin^2\left(\frac{\pi x}{L}\right) dx$$
$$= \frac{EI_d}{4}\left(\frac{\pi^4}{L^3}\right)(v_{so})^2$$

The strain energy of the pier is given as

$$U_p = \Sigma \frac{1}{2} k_p (v_p)^2$$
$$= \Sigma \frac{1}{2} k_p (v_{so})^2 \sin^2\left(\frac{\pi x_p}{L}\right)$$

The work done by the unit force, p_o, is given as

$$W_{po} = \int p_o\, v_s(x) dx$$
$$= \int p_o v_{so} \sin\left(\frac{\pi x}{L}\right) dx$$
$$= p_o\left(\frac{2L}{\pi}\right) v_{so}$$

Adding the quantities above results in π as function of v_{so} is as follows:

$$\pi = U_d + U_p - W_{po}$$

v_{so} can be found by taking the first derivative of π with respect to v_{so}

$$\frac{\partial \pi}{\partial v_{so}} = 0$$

The v_{so} that is found and the uniform distribution of mass (weight) that bridges usually have along their longitudinal axes $[w(x) = w_o]$ make it possible to calculate the parameters α, β, and γ as follows:

$$\alpha = \int v_s(x)\,dx = \int v_{so} \sin\left(\frac{\pi x}{L}\right) dx = \frac{2L}{\pi} v_{so}$$

$$\beta = \int w(x) v_s(x) dx = w_o v_{so} \int \sin\left(\frac{\pi x}{L}\right) dx = \frac{2L}{\pi} w_o v_{so}$$

$$\gamma = \int w(x) v_s(x)^2 dx = w_o (v_{so})^2 \int \sin^2\left(\frac{\pi x}{L}\right) dx = \frac{L}{2} w_o (v_{so})^2$$

Substituting the values above in the period and elastic force expressions results in

$$T_m = 2\pi \sqrt{\frac{\gamma}{p_o g \alpha}} = \sqrt{\frac{\pi^3 w_o v_o}{p_o g}}$$

$$p_e = \frac{\beta C_{sm}}{\gamma} w(x) v_o \sin\left(\frac{\pi x}{L}\right) = \frac{4 C_{sm}}{\pi} w_o \sin\left(\frac{\pi x}{L}\right)$$

$$= p_{eo} \sin\left(\frac{\pi x}{L}\right)$$

The force in the columns due to the earthquake loading, p_e, can also be found by similar simplified procedures that use the sinusoidal deflected shape.

By applying the earthquake load, $p_e = p_{eo} \sin(\pi x/L)$, to the bridge as shown in Fig. 9-18, the resulting deflected shape, v_e, due to p_e is also assumed to be sinusoidal

$$v_e = v_{eo} \sin\left(\frac{\pi x}{L}\right)$$

If we use the principle of minimum total potential energy to find v_{eo}, then

$$\pi = U - W$$

where $U = U_d + U_p$

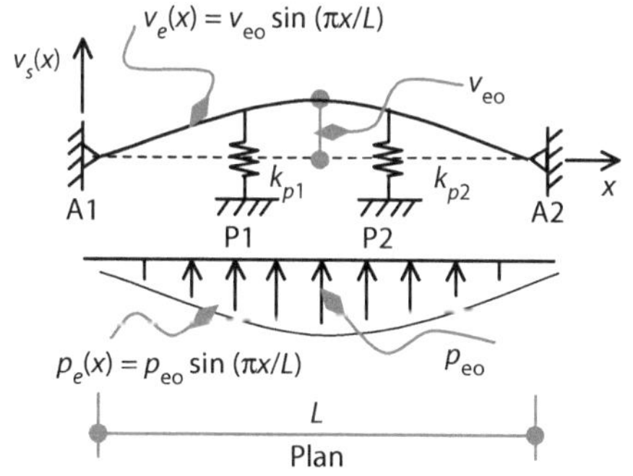

FIGURE 9-18 Application of load, p_e.

U_d is calculated previously for v_{so}. Thus,

$$U_d = \frac{EI_d}{4}\left(\frac{\pi^4}{L^3}\right)(v_{eo})^2 \qquad \text{(similar as before)}$$

$$U_p = \Sigma\frac{1}{2}\,k_p\,(v_{eo})^2 \sin^2\frac{\pi x_p}{L} \qquad \text{(similar as before)}$$

The work done by p_e is calculated as

$$W_{pe} = \int p_e v_e dx$$

$$= \int p_{eo} \sin\left(\frac{\pi x}{L}\right) = v_{eo} \sin\left(\frac{\pi x}{L}\right) dx$$

$$= p_{eo} v_{eo} \frac{L}{2}$$

Adding the quantities above results in

$$\pi = U_d + U_p - W_{pe}$$

v_{eo} can be found by taking the first derivative of π with respect to v_{eo}

$$\frac{\partial \pi}{\partial v_{eo}} = 0$$

Finally, the elastic forces in the pier (columns) are calculated as follows:

$$F_p = k_p v_e(x)$$
$$F_c = k_c v_e(x)$$

Example 9-3 The ease of application and the accuracy of the results may be demonstrated using the example given in Example 9-2, Fig. 1, which is repeated hear for clarity as shown in Example 9-3, Fig. 1.

Relevant parameters and properties are as follows.

Seismic parameters:

$$A_s = 0.24$$
$$S_{DS} = 0.60$$
$$S_{D1} = 0.32$$
$$T_s = 0.533,\ T_o = 0.107$$

Materials: $E_c = 23.5 \times 10^6\ \text{kN/m}^2$

Moment of inertia of the deck in the transverse direction

Deck: $I_d = \dfrac{tB^3}{12} = \dfrac{0.6(25)^3}{12} = 781.25\ \text{m}^4$

$$w(x) = \text{constant} = w_o = 375\ \text{kN/m}$$

Columns: $I_c = \dfrac{\pi D^4}{64} = \dfrac{\pi(1.2)^4}{64} = 0.101788\ \text{m}^4$

$$H = 8\ \text{m}$$

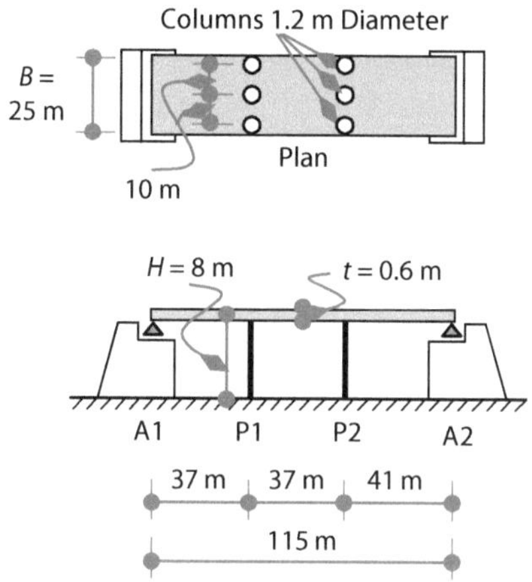

Example 9-3, Figure 1 Bridge layout.

Calculation of initial displacement due to uniformly distributed force, $p_o = 1$ kN/m, applied as shown in Example 9-3, Fig. 2

$$U_d = \frac{EI_d}{4}\left(\frac{\pi^4}{L^3}\right)(v_{so})^2 = \frac{23.5\times10^6(781.25)}{4}\left(\frac{\pi^4}{115^3}\right)(v_{so})^2$$

$$= 293{,}971\ (v_{so})^2\ \text{kN}\cdot\text{m}$$

$$k_c = \frac{12E_cI_c}{H^3} = \frac{12(23.5\times10^6)(0.101788)}{(8)^3} = 56{,}063\ \text{kN/m}$$

$$k_p = 3k_c = 3(56{,}063) = 168{,}189\ \text{kN/m}$$

$$U_p = \sum\frac{1}{2}\,k_c(v_{so})^2\,\sin^2\left(\frac{\pi x_c}{L}\right)$$

$$= \frac{1}{2}(56{,}063)(v_{so})^2\left[3\sin^2\frac{\pi(37)}{115} + 3\sin^2\frac{\pi(74)}{115}\right]$$

$$= 128{,}500\ (v_{so})^2\ \text{kN}\cdot\text{m}$$

$$W_{po} = p_o\left(\frac{2L}{\pi}\right)v_{so} = 1\left(\frac{2(115)}{\pi}\right)v_{so} = 73.211\ v_{so}\ \text{kN}\cdot\text{m}$$

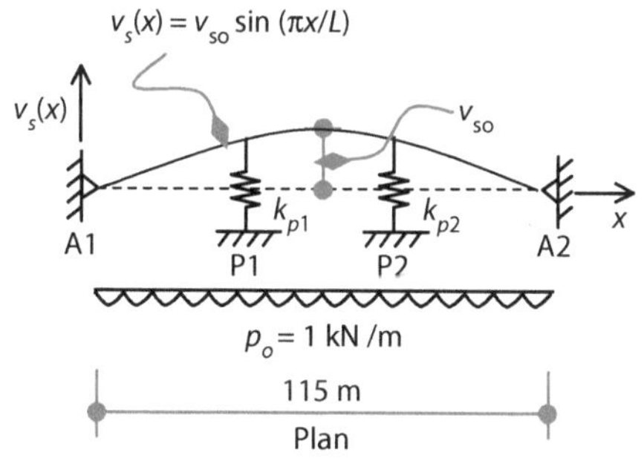

Example 9-3, Figure 2 Application of load, p_o.

Total potential energy, π, becomes

$$\pi = U_d + U_p - W_{po}$$

$$\pi = 293{,}971\,(v_{so})^2 + 128{,}500\,(v_{so})^2 - 73.211\,v_{so}$$

$$\frac{\partial \pi}{\partial v_{so}} = 844{,}942\,v_{so} - 73.211 = 0$$

Solving,

$$v_{so} = 87 \times 10^{-6}\ \text{m}$$

$$T_m = \sqrt{\frac{\pi^3 w_o v_o}{p_o g}} = \sqrt{\frac{\pi^3 (375)(87\times10^{-6})}{(1)(9.81)}} = 0.320\ \text{s}$$

Since T_m is between T_o and T_s, then

$$C_{sm} = S_{DS} = 0.60$$

As a result,

$$p_e = \frac{4C_{sm}}{\pi} w_o \sin\left(\frac{\pi x}{L}\right) = \frac{4(0.60)}{\pi}(375)\sin\left(\frac{\pi x}{115}\right)$$

$$= 287 \sin\frac{\pi x}{115}\ (\text{kN/m})$$

p_e is the earthquake-induced forces in the bridge that are proportional to the vibration mode. To find the displacements and the forces in the bridge due to this load, p_e, the minimum total potential is utilized again. In this case, the quantities U_d and U_p are readily available by replacing v_{so} by v_{eo}, in the previous v_{so} calculations, hence, and by referring to Example 9-3, Fig. 3

$$U_d = 293{,}971(v_{eo})^2\ \text{kN·m as before}$$

$$U_p = 128{,}500(v_{eo})^2\ \text{kN·m as before}$$

$$W_{pe} = p_{eo} v_{eo} \frac{L}{2} = 287\,v_{eo}\frac{115}{2} = 16{,}502\,v_{eo}\ \text{kN·m}$$

$$\pi = U_d + U_p - W_{pe}$$

$$\pi = 293{,}971(v_{eo})^2 + 128{,}500(v_{eo})^2 - 16{,}502\,v_{eo}$$

$$\frac{\partial \pi}{\partial v_{eo}} = 844{,}942\,v_{eo} - 16{,}502 = 0$$

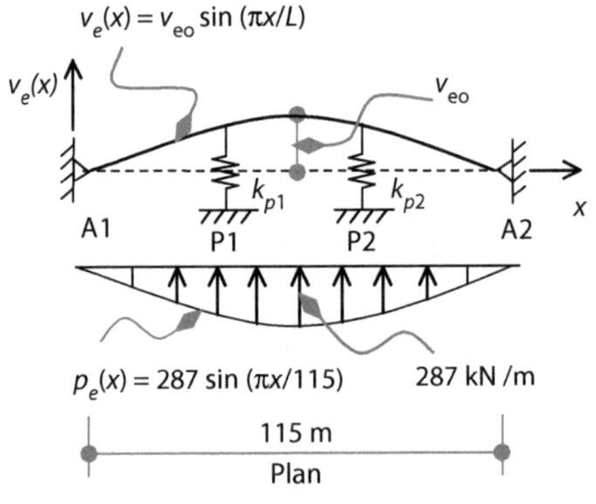

Example 9-3, Figure 3 Application of load, p_e.

Therefore,

$$v_{eo} = 0.019531 \text{ m}$$

The forces in the columns are calculated as a function of their stiffness

$$F_c = k_c v_e(x) = k_c v_{eo} \sin\left(\frac{\pi x}{L}\right)$$

For pier 1: $F_c = 56{,}063(0.019531) \sin\left(\frac{\pi(37)}{115}\right) = 928 \text{ kN}$

For pier 2: $F_c = 56{,}063(0.019531) \sin\left(\frac{\pi(74)}{115}\right) = 986 \text{ kN}$

These results and the associated effort can be compared with the results and effort of using AASHTO procedures by using the space frame model as given in Example 9-2.

Discontinuous Bridges

As noted in the uniform load method, discontinuous bridges are characterized by the presence of intermediate expansion joints. Thus, the deck of the bridge is divided into separate segments between the expansion joints. The bridge still needs to be analyzed in both longitudinal and transverse directions as noted earlier.

Longitudinal Direction The treatment of the longitudinal direction will be the same as it was in the uniform load method presented in the previous section.

Transverse Direction In the transverse direction, the bridge segments will not be independent at the location of the expansion joints as they are in longitudinal directions. The expansion joints are usually detailed to only accommodate the expansions in the longitudinal direction. The effect of the expansion joints in the transverse direction will be to reduce the stiffness of the bridge because they act as intermediate hinges. Because this effect is addressed in the software program structural model, the procedures can be followed exactly as outlined for continuous bridges and as illustrated in Example 9-1.

The implementation of this method can be demonstrated with the following example.

Example 9-4 A four-span bridge is supported by two abutments, A1 and A2, and three intermediate piers, P1, P2, and P3, with the dimensions shown in Example 9-4, Fig. 1. In the longitudinal direction, the deck is free to slide at both abutments. Each pier consists of two circular concrete columns of grade, $E_c = 25$ GPa. The columns are 1.2 m in diameter, and located 16 m apart as shown. The columns are fixed at the foundation level and free to rotate at the bridge level. The bridge has two expansion joints, J1 and J2, located as shown in the same figure. The bridge weight per unit length, $w(x)$, is constant and equal to, $w_o = 300$ kN/m.

If the bridge lies in a region with seismic coefficient, PGA = 0.20, $S_S = 0.50$, and $S_1 = 0.20$; and is supported on Soil Type C, use the single-mode spectral method to find the earthquake-induced elastic forces, elastic moments, and elastic displacements in the bridge. Remember that the bridge is treated as independent segments between expansion joints as explained in the section on Discontinuous Bridges.

Solution

Part 1: Longitudinal direction

Because the uniform load method and single-mode spectral method are the same, there is no need to use the single-mode spectral method for this direction. Refer to Example 9-1 for the analysis of this part.

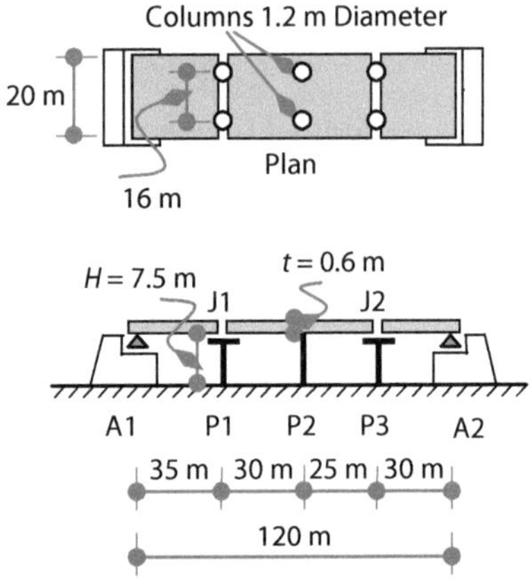

Example 9-4, Figure 1 Bridge layout.

Part 2: Transverse direction

1. Pier properties

$$E_c = 25{,}000 \text{ MPa}$$

$$I_c = \frac{\pi D^4}{64} = \frac{\pi(1.2)^4}{64} = 0.101788 \text{ m}^4$$

2. Displacement $v_s(x)$
The displacement in the transverse direction will not be as easy to evaluate as in the case of longitudinal excitation. In this case, the deck is considered to be flexible and is supported by two abutments and three piers as shown in Example 9-4, Fig. 2. Therefore, the system is highly statically indeterminate. To evaluate $v_s(x)$, the bridge needs to be modeled as a 3D frame structure. As a result, using structural analysis software becomes unavoidable.

The structure may be analyzed with the same space frame model as shown in Example 9-4, Fig. 3. The deck is modeled as a longitudinal beam with abutments as pinned supports at each end. The columns are modeled as vertical beams that are supported as required. To preserve the plane section hypothesis of beam theory in the deck, rigid links are required to connect the deck with the columns. The expansion joints are modeled by introducing end release at joints. These are shown as two hinges around nodes 3 and 7 in Example 9-4, Fig. 3. Because the deflected shape needs to be integrated to obtain the parameters α, β, and γ, it is important to insert as many nodes as practical along the bridge axis. In this example, one node is inserted between the piers and abutments for this purpose.

A unit force of p_o, equal to 1 kN, may be applied to the deck as shown in Example 9-4, Fig. 4. This will result in a deflected shape, $v_s(x)$, as also shown in the same figure.

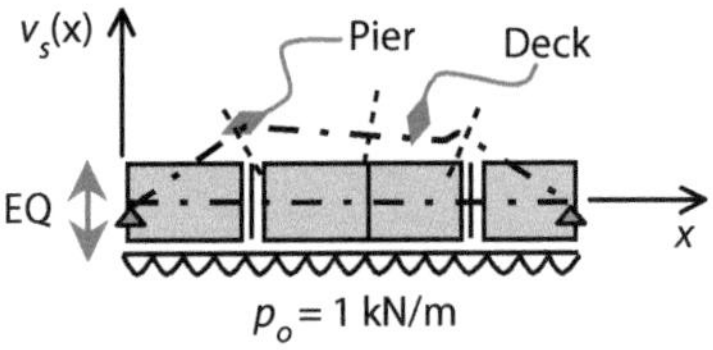

Example 9-4, Figure 2 Unit force application.

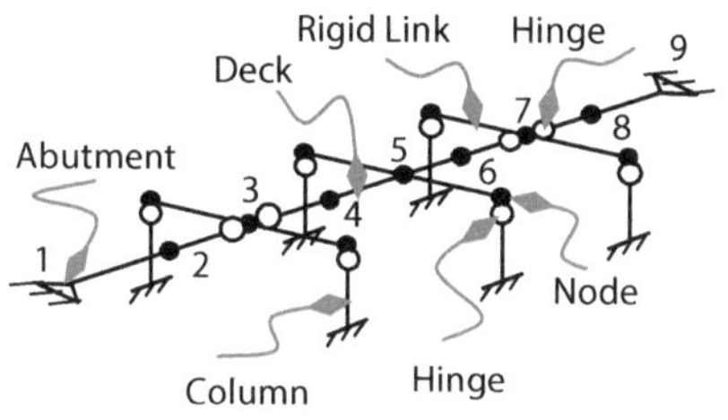

Example 9-4, Figure 3 Space frame model.

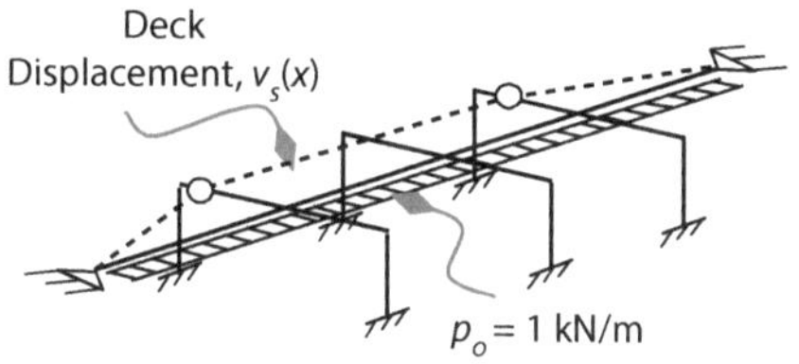

Example 9-4, Figure 4 Unit force application.

The space frame described above is entered into the SAP software program with the node arrangements shown in Example 9-4, Fig. 3, using the material and section properties that are defined earlier.

The resulting displacements from the SAP program output are shown in Example 9-4, Table 1.

3. Parameters α, β, and γ

   The parameters α, β, and γ may be obtained by numerical integration using the nodal displacements shown in Example 9-4, Table 1. Numerical integration results are obtained using the Excel spreadsheet as follows:

$$\alpha = \int v_s(x)dx = \text{numerical integration} = 0.071661\ \text{m}^2$$

$$\beta = \int w(x)v_s(x)dx = \text{numerical integration} = 21.498\ \text{kN}\cdot\text{m}$$

$$\gamma = \int w(x)v_s^2(x)dx = \text{numerical integration} = 0.015457\ \text{kN}\cdot\text{m}^2$$

4. Period

$$T_m = 2\pi\sqrt{\frac{\gamma}{p_o g \alpha}} = 2\pi\sqrt{\frac{0.015457}{1(9.81)(0.071661)}} = 0.932\ \text{s}$$

5. Elastic response spectrum coefficient, C_{sm}

   Site factors are obtained from tables in App. 9-1 for Soil Class C as follows:

$$F_{pga} = 1.2, \qquad F_a = 1.2, \qquad F_v = 1.6$$

Node Number	1	2	3	4	5	6	7	8	9
v_s ($\times 10^{-6}$ m)	0	445	884	884	810	779	746	375	0

Example 9-4, Table 1 Displacements, v_s, due to Unit Force, p_o = 1 kN/m

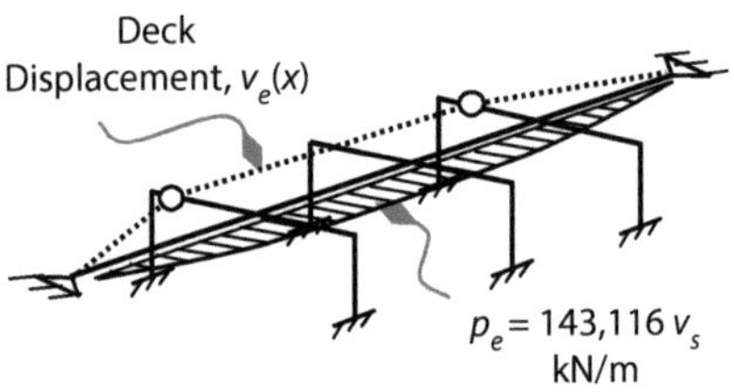

Example 9-4, Figure 5 Seismic force application.

hence,

$$A_s = F_{pga}\text{PGA} = 1.2(0.2) = 0.24$$
$$S_{DS} = F_a S_s = 1.2(0.5) = 0.60$$
$$S_{D1} = F_v S_1 = 1.6(0.2) = 0.32$$
$$T_s = \frac{S_{D1}}{S_{DS}} = \frac{0.32}{0.60} = 0.533$$
$$T_o = 0.2T_s = 0.2(0.533) = 0.107$$

Since $T_m > T_s$, then

$$C_{sm} = \frac{S_{D1}}{T_m} = \frac{0.32}{0.932} = 0.343$$

6. Force per unit length of the bridge

$$p_e = \frac{\beta C_{sm}}{\gamma} w(x) v_s(x) = \frac{21.498(0.343)}{0.015457}(300)v_s$$
$$= 143{,}116\ v_s(\text{kN/m})$$

7. Pier forces and displacements
The earthquake-induced forces and displacements are found by applying the force, p_e, to the bridge using the same space frame model used to find p_e as shown in Example 9-4, Fig. 5. The resulting displacements from the computer output at the nodes as defined in Example 9-4, Fig. 3, are given in Example 9-4, Table 2.
The resulting forces and displacements are obtained from the software program output provided in Example 9-4, Tables 2 and 3, as follows:

Support	**Pier 1**		**Pier 2**		**Pier 3**		**Abutment 1**	**Abutment 2**
Force (kN)	3,412		2,976		2,586		742	536
Column	**1**	**2**	**1**	**2**	**1**	**2**	–	–
Force (kN)	1,706	1,706	1,488	1,488	1,293	1,293	–	–

Example 9-2, Table 2 Forces and Reactions due to Force, $p_e = 143{,}116\ v_s$, kN/m

Node Number	1	2	3	4	5	6	7	8	9
v_e (mm)	0	48	95	89	83	78	72	36	0

Example 9-2, Table 3 Displacements, v_e, due to $p_e = 143{,}116\ v_s$, kN/m

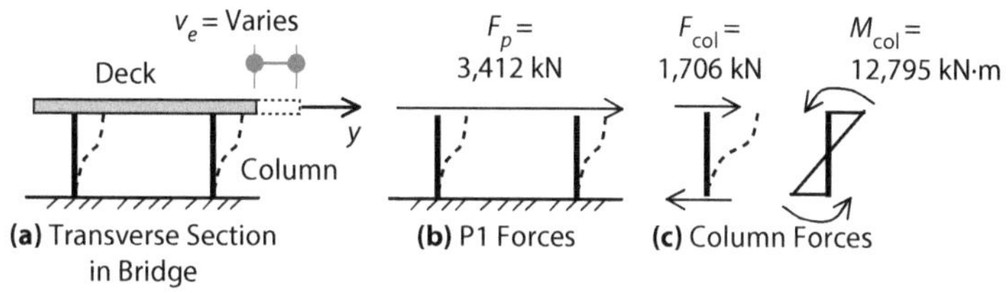

Example 9-4, Figure 6 Earthquake-induced forces.

The moments in the columns are also available from the software program's output. For this example, the moments may also be calculated for each column directly from statics as follows:

Pier P1: $M_{col} = F_{col}(7.5) = 1{,}706\ (7.5) = 12{,}795$ kN·m

Pier P2: $M_{col} = F_{col}(7.5) = 1{,}488\ (7.5) = 11{,}160$ kN·m

Pier P3: $M_{col} = F_{col}(7.5) = 1{,}293\ (7.5) = 9{,}698$ kN·m

The results for pier P1 are shown in Example 9-4, Fig. 6.

Rigid Deck Method for Discontinuous Bridges

The evaluation of earthquake-induced forces and displacements in the transverse direction of discontinuous bridges can also be cumbersome in a way that is similar to the case for continuous bridges. The procedures require the analyst to conduct three stages of analysis procedures that include two runs of a space frame program with an intermediate stage of numerical integration. These procedures may be simplified using the new rigid deck method developed by the author for discontinuous bridges. For more details on the development, see the author published paper on this subject as listed in the bibliography.

Similar to the sinusoidal method, this method uses the beam on elastic foundation model in conjunction with the minimum total potential concept to evaluate both v_s and v_e. The earthquake-induced forces are obtained by simplified procedures.

Because of the rigidity of the deck in the transverse direction with respect to the substructure, the deck segments between the expansion joints may be assumed to remain rigid under lateral displacement of the bridge as shown in Fig. 9-19. As a result, these joints do not store any strain energy. This behavior is similar to rigid diaphragm behavior in buildings. In this case, the bridge modeling is greatly simplified by limiting the degrees of freedom of each segment to two DOFs at its ends, where each segment is supported by elastic supports (the columns). Therefore, the bridge as a whole can be modeled as a series of rigid segments (rigid deck model) having a number of degrees of freedom equal to the number of the expansion joints.

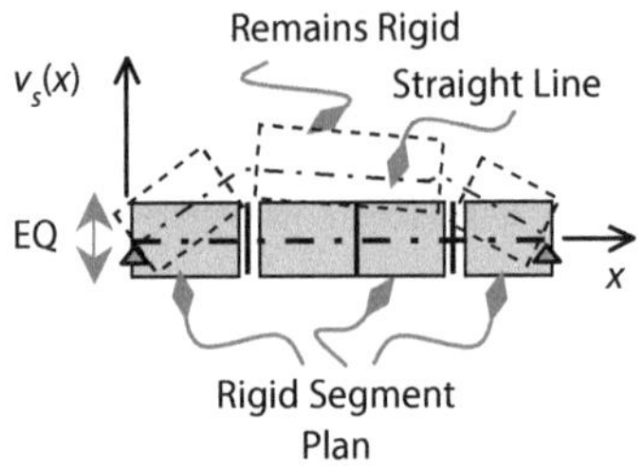

FIGURE 9-19 Rigid deck assumption.

For the development of this method, consider the bridge given in Example 9-4. This four-span bridge consists of a deck slab directly supported by two abutments at A1 and A2 and three intermediate piers, P1, P2, and P3, with height, H, as shown in Fig. 9-20. The bridge contains two expansion joints, J1 and J2, that are modeled as internal hinges. The bridge dimensions are given as total length, L, with four spans in between marked L_1, L_2, L_3, and L_4; breadth, B; and deck (slab) thickness, t. In the transverse direction, the bridge may be modeled as a beam on elastic foundation with rigid segments between expansion joints. Because the bridge is transversely restricted at the abutments, the bridge will have two degrees of freedom at the location of expansion joints, P1 and P3, and the rest of the bridge displacements will be expressed in terms of these two degrees of freedom.

With these assumptions, the equivalent stiffness of the pier can be evaluated as follows:

For fixed-fixed end: $$k_p = \Sigma \frac{12EI_c}{H^3}$$

For free-fixed end: $$k_p = \Sigma \frac{3EI_c}{H^3}$$

where I_c = moment of inertia of the columns
H = height of columns

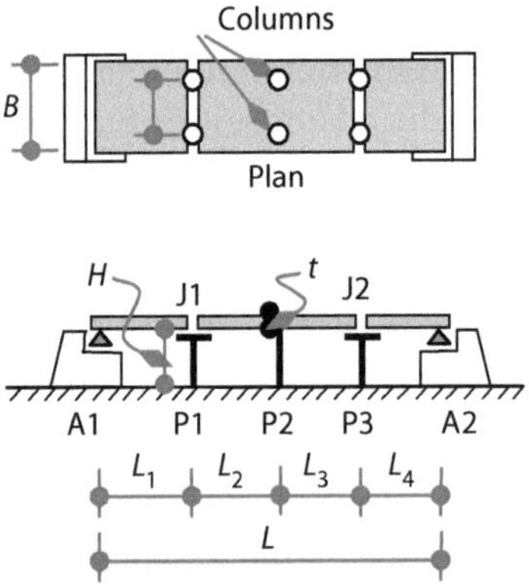

FIGURE 9-20 Bridge layout.

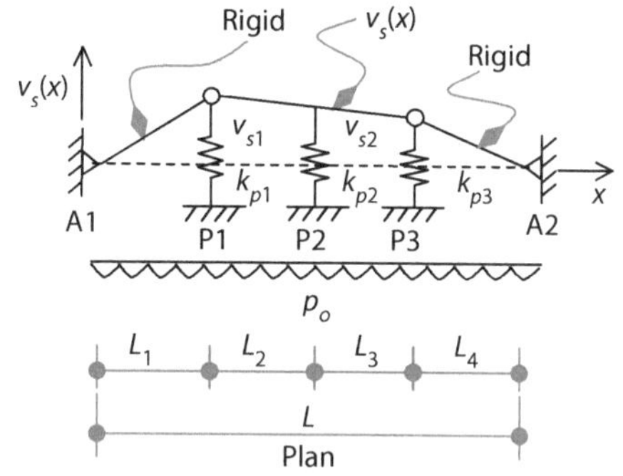

Figure 9-21 Application of load, p_o.

The procedures start by applying a uniform load to the bridge in the transverse direction. The resulting deflected shape will consist of three straight lines as shown in Fig. 9-21. The displacements of the expansion joints, v_{s1} and v_{s2}, are unknown quantities to be found.

The principle of minimum total potential is used to calculate v_{s1} and v_{s2}. The total potential, π, is given as the summation of the strain energy and the negative of the work done by external loads. Because the deck is considered to be rigid, the strain energy in this case is only the result of the deformation of the piers, U_p, while the work done is due to load p_o is W_{po}. As a result

$$\pi = U - W$$

where

$$\begin{aligned} U &= U_p \\ &= \Sigma \frac{1}{2} k_p (v_p)^2 \\ &= \frac{1}{2} k_{p1}(v_{s1})^2 + \frac{1}{2} k_{p3}(v_{s2})^2 + \frac{1}{2} k_{p2}\left(\frac{L_3 v_{s1}}{L_2 + L_3} + \frac{L_2 v_{s2}}{L_2 + L_3} \right)^2 \end{aligned}$$

and

$$\begin{aligned} W &= W_{po} = \int p_o v_s dx \\ &= p_o(\text{area of the deflected shape}) \\ &= p_o\left[\frac{1}{2} L_1 v_{s1} + \frac{1}{2}(L_2 + L_3)(v_{s1} + v_{s2}) + \frac{1}{2} L_4 v_{s2} \right] \end{aligned}$$

Adding the expressions above, the total potential energy becomes

$$\begin{aligned} \pi &= U_p - W_{po} \\ &= a_1(v_{s1})^2 + a_2(v_{s2})^2 + a_3 v_{s1} v_{s2} - a_4 v_{s1} - a_5 v_{s2} \end{aligned}$$

where a_1, a_2, a_3, a_4, and a_5 are constant quantities associated with the displacement terms.

v_{s1} and v_{s2} can be found by taking the first partial derivative of π with respect to v_{s1} and v_{s2}

$$\frac{\partial \pi}{\partial v_{s1}} = 2a_1 v_{s1} + a_3 v_{s2} - a_4 = 0$$

$$\frac{\partial \pi}{\partial v_{s2}} = 2a_2 v_{s2} + a_3 v_{s1} - a_5 = 0$$

For convenience, these equations may be presented in matrix form such that

$$\begin{bmatrix} 2a_1 & a_3 \\ a_3 & 2a_2 \end{bmatrix} \begin{Bmatrix} v_{s1} \\ v_{s2} \end{Bmatrix} = \begin{Bmatrix} a_4 \\ a_5 \end{Bmatrix}$$

The size of the coefficient matrix $[a_{ij}]$ corresponds to the number of expansion joints in the bridge.

Solving the matrix equation above yields the values of v_{s1} and v_{s2}. By noting that most bridges have uniform distribution of mass (weight) along their longitudinal axes $[w(x) = w_o]$, the parameters α, β, and γ as defined before can be calculated as

$$\alpha = \int v_s(x)dx = \text{area under the deflected shape}$$
$$= \frac{1}{2}L_1 v_{s1} + \frac{1}{2}(L_2 + L_3)(v_{s1} + v_{s2}) + \frac{1}{2}L_4 v_{s2}$$
$$\beta = \int w(x)v_s(x)dx = w_o\alpha$$
$$\gamma = \int w(x)v_s(x)^2\,dx = w_o\int (v_s)^2 dx$$
$$= w_o(\text{integration of two trapezoids: a standard operation})$$
$$= w_o\frac{1}{3}[L_1(v_{s1})^2 + (L_2 + L_3)\{(v_{s1})^2 + (v_{s2})^2 + (v_{s1}v_{s2})\} + L_4(v_{s2})^2]$$

Substituting the values above in the period and elastic force expressions given before results in

$$T_m = 2\pi\sqrt{\frac{\gamma}{p_o g \alpha}}$$

$$p_e = \frac{\beta C_{sm}}{\gamma} w(x)v_s(x)$$

that is, $$p_e \propto v_s$$

where C_{sm} is the elastic response spectrum as given before.

The force in the columns due to the earthquake loading, p_e, can also be found through similar simplified procedures by making use of the rigid segment assumption.

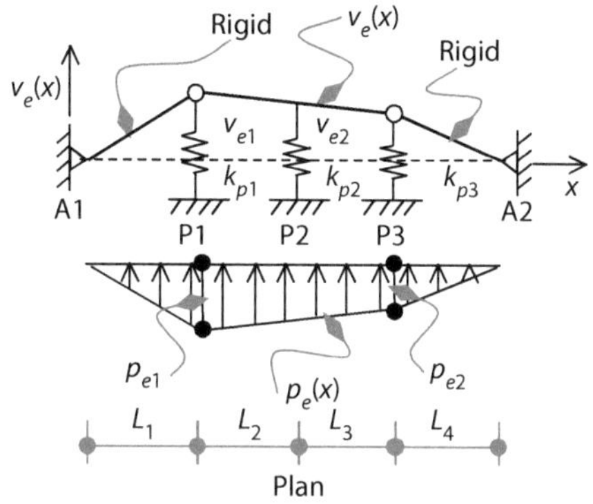

FIGURE 9-22 Application of load, p_e.

Because earthquake-induced force, p_e, is proportional to the vibration mode, which is the deflected shape, v_s, the force, p_e, is applied to the bridge as shown in Fig. 9-22. The resulting deflected shape, v_e, due to p_e, is found as before

$$\pi = U - W$$

where

$$U = U_p \quad \text{(as calculated previously for } v_s\text{)}$$

$$= \Sigma \frac{1}{2} k_p (v_p)^2$$

$$= \frac{1}{2} k_{p1}(v_{e1})^2 + \frac{1}{2} k_{p3}(v_{e2})^2 + \frac{1}{2} k_{p2}\left(\frac{L_3 v_{e1}}{L_2 + L_3} + \frac{L_2 v_{e2}}{L_2 + L_3}\right)^2$$

The work done by p_e is calculated as

$$W_{pe} = \int p_e v_e dx$$

$$= \text{integration of two trapezoids}$$

$$= \frac{1}{3} L_1 (p_{e1} v_{e1}) + \frac{1}{3} L_4 (p_{e2} v_{e2})$$

$$+ \frac{1}{6}(L_2 + L_3)(2 p_{e1} v_{e1} + 2 p_{e2} v_{e2} + p_{e1} v_{e2} + p_{e2} v_{e1})$$

Adding the expressions above, the total potential energy becomes

$$\pi = U_p - W_{pe}$$

$$= a_6 (v_{e1})^2 + a_7 (v_{e2})^2 + a_8 v_{e1} v_{e2} - a_9 v_{e1} - a_{10} v_{e2}$$

v_{e1} and v_{e2} can be found by taking the first partial derivatives of π with respect to v_{e1} and v_{e2}

$$\frac{\partial \pi}{\partial v_{e1}} = 2a_6 v_{e1} + a_8 v_{e2} - a_9 = 0$$

$$\frac{\partial \pi}{\partial v_{e2}} = 2a_7 v_{e2} + a_8 v_{e1} - a_{10} = 0$$

$$\begin{bmatrix} 2a_6 & a_8 \\ a_8 & 2a_7 \end{bmatrix} \begin{Bmatrix} v_{e1} \\ v_{e2} \end{Bmatrix} = \begin{Bmatrix} a_9 \\ a_{10} \end{Bmatrix}$$

Solving the matrix equation above yields the values of v_{e1} and v_{e2}.

Finally, the elastic forces in the pier (columns) are calculated as

$$F_p = k_p\, v_e(x)$$
$$F_c = k_c\, v_e(x)$$

Example 9-5 The ease of application and the accuracy of the results may be demonstrated using the bridge in Example 9-4. The bridge layout and dimensions are repeated in Example 9-5, Fig. 1 for convenience.

Relevant parameters and properties are as follows.

Seismic parameters:

$$A_s = 0.24$$
$$S_{DS} = 0.60$$
$$S_{D1} = 0.32$$
$$T_s = 0.533,\ T_o = 0.107$$

Materials: $E_c = 25 \times 10^6$ kN/m^2

$$I_c = \frac{\pi D^4}{64} = \frac{\pi (1.2)^4}{64}$$
$$= 0.101788 \text{ m}^4$$

$$k_c = \frac{3E_c I_c}{H^3} = \frac{3(25\times10^6)(0.101788)}{(7.5)^3} = 18{,}095 \text{ kN/m}$$

$$k_p = 2k_c = 2(18{,}095) = 36{,}191 \text{ kN/m}$$

The total potential energy of the bridge due to the application of load, $p_o = 1$ kN/m, to the bridge as shown in Example 9-5, Fig. 2, is calculated as follows:

$$U_p = \frac{1}{2}k_{p1}(v_{s1})^2 + \frac{1}{2}k_{p3}(v_{s2})^2 + \frac{1}{2}k_{p2}\left(\frac{L_3 v_{s1}}{L_2 + L_3} + \frac{L_2 v_{s2}}{L_2 + L_3}\right)^2$$

$$= \frac{1}{2}(36{,}191)\left[(v_{s1})^2 + (v_{s2})^2 + \left(\frac{25v_{s1}}{55} + \frac{30v_{s2}}{55}\right)^2\right]$$

$$= 21{,}834(v_{s1})^2 + 23{,}479(v_{s2})^2 + 8{,}973\, v_{s1}v_{s2}$$

$$W_{po} = p_o\left[\frac{1}{2}L_1 v_{s1} + \frac{1}{2}(L_2 + L_3)\ (v_{s1} + v_{s2}) + \frac{1}{2}L_4 v_{s2}\right]$$

$$= (1)\left[\frac{1}{2}(35)(v_{s1}) + \frac{1}{2}(55)(v_{s1} + v_{s2}) + \frac{1}{2}(30)(v_{s2})\right]$$

$$= 45v_{s1} + 42.5v_{s2}$$

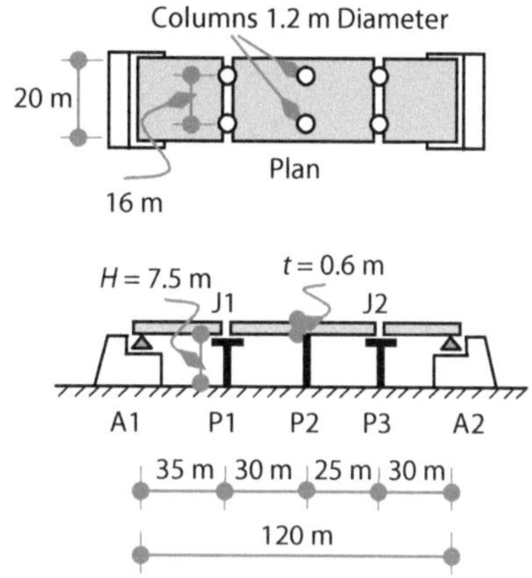

Example 9-5, Figure 1 Bridge layout.

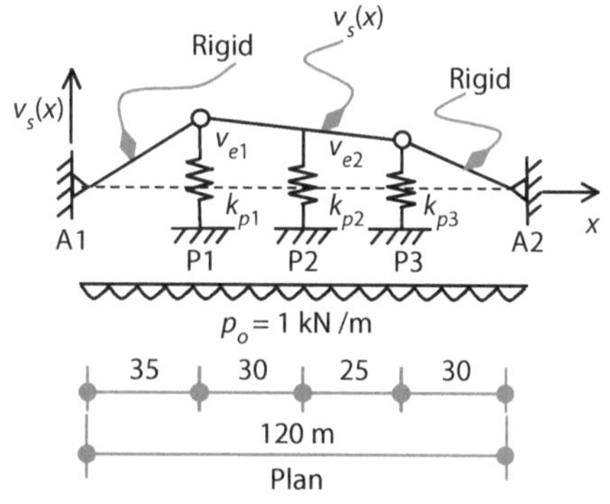

Example 9-5, Figure 2 Application of load, p_o.

Therefore,

$$\pi = U_p - W_{po}$$
$$= 21{,}834(v_{s1})^2 + 23{,}479(v_{s2})^2 + 8{,}973 v_{s1} v_{s2} - 45 v_{s1} - 42.5 v_{s2}$$

Taking partial derivatives of Π w.r.t. V_{s1} and V_{s2}, we get

$$\frac{\partial \pi}{\partial v_{s1}} = 43{,}668 v_{s1} + 8{,}973 v_{s2} - 45 = 0$$

$$\frac{\partial \pi}{\partial v_{s2}} = 8{,}973 v_{s1} + 46{,}958 v_{s2} - 42.5 = 0$$

In matrix form, the two equations above are given as

$$\begin{bmatrix} 43{,}668 & 8{,}973 \\ 8{,}973 & 46{,}958 \end{bmatrix} \begin{Bmatrix} v_{s1} \\ v_{s2} \end{Bmatrix} = \begin{Bmatrix} 45.0 \\ 42.5 \end{Bmatrix}$$

Solving the above matrix equation, v_{s1} and v_{s2} are obtained as

$$v_{s1} = 0.879 \text{ mm}$$

$$v_{s2} = 0.737 \text{ mm}$$

The parameters α, β, and γ are calculated as a function of v_{s1} and v_{s2}:

α = area under deflected shape which is evaluated for W_{po} calculations;

$$= 45v_{s1} + 42.5v_{s2}$$

$$= 45\,(0.879 \times 10^{-3}) + 42.5\,(0.737 \times 10^{-3}) = 0.070878 \text{ m}^2$$

$$\beta = w_o\,\alpha$$

$$= 300\,(0.070878) = 21.263 \text{ kN·m}$$

$$\gamma = w_o \frac{1}{3}[L_1(v_{s1})^2 + (L_2 + L_3)\{(v_{s1})^2 + (v_{s2})^2 + (v_{s1}\, v_{s2})\} + L_4(v_{s2})^2]$$

$$= 300\left(\frac{1}{3}\right)(10^{-6})\,[35(0.879)^2 + 55\,\{(0.879)^2 + (0.737)^2 + (0.879)(0.737)\} + 30(0.737)^2]$$

$$= 0.015133 \text{ kN·m}^2$$

Using parameters α, β, and γ, the period and the equivalent earthquake forces are calculated as

$$T_m = 2\pi\sqrt{\frac{\gamma}{p_o g \alpha}} = \sqrt{\frac{0.015133}{1(9.81)(0.070878)}} = 0.927 \text{ s}$$

Since $T_m > T_s$, then

$$C_{sm} = \frac{S_{D1}}{T_m} = \frac{0.32}{0.927} = 0.345$$

Equivalent earthquake force, p_e, is calculated as

$$p_e = \frac{\beta C_{sm}}{\gamma} w(x) v_s(x) = \frac{21.263(0.345)}{0.015133}(300)v_s$$

$$= 145{,}425\; v_s \text{ kN/m}$$

Substituting the value of v_{s1} and v_{s2} in the equation above, the intensity of the loads at the locations of expansion joints J1 and J2 can be calculated as

$$p_{e1} = 145{,}425\,(0.879 \times 10^{-3}) = 128 \text{ kN/m}$$

$$p_{e2} = 145{,}425\,(0.737 \times 10^{-3}) = 107 \text{ kN/m}$$

Applying the equivalent earthquake forces to the bridge as shown in Example 9-5, Fig. 3, the resulting deflected shape, v_e, is calculated by replacing v_s by v_e in the energy and the work done expressions, which result in

$$U_p = 21{,}834(v_{e1})^2 + 23{,}479(v_{e2})^2 + 8{,}973\, v_{e1} v_{e2}$$

$$W_{pe} = \frac{1}{3} L_1\,(p_{e1} v_{e1}) + \frac{1}{3} L_4\,(p_{e2} v_{e2})$$

$$+ \frac{1}{6}(L_2 + L_3)(2p_{e1}v_{e1} + 2p_{e2}v_{e2} + p_{e1}v_{e2} + p_{e2}v_{e1})$$

$$= \frac{1}{3}(35)(128)v_{e1} + \frac{1}{3}(30)(107)v_{e2}$$

$$+ \frac{1}{6}(55)[2(128)v_{e1} + 2(107)v_{e2} + 128v_{e2} + 107v_{e1}]$$

$$= 4{,}821v_{e1} + 4{,}205v_{e2}$$

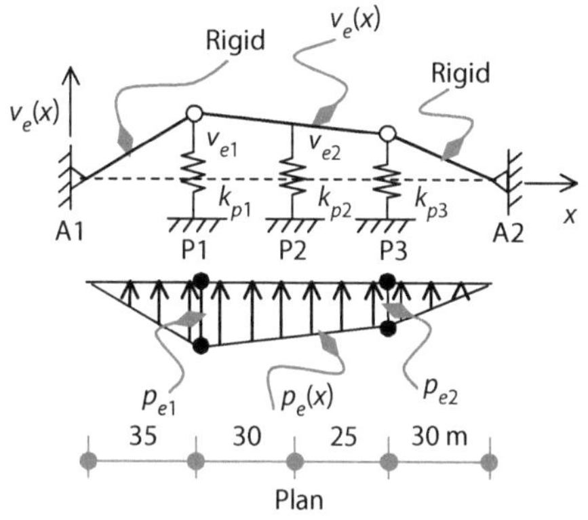

Example 9-5, Figure 3 Application of load, p_e.

Therefore, the total potential of the bridge under displacements v_{e1} and v_{e2} will be

$$\begin{aligned}\pi &= U_p - W_{pe} \\ &= 21{,}834(v_{e1})^2 + 23{,}479(v_{e2})^2 + 8{,}973 v_{e1} v_{e2} - 4{,}821 v_{e1} - 4{,}205 v_{e2}\end{aligned}$$

Taking partial derivatives of π w.r.t. v_{e1} and v_{e2}, we find that

$$\frac{\partial \pi}{\partial v_{e1}} = 43{,}668 v_{e1} + 8{,}973 v_{e2} - 4{,}821 = 0$$

$$\frac{\partial \pi}{\partial v_{e2}} = 8{,}973 v_{e1} + 46{,}958 v_{e2} - 4{,}205 = 0$$

In a procedure similar to solving for v_{s1} and v_{s2}, v_{e1}, and v_{e2} can be obtained by solving the two equations above to yield

$$v_{e1} = 96 \text{ mm}$$
$$v_{e2} = 72 \text{ mm}$$

Knowing the earthquake-induced displacements, the earthquake-induced forces at the top of columns are calculated as follows:

At pier P1: $v_{p1} = v_{e1} = 96 \text{ mm}$

Thus,

$$F_c = k_c v_{p1} = 18{,}096\,(0.096) = 1{,}737 \text{ kN}$$

At pier P2:

$$\begin{aligned}v_{p2} &= \frac{25}{55} v_{e1} + \frac{30}{55} v_{e2} \\ &= \frac{25}{55}(96) + \frac{30}{55}(72) = 83 \text{ mm}\end{aligned}$$

Thus,

$$F_c = k_c v_{p2} = 18{,}096\,(0.083) = 1{,}502 \text{ kN}$$

At pier P3: $v_{p3} = v_{e2} = 72 \text{ mm}$

Thus,

$$F_c = k_c v_{p1} = 18{,}096\,(0.072) = 1{,}303 \text{ kN}$$

These results and the associated effort can be compared with the results and effort of using AASHTO procedures by using the space frame model.

9.8.3 Multimode Spectral Method

AASHTO requires that the multimode spectral analysis method be used for bridges in which coupling occurs in more than one of the three coordinate directions within each mode of vibration. As a minimum, linear dynamic analysis using a 3D model shall be used to represent the structure.

This method is simply the modal superposition presented in Chap. 3. As mentioned earlier, modeling is considered to be an art and an engineering tool. Therefore, the bridge needs to be modeled properly, taking into consideration critical details such as expansion joints, the state of the supporting pads in different directions, and the connections among different components.

Examples of modeling the bridge in 3D are already given in the presentation of the previous two methods, namely, the uniform load method and the single-mode spectral method. (Refer to those sections for guidance and insight.)

The response spectra given in previous sections and shown in Fig. 9-6 should be used for this analysis. Refer to Example 6-3 for details on how to prepare the response spectrum for input in this analysis.

The elastic design forces are obtained by combining enough modes to yield reasonable accuracy. AASHTO requires that the number of modes included in this analysis is at least three times the number of spans in the model. The modes may be combined using the known complete quadratic combination method (CQC). If the CQC method option is not available, SRSS or SABS combination methods may be used as applicable. Within this context, SRSS is appropriate for well-separated modes, whereas, SABS is appropriate for closely spaced modes as explained in Chap. 3.

Because the resulting forces given above are the elastic forces, they should be divided by the response modification factor, R, for design purposes.

9.8.4 Time-History Method

The time-history method is the general method in dynamic analysis. Elastic time-history and inelastic time-history analysis may be used. If used, time-history analysis requires that earthquake records be explicitly included in the analysis.

According to AASHTO, time histories shall reflect the seismic environment and local site conditions at the site. Response-spectrum-compatible records shall be developed from representative recorded motions. However, if such records are not available, spectrum-compatible records must be generated according to the response spectra given in Fig. 9-6. More details on the specifics of the characteristics of such records may be found in AASHTO, Chap. 4, Sec. 4.7.4.3.4b. Generation of synthetic earthquake records is addressed in Chap. 11 in this book.

AASHTO requires that three sets of three records each shall be used as a minimum in any time-history analysis. All three components of each set shall be applied simultaneously to the structure in the three orthogonal directions (x, y, and z) in an 3D structural model of the bridge. If such application is conducted, the response quantities for design shall be taken as the maximum of the response in each principal direction from each run.

However, if seven sets of records are used instead of three sets, AASHTO allows the response quantities for design to be taken as the average of the seven record results rather than the maximum values as in the case of using only three sets of records mentioned above.

As noted earlier, if elastic time-history analysis is conducted, the resulting forces should be divided by the force modification factor, R. However, if explicit inelastic-time-history analysis is conducted, the resulting design forces and displacements will be readily available from the software program output.

Of course, performing inelastic-time-history analysis requires a realistic description of the hysteretic behavior of the structure. This is not an easy task.

9.8.5 Directional Effect

The directional effect of earthquakes is taken into consideration by treating the analysis in the longitudinal and transverse directions similar to mode shape combining criteria. Within this context, combining method, such as SABS, SRSS, or CQC, are theoretically valid for combining the effect of earthquakes from the two principal directions.

However, AASHTO considers this effect to be satisfied by adding 30 percent of the results from one direction to 100 percent of the results from the other direction to form two load cases as follows:

$$\mathrm{EQ} = \mathrm{EQ}_x + 0.3\,\mathrm{EQ}_y \qquad \text{(Case 1)}$$

$$\mathrm{EQ} = 0.3\,\mathrm{EQ}_x + \mathrm{EQ}_y \qquad \text{(Case 2)}$$

where EQ = combined earthquake effect
EQ_x = earthquake effect in the longitudinal direction
EQ_y = earthquake effect in the transverse direction

It should be pointed out that the combination criterion above is required when seismic analysis is conducted using two independent orthogonal directions separately as in the case of the uniform load method or the single-mode method. This directional combination does not apply to the multimode spectral method or time-history method as presented in AASHTO because the directional effect is inherently included in the required 3D analysis.

The combination procedures given above are waived for foundations and/or column connection forces if they are determined from plastic hinging of the columns.

9.9 Load Combinations

The load combinations in AASHTO depend on the presence of different loading conditions. The general load combination that includes earthquake effect is given in the following expression:

$$U = \gamma_p\, G + \gamma_{\mathrm{EQ}}\, L + 1.0\,\mathrm{WA} + 1.0\,\mathrm{FR} + 1.0\,\mathrm{EQ}$$

where U = ultimate factored design forces
G = gravity-related loads such as dead loads, superimposed dead loads, and earth pressure
γ_p = load factor that depends on the type of gravity load under consideration. For each type of load, this factor takes two values: a maximum and a minimum
L = live load-related loads, including vehicular effects such as impact, breaking forces, and centrifugal forces
WA = water-related loads
FR = friction forces
EQ = earthquake effect
γ_{EQ} = load factor of live loads, which must be considered on a project-by-project basis. In the absence of such information, AASHTO suggests in its commentary the use of $\gamma_{EQ} = 0.5$

For a complete range of values of load factors, refer to AASHTO's general load combinations, Sec. 3.4.1.

For example, in the presence of the dead load of structural components and nonstructural components, DC; the superimposed dead load of wearing surface and utilities, DW; and half of the live load effect and in the absence of water and friction loads, the earthquake load combination will be as follows:

$$U = 1.25\,\text{DC} + 1.5\,\text{DW} + 0.5\,L + 1.0\,\text{EQ}$$
$$U = 0.9\,\text{DC} + 0.65\,\text{DW} + 0.5\,L + 1.0\,\text{EQ}$$

9.10 Design Requirements

For the load combinations that include an earthquake event, the bridge design forces are obtained by dividing the elastic earthquake-induced forces by an appropriate response modification factor R, as given in Tables 9-4 and 9-5. These tables are also listed in App. 9-1 at the end of this chapter.

	Operational Category		
Substructure	**Critical**	**Essential**	**Other**
Pier walls	1.5	1.5	2.0
Reinforced concrete pile bents:			
Vertical piles only	1.5	2.0	3.0
With batter piles	1.5	1.5	2.0
Single columns	1.5	2.0	3.0
Steel or composite steel and concrete pile bents:			
Vertical piles only	1.5	3.5	5.0
With batter piles	1.5	2.0	3.0
Multiple column bents	1.5	3.5	5.0

TABLE 9-4 Response Modification Factor, R, for Substructures

Connection	All Operational Categories
Superstructure to abutment	0.8
Expansion joints within a span of the superstructure	0.8
Columns, piers, or pile bents to cap beam or superstructure	1.0
Columns or piers to foundations	1.0

Table 9-5 Response Modification Factor, *R*, for Connections

As shown in Tables 9-4 and 9-5, the response modification factor depends on the importance of the bridge. Note that there are no levels of detailing defined for levels of force reductions as in the case of building systems given in building codes. However, AASHTO defines R-values only for substructure main members as well as for connections. Consequently, the detailing requirements will be the same for all structural elements.

The requirements for beam-columns and pier walls are presented in the following sections.

9.11 Design Requirements of Reinforced Concrete Beam-Columns

The design requirements of reinforced concrete beam-columns are given in terms of force requirements and detailing requirements. The detailing basically follows the detailing of ACI 318 as presented in Chap. 7 for beam-columns of special moment frames. The minor changes required for reinforced concrete beam-columns are presented in the following sections.

Design requirements are given only for columns and pier walls. There are no special requirements for cap beams. The design requirements are given according to the seismic zones as follows.

9.11.1 Bridges in Seismic Zone 1

Design requirements in Zone 1 depend on the response acceleration coefficient, S_{D1}, as follows:

For $S_{D1} < 0.10$: There are no design requirements.

For $S_{D1} \geq 0.10$: There are no design requirements except at the plastic hinge regions at tops and bottoms of columns. The plastic hinge regions and their extension into the cap beams and into the foundations must satisfy the plastic hinge region requirements given in Seismic Zones 2, 3, and 4. These requirements include the length of the plastic hinge regions and their extension, their minimum amount of transverse reinforcement, and their detailing as presented in the following sections.

It is important to point out that the requirements above are required in addition to the design forces of the connections between the superstructure and the substructure as noted in previous sections.

In addition, the seismic seat requirements must also be satisfied as will be presented later in this chapter.

9.11.2 Bridges in Seismic Zone 2

The requirements of Seismic Zone 2 is elevated to Seismic Zones 3 and 4 in the recent AASHTO code (2012) as the earthquake hazard level is increased from 500 to 1,000 years return period.

The only difference that remains between them is the maximum gross amount of longitudinal reinforcement in the column, which is given as 6 percent for Zone 2, while kept at 4 percent in Zones 3 and 4.

9.11.3 Bridges in Seismic Zones 3 and 4

Seismic requirements in Zones 3 and 4 are given in terms of dimensions, strength requirements, longitudinal reinforcement detailing, and transverse reinforcement detailing as follows:

1. Dimensions

   To satisfy the beam theory hypothesis of plane section, to ensure flexure behavior, and to prevent premature shear failures, the beam-column span/depth ratio shall not be less than 2.5. The depth is defined as the maximum dimensions of the cross section. For flared columns, the section shall be taken as the smaller cross section along the column as shown in Fig. 9-23.

   These limitations are applied to both axes separately and should be treated as such, for example, the major axis is treated as pier while the minor axis is treated as a column in the following case:

   If $\frac{H_o}{h_1} < 2.5$, the member is treated as pier about y-axis

   and if $\frac{H_o}{h_2} \geq 2.5$, the member is treated as a column about x-axis

2. Strength

   Strength-reduction factor

   The strength-reduction factor, ϕ, shall be taken as 0.9 for both tied columns and spiral columns.

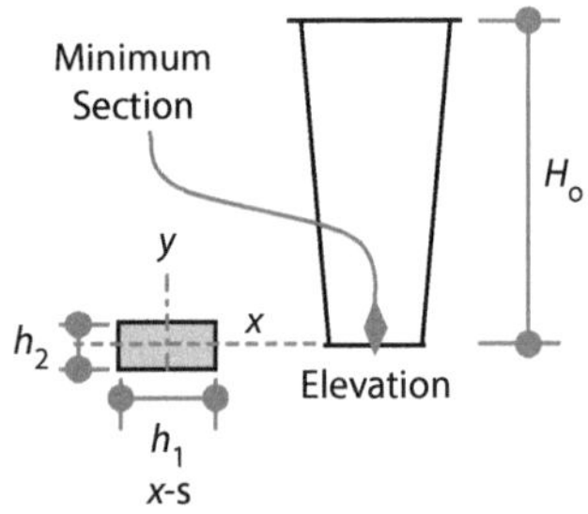

FIGURE 9-23 Flared column.

Shear in end regions (plastic hinge regions)

The shear strength in the end regions (plastic hinge regions), to be defined later in this chapter, shall be taken according to the presence of ultimate factored external load as follows:

1. For $P_u \geq 0.10 f_c' A_g$: $V = V_c$
2. For $P_u = 0$: $V = 0$

where V_c is the designated shear strength of concrete as defined by AASHTO for cases without earthquakes.

For values of P_u in between linear interpolation shall be used.

Shear capacity of concrete inside joints (connections)

The nominal shear resistance provided by concrete inside the joints, V_n, shall be limited to

$$V_n = 1.0\sqrt{f_c'}\,bd$$

where f_c' = characteristic strength of concrete in MPa
b = width of column web
d = effective depth of the column (distance from extreme compression fibers to centroid of tension steel

Note that the shear capacity limit V_n given above equals to six times the capacity of concrete in pure beams, $V_c = 0.165\sqrt{f_c'}\,bd$.

3. Longitudinal reinforcement

The gross reinforcement ratio, ρ_g, shall be limited between 1 percent and 4 percent of the gross-sectional area, that is,

$$A_{s,\max} \leq 0.04\,A_g$$

$$A_{s,\min} \geq 0.01\,A_g$$

Note that maximum reinforcement ratio in Seismic Zone 2 is 6 percent.

Strength-reduction factor

The strength-reduction factor, ϕ, shall be taken as 0.9 for both tied columns and spiral columns.

Development of reinforcement

The development length of all bars shall be 1.25 times the full development length in tension for nonseismic regions.

Splice

Lap splices shall not be used in seismic regions. If other splice types, such as full-welded or full-mechanical splices, are used, the spliced bars shall be alternate bars in each layer. In addition, the distance between adjacent bars shall not be less than 600 mm along the axis of the column.

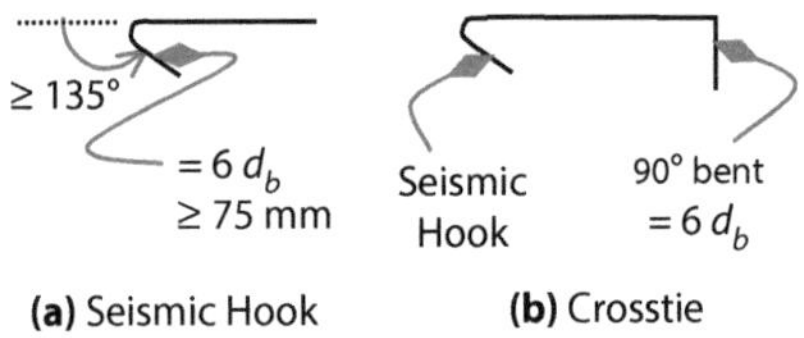

FIGURE 9-24 Seismic ties.

4. Transverse reinforcement

The requirements of transverse reinforcement are given in terms of their amount and their detailing. Before addressing these requirements, the following definitions are deemed necessary:

Seismic hooks

A seismic hook is a bar bent of at least 135 degrees with projecting length inside the core of a rectangular concrete section. The projecting length shall be at least six times the bar diameter, d_b, but not less than 75 mm as shown in Fig. 9-24(a).

Crossties

A crosstie is a straight tie that has a seismic hook on one side and a 90° bent on the other. The bent must have at least six times its diameter as shown in Fig. 9-24(b). The crossties are intended to ease the installation of the ties by engaging the seismic hook first and then pushing the 90° bent afterward.

Hoops

Hoops are special ties for seismic sections. Hoops are characterized as elements that are different than ties and are classified as either continuous hoops or composed (overlapping) hoops. Continuous hoops consist of one continuous bar, whereas composed hoops consist of an open stirrup and a crosstie arranged as shown in Fig. 9-25. Continuous hoops provide better support than composed hoops; however, composed hoops are easier to install. Both types of hoops must be detailed to conform to seismic hook requirements as shown in Fig. 9-25.

Detailing of Transverse Reinforcement

Detailing of transverse reinforcement is given in terms of cross-section details and layout details as follows:

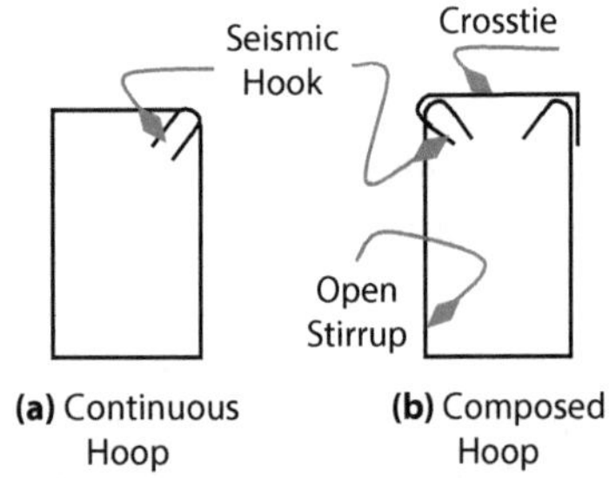

FIGURE 9-25 Hoops.

Cross-Section Detailing

1. The distance center to center of the outer hoop is designated as h_c in the direction under consideration.
2. The distance center to center between crossties in any direction, x_i, shall not exceed 350 mm as shown in Fig. 9-26.
3. Alternate bars must be supported. The maximum clear distance between supported bars and unsupported bars shall be limited to 150 mm as shown in Fig. 9-26.
4. All hooks shown in Fig. 9-26 shall be seismic hooks as defined earlier. The 90° bent of consecutive crossties shall be on opposite sides of the column in each direction of the cross section, or along the axis of the column.
5. Maximum spacing of longitudinal reinforcement in a circular column shall be limited to 200 mm as shown in Fig. 9-27.

Layout and Loacation Detailing Refer to the following details in Fig. 9-28

1. The potential plastic hinge region is assumed to be within a distance (ℓ_o) from the face of the support. The minimum length of the plastic hinge region (ℓ_o) is given as a function of the clear span of the beam-column, the dimensions of the section, and 450 mm, as shown in Detail A. Therefore, over this distance (ℓ_o), the maximum spacing of stirrups is given in terms of the sides of the section and is limited to 100 mm as defined in Detail B.

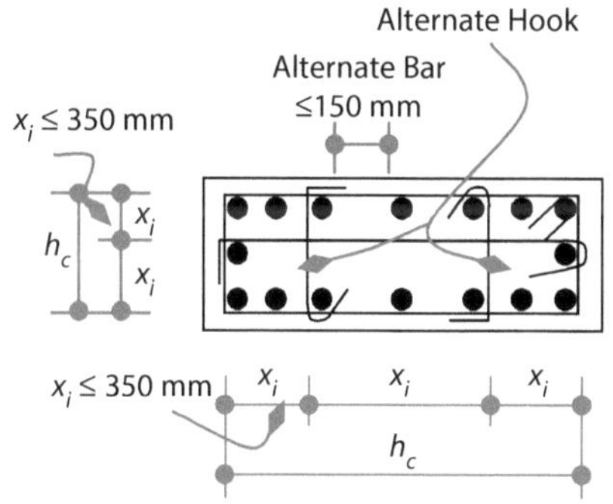

FIGURE 9-26 Hoop dimensions.

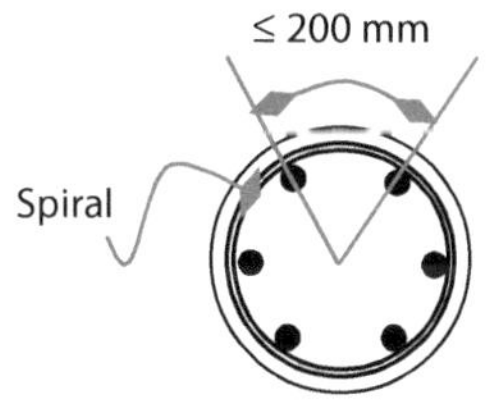

FIGURE 9-27 Spacing of main reinforcement.

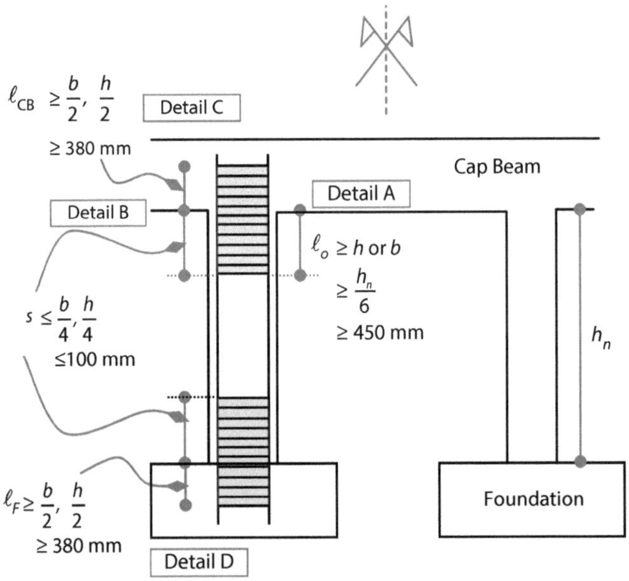

FIGURE 9-28 Beam-column requirements and detailing.

2. The amount and spacing of stirrups in the plastic hinge region must extend through the cap beam and through the foundation as shown in Details C and D.

Minimum Transverse Reinforcement within the Plastic Hinge Region The minimum amount of hoops in the plastic hinge region depends on the shape of the section. The minimum hoops for rectangular and circular sections are given in the following sections. In either case, the yield stress of transverse reinforcement shall not exceed the yield stress of the longitudinal reinforcement.

Rectangular Sections The minimum area of hoops, A_{sh}, in any direction of the rectangular section shall be as follows:

$$A_{sh} \geq 0.30\, sh_c \frac{f'_c}{f_y}\left(\frac{A_g}{A_c}-1\right)$$

$$\geq 0.12\, sh_c \frac{f'_c}{f_y}$$

where A_{sh} = Area of hoops for shear
A_g = gross area of the concrete section
A_c = area of concrete core, including hoops
f'_c = characteristic strength of concrete
f_y = yield stress of hoop reinforcement
h_c = out-to-out dimension of the peripheral hoop in the direction under consideration as shown in Fig. 9-26
s = spacing of hoops along the axis of the member

Note that the coefficient of the second equation is 0.12, which is higher than ACI corresponding coefficient of 0.09.

Circular and Spiral Sections The minimum volume of hoops, ρ_s, in circular sections with hoops or with spirals shall be as follows:

$$\rho_s \geq 0.12 \frac{f'_c}{f_y}$$

$$\geq 0.45 \frac{f'_c}{f_y}\left(\frac{A_g}{A_c} - 1\right)$$

where A_c = area of concrete core, including hoops
A_g = gross area of the concrete section
f'_c = characteristic strength of concrete
f_y = yield stress of hoop reinforcement

9.12 Design Requirements of Reinforced Concrete Pier Walls

A member is defined as a pier wall if its span/depth ratio does not satisfy the requirements of the beam-column of 2.5 as given in the previous section. However, if one side of the pier wall satisfies beam-column requirements (the weak axis), it may be designed as a column, while the other axis (the strong axis) is designed as a pier wall. The requirements of pier walls are given for shear strength and for reinforcement detailing as follows:

Reinforcement Requirements The horizontal reinforcement ratio, ρ_h, and the vertical reinforcement ratio, ρ_v, as shown in Fig. 9-29 shall satisfy the following minimum criterion:

$$\rho_h \geq 0.0025$$

$$\rho_v \geq \rho_h$$

Both horizontal and vertical reinforcements shall be installed in two layers close to the face of the pier wall. The maximum spacing, s, of both horizontal and vertical reinforcement shall be limited to 450 mm

$$s_{max} = 450 \text{ mm}$$

The reinforcement shall be properly developed and spliced. Splices shall be staggered.

Shear Strength Requirements The shear resistance of the pier wall shall be taken as follows:

If the nominal shear strength of the pier is defined as V_n, such that

$$V_n = [0.165\sqrt{f'_c} + \rho_h f_y]\, bd$$

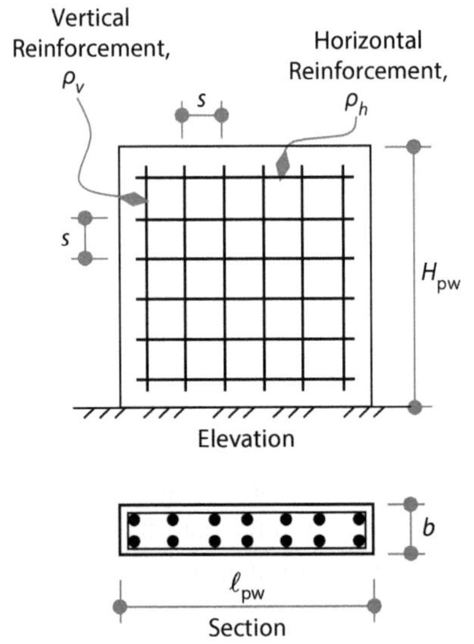

FIGURE 9-29 Pier wall.

then the factored shear strength of the pier wall, V_r, will be given as

$$V_r = \phi V_n$$
$$\leq 0.66\sqrt{f'_c}\,bd \quad \text{Note} = 4(0.165\sqrt{f'_c}\,bd)$$

where b = width of column web as shown in Fig. 9-29
d = effective depth of the column (distance from extreme compression fibers to centroid of tension steel)
f'_c = characteristic strength of concrete
f_y = yield stress of steel
ρ_h = horizontal steel ratio as defined before

9.13 Special Topics

The following complementary issues must be satisfied to ensure the safety of the bridge. Such issues include P-Δ requirements, displacement requirements (seismic seats), longitudinal restrainer, hold-down devices, and liquefaction.

9.13.1 P-Δ Requirements

In treatment of second-order effect, or P-Δ effect, AASHTO philosophy is to eliminate the second-order effect rather than dealing with it. Accordingly, AASHTO limits the relative inelastic displacement of the columns and piers to a value that render P-Δ effect insignificant.

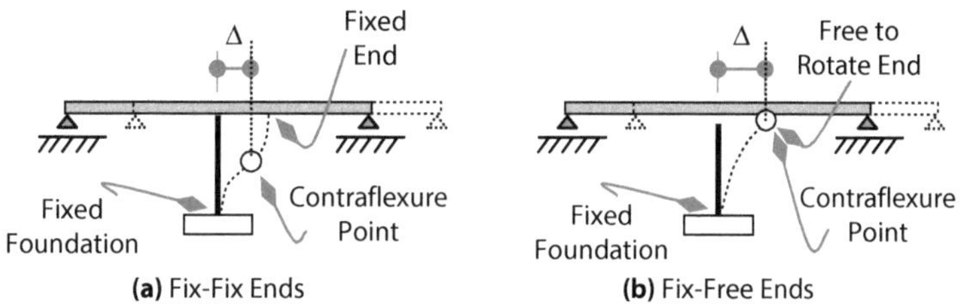

FIGURE 9-30 Definition of displacement, Δ, for *P*-Δ, effect limitations.

The reference displacement, Δ, in this context is the relative displacement between the point of foundation fixity and the point of contraflexure in the column, or in the pier, as shown in Fig. 9-30.

Accordingly, the displacement limitations are given follows:

$$\Delta_{\text{inel}} \le \Delta_{\text{limit}}$$

where Δ_{limit} = inelastic displacement limit at the location defined in Fig. 9-30, which is given as

$$\Delta_{\text{limit}} = 0.25\,\frac{\phi M_n}{P_u}$$

Δ_{inel} = actual inelastic displacement at the location defined in Fig. 9-30, which is given as

$$\Delta_{\text{inel}} = R_d\,\Delta_e$$

Δ_e = elastic displacement at the location defined in Fig. 9-30. This displacement is due to elastic strength demand before dividing the elastic forces by *R*. This displacement is not the yield displacement due to forces after reduction by *R*.

P_u = factored axial load in the column (pier)

M_n = nominal flexure strength of the column (pier)

ϕ = strength reduction factor for column (pier) under seismic forces, which is given as $\phi = 0.9$ for both circular and rectangular column

R_d = ductility demand coefficient depends on the period of the bridge, which is given as follows:

- For short-period bridges where $T < 1.25\,T_s$, R_d is given as

$$R_d = \left(1 - \frac{1}{R}\right)\frac{1.25T_s}{T} + \frac{1}{R}$$

- For long-period bridges where $T \ge 1.25\,T_s$, R_d is given as

$R_d = 1$

R = Response modification factor defined earlier

Recall that the relationship between the force reduction factor, R, and the ductility demand, μ_d, is initially given by Newmark as illustrated in Chap. 5. Newmark has shown that the R-μ_d relationship depends on the period of the structure where he proposed the equal energy criterion, EEC, for short-period structures, and the equal displacement criterion, EDC, for long-period structures. It should be pointed out that this dependency on the period is recognized by AASHTO through its coefficient, R_d, which is given as function of the period, while it is not recognized by IBC-code in its coefficient of displacement, C_d, as illustrated in Chap. 6.

In order to draw comparison between AASHTO criterion and Newmark criteria, AASHTO relationship between R and μ_d may be constructed as follows:

Recall that
$$R=\frac{\Delta_e}{\Delta_y} \text{ (see Chap. 5)}$$

or,
$$\Delta_e = R\,\Delta_y$$

and also recall that
$$\Delta_{\text{inel}} = \mu_d \Delta_y$$

Substituting Δ_e from the expression above into AASHTO equation for Δ_{inel}, then
$$\Delta_{\text{inel}} = R_d\Delta_e = R_d(R\Delta_y) = \mu_d\Delta_y$$

hence,
$$\mu_d = R_d R$$

Substituting $R_d=\left(1-\frac{1}{R}\right)\frac{1.25T_s}{T}+\frac{1}{R}$, into the above expression, then
$$\mu_d = \left\{\left(1-\frac{1}{R}\right)\frac{1.25T_s}{T}+\frac{1}{R}\right\} R$$
$$\mu_d = (R-1)\frac{1.25T_s}{T}+1$$

or,
$$R=(\mu_d-1)\frac{T}{1.25T_s}+1$$

The relationship between R and μ_d given above is plotted against Newmark criteria for EEC and EDC as shown in Fig. 9-31. The parameters selected for this plot are as follows:
$$T_s = 0.5 \text{ s}$$

T varies as 0.0, 0.1, 0.2, 0.3, 0.4, 0.5, and the border line for long period, T = 1.25 and T_s = 0.625.

Figure 9-31 shows that AASHTO relationship between R and μ_d is a fan of straight lines that starts at 1 and spreads between zero period and Newmark, EDC. Note that

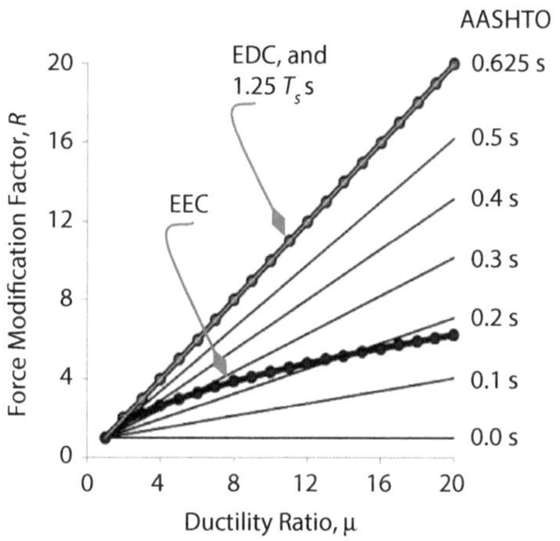

FIGURE 9-31 AASHTO relationship between R and ductility ratio, μ.

EDC coincides with AASHTO, $R_d = 1$, as it should be. Note also that EEC matches AASHTO criterion around the period of 0.2 s., which is the reference period specified in AASHTO as the short period in spectral acceleration coefficient, S_S. Note also that AASHTO curves drop below EEC for periods less than 0.2 s. It should be pointed out that at the limit when the period equals to zero, the structure cannot sustain any ductility demand regardless of the value of R, as the R-μ_d curve becomes horizontal. This result is in line with research on this subject.

It is the opinion of this author that IBC-code should revisit its inelastic displacement criteria, and must relate that to the period of the structure, especially that ductility demand becomes extremely high at very short periods as obvious from Fig. 9-31. In addition, the author has shown through his published research that, for short-period structures, ductility alone cannot reduce seismic demand to acceptable levels where he indicates that other means, such as seismic isolation, must be utilized to achieve survival requirements at very short periods.

9.13.2 Displacement Requirements (Seismic Seats)

Seismic seats are required in bridges at the expansion bearings if restrainers are not used in the longitudinal directions.

Therefore, if restrainers are not used, seismic seats shall be provided between the expansion joints even if there are supports in between as illustrated in Fig. 9-32. In this

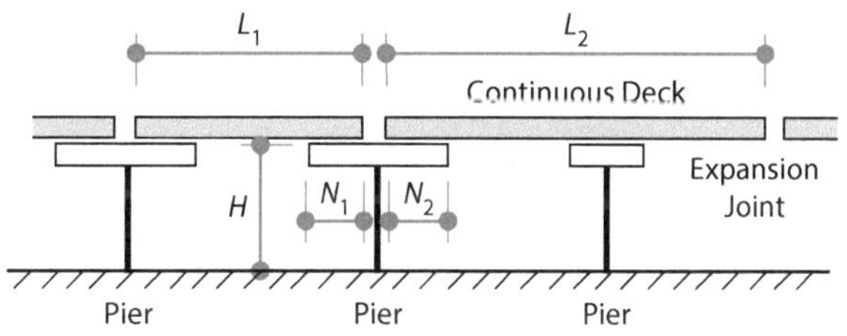

FIGURE 9-32 Seismic seats in multiple span bridges.

case, seismic seats may differ on each side of the expansion joint, which are defined as N_1 and N_2 as shown in Fig. 9-32.

These seismic seats depend on the seismic zone and the acceleration coefficient, A_s. Accordingly, seismic seats, N_1 and N_2, are determined in terms of basic seismic seats, N_{b1} and N_{b2}, which are given by the following expressions:

$$N_{b1} = (203\text{ mm} + 1.67\,L_1 + 6.67\,H)\,(1 + 0.000125\,S^2)$$

$$N_{b2} = (203\text{ mm} + 1.67\,L_2 + 6.67\,H)\,(1 + 0.000125\,S^2)$$

where N_{b1} = minimum basic seismic seat distance to the left of the pier under consideration in millimeters

N_{b2} = minimum basic seismic seat distance to the right of the pier under consideration in millimeters

L_1 = length of continuous deck segment in meters

L_2 = length of continuous deck segment in meters

H = height of substructures. This is given as follows:

For *single-span* bridge:

$H = 0.0$

For *abutments*:

H = distance from foundation to top of substructure in meters

For *piers*:

H = average distance from foundation to top of all piers that appear in the deck segment to the next expansion joint in meters

S = skew angle of the support measured from normal to span in degrees

Once N_{b1} and N_{b2} are determined, the actual minimum seismic seats shall be provided as follows:

Zone 1:	for $A_s < 0.05$ →	$N_1 = 0.75\,N_{b1}$
		$N_2 = 0.75\,N_{b2}$
	for $A_s \geq 0.05$ →	$N_1 = N_{b1}$
		$N_2 = N_{b2}$
Zones 2, 3, and 4:		$N_1 = 1.50\,N_{b1}$
		$N_2 = 1.50\,N_{b2}$

9.13.3 Longitudinal Restrainers

The bridge spans may be restrained by restraining devices to tie the superstructure to the substructure at the expansion joints. If used, the restrainer must have sufficient slack to accommodate the expected inelastic design displacement such that the restrainer becomes active only when this displacement is reached.

9.13.4 Hold-Down Devices

This section applies to Seismic Zones 2, 3, and 4. Hold-down devices are required to tie the superstructure to the substructure at the support locations under the following conditions.

If the seismic-induced vertical force due to longitudinal seismic force that opposes the permanent loads is designated as Q_v, and the vertical reaction due to permanent loads is designated as R_{perm}, then the hold-down devices are required if

$$Q_v \geq 0.5\, R_{perm}$$

If, $Q_v \leq 1.0\, R_{perm}$, the design force for hold-down devices, Q_{HD}, is given as

$$Q_{HD} = 0.1\, R_{perm} \qquad \text{(calculated as if it is simply supported)}$$

If on the other hand, $Q_v > 1.0\, R_{perm}$, the design force for hold-down devices, Q_{HD}, is given as

$$Q_{HD} = 1.20\,(Q_v - R_{perm})$$
$$\geq 0.1\, R_{perm} \qquad \text{(calculated as if it is simply supported)}$$

9.13.5 Liquefaction

Liquefaction is considered to be one of the most dangerous and dramatic events that usually cause disastrous collapse in structures. Therefore, the site must be treated to prevent such a phenomenon.

The treatment of liquefaction will be addressed in Chap. 10, Geotechnical Aspects and Foundations.

APPENDIX 9-1

Appendix 9-1 contains information on seismic parameters presented in tables based on the AASHTO 2012.

Seismic Zone	1	2	3	4
Acceleration coefficient, S_{D1}	≤ 0.15	≤ 0.30 > 0.15	≤ 0.50 > 0.30	> 0.50

Table 9-A1 AASHTO Seismic Zones and Coefficients

Parameter	Value				
Number of spans	2	3	4	5	6
Maximum subtended angle for a curved bridge	90°	90°	90°	90°	90°
Maximum span length ratio from span to span	3	2	2	1.5	1.5
Maximum bent/pier stiffness ratio from span to span	–	4	4	3	2

Table 9-A2 Regular Bridge Requirements of Straight Bridges

Seismic Zone	Single-Span Bridges	Multispan Bridges					
		Other		Essential		Critical	
		Regular	Irregular	Regular	Irregular	Regular	Irregular
1	No seismic analysis required	No seismic analysis required					
2		SM/UL	SM	SM/UL	MM	MM	MM
3		SM/UL	MM	MM	MM	MM	TH
4		SM/UL	MM	MM	MM	TH	TH

Table 9-A3 Method of Analysis for Seismic Effect

Site Class	Peak Ground Acceleration (PGA) Coefficient				
	PGA < 0.10	PGA = 0.20	PGA = 0.30	PGA = 0.40	PGA > 0.50
A	0.8	0.8	0.8	0.8	0.8
B	1.0	1.0	1.0	1.0	1.0
C	1.2	1.2	1.1	1.0	1.0
D	1.6	1.4	1.2	1.1	1.0
E	2.5	1.7	1.2	0.9	0.9
F	Site-specific investigation is required				

Table 9-A4 AASHTO Site Factor at Zero Period, F_{pga}

Site Class	Spectral Acceleration Coefficient at 0.2 s				
	$S_s < 0.25$	$S_s = 0.50$	$S_s = 0.75$	$S_s = 1.00$	$S_s > 1.25$
A	0.8	0.8	0.8	0.8	0.8
B	1.0	1.0	1.0	1.0	1.0
C	1.2	1.2	1.1	1.0	1.0
D	1.6	1.4	1.2	1.1	1.0
E	2.5	1.7	1.2	0.9	0.9
F	Site-specific investigation is required				

Table 9-A5 AASHTO Site Factor for Short Period, F_a

Site Class	Spectral Acceleration Coefficient at 1.0 s				
	$S_1 < 0.1$	$S_1 = 0.2$	$S_1 = 0.3$	$S_1 = 0.4$	$S_1 > 0.5$
A	0.8	0.8	0.8	0.8	0.8
B	1.0	1.0	1.0	1.0	1.0
C	1.7	1.6	1.5	1.4	1.3
D	2.4	2.0	1.8	1.6	1.5
E	3.5	3.2	2.8	2.4	2.4
F	Site-specific investigation is required				

Table 9-A6 AASHTO Site Factor for Long Period, F_v

Substructure	Operational Category		
	Critical	Essential	Other
Pier Walls	1.5	1.5	2.0
Reinforced concrete pile bents:			
Vertical piles only	1.5	2.0	3.0
With batter piles	1.5	1.5	2.0
Single columns	1.5	2.0	3.0
Steel or composite steel and concrete pile bents:			
Vertical piles only	1.5	3.5	5.0
With batter piles	1.5	2.0	3.0
Multiple column bents	1.5	3.5	5.0

Table 9-A7 Response Modification Factor, *R*, for Substructures

Connection	All Operational Categories
Superstructure to abutment	0.8
Expansion joints within a span of the superstructure	0.8
Columns, piers, or pile bents to cap beam or superstructure	1.0
Columns or piers to foundations	1.0

Table 9-A8 Response Modification Factor, *R*, for Connections

CHAPTER 10

Geotechnical Aspects and Foundations

10.1 Introduction

Geotechnical engineering is as important for seismic design as it is for nonseismic design. Most of the geotechnical aspects in normal design must be observed for seismic design, along with additional requirements. Because foundation failures are catastrophic and usually result in loss of human life, they must be carefully designed to be the last structural element to fail.

Experience from earthquakes has shown that liquefaction is the most dangerous type of foundation failure, which can result in many catastrophic structural failures. Therefore, every effort must be made to avoid such failures. Stability of slopes and lateral soil pressure are also affected by seismic activities and must be treated with dynamic aspects in the case of ground motions. Because earthquakes generally increase soil pressure and decrease resistance, these must be evaluated accordingly.

Earthquake excitation is transferred to the building through the ground. Because the dynamic characteristics at the site greatly affect the ground acceleration and its transfer to the structure, dynamic analysis of soil is necessary. In addition, the type of soil alters the characteristics and intensity of the seismic waves as they travel through the soil. This effect is reflected in the types of response spectra given by various codes and presented in previous chapters.

This chapter neither intends to cover every aspect of geotechnical engineering nor to replace the expertise of professional geotechnical engineers. However, it will address important issues on the geotechnical side that must be considered when evaluating structural safety under earthquake excitation. It will also address some of the problems that are encountered in foundation design.

10.2 Wave Propagation

In its simplest form, soil vibration is similar to the vibration of a thin rod. The modes of vibration relevant to soil behavior are the axial and the torsional modes of vibrations. The flexure mode of vibration has little application in soil dynamics. Therefore, only axial and torsional modes will be examined in this section. The flexure mode is extensively analyzed in Chap. 3.

The differential equation of the axial vibration is derived in Sec. 3.8. The differential equation is given in terms of modulus of elasticity (E), mass per unit length (m), and cross-sectional area (A) as follows:

$$m\ddot{u} - EAu'' = 0$$

Dividing the equation above by m yields the following:

$$\ddot{u} - \left(\frac{EA}{m}\right)u'' = 0$$

$$\ddot{u} - v_a^2 u'' = 0$$

v_a has units of velocity (m/s) and is known as the wave propagation velocity. This is given in terms of modulus of elasticity and mass density as

$$v_a = \sqrt{\frac{EA}{m}} = \sqrt{\frac{E}{\rho}}$$

The propagation velocity expression above implies that the propagation velocity depends on both stiffness and density of the material.

The torsional vibration mode can be shown to have a differential equation that is similar to the development above. The differential equation is given in terms of rotational angle θ as follows:

$$\ddot{\theta} - v_s^2 \theta'' = 0$$

v_s also has units of velocity (m/s) and is known as the shear wave propagation velocity. This is given in terms of the shear modulus of elasticity and mass density as

$$v_s = \sqrt{\frac{G}{\rho}}$$

Because the objectives of wave propagation analysis are different than the objectives of structural dynamic analysis, the solution of the differential equations of wave propagation may be treated differently than the approach used in the sections of structural dynamics. Therefore, the general differential equation of wave propagation may be given in the following form:

$$\ddot{u} - v^2 u'' = 0$$

The solution of the differential equation above as a function of space and time may be given in the following form:

$$u(x,t) = f(vt - x) + g(vt + x)$$

where f and g can be any arbitrary functions of $(vt - x)$ and $(vt + x)$ that satisfy the differential equation given above. Because the argument of $\{f(vt - x)\}$ remains constant as x increases with time, its solution describes a displacement wave that is traveling with velocity v in the positive x-direction. Similarly, the argument of $\{g(vt + x)\}$ describes a displacement wave that is traveling with velocity v in the negative x-direction.

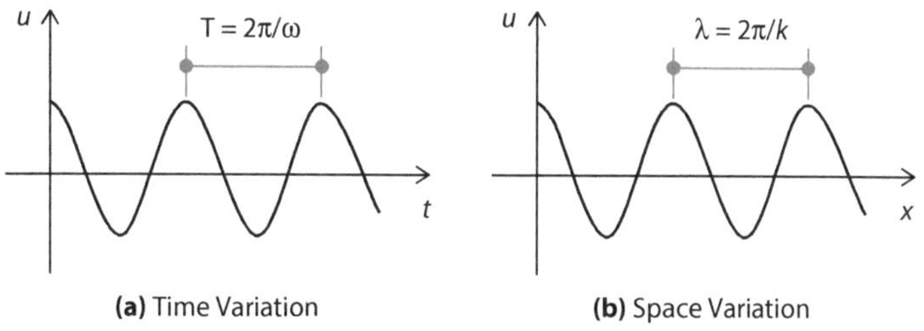

FIGURE 10-1 Wave propagation progress.

When subjected to harmonic stress forcing function of the form

$$\sigma = \sigma_o \cos \omega_f t$$

the wave propagation of the stress in the positive and negative directions in the rod can be shown to take the following form:

$$u(x,t) = A\cos(\omega_f t - kx) + B\cos(\omega_f t + kx)$$

where k = wave number (ω_f/v)
ω_f = frequency of the forcing function

Consequently, the wave length λ is given as follows:

$$\lambda = vT_f = \frac{v}{f_f} = \frac{2\pi}{\omega_f}v = \frac{2\pi}{k}$$

where the subscript f stands for the forcing function.

The wave propagation may be plotted with respect to time and space as shown in Fig. 10-1. The solution above may also be expressed in its complex form equivalence as

$$u(x,t) = Ce^{j(\omega ft - kx)} + De^{j(\omega ft + kx)}$$

Remember that the analysis above is given for a simplified case in a 1D state. This analysis may be refined to include 3D wave propagation cases. Such analysis may be found in the bibliography at the end of the book.

10.3 Ground Response

This section develops an expression that can describe the transfer of ground motion at the bedrock to the ground surface as a function of soil properties. As in the previous section, the analysis of this section takes the simplest form to give the reader an indication of the nature of behavior. For a more refined analysis, see the bibliography at the end of the book.

For the objectives mentioned above, the analysis considers linear behavior with uniform soil layers above the bedrock that have the height, H, as shown in Fig. 10-2. Any horizontal motion of the bedrock produces vertical propagating shear waves that

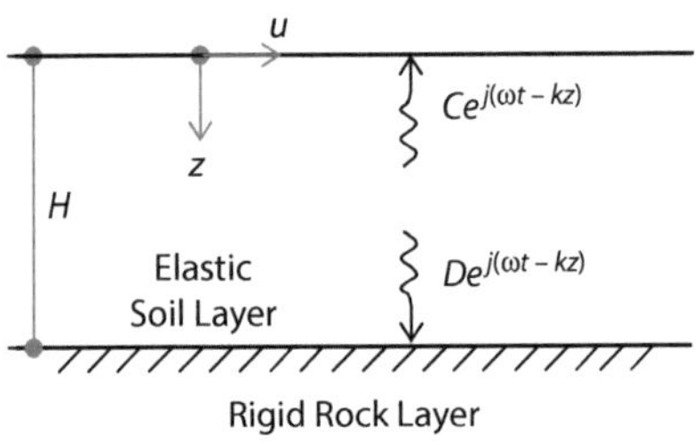

FIGURE 10-2 Ground motion transfer.

result in horizontal displacement that may be described with the expression developed in the previous section

$$u(z,t) = Ce^{j(\omega t - kz)} + De^{j(\omega t + kz)}$$

where C = the amplitude of waves traveling upward (−z)
D = the amplitude of waves traveling downward (+z)
k = wave number (ω/v_s)
ω = circular frequency of the ground shaking

At the free surface, where $z = 0$, the shear stress and shear strain are zero. Therefore,

$$\tau(0,t) = G\gamma = G\frac{\partial u}{\partial z} = 0$$

Substituting $u(z,t)$ into the shear equation above yields

$$Gjk(Ce^{jk(0)}) - (De^{-jk(0)})e^{j\omega t} = Gjk(C - D)e^{j\omega t} = 0$$

The solution above is satisfied when $C = D$. Thus,

$$u(z,t) = 2C \cos kz\, e^{j\omega t}$$

The equation describes a standing wave of amplitude, $2C \cos kz$, and has a fixed shape with respect to depth. This equation may be used to define a transfer function that relates the amplitude of displacements at any two points in the soil layer. This is the primary objective of the analysis. To find the ratio of the amplitude at the surface to the amplitude at the bedrock, z is substituted as zero at the surface and H at the bedrock to yield

$$F(j\omega) = \frac{u_{\max}(0,t)}{u_{\max}(\mathrm{H},\mathrm{t})} = \frac{2C\, e^{j\omega t}}{2C \cos kH\, e^{j\omega t}} = \frac{1}{\cos kH} = \frac{1}{\cos(\omega H/v_s)}$$

The modulus of the transfer function above gives the amplification of the amplitude of the bedrock at surface, which is expressed as follows:

$$|F(j\omega)| = \frac{1}{|\cos(\omega H/v_s)|}$$

The expression above will always be one or more than one because the maximum value of the cosine function is one. This means that the bedrock amplitude will always be amplified and will be at least equal to the amplitude of the bedrock itself. In addition, the amplitude amplification is a function of the frequency of the bedrock, the height of the overlaying layer, and the shear velocity of the overlaying layer.

In general, seismic codes give the amplification factors of the ground response according to the type of the soil layers above the bedrock, their height, and their shear velocities. As an example, Chap. 6 in the International Building Code (IBC) divides the soil profile above bedrock into six categories that are classified according to the average properties of the top 30 m (100 ft) below the ground surface. The average properties are indicated by the type of soil particles and their shear velocity. For demonstration purposes, the IBC classification is given in Table 10-1. As noted in Chap. 6, the response amplifications are defined according to this classification.

Soil Profile Type	Description	Average Properties for Top 30 m (100 ft)		
		Shear Wave Velocity, vs. m/s (ft/s)	Standard Penetration Test (blows/ft)	Undrained Shear Strength kPa (psf)
A	Hard rock	> 1,500 (5,000)	N/A	N/A
B	Rock	760 to 1,500 (2,500 to 5,000)		
C	Very dense soil and soft rock	360 to 760 (1,200 to 2,500)	> 50	> 100 (2,000)
D	Stiff soil profile	180 to 360 (600 to 1,200)	15 to 50	50 to 100 (1,000 to 2,000)
E	Soft soil profile	< 180 (600)	< 15	< 50 (1,000)
	—	Any profile with more than 3.28 m (10 ft) of soil having the following characteristics: 1. Plasticity index, $PI > 20$, 2. Moisture content, $w \geq 40\%$, and 3. Undrained shear strength < 24 kPa (500 psf).		
F	—	Any profile containing soils having one or more of the following characteristics: 1. Soils vulnerable to potential failure of collapse under seismic loading such as liquefiable soils; quick and high sensitive clays; and collapsible, weakly cemented soils. 2. Peats and/or highly organic clays [$H > 3.28$ m (10 ft)] of peat and/or highly organic clay where H = thickness of soil. 3. Very high plasticity clays [$H > 7.5$ m (25 ft)] with plasticity index of $P > 75$. 4. Very thick soft/medium stiff clays [$H > 36$ m (120 ft)].		

Table 10-1 Soil Profile Type

10.4 Liquefaction

As noted earlier, liquefaction is one of the most dangerous phenomena that can result in devastating damage. Liquefaction is associated with saturated cohesionless soil under undrained conditions. If such soil conditions exist in their loose state and are subjected to sudden application of forces as in the case of an earthquake, the pore pressure will develop immediately and leave little for the effective stress in the soil mass. As a consequence, the shear strength drops suddenly. Before the transfer of the stress from water to soil takes place, the soil mass behaves as liquid (with zero shear strength) and liquefaction occurs.

The susceptibility of a site to liquefaction is related to the site's critical void ratio (CVR). Critical void ratio is the void ratio that marks the borderline between contraction and dilation of the soil mass under external pressure. In general, a dense sample of sand dilates under pressure, whereas a loose sample of sand contracts under pressure. The void ratio of the state of densification that causes no change in the volume of the sample under external pressure is the CVR.

In general, soil mass with a void ratio larger than the CVR is considered susceptible to liquefaction, whereas soil mass with a void ratio smaller than the CVR is considered unsusceptible to liquefaction. This susceptibility condition is marked by the CVR line, shown in Fig. 10-3.

As a general recommendation, loose to medium-dense saturated sites should be avoided in practice. If one must use such sites, the liquefaction hazard can be reduced by proper drainage and/or proper densification of the soil at the site.

Equivalent Uniform Cyclic Shear Stress Method

Practical evaluation of liquefaction hazard may be related to a quantity known as equivalent uniform cyclic shear stress, τ_{cyc}. The equivalent uniform cyclic shear stress induced by earthquake excitation may be calculated as a function of the total overburden pressure, σ_o, by the following expression:

$$\tau_{cyc} = 0.65\left(\frac{a_{max}}{g}\right)\sigma_o r_d$$

where a_{max} = effective peak ground acceleration at the site
σ_o = total overburden pressure
r_d = stress-reduction factor varying from 1.0 at the surface to 0.9 at a depth of 9 m from the surface
τ_{cyc} = equivalent uniform cyclic shear stress induced by earthquakes

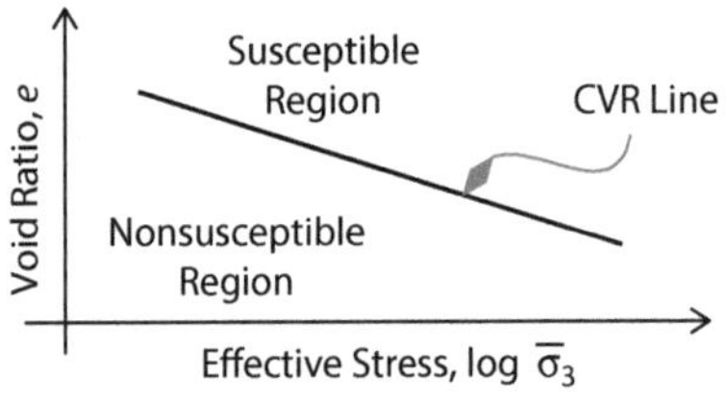

FIGURE 10-3 Liquefaction susceptibility.

The equivalent uniform cyclic shear stress may be divided by the effective overburden vertical pressure, $\bar{\sigma}_o$, to yield a normalized quantity identified as cyclic stress ratio, C_R. The cyclic stress ratio induced by earthquakes is designated as C_{RE} and is expressed in the following form:

$$C_{\mathrm{RE}} = \frac{\tau_{\mathrm{cyc}}}{\bar{\sigma}_o} = 0.65\left(\frac{a_{\max}}{g}\right)\frac{\sigma_o r_d}{\bar{\sigma}_o}$$

Based on historical field measurements, we know that liquefaction has been triggered in previous earthquakes when the C_{RE} exceeded a certain threshold value. This threshold cyclic stress ratio C_{RT} can be correlated to the corrected standard penetration test reading, N_1.

The corrected standard penetration test value, N_1, may obtained from the field standard penetration test, N, using the following expression:

$$N_1 = C_N N$$

where C_N is a correction coefficient that depends on the effective overburden pressures as given in Fig. 10-4.

The threshold cyclic stress ratio that triggers liquefaction (based on historical field measurements) is given in Fig. 10-5 for earthquake magnitudes of 6, 7.5, and 8.5, respectively. To reduce the likely occurrence of liquefaction, the values of the cyclic stress ratio must satisfy the following condition:

$$C_{\mathrm{RE}} \le C_{\mathrm{RT}}$$

The threshold cyclic stress ratio that can trigger liquefaction may also be determined by laboratory tests. If such tests are available, the potential of liquefaction should be evaluated and compared with the earthquake-induced cyclic stress ratio at different depths as shown in Fig. 10-6.

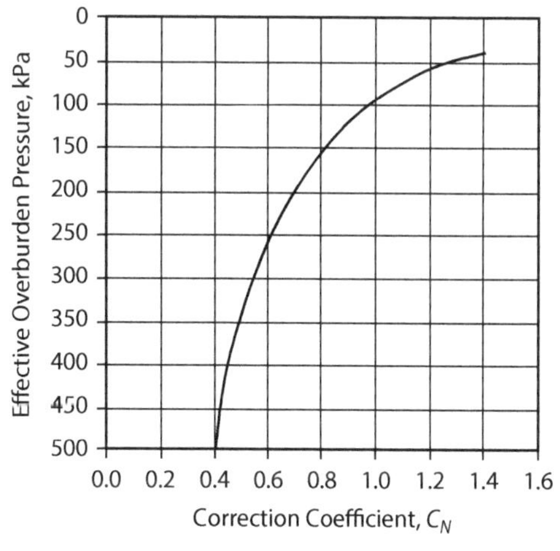

Figure 10-4 Correction coefficient of standard penetration test.

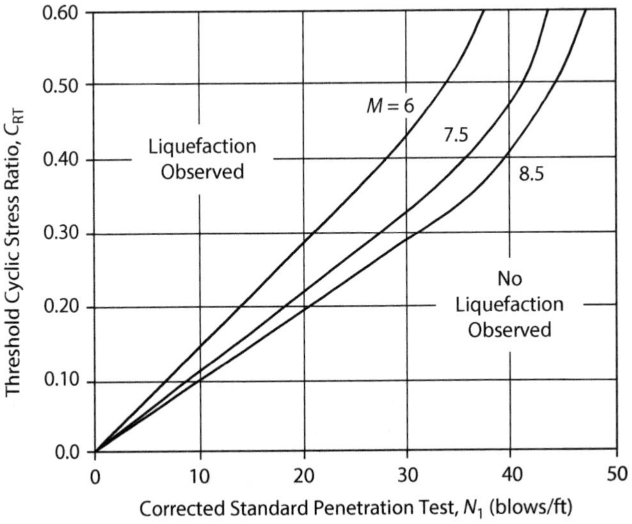

FIGURE 10-5 Historical relationship between C_{RT} and N_1.

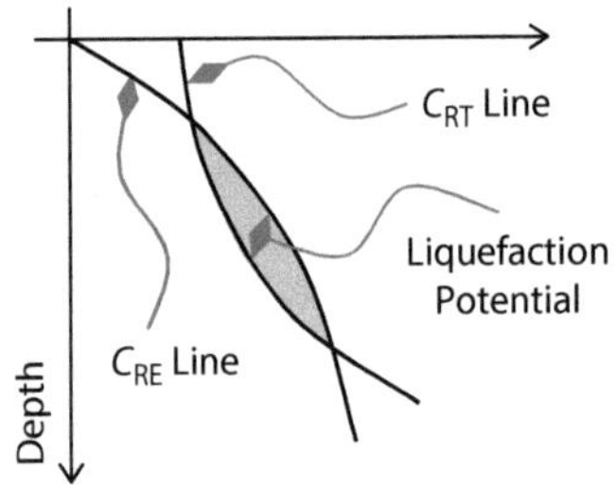

FIGURE 10-6 Theoretical-experimental liquefaction potential.

Example 10-1 A deposit layer of sand with effective unit weights as shown in Example 10-1, Fig. 1, extends 10 m below the ground surface. The ground water table (GWT) extends 3 m below the ground surface. The corrected standard penetration test, N_1, is evaluated as 20 blows per ft.

The site exists in a seismic zone that has an effective peak acceleration coefficient, a_{max}, equal to 0.2 g.

Examine the potential of liquefaction at the midheight of the sand layer, using historical data for earthquakes with magnitude M equal to 8.5.

Solution Total overburden pressure at midheight (at a depth of 5 m)

$$\sigma_o = \gamma_d h_d + \gamma_b h_b + \gamma_w h_w$$
$$= 17(3) + 11(2) + 10(2) = 93 \text{ kN/m}^2$$

Pore pressure

$$\sigma_w = \gamma_w h_w = 10(2) = 20 \text{ kN/m}^2$$

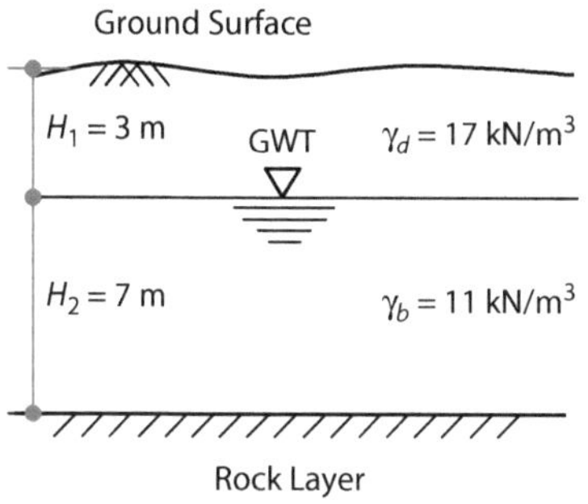

Example 10-1, Figure 1 Sand deposit.

Effective overburden pressure

$$\bar{\sigma}_o = \sigma_o - \sigma_W = 93 - 20 = 73 \text{ kN/m}^2$$

Stress-reduction factor, r_d, is the interpolation between 1.0 at surface and 0.9 at a depth of 9.0 m

$$r_d = 1.0 - \left(\frac{0.1}{9}\right)(5) = 0.94$$

Equivalent uniform cyclic shear stress

$$\tau_{\text{cyc}} = 0.65\left(\frac{a_{\max}}{g}\right)\sigma_o r_d$$

$$\tau_{\text{cyc}} = 0.65\left(\frac{0.2g}{g}\right)93(0.94) = 11.36 \text{ kN/m}^2$$

Induced cyclic stress ratio

$$C_{\text{RE}} = \frac{\tau_{\text{cyc}}}{\bar{\sigma}_o} = \frac{11.36}{73} = 0.16$$

Threshold cyclic stress ratio at $N_1 = 20$ blows per ft

$$C_{\text{RT}} = 0.19 \text{ (from chart, Fig. 10-5)}$$

Since $C_{\text{RE}} < C_{\text{RT}}$, the sand layer at midheight is unlikely to liquefy.

10.5 Slope Stability

Normal slope stability is usually checked by considering the equilibrium of the soil mass bounded by a potential failure circle shown in the shaded area in Fig. 10-7. The soil mass driven by its weight, *W*, is kept in place by the cohesion and friction forces, *T*, along the circle perimeter as shown in the same figure. The slope is usually required to maintain a 1.5 factor of safety against sliding.

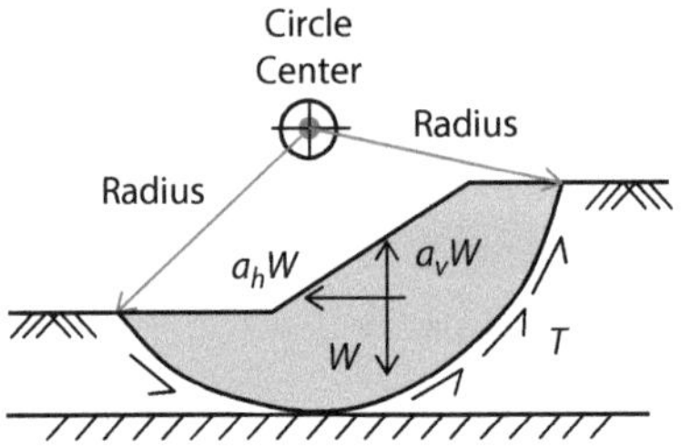

FIGURE 10-7 Slope stability.

For ground motion, the same procedures are still valid with consideration of the effect of ground acceleration. It can be noted from Fig. 10-7 that the ground acceleration has two components, namely, a horizontal component with acceleration coefficient a_h and a vertical component with acceleration coefficient a_v. The horizontal force component (a_hW) will always tend to add to the driving forces of the slope mass, thus worsening the situation. The vertical force component (a_vW) may or may not have an adverse effect on the stability of the driven circle. Its effect is decided according to its point of application with respect to the center of the driven circle. The vertical component also has an adverse effect by reducing the vertical force at the failure surface, thereby reducing the resisting friction forces.

10.6 Lateral Earth Pressure

The effect of ground acceleration on earth pressure is somewhat similar to the effect on slope stability, that is, the horizontal and vertical components of the ground accelerations need to be considered in the failure of the wedge behind the retaining structures. Among the popular methods used to evaluate seismic lateral earth pressure is the Mononobe-Okabe method, which is considered an extension to Coulomb failure criterion.

The Mononobe-Okabe method considers both active and passive lateral seismic earth pressure. The total active lateral force, including seismic effect E_{AE}, behind retaining walls may be expressed as follows:

$$E_{AE} = \frac{1}{2}\gamma H^2 (1 - k_v) K_{AE}$$

where the seismic active pressure coefficient, K_{AE}, is given with reference to Fig. 10-8 as follows:

$$K_{AE} = \frac{\cos^2(\phi - \theta - \beta)}{\psi \cos\theta \cos^2\beta \cos(\delta + \beta + \theta)}$$

where β = slope of soil face
δ = angle of friction between soil and retaining wall
γ = unit weight of soil
H = height of soil face
i = backfill slope angle
k_h = horizontal acceleration coefficient

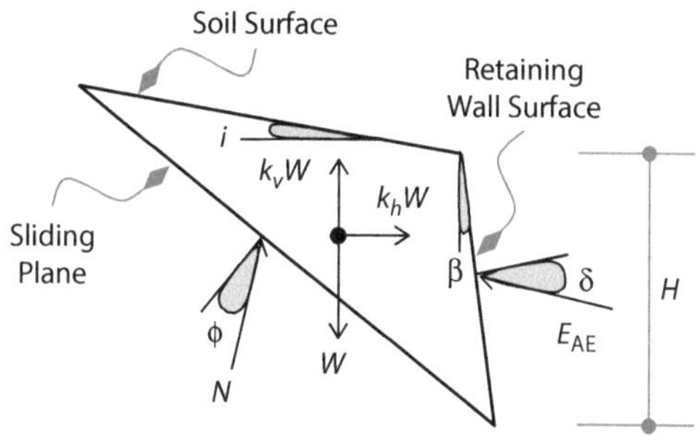

FIGURE 10-8 Equilibrium of failure wedge.

k_v = vertical acceleration coefficient
ϕ = angle of friction soil

$$\psi = \left[1 + \sqrt{\frac{\sin(\phi+\delta)\sin(\phi-\theta-i)}{\cos(\delta+\beta+\theta)\cos(i-\beta)}}\right]^2$$

θ = seismic inertial angle, $\tan^{-1}\left(\frac{k_h}{1-k_v}\right)$

The assumed wedge failure is the same as in the static case. The additional forces that are considered in seismic analysis are the horizontal seismic force (k_hW) and the vertical seismic force (k_vW).

The equivalent expression for passive earth pressure is given as follows:

$$E_{PE} = \frac{1}{2}\gamma H^2(1-k_v)K_{PE}$$

where the seismic active pressure coefficient, K_{PE}, is given as follows:

$$K_{PE} = \frac{\cos^2(\phi-\theta+\beta)}{\Gamma\cos\theta\cos^2\beta\cos(\delta+\beta+\theta)}$$

where

$$\Gamma = \left[1 - \sqrt{\frac{\sin(\phi-\delta)\sin(\phi-\theta+i)}{\cos(\delta-\beta+\theta)\cos(i-\beta)}}\right]^2$$

There have been suggestions on how to apply this force. For example, some suggest one should consider the static component of the force to act at 0.33 H from the base, as is usually done, whereas the dynamic component is taken at 0.6 H from the base. For most practical solutions, the American Association of State Highway and Transportation Officials (AASHTO) code provisions suggest that one considers the total force, including both static and dynamic components, to act at 0.5 H from the base with uniform distribution of pressure as shown in Fig. 10-9.

It is important to emphasize that the philosophy of designing retaining walls under seismic excitations is to force sliding failures rather than overturning failures. The Newmark sliding block concept has shown that sliding of a wall will alleviate seismic

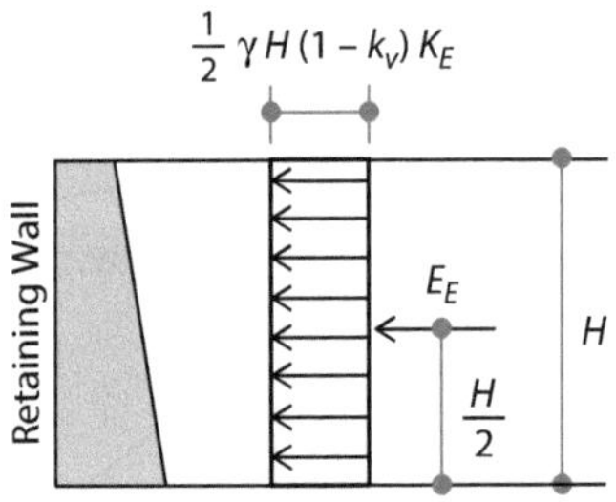

FIGURE 10-9 Nominal pressure distribution.

forces in a way similar to the influence force reduction factors have on structural behavior under excitation. Furthermore, if sliding of the wall occurs, it can often be accommodated by the wall and will usually not result in catastrophic failures. In contrast, overturning usually results in catastrophic failure and loss of the structure.

10.7 Foundations

A foundation is a very important element for the safety of a structure because its failure usually causes catastrophic failure of the structure. In general, foundation failures should be avoided. For example, seismic codes require conditions to force the formation of plastic hinges in the superstructure, allowing the foundations to remain in their elastic range as explained in previous sections. Consequently, foundations are designed with the same concepts and procedures for regular nonseismic cases. However, the large overturning moments that result from seismic forces need special treatment.

Owing to large overturning moments, especially in nonbearing shear walls, the vertical gravity load is not enough to hold the foundation down as shown in Fig. 10-10(a). As a result, the foundation cannot carry the load by itself. Two solutions may be used to rectify this problem. One solution is to use a pile foundation to

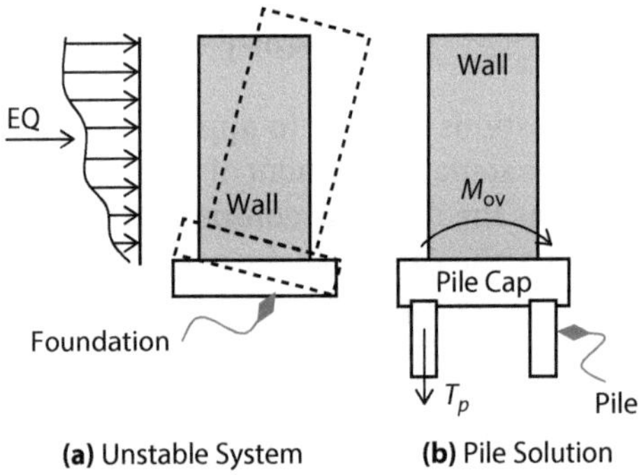

FIGURE 10-10 Stability of foundations.

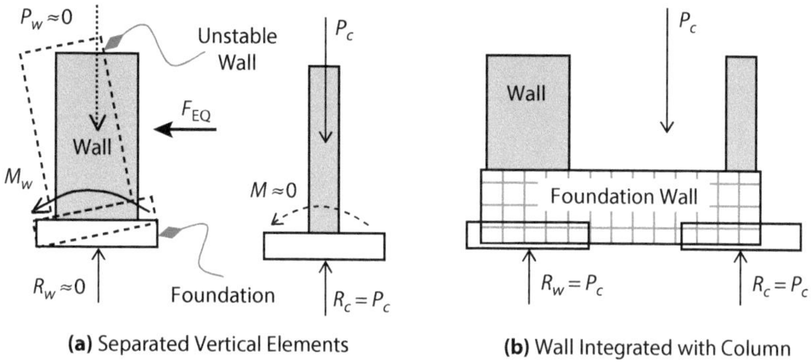

FIGURE 10-11 Foundation walls (strap girders).

take the tension forces that are developed at the bottom of the foundation as shown in Fig. 10-10(b).

The other solution is to tie the foundation to the building by stiff girders known as foundation walls or strap girders as shown in Fig. 10-11(b). These foundation walls are similar in principle to the strap beams used to tie the property line footings with an inside footing. Fig. 10-11(a) shows a wall under large moment but with a minimal axial load. As a result, this wall will be unstable under overturning moment. However, the adjacent column is subjected to heavy axial load with or without moment. Figure 10-11(b) shows the arrangement of using a stiff strap girder (foundation wall) to tie the two elements together and force them to behave as one unit. In this case, the system will act as one unit with a vertical load equal to the column load. The effect of the strap girder will be to shift the resultant to the middle region and provide stability of the system as a whole.

Obviously, the solution of such problems depends largely on the distribution of the structural system of the building. Note the difference between wall foundations and foundation walls. Wall foundations are the strip footings used to carry wall loads whether they are shear walls or regular bearing walls. Foundation walls are the stiff girder systems that tie unstable elements with adjacent walls or columns.

The design forces of foundations are required by seismic codes to exceed the strength that is developed due to plastic hinging of the superstructure. This is intended to guarantee their integrity and prevent any premature failures in the foundations. In addition, the codes require special detailing of the connections between the foundations and the superstructure. Such evaluation and detailing are given in Chap. 8.

Example 10-2 A nonbearing shear wall is subjected to a seismic lateral static force equal to 200 kN acting at a height as shown in Example 10-2, Fig. 1. The weight of the wall is 250 kN (56.21 kip).

1. Check the stability of the wall under these forces. If the wall is not stable, examine alternatives (2) and (3) below to stabilize the wall.
2. Use four piles under the wall foundation.
3. Use a strap girder (foundation wall) to tie the wall to the adjacent column as shown in Example 10-2, Fig. 1.

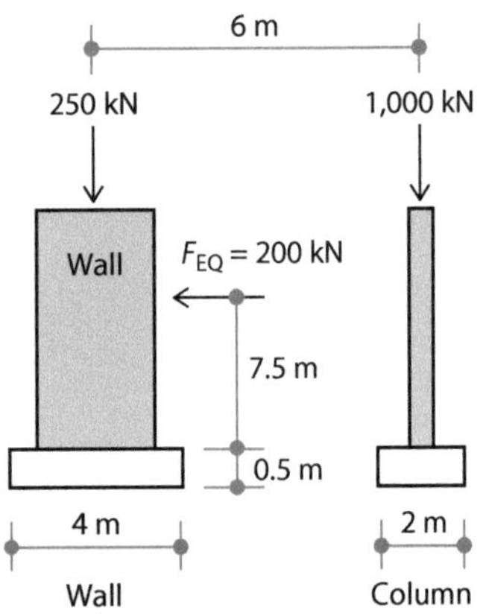

Example 10-2, Figure 1

Solution

Part 1. Wall stability

Taking moments about point "O" shown in Example 10-2, Fig. 2, the overturning and stabilizing moments are calculated as follows:

$$M_{ov} = 200(8) = 1{,}600 \text{ kN·m}$$

$$M_{st} = 250(2) = 500 \text{ kN·m}$$

Because $M_{ov} > M_{st}$, the wall is unstable under these forces and must be stabilized by other means.

The stability of the wall may also be checked by calculating the eccentricity, e, of all forces from the centroid of the foundation

$$e = \frac{\Sigma M}{\Sigma P} = \frac{1{,}600}{250} = 6.4 \text{ m}$$

Because the eccentricity is outside of the foundation limits, the wall is unstable.

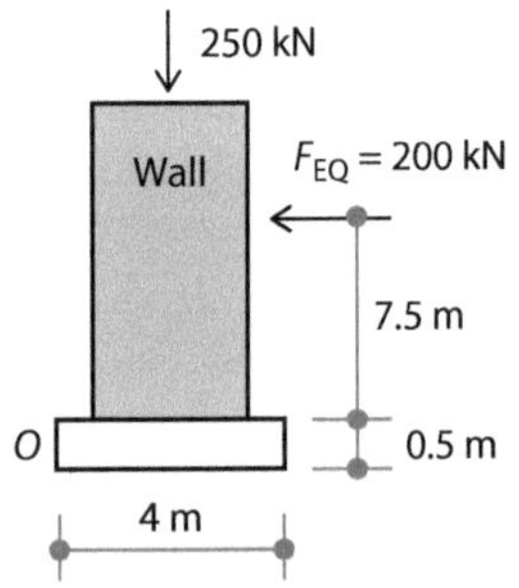

Example 10-2, Figure 2 Wall stability.

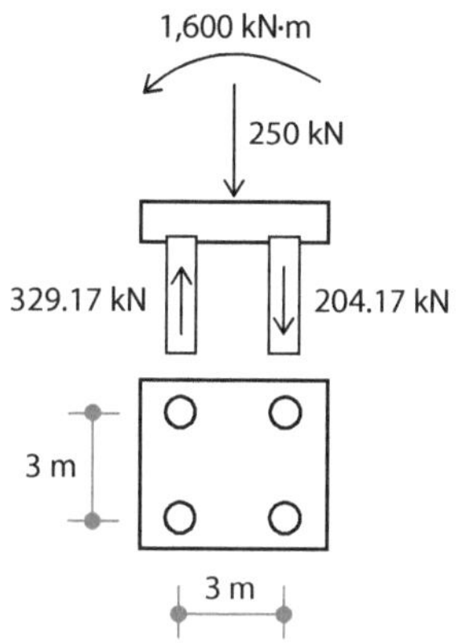

Example 10-2, Figure 3 Pile solution.

Part 2. Pile solution

Forces and moments acting on the top of the pile group are as shown in Example 10-2, Fig. 3, as follows:

$$Q = 250 \text{ kN}$$

$$M = 200(8) = 1{,}600 \text{ kN·m}$$

The force in each pile, F_p, is found from the following expression:

$$F_i = \frac{Q}{N} + \frac{M}{\Sigma r_j^2} r_i$$

where N = number of piles = 4 piles

Σr_j^2 = polar moment of inertia, which is given as

$$2(1.5)^2 + 2(1.5)^2 = 9 \text{ m}^2$$

Hence,

$$F_i = \frac{250}{4} \pm \frac{1{,}600}{9} r_i$$
$$= 62.5 \pm 177.78(1.5)$$

Compression piles: $F_{comp} = 62.5 + 266.67 = 329.17$ kN

Tension piles: $F_{tens} = 62.5 - 266.67 = -204.17$ kN

Part 3. Strap girder solution

The system will be tied together by a stiff strap girder (SG) as shown in Example 10-2, Fig. 4(a). The statical system of this arrangement consists of a beam (strap girder) with a length equal to 9 m as shown in Example 10-2, Fig. 4(b), supported at the centroids of the wall and the column, and loaded by the pressure under the footings. The wall is subjected to moment equal to1,600 kN·m at its base as calculated in Part 2.

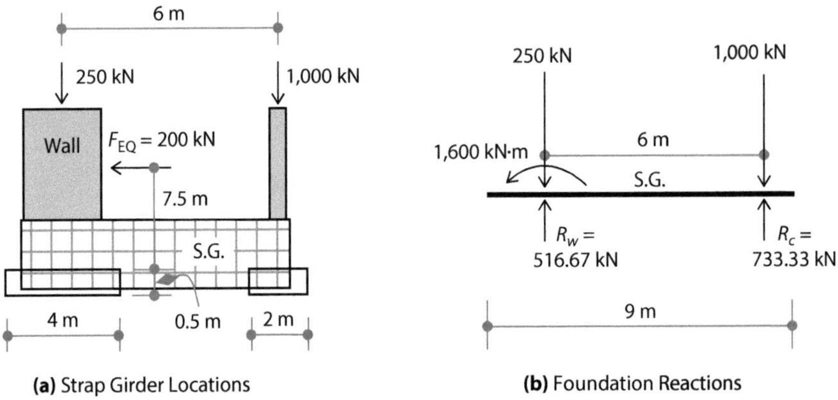

EXAMPLE 10-2, FIGURE 4 Statical system of strap girder.

The reactions of the foundations are found by taking moments about any element. For example, if moments are taken about the centroid of the wall as in Example 10-2, Fig. 4(b), the reactions are given as follows:

$$R_c = \frac{1{,}000(6) - 1{,}600}{6} = 733.33 \text{ kN}$$

$$R_w = 1{,}250 - 733.33 = 516.67 \text{ kN}$$

The pressure under each footing is calculated by dividing the reaction force by the length of the footing

$$q_w = \frac{516.67}{4} = 129.17 \text{ kN/m}$$

$$q_c = \frac{733.33}{2} = 366.67 \text{ kN/m}$$

The final loading diagram of the strap girder is shown in Example 10-2, Fig. 5.

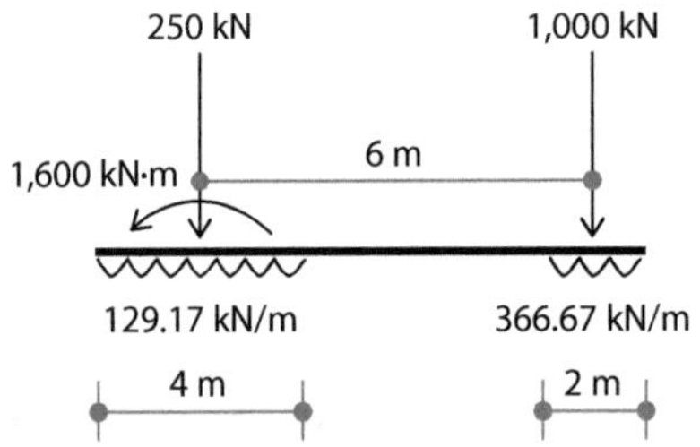

EXAMPLE 10-2, FIGURE 5 Loading diagram.

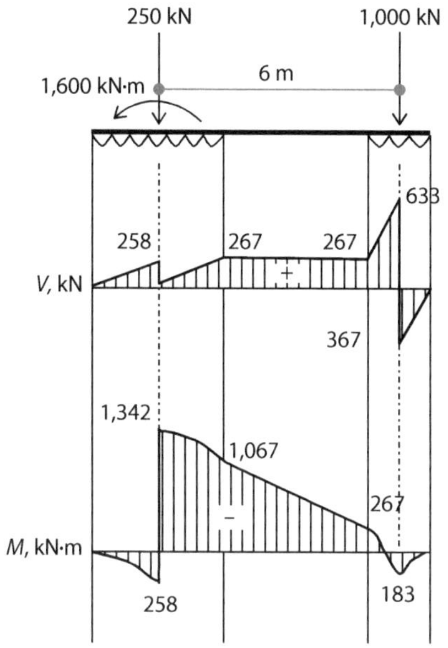

Example 10-2, Figure 6 Shear and bending diagrams.

The bending moments and shearing forces acting on this girder can now be drawn as a result of the loading shown in Example 10-2, Fig. 5. The resulting diagrams are shown in Example 10-2, Fig. 6.

The depth of the girder may be calculated using the expression of convenient depth given in App. 7-1. The convenient depth is the depth that covers deflection limitations by providing the section with a reinforcement index ω, equal to 0.18. Using such an expression for an assumed width of this girder, $b = 500$ mm, $f_c' = 25$ MPa, and for the sake of illustration, consider this loading scheme is already a factored loading for dead and seismic effect. As a result

$$d_{\text{conv}} = 2.5\sqrt{\frac{M_n}{f_c' b}}$$

where

$$M_n = \frac{M_u}{\phi} = \frac{1,342}{0.9} = 1,499 \text{ kN·m}$$

Therefore,

$$d_{\text{conv}} = 2.5\sqrt{\frac{1,499 \times 10^6}{25(500)}} = 866 \text{ mm}$$

Use a total depth of the strap girder equal to 1.0 m.

APPENDIX 10-1

Sheet 1 Soil Classification

Soil Profile Type	Description	Average Properties for Top 30 m (100 ft)		
		Shear Wave Velocity, vs. m/s (ft/s)	**Standard Penetration Test (blows/ft)**	**Undrained Shear Strength kPa (psf)**
A	Hard rock	> 1,500 (5,000)	N/A	N/A
B	Rock	760 to 1,500 (2,500 to 5,000)		
C	Very dense soil and soft rock	360 to 760 (1,200 to 2,500)	> 50	> 100 (2,000)
D	Stiff soil profile	180 to 360 (600 to 1,200)	15 to 50	50 to 100 (1,000 to 2,000)
E	Soft soil profile	< 180 (600)	< 15	< 50 (1,000)
	—	Any profile with more than 3.28 m (10 ft) of soil having the following characteristics: 1. Plasticity index, $PI > 20$, 2. Moisture content, $w \geq 40\%$, and 3. Undrained shear strength < 24 kPa (500 psf).		
F	—	Any profile containing soils having one or more of the following characteristics: 1. Soils vulnerable to potential failure of collapse under seismic loading such as liquefiable soils, quick and high sensitive clays, and collapsible weakly cemented soils. 2. Peats and/or highly organic clays [$H > 3.28$ m (10 ft)] of peat and/or highly organic clay where H = thickness of soil. 3. Very high plasticity clays [$H > 7.5$ m (25 ft)] with plasticity index of $P > 75$. 4. Very thick soft/medium stiff clays [$H > 36$ m (120 ft)].		

Table 10-A1 Soil Profile Type According to IBC Classification

Sheet 2 Correction of Standard Penetration Test, SPT

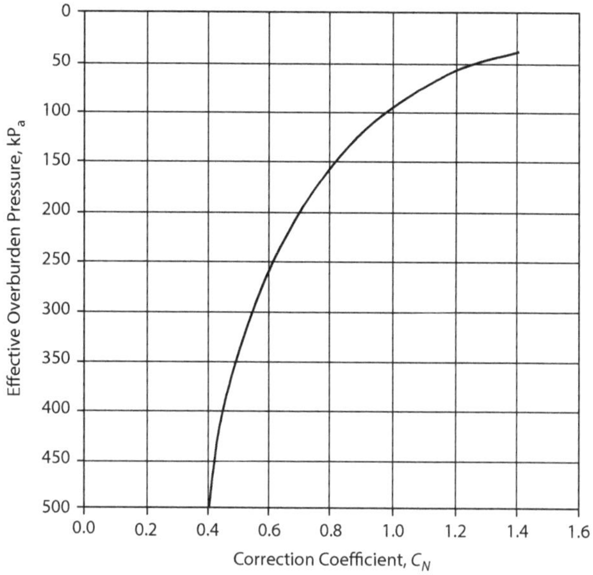

Correction coefficient of standard penetration test.

Sheet 3 Historical Liquefaction Potential Curves

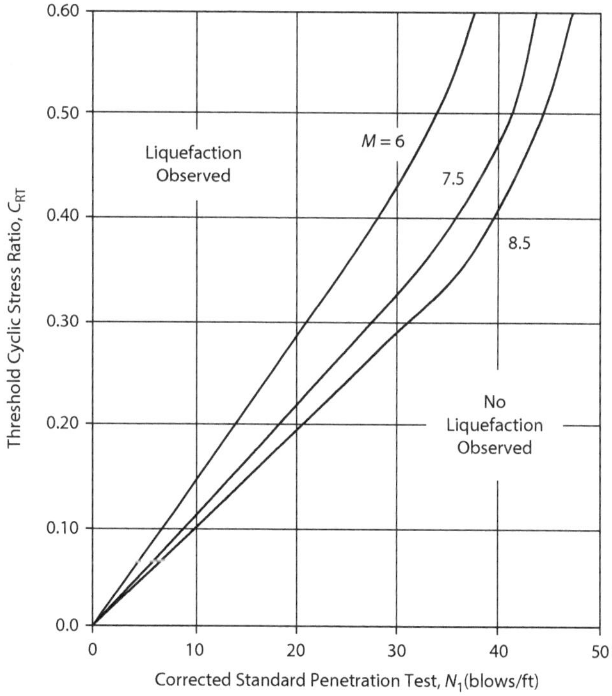

Historical relationship between C_{RT} and N_1.

CHAPTER 11

Synthetic Earthquakes

11.1 Introduction

Although earthquake records are commonly used in the dynamic analysis of structures, they do not reflect site conditions nor do they represent any future earthquake prediction. Earthquakes are characterized by initiation of seismic waves by irregular slippage along faults, followed by numerous random changes in their travel through random media properties. Research shows that if strong ground motions have anything in common, it is their randomness. Therefore, it seems logical to use this property of randomness in the analysis of earthquakes.

A probabilistic approach to engineering problems and structural safety is now widely accepted. Even for loading conditions with much higher certainty than earthquake loading, modern codes use a probabilistic approach and reliability indices as the basis for their provisions. Thus, it seems realistic to use a probabilistic approach for a less certain process, the earthquake loading.

Fourier analysis of earthquakes reveals that earthquakes can be described as noise signals. Such signals contain frequency bands of high Fourier amplitudes according to the site characteristics of the ground. Because the power spectral density function, which is defined as the energy per unit frequency, is proportional to Fourier amplitude, the power spectral density of specific sites can be constructed from a statistical database of actual records.

In order to incorporate randomness as well as power spectral distribution of the earthquake, ground motions are best simulated by nonstationary filtered white noise. Site characteristics are introduced in this simulation by the application of filters containing those characteristics, whereas the sensitivity of structural response to inelastic properties is addressed by the application of nonstationary envelope functions.

11.2 Fourier Transform

In review, it is known that any periodic function of time can be approximated by a Fourier series as a summation of sinusoidal functions. The trigonometric form of a Fourier series takes the form

$$f(t) = a_o + \Sigma_1^n a_n \cos(n\omega_o t) + \Sigma_1^n b_n \sin(n\omega_o t)$$

The constants a_o, a_n, and b_n are given as follows:

$$a_o = \frac{1}{T}\int_o^T f(t)dt$$

$$a_n = \frac{2}{T}\int_o^T f(t)\cos(n\omega_o t)dt$$

$$b_n = \frac{2}{T}\int_o^T f(t)\sin(n\omega_o t)dt$$

The complex form of a Fourier series takes the form

$$f(t) = \sum_{-\infty}^{\infty} c_n e^{jn\omega_o t}$$

where

$$c_n = \frac{1}{T}\int_{-T/2}^{T/2} f(t)e^{-jn\omega_o t}\, dt$$

However, for nonperiodic functions, the period approaches ∞, and ω_o becomes vanishingly small ($\omega_o = 2\pi/T$). This condition may be represented by differential form such that, $\omega_o \rightarrow d\omega_o$. Therefore, a Fourier series takes a differential form as follows:

$$F(j\omega) = \int_{-\infty}^{\infty} f(t)\, e^{-j\omega t}\, dt$$

$$f(t) = \frac{1}{2\pi}\int_{-\infty}^{\infty} F(j\omega)\, e^{j\omega t}\, d\omega$$

The two equations above are known as Fourier transform pairs. This implies that any function in the time domain may be transformed into the frequency domain and vice versa. Fourier transform pairs have wide application in engineering circuits and communication fields. In civil engineering, they are useful in analyzing earthquake signals that are similar to noise signals in communication systems.

It is important for us to understand the physical meaning of the Fourier transform. For this purpose, it is necessary to define the pulse function as a function that has a unit area over an infinitesimal interval. To illustrate, consider the rectangular function shown in Fig. 11-1. In Fig. 11-1(a), a rectangular function has a width of Δ with a height of $1/\Delta$. Thus, the area bounded by this rectangle is one.

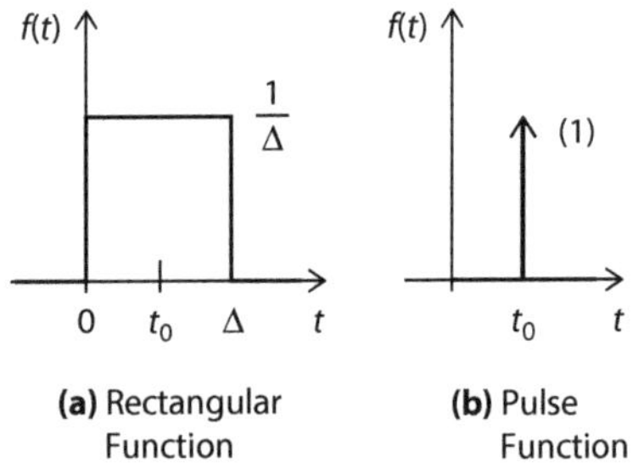

FIGURE 11-1 Pulse function.

Let the width of this rectangular function shrink without changing the area. In the limit, when $\Delta \to 0$, the height of this function $\to \infty$. The new function at this limit is called the pulse function, as shown in Fig. 11-1(b). The pulse function may be represented mathematically as follows:

$$\delta(t - t_0) = 0 \qquad (\text{for } t \neq t_0)$$

The area of one is called the strength of this function and is written as (1). Consequently, a function that is given as

$$A(t) = B\delta(t - t_0) = 0$$

is said to have a strength of B, written as (B).
Using the pulse function notation, it can be shown that Fourier transform pairs of the cosine function ($\cos \omega_o t$) are given as

$$f(t) = \cos \omega_o t$$

$$F(j\omega) = \pi[\delta(\omega + \omega_0) + \delta(\omega - \omega_0)]$$

This result shown in Fig. 11-2 implies that the cosine function that has one frequency in the time domain appears with two frequencies in the frequency domain that are symmetric about the vertical axis with a strength of π. This is a property of Fourier transform functions, which can be stated as follows: the Fourier transform of real time functions will always be symmetric about the vertical axis in the frequency domain, known as double-sided functions. Figure 11-3 shows an example of popular functions in the time domain and their transform in the frequency domain.

The transform of the pulse function appears with constant value for all frequencies as shown in Fig. 11-3(a). Because this function contains all frequencies, it is termed white noise similar to the term white color that contains all frequencies. All real functions in the time domain appear symmetric in the frequency domain. Therefore, because the pulse of Fig. 11-3(c) is not symmetric, its function is not real in the time domain. It is important to note that a general function that appears in Fig. 11-3(d) transforms into another symmetric general function. This function is of interest because it represents the acceleration signal of the earthquake in the time domain.

Note that the transform process is carried out through mathematical tools as the discrete Fourier transform, or in an even more efficient scheme, as the fast Fourier transform. For more on this subject, consult textbooks in mathematics and in structural dynamics.

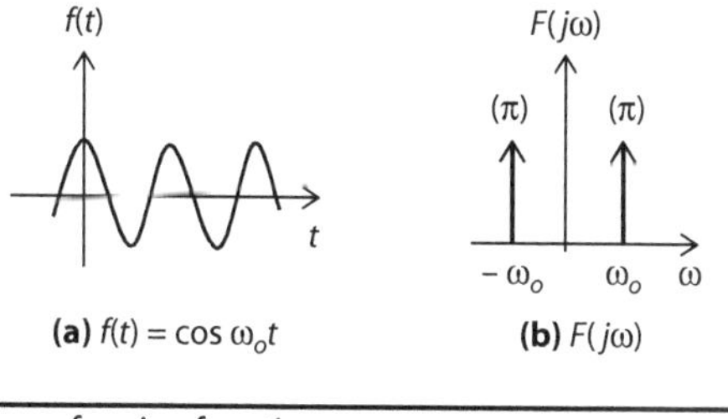

FIGURE 11-2 Fourier transform of cosine function.

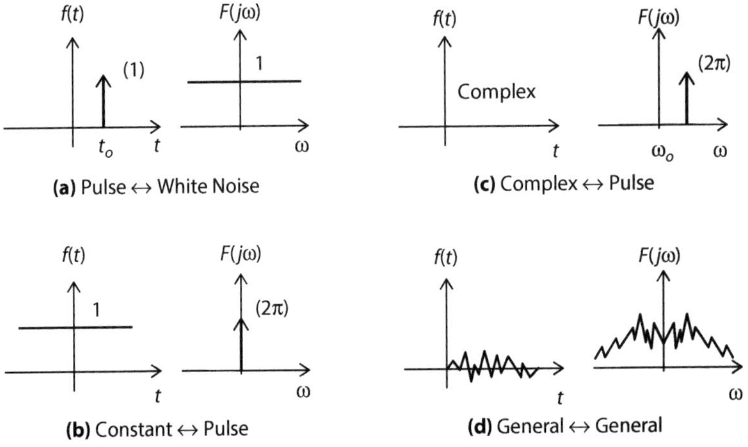

FIGURE 11-3 Fourier transform of four popular functions.

11.3 Power Spectral Density

Power spectral density (PSD) is a measure of the power associated with the frequency content of a function. This concept is best explained using circuit analysis. Consider a resister with resistance equal to 1 ohm. If a current, $f(t)$, is passing through this resister, the voltage will also be equal to 1 insofar as

$$V = iR = f(t)(1) = f(t)$$

Because the power that is delivered to this resistor is the product of the voltage by the current, the power P is expressed as

$$P = vi = f(t)^2$$

Accordingly, the total energy, E, delivered to this resistor will be the integration of the power over the entire time interval

$$E = \int_{-\infty}^{\infty} f(t)^2\,dt$$

The energy delivered to this resistor may also be expressed in the frequency domain if the expression above is used with consideration such that, $f(t)^2 = f(t)\cdot f(t)$. As a result

$$E = \int_{-\infty}^{\infty} f(t)^2\,dt = \int_{-\infty}^{\infty} f(t)f(t)dt$$

Using the Fourier transform from the section before, $f(t)$ may be replaced by its transform as follows:

$$\begin{aligned} E &= \int_{-\infty}^{\infty} f(t)f(t)dt \\ &= \int_{-\infty}^{\infty} f(t)\left[\frac{1}{2\pi}\int_{-\infty}^{\infty} F(j\omega)e^{j\omega t}d\omega\right]dt \end{aligned}$$

By rearranging the terms above, the expression may be reduced to the following:

$$
\begin{aligned}
E &= \frac{1}{2\pi}\int_{-\infty}^{\infty} F(j\omega)\left[\int_{-\infty}^{\infty} f(t)e^{j\omega t}\,dt\right]d\omega \\
&= \frac{1}{2\pi}\int_{-\infty}^{\infty} F(j\omega)\left[F(-j\omega)\right]d\omega \\
&= \frac{1}{2\pi}\int_{-\infty}^{\infty} \left|F(j\omega)\right|^2 d\omega
\end{aligned}
$$

The expression above indicates that the energy of the system may be expressed in the frequency domain as the square of Fourier amplitude of Fourier transform of the time function.

The average power can now be evaluated by dividing the energy by the period when the period tends to infinite

$$
\begin{aligned}
P &= \lim_{T\to\infty}\frac{1}{T}\cdot\frac{1}{2\pi}\int_{-\infty}^{\infty} \left|F(j\omega)\right|^2 d\omega \\
&= \int_{-\infty}^{\infty} \lim_{T\to\infty}\frac{\left|F(j\omega)\right|^2}{2\pi T}\,d\omega = \int_{-\infty}^{\infty} S(\omega)\,d\omega
\end{aligned}
$$

The quantity $S(\omega)$ is known as the power spectral density because it represents some form of power density of the function $f(t)$. As a result, the PSD may now be expressed as follows:

$$
S(\omega) = \lim_{T\to\infty}\frac{|F(j\omega)|^2}{2\pi T}
$$

At this stage, the power associated with the frequency is proportional to the square of Fourier amplitude, $|F(j\omega)|^2$.

The response spectrum represented in many chapters of this book represents a kind of amplification of the ground parameters, namely ground displacement, ground velocity, and ground acceleration. Therefore, the response spectrum can be seen as representing some degree of the power associated with the frequency of the ground because the amplification is high at resonance when the frequency of the system and the forcing function coincide. This is an important result because the power spectral density and the response spectrum can be considered to be two sides of the same coin. Because they both measure the energy associated with the frequency of the system, the PSD may be used to develop the time domain earthquake signal. This will be discussed in the next section.

11.4 Stationary Random Processes

A random process is defined as a family, or ensemble, of n random variables, where each variable is a function of other independent variables. The process is said to be random because it can only be known in a probabilistic sense. For synthetic earthquakes, we are interested in a harmonic random process that is a function of only one independent random variable, θ_n. This random process is defined by the relation

$$x_n(t) = A_n \sin(\omega_n t + \theta_n), \ (n = 1, 2, \ldots \infty)$$

where A_n = amplitude of each individual harmonic wave form
ω_n = circular frequency of each individual harmonic
θ_n = nth sampled value of a random phase angle between 0 and 2π
$x_n(t)$ = nth member of the ensemble

This process is considered stationary because the averages of each individual sample of its family are independent of time. Such averages include mean values, mean square values, variances, covariances, and correlation coefficients. However, the average of greatest interest in a stationary random process is the mean square value. For one sample of periodic function, the time mean square value denoted as $< x_n(t)^2 >$ is given as

$$< x_n(t)^2 > = \frac{1}{T}\int_{-T/2}^{T/2} x_n(t)^2 \, dt$$

A Fourier series can be used to show that the mean square value, $< x_n(t)^2 >$, for a periodic function can be expressed in the frequency domain as follows:

$$< x_n(t)^2 > = \sum_{r=1}^{\infty} \frac{A_{\mathrm{nr}}{}^2}{2}$$

where A_n is amplitude of the rth term in Fourier series of the nth harmonic function. Because the PSD function is the most representative and useful measure of a random process, it is helpful to express the mean square in terms of the PSD. The mean square value, $< x_n(t)^2 >$, can also be expressed in terms of the PSD using Fourier transform pairs as follows:

$$< x_n(t)^2 > = \int_{-\infty}^{\infty} S_{\mathrm{xn}} \, d\omega = \sum_{r=1}^{\infty} 2S_{\mathrm{xnr}}(\omega)\Delta\omega$$

where S_{xn} = PSD function of the nth harmonic function
S_{xnr} = PSD value of the rth term in the summation above

When $x_n(t)$ is a real function, the PSD is symmetric. Therefore, a factor of 2 appears in the above summation over the range, $r = 1 \rightarrow \infty$.

11.5 Random Ground Motion Model

As noted earlier, ground motions are best simulated by nonstationary filtered white noise that allows the incorporation of the energy associated with the frequency content while preserving the randomness and sensitivity to inelastic dynamic analysis. These elements can be achieved by subjecting a stationary filtered white noise to a nonstationary envelope function. Accordingly, the simulated ground acceleration, $\ddot{u}_g(t)$, may take the following form:

$$\ddot{u}_g(t) = \mathrm{PGA} \bullet e(t) \bullet X(t)$$

where $e(t)$ = normalized nonstationary envelope function required to describe the buildup and subsequent decay of the ground amplitude
PGA = peak ground acceleration
$X(t)$ = stationary filtered white noise ensemble (family)

$X(t)$ can be expressed by a family of harmonics in the form

$$X(t) = \Sigma x_n(t) = \Sigma A_n \cos(\omega_n t + \theta_n), (n = 1, 2, \ldots m) \tag{11-1}$$

As noted in the previous section, the mean square value of the sample (n) in terms of amplitudes is given as

$$< x_n(t)^2 > = \frac{1}{2}\Sigma A_{nr}^2 \quad (n = 1, 2, \ldots m; r = 1, 2, \ldots \infty) \tag{11-2}$$

Also, the mean square value of the sample (n) in terms of PSD is given as

$$< x_n(t)^2 > = 2\, \Sigma S_{xnr}(\omega_n)\Delta\omega, \quad (n = 1, 2, \ldots m; r = 1, 2, \ldots \infty) \tag{11-3}$$

Therefore, by comparison of results in Eqs. (11-2) and (11-3), we can conclude that

$$A_n^2 = 4S_{xn}(\omega)\Delta\omega \quad (n = 1, 2, \ldots m) \tag{11-4}$$

As mentioned earlier, $S_{xn}(\omega)$ is symmetric and with its two mirror sides about the y-axis, is said to be double sided. However, if only one side is considered in the analysis, the PSD is said to be one sided. Accordingly, if one-sided PSD is considered, the amplitude, A_n, will be given as

$$A_n^2 = 2S_{xn}(\omega)\Delta\omega \quad (n = 1, 2, \ldots m) \tag{11-5}$$

When we substitute (A_n) from Eq. (11-5) into the original expression in Eq. (11-1) and use S_g to refer to the ground PSD, $X(t)$ can now be expressed as

$$X(t) = \Sigma\sqrt{2S_g(\omega_n)\,\Delta\omega}\cdot\cos(\omega_n t + \theta_n) \quad (n=1,2,\ldots m) \tag{11-6}$$

where $\Delta\omega$ = incremental frequency (ω_{max}/m)
m = number of samples, which is usually taken as 250
n = the nth generated sample of the ensemble
ω_n = frequency of the nth generated sample ($n\,\Delta\omega$)
ω_{max} = cutoff value in the PSD distribution (e.g., ω_{max} = 60 in Fig. 11-4)
$S_g(\omega_n)$ = earthquake one-sided PSD function with dimensions, L^2/T^3
θ_n = random phase angle of the, nth, generated sample between 0 and 2π

$S_g(\omega_n)$ is needed to incorporate site characteristics, which can be constructed as a filtered white noise. Statistical analysis of actual field earthquake records provides both filter shape and white noise amplitude as functions of site characteristics. Site characteristics are given in terms of dominant ground frequency, ω_g, and ground damping ratio, ζ_g. Table 11-1 provides a number of recommended values for the parameters ω_g and ζ_g that reflect the characteristics of the major types of site conditions (rock, deep cohesionless, and soft sites).

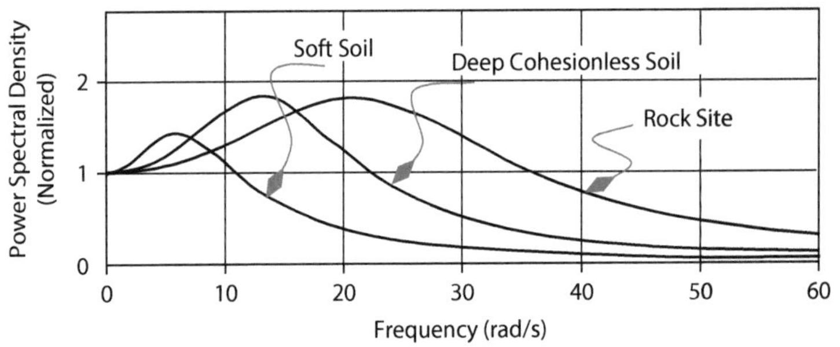

FIGURE 11-4 Site-dependent power spectral density of earthquakes.

Soil Type	ω_g	ζ_g
Rock	8π	0.6
Deep cohesionless	5π	0.6
Soft	2.4π	0.85

TABLE 11-1 Site Condition Parameters

Knowing, ω_g and ζ_g, the PSD can be expressed in the following form:

$$S_g(\omega) = |H(\omega)|^2 \bullet S_o \tag{11-7}$$

where

$$|H(\omega)|^2 = \text{Stationary filter}$$

An appropriate filter may be given as

$$|H(\omega)|^2 = \frac{1+4\zeta_g^2(\omega/\omega_g)^2}{[1-(\omega/\omega_g)^2]^2 + 4\zeta_g^2(\omega/\omega_g)^2}$$

S_o is the amplitude of white noise, which may be given as

$$S_o = \frac{4\zeta_g}{\pi(1+4\zeta_g^2)\omega_g}$$

In earthquake engineering, the peak ground acceleration is established independently in zoning maps. As a result, the white noise amplitude, S_o, becomes irrelevant as the generated records are normalized. Using the values of ω_g and ζ_g from Table 11-1 in the above expression for $S_g(\omega)$, three normalized PSD functions can be constructed as shown in Fig. 11-4. The PSD shapes in Fig. 11-4 resemble the measured field response spectra shown in Fig. 3-13, and these are our target distribution.

Even though the elastic dynamic response can be directly carried out in the frequency domain with a given $S_g(\omega)$, this is not possible in the case of inelastic dynamic analysis because the assumed superposition in the frequency domain does not apply. For nonlinear systems, the structure resistance is history dependent. Therefore, the analysis must be completed in the time domain.

Ground motion simulation models are successfully used to model earthquakes for elastic dynamic analysis, and this has proven to be useful. However, response agreement with elastic analysis does not directly lead to an acceptable response in the inelastic analysis.

In the inelastic dynamic analysis, the earthquake duration is of concern as much as the deformation of the ground. Research has shown that inelastic response of structures is even more sensitive to the excitation duration than to the peak deformation of the ground. The Amin-Ang envelope function is applicable to inelastic dynamic analysis. The Amin-Ang envelope function depends on a parameter known as the equivalent earthquake duration, T_n, which is also site dependent. Statistical analysis of actual records suggests typical values for T_n as 10 s for rock sites, 15 s for deep cohesionless sites, and 20 s for soft sites.

As shown in Fig. 11-5, the Amin-Ang envelope function, $e(t)$, is given as

$$e(t)=\left(\frac{t}{t_1}\right)^2 \quad (\text{for } t<t_1)$$

$$e(t)=1 \quad (\text{for } t_1<t<t_2)$$

$$e(t)=e^{-c(t-t_2)} \quad (\text{for } t>t_2)$$

t_1, t_2 (seconds), and c values depend on T_n, which are given for $T_n > 3$ s, by the following expression:

$$t_1 = 1.5 \text{ s}$$

$$t_2 = T_n - 2.18 \text{ s}$$

$$c = 0.18$$

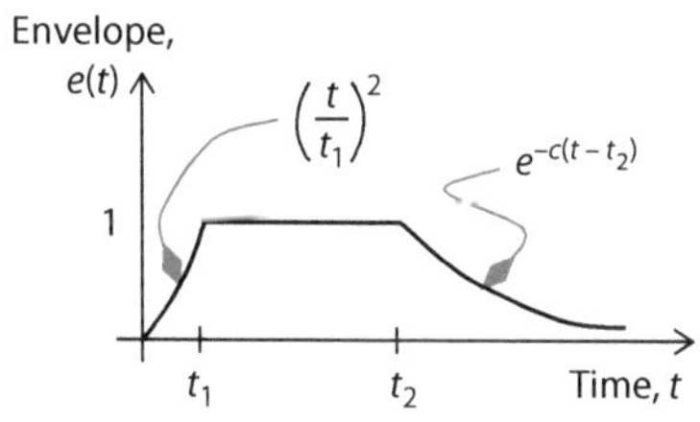

FIGURE 11-5 Amin-Ang envelope.

11.6 Implementation of Ground Motion Model

The concepts above can easily be implemented with computers. Figure 11-6 shows the results from such an implementation with three generated samples of synthetic records for the major site conditions. Each sample is based on a corresponding target PSD distribution as given in Fig. 11-4.

The target PSD functions in Fig. 11-4 are smooth curves because they represent an average results over a large number of records. However, the PSD shape for each individual record will be jagged. As an illustration, Fig. 11-7 shows the Fourier amplitude spectrum (proportional to the PSD) for each individual record given in Fig. 11-6.

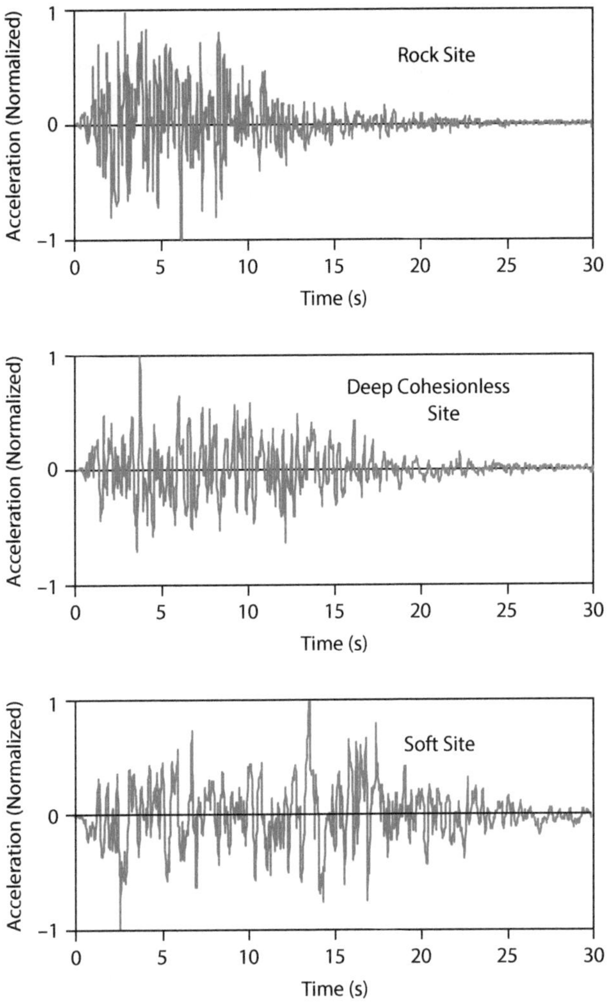

Figure 11-6 Samples of synthetic records.

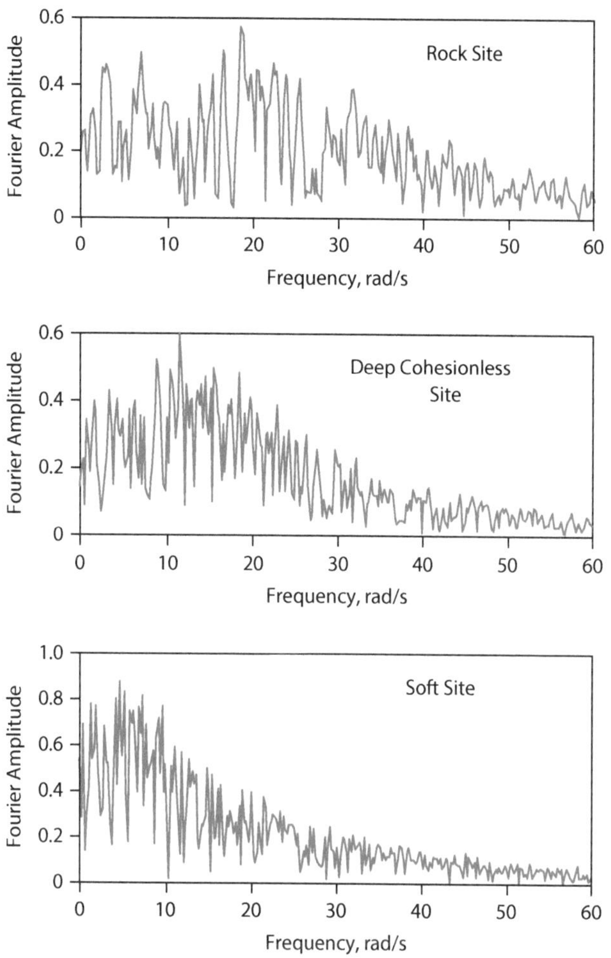

FIGURE 11-7 Fourier amplitude spectrum of synthetic records.

Nevertheless, the trend of each individual PSD distribution follows the general shape of the corresponding target PSD distribution as given in Fig. 11-4.

11.7 Validity of Synthetic Earthquakes

The ground motion model that is based on a stochastic approach has a solid mathematical basis. The validity of such models has been proven to be sound and well representative in their effect on the structural response. In order to examine the validity of synthetic earthquakes, the following paragraphs will present the results of a parametric study of an equivalent single degree of freedom (ESDOF) system that was subjected to the excitation of nine actual earthquake records and nine synthetic earthquake records.

The parametric study used inelastic dynamic analysis and was performed for each set of records by varying the period and the yielding level of a bilinear hysteresis model

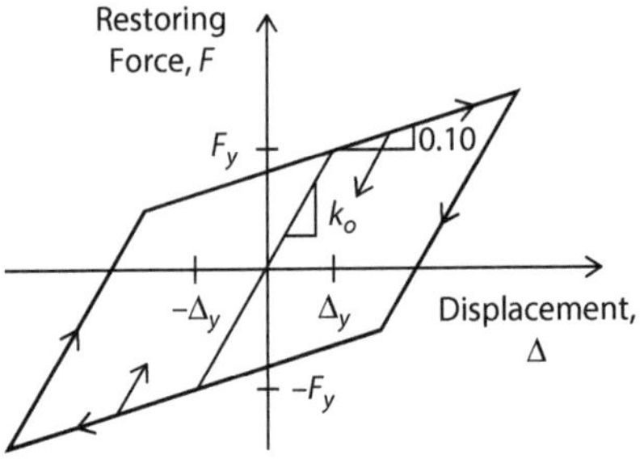

FIGURE 11-8 Bilinear hysteresis model.

shown in Fig. 11-8. The parameter variation included six periods and six levels of yielding of the hysteresis model for each set of earthquake records. This parameter variation resulted in 324 pairs of the force reduction factor, R, and the global ductility demand, μ_d. Values for each set of records were grouped and statistically analyzed and examined.

The inelastic dynamic analysis results are compiled into two groups as shown in Figs. 11-9 and 11-10, which plot all data points of R versus μ_d at different periods. Figure 11-9 shows the results of the response of the ESDOF system to the excitation of the nine actual earthquakes, while Fig. 11-10 shows the response of the same system to the excitation of the nine synthetic earthquakes.

Figures 11-9 and 11-10 clearly indicate that the R versus μ_d relationship is extremely random in both earthquake groups as shown by the scatter of the data points. However, the general shape of the trend of data is similar for both cases. Note that the trend of excessive ductility demand at short periods applies to both cases, whereas relatively moderate ductility demand is required at higher periods.

Statistical comparison between the two cases is also presented by performing regression analysis and evaluation of dispersion measures. Figure 11-11 shows the results

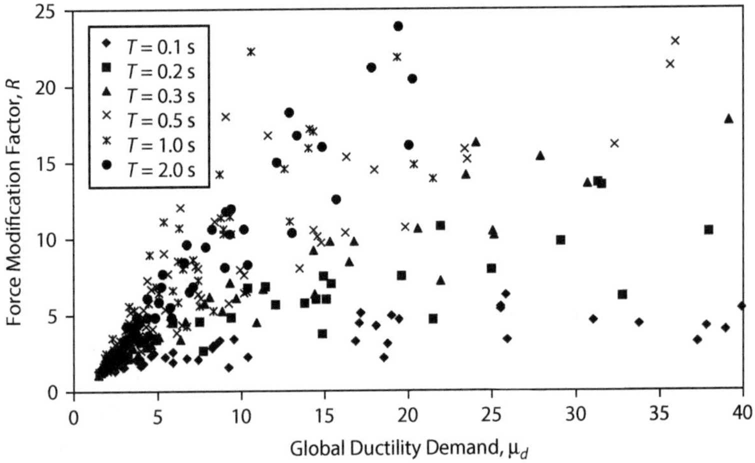

FIGURE 11-9 Response to nine actual earthquakes.

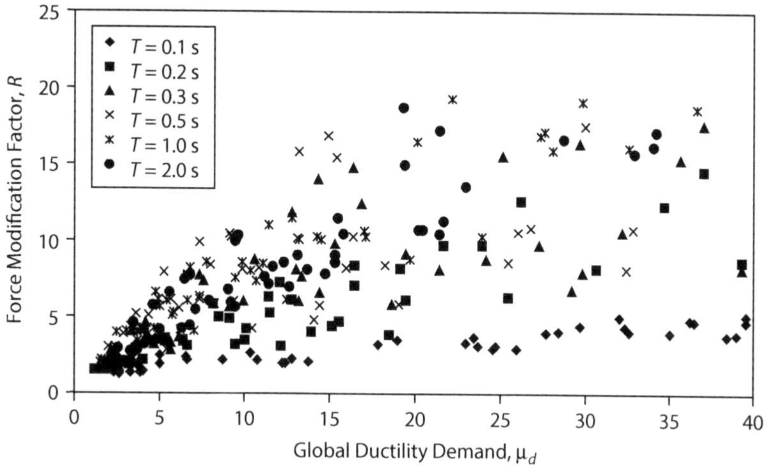

FIGURE 11-10 Response to nine synthetic earthquakes.

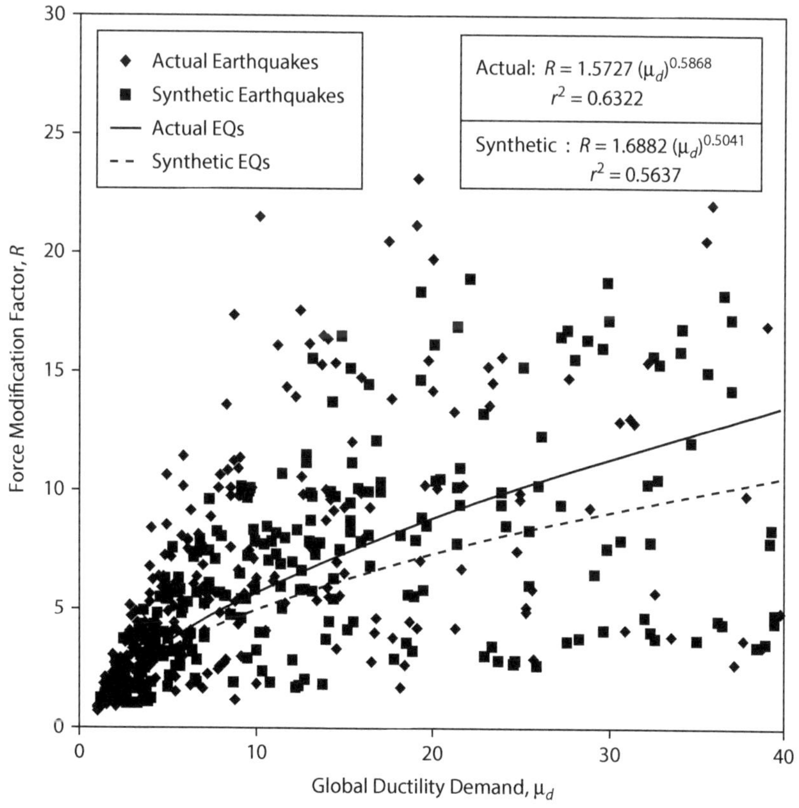

FIGURE 11-11 Comparison between actual and synthetic earthquakes.

Statistical Parameter	Actual Earthquake Records		Synthetic Earthquake Records	
	Global Ductility Demand (μ_d)	Force Modification Factor (*R*)	Global Ductility Demand (μ_d)	Force Modification Factor (*R*)
Mean	10.90	6.30	17.50	7.00
Standard deviation	12.24	5.23	18.71	4.81
Variance	149.90	27.34	350.09	23.07
Covariance	38.65		54.01	
Correlation coefficient	0.606		0.602	

Table 11-2 Statistical Measures within Each Group of Earthquake Records

Statistical Parameter	Force Modification Factor (*R*)		Global Ductility Demand (μ_d)	
	Actual Earthquake Records	Synthetic Earthquake Records	Actual Earthquake Records	Synthetic Earthquake Records
Mean	6.30	7.00	10.90	17.50
Standard deviation	5.23	4.81	12.24	18.71
Variance	27.34	23.07	149.90	350.09
Covariance	23.07		179.64	
Correlation coefficient	0.920		0.787	

Table 11-3 Statistical Measures Across Groups of Earthquake Records

of all data points plotted together with two trend lines drawn for each group of earthquakes. Figure 11-11 shows that the curve that represents synthetic records lies slightly below the curve that represents the actual records. However, the trend curves are not too far from each other, as shown by a comparison of the trend line equations with their associated coefficients of determination (*r* squared values) that appear in the same figure. For the same reduction of elastic strength demand (for the same *R* value), synthetic earthquakes demand higher ductility ratios than their actual counterparts.

Statistical measures are also presented in Tables 11-2 and 11-3. Table 11-2 shows basic statistical measures within each group of earthquake records. The correlation coefficient within each group is almost identical, 0.606 versus 0.602, implying that each group exhibits the same effect on the *R* versus μ_d relationship. The correlation agreement also indicates that synthetic records can be considered to be equivalent to the actual records in their dispersion of characteristics and effect on structural response.

In contrast, Table 11-3 shows basic statistical measures across both groups of records. Here, the correlation coefficient between the *R* values across the two groups is very high. This is anticipated because the *R* values are selected beforehand to establish the parameter variation mentioned earlier. However, the correlation coefficient between the μ_d values across the two groups is also high, indicating that ductility demand under the excitation of both groups is highly correlated.

CHAPTER 12
Seismic Isolation

12.1 Introduction

Seismic isolation is a modern technique used to seismically isolate structures from the ground at the base level. (It is also called base isolation.) Isolation is achieved by introducing the isolating system at the base or at selected locations in order to reduce the transfer of the ground motion effects to the structure. The significant advantage of seismic isolation is its ability to minimize the damage usually associated with earthquake events. However, the disadvantage of seismic isolation is that it may need to be replaced during the lifetime of the structure.

Over the past several decades, seismic isolation has gained in popularity and frequency of use, especially in bridges. Seismic isolation is considered the only practical solution for historical monuments if the superstructure itself cannot be modified or strengthened for obvious reasons.

Today, *isolators* for seismic isolation systems come in many different forms and designs. Lead-rubber bearing isolators are the most popular and perhaps the most economical. In addition, because elastomeric bearings are installed in bridges for nonseismic performance, including lead-rubber bearing isolators would only be a minor modification to the original bearings. Therefore, this chapter will focus on seismic isolation systems that use lead-rubber bearing isolators.

12.2 Concept of Seismic Isolation

The basic concept of seismic isolation can be illustrated by a helicopter-pier analogy. If a helicopter lands on a rigid pier as shown in Fig. 12-1(a), the helicopter experiences the same acceleration of the ground as it is transferred through the supporting rigid pier. However, if the helicopter takes off while connected to the ground by a wire (soft support), the helicopter does not feel any of the ground acceleration as shown in Fig. 12-1(b). Even though the helicopter does not feel the acceleration in the second case, it will experience the same displacement of the ground relative to its position.

The same conclusion is reached when we examine the response spectrum properties. Recall that the response spectrum acceleration approaches zero as the period of

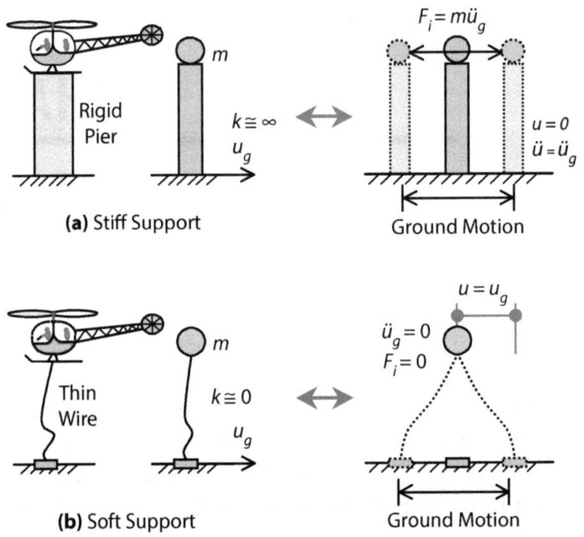

FIGURE 12-1 The concept of seismic isolation.

the structure approaches infinity (soft support), while the response spectrum acceleration approaches the ground acceleration as the period approaches zero (stiff support). Figure 12-2 shows these limits for the El Centro earthquake. Note that isolators usually have large energy-dissipating elements that further reduce the forces and displacements of the system (see the effect of increased damping ratios in Fig. 12-2).

In summary, the target of any seismic isolation system is to elongate the period of the structure to limit the transfer of the acceleration from the ground to the structure. This also implies that the structure experiences little force, if any. However, this action is associated with large displacements in the isolators that must be accommodated. Remember that seismic isolation is a tradeoff between large forces and large displacements.

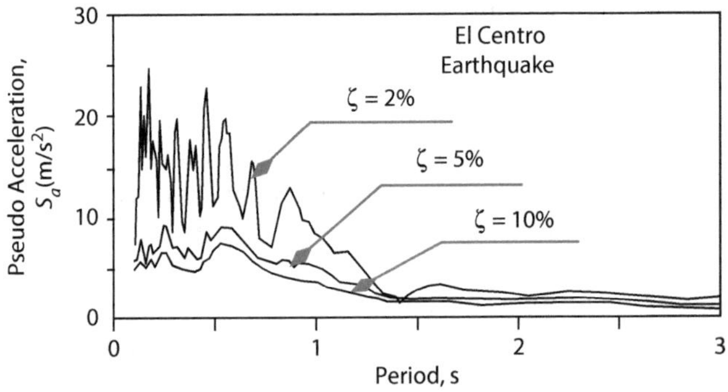

FIGURE 12-2 Response spectrum.

12.3 Lead-Rubber Bearing Isolators

As mentioned earlier, this book will consider the lead-rubber bearing isolation system as a model isolator to explore the concept and the design requirements for isolators in general. The lead-rubber bearing isolator consists of a rubber pad reinforced with laminated steel sheets as shown in Fig. 12-3. A circular lead core is inserted in the middle of the rubber pad as shown in the same figure. The pad is then attached to two steel plates at both faces as an interface to connect the isolator with the structure.

The rubber used in these isolators consists of the same elastomeric material used in bridges with special specifications to accommodate seismic requirements. The rubber in this case serves as a soft material that provides soft support for isolation purposes. The rubber itself is not a good energy dissipater, as it is a purely elastic material. Furthermore, under regular loads, the structure may experience large uncomfortable displacements that are due to its low stiffness. In addition, the structure may be subjected to low-frequency vibration under normal conditions. Therefore, lead core is inserted in the middle of the rubber pad to handle these shortcomings.

As a material, lead has moderate stiffness and a large capacity of energy dissipation. As a result, it is considered ideal for this system to achieve three important goals:

1. Limit large displacements that are due to lateral forces under normal conditions because of its elastic stiffness.
2. Yield under seismic excitations at low force levels to activate the function of the rubber as an isolator.
3. Serve as a damper on account of its large energy dissipation capacity.

The elastomer as a material is considered to be linear elastic. The properties of the elastomer depend on its hardness, which is usually specified by the manufacturer. In the absence of manufacturer data, the American Association of State Highway and Transportation Officials (AASHTO) code provides the shear modulus of elasticity and creep properties for elastomer at different hardness values. This information is also provided in Table 12-A1 at the end of this chapter. In contrast, lead as a material exhibits ideal elastoplastic behavior in shear. The lead properties in shear are also given in Table 12-A2.

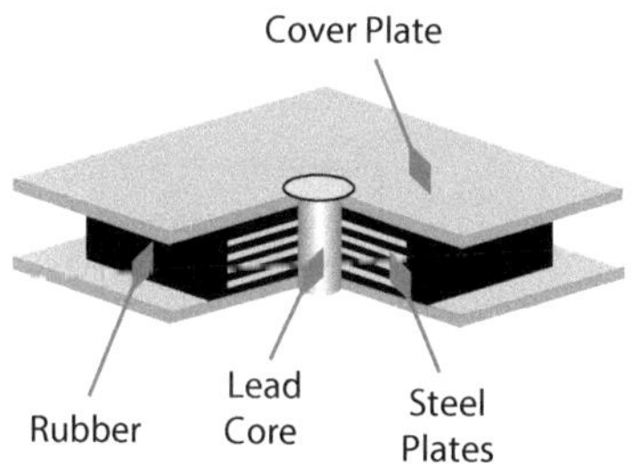

FIGURE 12-3 Lead-rubber bearing.

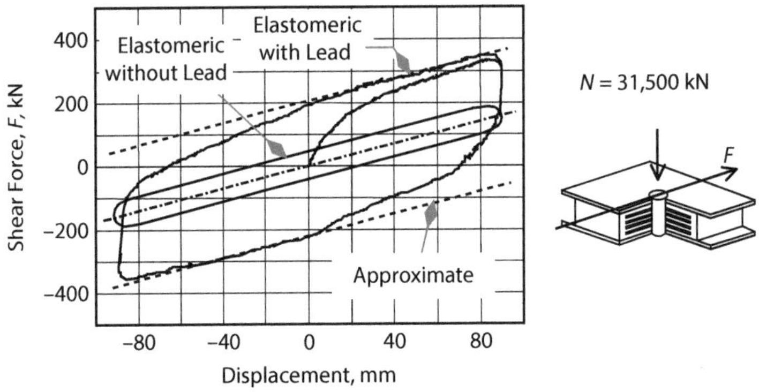

FIGURE 12-4 Lead-rubber bearing hysteretic behavior.

When combined, elastomer and lead form an ideal system that exhibits ideal bilinear hysteretic behavior as shown in Fig. 12-4. Extensive experimental testing was conducted on lead-rubber bearings under constant vertical load, N, and subjected to varying cyclic shearing force. Some test results (adopted from Skinner et al., 1993) are shown in Fig. 12-4, which illustrates the contribution of both rubber and lead material to stiffness and to energy dissipation. The large elastic stiffness and large hysteretic loop are due to lead contribution. Upon yielding of lead, the overall effective stiffness of the bearing will only be the stiffness of the elastomer.

12.4 Analysis of Seismically Isolated Structures

Because of the relatively high stiffness ratio between the structure and the isolator, the structure is considered to behave as a rigid body with all deformations concentrated in the isolator as shown in Fig. 12-5. With this assumption, the system may be treated as an equivalent single degree of freedom system with its mass concentrated in the building; the stiffness and displacements of the system are concentrated in the isolator.

Because the isolator properties are known through their hysteretic behavior, inelastic dynamic analysis can be used to analyze the system under any excitation as illustrated in Chap. 4. Indirect analysis using force reductions and ductility demand/capacity will also be valid.

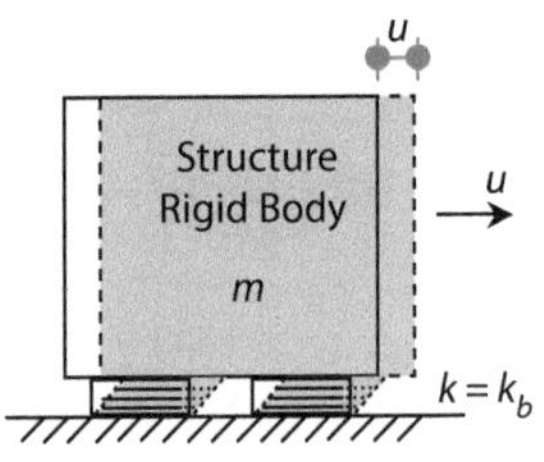

FIGURE 12-5 Seismic isolation model.

In general, the system must be designed to meet the required specifications and the conditions of serviceability. Such requirements include, for example, the design of the bearing to carry the vertical gravity loads and to meet the strain limitations, a minimum yield force to resist wind during normal conditions, and a minimum displacement capacity (ductility capacity) to meet the expected large inelastic displacements during earthquake excitations.

12.5 Design of Seismically Isolated Structures

Design requirements of this section will be given according to AASHTO specifications for design of elastomeric bearings. According to AASHTO, elastomeric bearings with reinforcement steel sheets may be designed using the allowable stress design method. As indicated earlier, the bearing consists of a stack of elastomer layers and steel sheet layers bonded together.

The bearing components are typically arranged as shown in Fig. 12-6, which are usually optimized by the manufacturer. To get a sense of the bearing details, view the following dimensions as general dimensions:

1. The elastomer layer thickness (around 12 mm) between steel sheets is designated as h_{ri}.
2. The total thickness of rubber layers is $h_{rt} = \Sigma h_{ri}$.
3. The total thickness of bearing, including steel sheets and end plates, is designated as h_b.
4. The thickness of steel sheets is typically taken as 3 mm.
5. The end steel plates are typically taken as around 20 mm.
6. The cover of elastomer around steel sheets is taken as around 12 mm.

The area of steel sheets is designated as the bonded area, which is considered the effective area in resisting bearing deformations. The overall dimensions of the bearing in plan are given as shown in Fig. 12-7 as follows:

L = dimension along the axis of the bridge
W = dimension perpendicular to the axis of the bridge

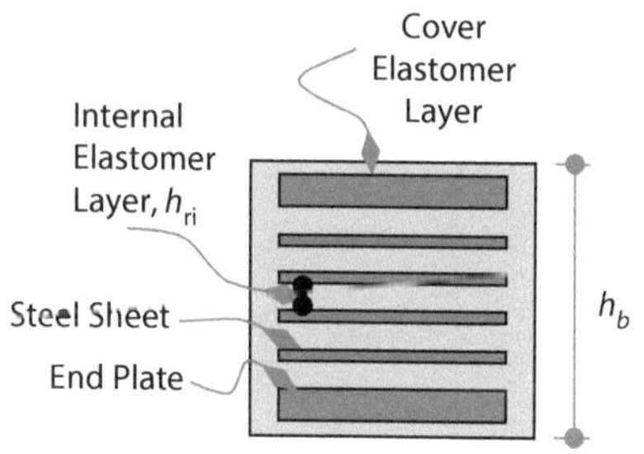

FIGURE 12-6 Typical elastomeric bearing without lead core.

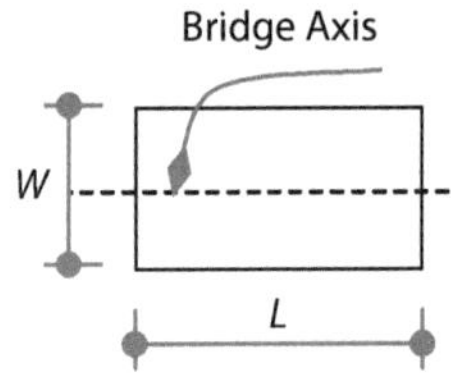

Figure 12-7 Bearing dimensions in plan.

A shape factor, S, is defined as follows:

$$S = \frac{\text{plan area}}{\text{free to bulge area}} = \frac{LW}{(L+W)(2)(h_{ri})}$$

where h_{ri} is the thickest rubber layer.

Allowable Compressive Stress

Allowable compressive stress of elastomer is given as a function of its shape factor defined earlier. The allowable compressive stress for steel-reinforced bearings under dead and live loads, $\sigma_{c,\text{all}}$, is given as follows:

$$\sigma_{c,\text{all}} = \frac{GS}{\beta}$$

$$\sigma_{c,\text{all}} \leq 7\ \text{MPa}$$

where G = shear modulus of elasticity of the elastomer as given in Table 12-A1
S = shape factor as defined earlier
β = factor taken as 1.0 for internal layers and as 1.4 for cover layer as defined in Fig. 12-6

Bearings (as rubber material) are excluded from the impact effect of live loads.

Allowable Shear Deformation (Shear Strain)

The maximum shear strain in the bearing measured over pure rubber material is given as 0.5. This limitation implies that the total rubber thickness, h_{rt}, excluding steel sheets and end plates, must be at least

$$\frac{\Delta_s}{h_{rt}} \leq 0.5$$

where Δ_s = maximum top displacement of the bearing in the lateral direction (see Fig. 12-9)
h_{rt} = total rubber height of the bearing ($h_{rt} = \Sigma h_{ri}$)

Allowable Rotation

The maximum rotation of the bearing plan surface, q, under the dead and live loads is limited to the following:

$$\theta \leq 2\frac{\Delta_c}{L}$$

$$\theta \leq 2\frac{\Delta_c}{W}$$

where Δ_c = instantaneous vertical compressive deflection of the bearing
L, W = bearing dimensions as given in Fig. 12-7
θ = rotation of the bearing, which is due to the rotation of the supported beam end

The instantaneous vertical compressive deflection of the bearing is a function of the vertical stress. Unfortunately, the axial stress-strain relationship of rubber is nonlinear. Thus, the deflection is given in AASHTO by means of relation charts. Appendix 12-1 presents two charts of vertical stress versus vertical strain relationships for elastomer of two hardness grades of 50 and 60.

Stability Requirements

The total thickness of the bearing is limited to a maximum to prevent the toppling and rolling of the bearing under large deformations. These limits are given for reinforced bearings as follows:

Rectangular bearings: $$h_b \leq \frac{L}{3}, \frac{W}{3}$$

Circular bearings: $$h_b \leq \frac{D}{4}$$

where h_b = total height of the bearing, including steel sheets and end plates
L, W = bearing dimensions as given in Fig. 12-7
D = bearing diameter in case of circular bearings

The requirements above may be waived if the manufacturer provides the bearing with positive mechanical connections to the structure that prevent the instability of the bearing. For example, an interface plate bolted to the superstructure may be considered as a waiver from this stability requirement as long as the manufacturer proves that his or her product is stable under loads.

Lead Core Dimensions

The lead core diameter, D_L, is recommended to be limited to the least bearing dimensions as follows:

$$D_L \leq \frac{B}{3}, \quad D_L \geq \frac{B}{6}$$

where B is the least dimension of the bearing (a minimum of L, W, or D).

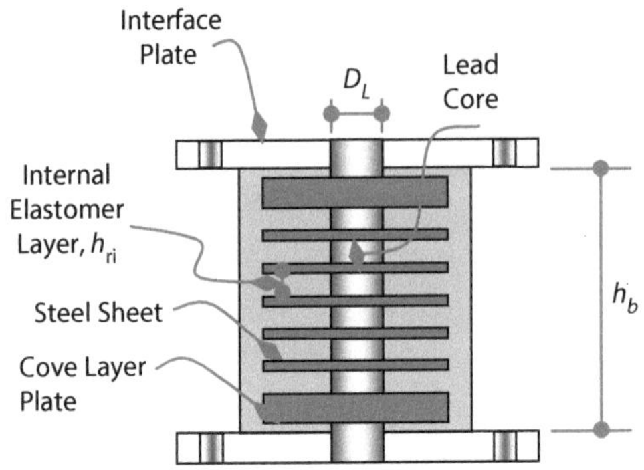

FIGURE 12-8 Typical elastomeric bearing with lead core.

This is not an AASHTO requirement. Instead, it is a recommendation given by lead-rubber bearing manufacturers. Figure 12-8 shows a typical arrangement of bearings with a lead core.

Shear Stiffness

The bearing stiffness is given as a function of the shear stiffness because its response is basically in the shear mode as shown in Fig. 12-9. For a given force, F_s, with displacement, Δ_s, the stiffness is evaluated as follows:

$$\begin{aligned} F_s &= \tau A_r \\ &= (G_r \gamma) A_r \\ &= G_r A_r \left(\frac{\Delta_s}{h_r} \right) \\ &= \left(\frac{G_r A_r}{h_r} \right) \end{aligned}$$

Consequently, the shear stiffness of the bearing is given as

$$k_r = \frac{G_r A_r}{h_r} \Delta s$$

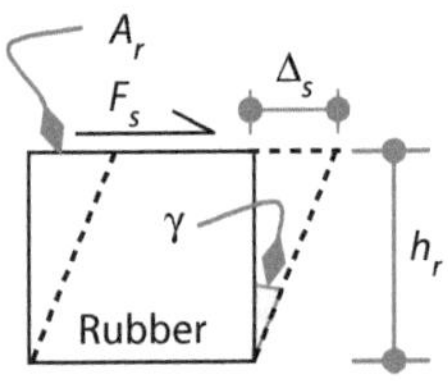

FIGURE 12-9 Bearing stiffness.

Similarly, the shear stiffness of the lead is given as

$$k_L = \frac{G_L A_L}{h_b}$$

where G = shear modulus of elasticity
L = lead
r = rubber

The design requirements given above are best illustrated by the following example.

Example 12-1 A single-span bridge is shown in Example 12-1, Fig. 1. The deck is supported by two longitudinal girders. The bridge lies in a seismic zone with a seismic coefficient, $S_S = 0.5$ and $S_1 = 0.2$. The site is of Soil Type F. The total weight of the bridge is 1,000 kN. The bridge is subjected to a total live load of 500 kN.

Other relevant data:
The short-term forces that are due to wind, breaking forces, and others are 20 kN. Shrinkage and temperature displacement is 5 mm.

The available isolators are lead-rubber bearings. The elastomer has a shear modulus of elasticity of 1.0 MPa layered at 12 mm.

Since the soil is Type F, a site-specific response spectrum is provided by the geotechnical engineer as given in Example 12-1, Fig. 2.

Recall also that the isolators behave according to the equal displacement criterion given in Chap. 5, that is, elastic and inelastic displacements are equal under earthquake excitation.

Use these rubber-lead bearings to isolate the bridge.

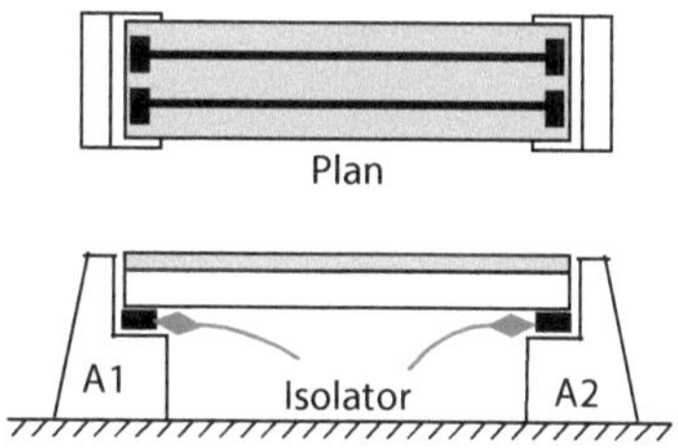

Example 12-1, Figure 1 Bridge layout.

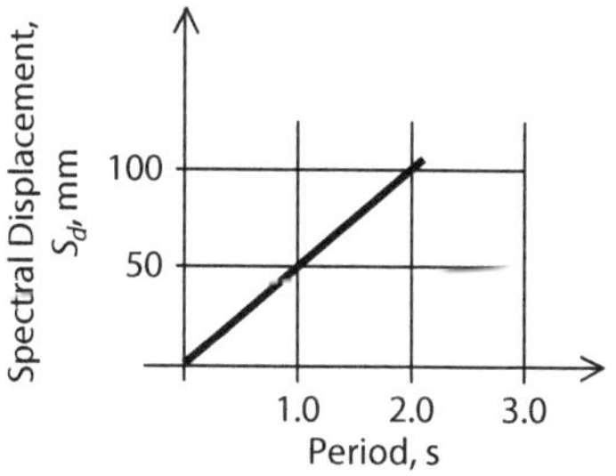

Example 12-1, Figure 2 Site-specific response spectrum.

Solution Because the bridge has two main girders, two isolators are needed for each girder, resulting in a total of four isolators placed as shown in Example 12-1, Fig. 1. The design must be carried out in two stages. First, the bearings must be designed for dead loads, live loads, and any other nonseismic loading specified by the code; and second, the bearings must be designed for seismic requirements specified by the code.

In this example, and for our objectives, our design will be limited to treatment of dead, live, and seismic load effect.

The bearings will be designed of the same height. Therefore, all loads will be equally divided by the number of bearings, that is, by 4.

Part 1: Nonseismic dimensioning

1. Each bearing share of gravity loads

$$D = \frac{1{,}000}{4} = 250 \text{ kN}$$

$$L = \frac{500}{4} = 125 \text{ kN}$$

2. Plan dimensions by trial and error since allowable stresses are a function of the shape factor. Start with $\sigma_{c,\text{all}} = 7$ MPa

$$A_{\text{req}} = \frac{D+L}{\sigma_{c,\text{all}}} = \frac{(250+125)(1{,}000)}{7} = 53{,}571 \text{ mm}^2$$

Try 231×231 mm to find the new allowable stress.

Shape factor:

$$S = \frac{\text{LW}}{(L+W)(2)(h_{\text{ri}})} = \frac{231(231)}{(231+231)(2)(12)} = 4.82$$

$$s_{c,\text{all}} = \frac{\text{GS}}{\beta} = \frac{1.0(4.82)}{1} = 4.82 \text{ MPa}$$

Therefore,

$$A_{\text{req}} = \frac{(250+125)(1{,}000)}{4.82} = 77{,}800 \text{ mm}^2$$

Try 279×279 mm to find the new allowable stress.

Shape factor:

$$S = \frac{\text{LW}}{(L+W)(2)(h_{\text{ri}})} = \frac{279(279)}{(279+279)(2)(12)} = 5.81$$

$$s_{c,\text{all}} = \frac{\text{GS}}{\beta} = \frac{1.0(5.81)}{1} = 5.81 \text{ MPa}$$

Therefore,

$$A_{\text{req}} = \frac{(250+125)(1{,}000)}{5.81} = 64{,}544 \text{ mm}^2$$
$$= 254 \times 254 \text{ mm}$$

Use 280 × 280 mm for overall dimensions. With 20-mm horizontal cover on each side, the bonded area becomes 240 × 240 mm.

3. Shear deformations limitations

$$\frac{\Delta_s}{h_{rt}} \leq 0.5 \rightarrow h_{rt} \geq 2\Delta_s = 2(5) = 10 \text{ mm}$$

Use 1-layer, 12 mm each; and 2-layers cover, 3 mm each

$$h_{rt} = 12 + 2(3) = 18 \text{ mm}$$

4. Rotation should also be checked. This check would be completed in a straightforward fashion as a function of the rotation of the girder ends.
5. Stability check

Because this type of bearings has a positive interface connection, the stability limits may be waived. If desired, the stability check for such bearing may be done as follows:

$$\text{Height limit} = \frac{L}{3} = \frac{280}{3} = 93 \text{ mm}$$

Total bearing height, h_b, is given as follows:

1 - layer rubber:	12 mm
0 - layer of steel sheets, 3 mm each:	0 mm
2 - end plates, 12 mm each:	24 mm
2 - cover rubber layers, 3 mm each:	6 mm

Therefore,

$$h_b = 12 + 0 + 24 + 6 = 42 \text{ mm} < 93 \text{ mm ok}$$

Part 2: Seismic dimensioning

6. Lead core diameter

$$D_L \leq \frac{B}{3} = \frac{280}{3} = 93 \text{ mm}$$
$$D_L \geq \frac{B}{6} = \frac{280}{6} = 47 \text{ mm}$$

Use 50-mm diameter core. As a result, the cross-sectional area of the lead core, A_L, and the yield force of the core, F_y, becomes

$$A_L = \frac{\pi D_L^2}{4} = \frac{\pi(50)^2}{4} = 1{,}963 \text{ mm}^2$$
$$F_y = \tau_y A_L = 10(1{,}963) = 19{,}630 \text{ N (19.63 kN)}$$

7. Seismic displacements

It should be pointed out that since the shear deformation requirements are function of the height bearing and the displacement due to earthquake excitation, the final dimensions are obtained by trial and error iterations as follows:

Bonded area of rubber: $A_r = A_{\text{bond}} - A_L = (240)^2 - (1{,}963) = 55{,}637\ \text{mm}^2$

Rubber stiffness: $k_r = \dfrac{G_r A_r}{h_{\text{rt}}} = \dfrac{1.0(55{,}637)}{18} = 3{,}091\ \text{N/mm}$

Lead stiffness: $k_L = \dfrac{G_L A_L}{h_b} = \dfrac{130(1{,}963)}{42} = 6{,}076\ \text{N/mm}$

Total stiffness: $k_b = k_r + k_L = 3{,}091 + 6{,}076 = 9{,}167\ \text{N/mm}$

Period: $T = 2\pi\sqrt{\dfrac{W}{gK_b}} = 2\pi\sqrt{\dfrac{250}{9.81(9{,}167)}} = 0.331\ \text{s}$

Using the response spectrum given in Example 12-1, Fig. 2, the maximum displacement due to earthquake is given as

$$\Delta_{\text{EQ}} = S_d = 0.331(50) = 16.55\ \text{mm}$$

Shear deformations limitations

$$\frac{\Delta_{\text{EQ}}}{h_{\text{rt}}} = \frac{16.55}{18} = 0.9 > 0.5\ \text{NG}$$

Since shear deformation is unsafe, the bearing rubber thickness must be increased, and checked against earthquake displacement again. Since the rubber thickness and the earthquake displacement are dependent on each other, this task is accomplished by trial and error procedures. After few trials, it can be shown that 4 layers of internal rubber layers satisfy these requirements as follows:

Use 4-layer, 12 mm each; and 2-layers cover, 3 mm each

$$h_{\text{rt}} = 4(12) + 2(3) = 54\ \text{mm}$$

Accordingly, the final total bearing height, h_b, becomes

4 - layer of rubber, 12 mm each:	48 mm
3 - layer of steel sheets, 3 mm each:	9 mm
2 - end plates, 12 mm each:	24 mm
2 - cover rubber layers, 3 mm each:	6 mm

Therefore,

$$h_b = 48 + 9 + 24 + 6 = 87\ \text{mm} < 93\ \text{mm ok}$$

Bonded area of rubber: $A_r = A_{\text{bond}} - A_L = (240)^2 - (1{,}963) = 55{,}637\ \text{mm}^2$

Rubber stiffness: $k_r = \dfrac{G_r A_r}{h_{\text{rt}}} = \dfrac{1.0(55{,}637)}{54} = 1{,}030\ \text{N/mm}$

Lead stiffness: $k_L = \dfrac{G_L A_L}{h_b} = \dfrac{130(1{,}963)}{87} = 2{,}933 \text{ N/mm}$

Total stiffness: $k_b = k_r + k_L = 1{,}030 + 2{,}933 = 3{,}963 \text{ N/mm}$

Period: $T = 2\pi\sqrt{\dfrac{W}{gK_b}} = 2\pi\sqrt{\dfrac{250}{9.81(3{,}963)}} = 0.504 \text{ s}$

Using the response spectrum from Example 12-1, Fig. 2, the maximum displacement demand is given as

$$\Delta_{EQ} = S_d = 0.504(50) = 25.19 \text{ mm}$$

Shear deformations limitations

$$\frac{\Delta_{EQ}}{h_{rt}} = \frac{25.19}{54} = 0.466 < 0.5 \text{ ok}$$

8. Seismic forces

Yield displacement of the lead core

$$\Delta_y = \frac{F_y}{k_L} = \frac{19{,}630}{2{,}933} = 6.69 \text{ mm}$$

Since $\Delta_{EQ} > \Delta_y$, the lead yields under earthquake excitation. The earthquake-induced force may now be calculated as the summation of the force in the lead core and in the rubber as follows:

$$F_r = k_r \Delta_{EQ} = 1{,}030(25.19) = 25{,}946 \text{ N}$$

$$F_{EQ} = F_L + F_r = 19{,}630 + 25{,}946 = 45{,}576 \text{ N (46 kN)}$$

9. Service displacements
   a. Lead yield displacement: $\Delta_y = 6.69$ mm (see previous step)
   b. Shrinkage and temperature: $\Delta_s = 5.0$ mm
   c. Displacement due to wind, breaking, and other forces as given in the example statement, $F_{wt} = 20$ kN

Force per each bearing: $F_w = \dfrac{F_{wt}}{4} = \dfrac{20}{4} = 5 \text{ kN}$

$$\Delta_w = \frac{F_w}{k_b} = \frac{5{,}000}{3{,}963} = 1.26 \text{ mm}$$

   d. Total service displacement:

$$\Delta_{total} = \Delta_s + \Delta_w = 5.0 + 1.26 = 6.26 \text{ mm} < \Delta_y = 6.69 \text{ mm}$$

Note that the total service displacement 6.26 is less than the lead yield displacement, and hence, the lead core remains elastic under service conditions. If the total service displacements exceed the lead core yield displacement, the lead core diameter must be increased to keep the lead core in the elastic range under service conditions.

APPENDIX 12-1

Sheet 1 Material Properties of Elastomeric Bearings

Hardness	50	60	70
Shear modulus, G at 23°C (MPa)	0.68–0.93	0.93–1.43	1.43–2.14
Creep deflection at 25 years as percentage of instantaneous deflection	25%	35%	45%

TABLE 12-A1 Elastomer Properties at Different Hardness Levels According to the AASHTO Code

As a material, lead exhibits ideal elastoplastic behavior in shear. The lead properties in shear are given in Table 12-A2.

Lead	Shear modulus, *G*	130 MPa
	Yield stress in shear, τ_y	10 MPa
Elastomer	Shear modulus, *G*	See Table 12-A1
	Allowable compressive stress, $\sigma_{c,all}$	$\sigma_{c,all} = \frac{GS}{\beta}$ * $\sigma_{c,all} \leq 7$ MPa

*See details in the main chapter which are summarized as follows:
G = shear modulus of elasticity of the elastomer as given in Table 12-A1
S = shape factor defined as follows:

$$S = \frac{\text{plan area}}{\text{free to bulge area}} = \frac{LW}{(L+W)(2)(h_{ri})}$$

β = factor taken as 1.0 for internal layers and as 1.4 for cover layer

TABLE 12-A2 Mechanical Properties of Lead and Elastomer

Sheet 2 Stress-Strain Relationship of Elastomer at Hardness 50 and 60

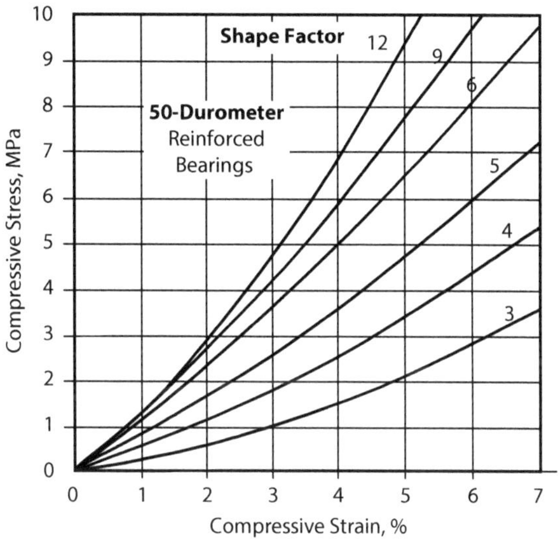

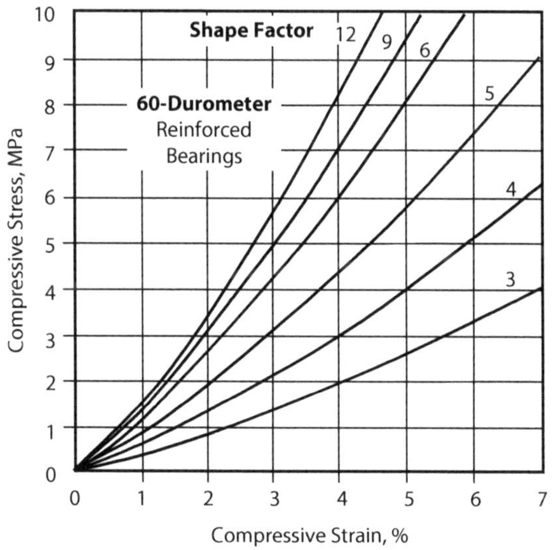

Bibliography

Chapter 1

FEMA, *NEHRP Recommended Provisions for the Development of Seismic Regulations for New Buildings, Part 1 Provisions*, Federal Emergency Management Agency, Washington, DC, 2009.

Chapter 2

Kramer, S. L., *Geotechnical Earthquake Engineering*, Prentice Hall, New York, 1996.

Paulay, T., and Priestly, M. J. N., *Seismic Design of Reinforced Concrete and Masonry Buildings*, Wiley, New York, 1992.

Wakabayashi M., *Design of Earthquake-Resistant Buildings*, McGraw-Hill, New York, 1986.

Chapter 3

Clough, R. W., and Penzien, J., *Dynamics of Structures*, 2nd ed., McGraw-Hill, New York, 1993.

Fleming, J. F., *Computer Analysis of Structural Systems*, 1st ed., McGraw-Hill, New York, 1989.

Huebner, K. H., and Thornton, E. A., *The Finite Element Method for Engineers*, 2nd ed., John Wiley & Sons, New York, 1981.

Seed, H. B., Ugas, C., and Lysmer, J., "Site-Dependent Spectra for Earthquake Resistant Design," *Bulletin of the Seismological Society of America*, 66 (1): 221–243, 1976.

Wakabayashi M., *Design of Earthquake-Resistant Buildings*, McGraw-Hill, New York, 1986.

Chapter 4

Clough, R. W., and Penzien, J., *Dynamics of Structures*, 2nd ed., McGraw-Hill, New York, 1993.

Wakabayashi M., *Design of Earthquake-Resistant Buildings*, McGraw-Hill, New York, 1986.

Chapter 5

Aktan, A. E., Karlsson, B. I., and Sozen, M. A., "Stress-Strain Relationships of Reinforcing Bars Subjected to Large Strain Reversals," Civil Engineering Studies, Structural Research Series No. 397, University of Illinois, Urbana, 1973.

Armouti, N. S., "Seismic Performance of Precast Concrete Structural Walls," Ph.D. Dissertation, Civil Engineering Department, Lehigh University, Pennsylvania, 1993.

Bresler, B., and Bertero, V. V., "Influence of High Strain Rate and Cyclic Loading on Behavior of Unconfined and Confined Concrete in Compression," Proceedings of the 2nd Canadian Conference on Earthquake Engineering, McMaster University, Ontario-Canada, June 1975.

Chen, W. F., and Han, D. J., *Plasticity for Structural Engineers*, Springer-Verlag, New York, 1988.

Chen, W. F., and Sohal, I., *Plastic Design and Second-Order Analysis of Steel Frames*, Spring-Verlag, New York, 1995.

FEMA, *NEHRP Guidelines for the Seismic Rehabilitation of Buildings*, FEMA 273, Federal Emergency Management Agency, Washington, DC, 2009.

FEMA, *NEHRP Recommended Provisions for the Development of Seismic Regulations for New Buildings, Part 1 Provision,"* Federal Emergency Management Agency, Washington, DC, 2009.

Nassar, A. A., and Krawinkler H., "Seismic Demands for SDOF and MDOF Systems," Report No. 95, The John A. Blume Earthquake Engineering Center, Stanford University, California, 1991.

Newmark, N. M., and Rosenblueth, E., *Fundamentals of Earthquake Engineering*, Prentice Hall, New York, 1971.

Park, R., and Paulay, T., *Reinforced Concrete Structures*, Wiley, New York, 1975.

Paulay, T., and Priestly, M. J. N., *Seismic Design of Reinforced Concrete and Masonry Buildings*, Wiley, New York, 1992.

PCI, "Connections for Precast Prestressed Concrete Buildings, including Earthquake resistance," Prestressed Concrete Institute, Chicago, IL, 1982.

Richart, F. E., Brandtzaeg, A., and Brown, R. L., "A Study of the Failure of Concrete under Combined Compressive Stresses," Bulletin No. 185, University of Illinois Engineering Experimental Station, 1928.

Sinha, B. P., Gerstle, K. H., and Tulil n, L. G., "Stress-Strain Relationships for Concrete under Cyclic Loading," *ACI Journal*, 61 (2): 195–211, 1964.

Wakabayashi M., *Design of Earthquake-Resistant Buildings*, McGraw-Hill, New York, 1986.

Chapter 6

ASCE 7, *Minimum Design Loads for Buildings and Other Structures*, American Society of Civil Engineers, Virginia, 2010.

CSI, *Structural Analysis Program, SAP2000 Nonlinear Version 14*, Computers and Structures, Inc., Berkeley, California, 2008.

FEMA, *NEHRP Recommended Provisions for the Development of Seismic Regulations for New Buildings, Part 1 Provisions*, Federal Emergency Management Agency, Washington, DC, 2009.

IBC, *International Building Code*, International Code Council, Washington, DC, 2015.

Seed, H. B., Ugas, C., and Lysmer, J., "Site-Dependent Spectra for Earthquake Resistant Design," *Bulletin of the Seismological Society of America*, 66 (1): 221–243, 1976.

Chapter 7

ACI, *Building Code Requirements for Structural Concrete (ACI 318-11) and Commentary*, American Concrete Institute, Detroit, MI, 2011.

Nawy, E. G., *Reinforced Concrete: A Fundamental Approach*, 4th ed., Prentice Hall, New York, 2000.

Nilson, A. H., Darwin, D., and Dolan, C. W., *Design of Concrete Structures*, 14th ed., McGraw-Hill, New York, 2010.

Park, R., and Paulay, T., *Reinforced Concrete Structures*, Wiley, New York, 1975.

Paulay, T., and Priestly, M. J. N., *Seismic Design of Reinforced Concrete and Masonry Buildings*, New York, Wiley, 1992.

Wang, C., and Salmon, C. G., *Reinforced Concrete Design*, Addison Wesley, New York, 1998.

Chapter 8

AISC, *Manual of Steel Construction, Load & Resistance Factor Design*, 14th ed., American Institute of Steel Construction, Inc., Chicago, 2011.

AISC, *Seismic Provisions for Structural Steel Buildings*, American Institute of Steel Construction, Inc., Chicago, 2010.

Beedle, L. S., *Plastic Design of Steel Frames*, John Wiley & Sons., New York, 1958.

Chen, W. F., and Han, D. J., *Plasticity for Structural Engineers*, Springer-Verlag, New York, 1988.

Chen, W. F., and Lui, E. M., *Structural Stability*; 1st ed., Elsevier, New York, 1987.

Chen, W. F., and Sohal, I., *Plastic Design and Second-Order Analysis of Steel Frames*, Spring-Verlag, New York, 1995.

Timoshenko, S. P., and Gere, J. M., *Theory of Elastic Stability*, 2nd ed., McGraw-Hill, New York, 1961.

Chapter 9

AASHTO, *LRFD Bridge Design Specifications*, 6th ed., American Association of State Highway and Transportation Officials, Washington, DC, 2012.

AASHTO, *Standard Specifications for Highway Bridges*, Division I-A, 15th ed., American Association of State Highway and Transportation Officials, Washington, DC, 1992.

ACI, *Building Code Requirements for Structural Concrete (ACI 318-11) and Commentary*, American Concrete Institute, Detroit, MI, 2011.

Armouti, N. S., "Transverse Earthquake-Induced Forces in Continuous Bridges," *Structural Engineering and Mechanics, An International Journal*, 14 (6): 733–738, 2002.

Armouti, N. S., "Transverse Earthquake-Induced Forces in Discontinuous Bridges," *Journal of Structural Engineering* 132 (2): 75–84, 2004.

Buckle, I. G., Mayes, R. L., and Button, M. R., "Seismic Design and Retrofit Manual for Highway Bridges," Report No. FHWA-IP-6, Final Report, National Technical Information Service, Springfield, Virginia, 130–136, 1987.

Chen, W. F., and Lui, E. M., *Structural Stability*, Elsevier, New York, 1987.

FEMA, *NEHRP Recommended Provisions for the Development of Seismic Regulations for New Buildings*, Part 1 Provisions, Federal Emergency Management Agency Washington, DC, 2009.

STAAD, *Structural Analysis and Design Program, STAAD pro V8i*, Bentley Systems, Incorporated, Exton, Pennsylvania, 2014.

Chapter 10

AASHTO, *Standard Specifications for Highway Bridges,* Division 1-A, 15th ed., Washington, DC, 1992.

Bowles, J. E., *Foundation Analysis and Design,* New York, McGraw-Hill, 1988.

ICC, *International Building Code*, International Code Council, Washington, DC, 2015.

Kramer, S. L., *Geotechnical Earthquake Engineering,* Prentice Hall, New York, 1996.

Chapter 11

Amin, M., Ang, A. H. -S., "Nonstationary Stochastic Model of Earthquake Motions," *Journal of the Engineering Mechanics Division,* ASCE, 94 (EM2): 559–583, 1968.

Armouti, N. S., "Random Ground Motion Model for Synthetic Earthquakes," *Dirasat: Engineering Sciences,* 29 (1): 20–30, 2002.

Armouti, N. S., "Response of Structures to Synthetic Earthquakes," Proceedings of the 9th Arab Structural Engineering Conference, Abu Dhabi, UAE, Vol. 1, 331–339, 2003.

Armouti, N. S., "Evaluation of Structural Response Subjected to Synthetic Earthquake Excitation," *Journal of Structural Engineering,* 31 (3): 175–180, 2004.

Clough, R. W., and Penzien, J., *Dynamics of Structures,* 2nd ed., McGraw-Hill, New York, 1993.

FEMA, *NEHRP Recommended Provisions for the Development of Seismic Regulations for New Buildings, Part 1 Provisions,* Federal Emergency Management Agency, Washington, DC, 2009.

Housner, G. W., Characteristics of strong motion earthquakes, *Selected Earthquake Engineering Papers of George W. Housner,* ASCE, New York, 1990, 8–20.

Hwang, H. H., and Jaw, J. W., "Statistical Evaluation of Response Modification Factors for Reinforced Concrete Structures," Report No. NCEER-89-0002, National Center for Earthquake Engineering Research, State University of New York at Buffalo, Buffalo, New York, 1989.

Hyat, Jr., W. H., Kemimerly, J. E., and Durbin, S., M., *Engineering Circuit Analysis,* McGraw-Hill, New York, 2002.

Kameda, H., "Stochastic Process Model of Strong Earthquake Motions for Inelastic Structural Response," Proceedings of the US-Southeast Asia Symposium on Engineering for Natural Hazard Protection, Manila, Philippines, 71–85, 1977.

Kanai, K., "Semi-empirical Formula for the Seismic Characteristics of the Ground," Bulletin of the Earthquake Research Institute, University of Tokyo, Vol. 35, pp. 309–325, 1957.

Lathi B. P., *Modern Digital and Analog Communication Systems,* Oxford University Press, USA, 1998.

Seed, H. B., Ugas, C., and Lysmer, J., "Site-Dependent Spectra for Earthquake Resistant Design," *Bulletin of the Seismological Society of America,* 66 (1) : 221–243, 1976.

Tajimi, H. A., "Statistical Method of Determining the Maximum Response of a Building Structure during an Earthquake," Proceedings of the 2nd World Conference on Earthquake Engineering, Tokyo and Kyoto, Vol. II, 781–798, 1960.

ICBO, *"Uniform Building Code,"* International Conference of Building Officials, Whittier, California, 1997, copyright International Code Council, Washington, DC.

Wakabayashi M., *Design of Earthquake-Resistant Buildings,* McGraw-Hill, New York, 1986.

Yang C. Y., *Random Vibration of Structures,* Wiley, New York, 1986.

Chapter 12

AASHTO, "*Standard Specifications for Highway Bridges,*" Division I-A, 15th ed., American Association of State Highway and Transportation Officials, Washington, DC, 1992.

ICC, "*International Building Code,*" International Code Council, Washington, DC, 2015.

Skinner, R. I, Robinson, W. H., and McVerry, G. H., *An Introduction to Seismic Isolation*, Wiley, New York, 1993.

Index

Q

R

S

9 781259 587122

Printed by Libri Plureos GmbH in Hamburg,
Germany